Low-Grade Metamorphism

Low-Grade Metamorphism

Martin Frey

Mineralogisch-Petrographisches, Institut der Universität Basel
Bernoullistrasse 30, CH-4056 Basel, Switzerland

Doug Robinson

Department of Earth Sciences, University of Bristol
Wills Memorial Building, Queens Road, Bristol BS8 1RJ, UK

Blackwell
Science

© 1999 by
Blackwell Science Ltd
Editorial Offices:
Osney Mead, Oxford OX2 0EL
25 John Street, London WC1N 2BL
23 Ainslie Place, Edinburgh EH3 6AJ
350 Main Street, Malden
  MA 02148 5018, USA
54 University Street, Carlton
  Victoria 3053, Australia
10, rue Casimir Delavigne
  75006 Paris, France

Other Editorial Offices:
Blackwell Wissenschafts-Verlag GmbH
  Kurfürstendamm 57
  10707 Berlin, Germany

Blackwell Science KK
  MG Kodenmacho Building
  7–10 Kodenmacho Nihombashi
  Chuo-ku, Tokyo 104, Japan

The right of the Authors to be
identified as the Authors of this Work
has been asserted in accordance
with the Copyright, Designs and
Patents Act 1988.

First published 1999

Set by Setrite Typesetters, Hong Kong
Printed and bound in
Great Britain at the
University Press, Cambridge

A catalogue record for this title
is available from the British Library

ISBN 0-632-04756-9

Library of Congress
Cataloging-in-publication Data

Low-grade metamorphism/[edited by]
Martin Frey, Doug Robinson.
        p.   cm.
    Includes bibliographical
    references.
    ISBN 0-632-04756-9
    1. Metamorphism (Geology)
I. Frey, Martin, 1940–. II. Robinson,
Doug.
QE475.A2L668—1998
552'.4—dc21   98-18900
                               CIP

DISTRIBUTORS

Marston Book Services Ltd
PO Box 269
Abingdon, Oxon OX14 4YN
(*Orders*: Tel: 01235 465500
           Fax: 01235 465555)

USA
Blackwell Science, Inc.
Commerce Place
350 Main Street
Malden, MA 02148 5018
(*Orders*: Tel: 800 759 6102
           781 388 8250
           Fax: 781 388 8255)

Canada
Login Brothers Book Company
324 Saulteaux Crescent
Winnipeg, Manitoba R3J 3T2
(*Orders*: Tel: 204 837–2987)

Australia
Blackwell Science Pty Ltd
54 University Street
Carlton, Victoria 3053
(*Orders*: Tel: 3 9347 0300
           Fax: 3 9347 5001)

For further information on
Blackwell Science, visit our website:
www.blackwell-science.com

Contents

List of contributors

J. C. Alt
Department of Geological Sciences, University of Michigan, Ann Arbor, MI 48109-1063, USA

R. E. Bevins
Department of Geology, National Museum of Wales, Cathys Park, Cardiff CF1 3NP, UK

S. Chaudhuri
Department of Geology, Kansas State University, Manhattan, KS 66506, USA

N. Clauer
Centre de Géochimie de la Surface (CNRS-LILP), Université Louis Pasteur, 1 rue Blessig, 67084 Strasbourg Cedex, France

H. W. Day
Department of Geology, University of California, Davis, CA 95616, USA

M. Frey
Mineralogisch-Petrographisches, Institut der Universität Basel, Bernoullistrasse 30, CH-4056 Basel, Switzerland

R. J. Merriman
British Geological Survey, Keyworth, Nottingham NG12 5GG, UK

D. R. Peacor
Department of Geological Sciences, University of Michigan, Ann Arbor, MI 48109-1063, USA

D. Robinson
Department of Earth Sciences, University of Bristol, Wills Memorial Building, Queens Road, Bristol BS8 1RJ, UK

P. Schiffman
Department of Geology, University of California, Davis, CA 95616, USA

Z. D. Sharp
Department of Earth and Planetary Sciences, Northrup Hall, The University of New Mexico, Albuquerque, NM 87131-1116, USA

Preface

Metamorphism is broadly a process of solid-state change in the mineralogy, texture and chemistry of rocks in the Earth's crust. It provides a powerful record of change in the Earth's most active settings where metamorphism has proceeded in response to global tectonics. Metamorphic rocks are defined in most petrological texts as having mineralogies and textures that are a consequence of recrystallization under $P–T$ regimes different from the diagenetic realm. Metamorphism 'proper' can thus be clearly recognized where rocks have been fully recrystallized and no direct evidence of the protolith remains. It is in such rocks that the principles of equilibrium thermodynamics have allowed, over the past two decades or so, an increasingly quantitative assessment of metamorphic processes.

In reality, however, there is no clear or sharp break between the diagenetic and metamorphic fields; instead there is a continuum of increasing recrystallization in the body of the protolith. Indeed, the extent of any transition zone will vary depending on features such as the type of earth material investigated, or the resolution of investigative technique. For example, vitreous and organic materials undergo change at much lower temperature than silicate materials, while Transmission Electron Microscopy provides greater resolution of the finest grained layer silicates than does X-ray diffraction. Although there is good coverage of the mainstream parts of the subject in a variety of textbooks, there is often only passing treatment of this transitional zone, suggesting that it is of only marginal interest to the subject as a whole. This book came to fruition with an aim of dispelling any such view by giving a specific analysis of the low-temperature part of the metamorphic spectrum. We believe that this field of metamorphism offers somewhat different perspectives on metamorphic processes, for which there are scientific questions that provide equally as exciting and challenging topics as at higher grades of metamorphism. We see research in this field as having reached a threshold at which several of the broad qualitative questions about the nature of the transformation and related processes have been successfully addressed over the last decade. This has provided a solid basis for moving forward to test these latest ideas and to provide for a more quantitative assessment of these processes, as has taken place in the mainstream of the subject. Low-temperature metamorphism is a widespread phenomenon associated with major processes of crustal growth and destruction, whose implications are perhaps underappreciated in terms of major Earth systems, such as arc volcanism and recycling of crustal materials into the mantle. Also, at the turn of the millennium, Earth Science is increasingly concentrating on processes at shallower levels in the Earth's crust, including low-temperature metamorphism, with their relevance to mass transfer and fluid–rock interaction aimed at better utilization of Earth resources and assessment of environmental impact. In the light of these considerations we believe that the field of low-temperature metamorphism has much of interest to offer to a broad audience and that the highlights of these features are to be found in the contents of this book.

The book does not offer a comprehensive coverage of all aspects of the subject; features such as organic maturation and fluid inclusions are not covered. Concentration is given to topics that we believe are most relevant to a wider audience and in which major advances have been made in recent years. The book starts with an overview of low-temperature metamorphism, highlighting advances made in the last decade and outlining its context to global settings and its relevance to other Earth systems. This is followed by detailed analysis in Chapters 2 and 3 of low-temperature metamorphism of metapelitic rocks. Chapter 2 assesses the remarkable advances made in our understanding of the continuum of prograde processes that occur in layer silicate minerals during the transformation from diagenesis to metamorphism, in particular those that have been provided by Electron Microscopy techniques. Chapter 3 examines regional patterns in metapelitic rocks from a global tectonic setting recognizing characteristic styles associated with contrasting settings.

Metabasites form the subject of Chapters 4, 5 and 6, commencing with an assessment and review of petrological methods that can be applied to low-grade metabasites of the style widely used at higher grades of metamorphism. Chapter 5 examines regional patterns of low-grade metamorphism tackled, as with metapelitic rocks, in terms of global tectonic settings. Chapter 6 covers the most widespread occurrence of low-temperature metamorphism, namely that in the ocean crust, resulting from hydrothermal metamorphism at ocean ridge settings, and identifies characteristics that distinguish its style from that of other settings.

Chapters 7 and 8 cover isotopic aspects of low-temperature metamorphism, the problems and potential of radiogenic and stable isotopes for resolving various problems and questions.

The Editors thank all authors for their efforts in producing their contributions to meet the deadline requirements and also for those reviewing chapters other than their own. Dr S. Th. Schmidt is thanked for reading and valuable comments on Chapters 4 and 5. Chapters 2 and 3 have benefited from the field and laboratory observations of Dr B. Roberts and the enthusiasm that he has generated over many years of fieldwork and discussion. R. J. Merriman publishes with permission of the Director, British Geological Survey.

D.R.

1 Low-temperature metamorphism: an overview

D. Robinson and R. J. Merriman

1.1 What is it?

Metamorphism is a process of change at elevated temperatures and pressures in rocks occupying the outer skin of the Earth. The challenge facing the metamorphic petrologist is to understand and interpret the record of changes that are imprinted in the texture, mineralogy and chemistry of the metamorphic rock, with relevance to understanding the evolution of Earth systems. In this book we investigate changes in terms of metamorphic phase transformations and the processes driving these changes in the broad pressure–temperature field that lies at subgreenschist and subblueschist facies conditions. This covers the temperature and pressure ranges up to $c.$ 400°C and $c.$ 4–5 kbar. Rocks affected by such conditions occupy large tracts of the Earth's upper continental and oceanic crust, and on this basis it must be expected that the processes involved in their formation are varied and diverse. Despite the widespread occurrence of such rocks, much about the actual mineralogical transitions and the processes involved are poorly understood compared to metamorphism at higher pressure–temperature conditions.

What is the reason for the relative lack of understanding of lower grade metamorphic rocks compared to their higher temperature counterparts? One reason is that most metamorphic petrologists have traditionally concentrated their attention on the changes taking place at elevated temperature ($+ c.$ 300°C) and pressures ($+ c.$ 3 kbar), reflecting changes in deeper parts of the Earth's crust and upper mantle. This has occurred because from the earliest days of the study of metamorphic rocks it was realized that regularity, recognized initially in these rocks in terms of their mineralogies at the hand specimen and petrographic scales, in time and space, meant that there were systematic controls that had the potential to be explored and understood. The mainstay of this approach was in the application of equilibrium thermodynamics to define the controlling features of the metamorphic process.

In terms of low-temperature metamorphism, the recrystallization is often not immediately obvious because of the fine-grained scale of the products. Also, the apparent lack of regularity in the metamorphic development and the partial recrystallization in the rocks with many relict features of the protolith preserved, point to non-equilibrated systems without obvious overall controls. Thus the likelihood of understanding such metamorphic evolution and predicting outcomes appeared to be low. However, the first sign that such a view was over-pessimistic was from the work of Coombs (1960). He showed a regularity in the distribution of zeolite minerals in South Island, New Zealand from which he proposed the establishment of the zeolite facies of metamorphism and since then, subgreenschist facies metamorphism has been recognized in many parts of the world. Documentation of subgreenschist facies metamorphism, however, still remains of a cursory nature in most traditional textbooks of metamorphic petrology that typically envisage the metamorphic process proper as one commencing at the greenschist/blueschist boundaries. It is one of the goals of this book to show that low-temperature metamorphic processes are a mainstream part of the subject with exciting scientific challenges and scope to advance the understanding of the metamorphic process in an equal manner as at higher grade conditions. Indeed, the features examined in this book relating principally to metapelite and metabasite rocks are representative of the very great majority of rocks at or near to the Earth's surface, and their relevance to and implication for many other earth systems are probably underestimated.

In the context, readers are perhaps expecting some definition as to what are considered to be the boundaries of low-temperature metamorphism. We do not, however, examine or attempt to rationalize features about the various definitions and correlations of this branch of metamorphism—such matters were discussed in detail by Frey and Kisch (1987) and Kisch (1987). Here we accept that depending on the type of Earth materials investigated (e.g. organic matter/silicate/evaporite/vitreous), different terminologies and divisions will apply, and that there can still be only broad correlation between them so that any proposed boundary will be of a markedly transitional character. Some of the different methods used to determine grade at low temperature and their respective terminologies to designate the progress of transformation are given in Table 1.1. It should be noted that use of a specific terminology implies the use of the appropriate methodology and should not be intermixed; for example, the anchizone should not be delimited other than with illite crystallinity data.

At high grades of metamorphism there is no sharp break between the metamorphic (solid state $<c.$ 700°C) and the igneous (molten) states, but it has still been a field that has attracted the attention of both metamorphic and igneous petrologists for the mutual benefit of both disciplines. There is similarly no sudden onset of metamorphism at low temperature with a transition between surface to near-surface processes of weathering and diagenesis and those regarded as truly metamorphic.

Table 1.1 Terminologies used to designate metamorphic grade in low-temperature metamorphism with reference to method used.

Method	Terminology for low-temperature grade
Illite crystallinity	Late diagenetic zone/Anchizone/ Epizone
Mineral assemblages	Zeolite/prehnite–pumpellyite/ prehnite–actinolite/ pumpellyite–actinolite
Vitrinite reflectance	Coal rank stages (Teichmüller, 1987, Table 4.2)
Fluid inclusion microthermometry	Various fluid zones: higher hydrocarbons/methane/water (see Mullis, 1987)

Accordingly, this domain of Earth processes benefits from the attentions of those workers who identify themselves as metamorphic or sedimentary petrologists. We certainly hope and believe that there is material contained herein to attract the attention of both these types of workers.

With reference to this book, this means that there is a selective coverage of the response of Earth materials to the specified pressure–temperature range. The omissions are a function related to space availability linked to coverage of items not traditionally seen as within the 'metamorphic' field and/or of materials that have been recently covered to greater depth than we could achieve. Also, the coverage is linked to areas in which there have been major advances in the understanding of the distribution of metamorphic patterns, the mineralogical changes and/or of the metamorphic processes involved. This does not mean, however, that the book covers a restricted field, as there have been many and major advances in a broad range of topics relevant to some of the most abundant materials of the Earth's crust, namely pelitic and metabasic rocks.

When considering pelitic rocks, much detail concerning changes in sedimentary rocks that come under the general umbrella of 'early diagenesis' (0 to $c.$ 100°C) is omitted. More advanced 'late diagenetic' changes in these rocks, such as the smectite to illite transition, have been covered particularly in relation to the development of bedding-parallel microfabric, which is described here for the first time. In terms of metabasite rocks, however, there is no traditional discipline break at an equivalent 'diagenetic/metamorphic' (early/ late diagenetic?) boundary and so those changes occurring progressively from 0 to $c.$ 400°C are covered for these rocks.

1.2 A decade of progress

The only previous book on low-temperature metamorphism (Frey, 1987a) provided the first comprehensive framework of the subject, reporting on the character of low-temperature metamorphism across a variety of rock types as well as organic maturation, fluid inclusions, radiogenic isotopes and correlation between different indices of metamorphism. This book offers a more focused assessment, concentrating purposely on pelitic and metabasic rock types where there have been major advances in terms of phase transitions, processes, and also in being able to link the metamorphic style with different geodynamic settings. These advances have been made through a greater interest in this field arising partly from an International Geological Correlation Project (IGCP 294) promoting the subject and from technological advances, particularly in application of transmission electron microscopy (TEM) techniques.

There have been advances in the understanding of various mineralogical transformations that for example have at last resolved the exact nature of the widely used illite crystallinity technique for pelitic rocks (Merriman *et al.*, 1990, 1995b). This advance has come about by application of high-resolution TEM techniques as described in Chapter 2. This method has identified that the changes in the 10 Å X-ray diffraction profile are the result of an increase in the number of the 10 Å layers in a diffracting white mica crystallite. The change in illite crystallinity was originally attributed to some 'ordering' of the crystalline lattice (Kübler 1967a), and Frey (1987a) identified temperature, fluid pressure, stress, time, lithology and illite chemistry as possible factors that might have an influence on this parameter. As a result of the rather vague understanding of the nature of illite crystallinity it was realized that its use had to be of a pragmatic nature (Kübler, 1984). Now, with a much clearer understanding of its fundamental cause, its application can become less pragmatic and the influence of the six possible controls listed above in relation to the actual setting of rocks being examined, can be more realistically treated in terms of their potential effect.

Such studies on illite crystallinity and related studies on layer silicate minerals, principally involving the smectite–illite–muscovite and tri-smectite–corrensite–chlorite series, have given rise to the concept that a given mineralogical assemblage records the state of reaction progress. Accordingly, these types of reaction pathways may be better interpreted in terms of reaction progress whose main controls are kinetic (Essene & Peacor, 1995). Certainly the revelation that changes in illite crystallinity are fundamentally due to changes in crystallite size means that the principal forces driving these types of change in fine-grained layer silicates—

grain boundary and strain energies—would appear to be clearly linked to kinetic effects. The prominence of reaction progress and its kinetic control will be more evident at lowest grades in connection with the very fine-grained nature of reacting materials than coarser grained products at higher grade. There is still, however, active debate about the merits of these latest proposals, and some examples of applications of this approach are detailed in Chapters 2, 3 and 5. Reaction progress in clay mineral assemblages can be used to establish a sequence of metapelitic zones that cover the transition from diagenesis to low grade (greenschist facies) metamorphism. In Chapter 3 the relationship between patterns of metapelitic zones and tectonostratigraphy is used to characterize the different geotectonic setting of very low-grade metamorphism. For example, as detailed in Chapter 3, there are differences in illite crystallinity between samples from high-strain zones at high- or deep-crustal levels that can now be explained in terms of kinetic effects driven by strain-related processes affecting crystallite size. Other examples in Chapter 5 relate to the distribution and type of mafic phyllosilicates in metabasites that appear to be controlled by kinetic effects related to the nature and character of fluid flow during the metamorphic process (Schmidt & Robinson, 1997). Application of this kinetic concept seems to offer a better explanation of certain aspects of mineralogical development at low temperature than the effects of temperature alone.

In recent years the intensity of studies involving application of radiogenic and stable isotope methods to low-grade metamorphic processes has not been as great as in other areas. As the respective reviews by Clauer and Chaudhuri (Chapter 7) and Sharp (Chapter 8) demonstrate, there has been a fuller understanding of the interrelationship between isotopic data from separated size fractions and whole-rock systems particularly in the case of pelitic rocks. In both radiogenic and stable isotope systems in pelitic rocks, metamorphic recrystallization has to reach levels of at least low greenschist facies (c. 400°C) before all size fractions in the rock have equilibrated. Below this grade, however, it is shown that results from separated size fractions can be used to provide valuable information to test the degree to which the different fractions might have equilibrated with each other. The extent to which this occurs is often related to the degree of fluid/ rock interaction and the ease or otherwise by which fluid is able to pervasively interact with all portions of the rock.

Metabasite rocks have also been the subject of much interest in recent years, particularly in the increasing use of thermodynamic-based approaches to obtain better understanding of the metamorphic processes. The derivation, by Frey *et al.* (1991), of the first model petrogenetic grid for low-grade metabasites based on an internal consistent thermodynamic database, has proved to be a valuable template against which to compare natural facies distributions. This work also demonstrated that the subgreenschist facies assemblages show extensive overlap in the derived pressure–temperature space as a consequence of variation in whole-rock chemistry. One of the greatest difficulties at low grades of metamorphism has been in establishing whether the mineral assemblages found in low-grade metabasites do indeed represent equilibrated relationships. This has been aided following the first application by Springer *et al.* (1992) of thermodynamically-valid chemographic projection techniques to the study of compositional relationships to such rocks. This showed that there can be consistent element partitioning relationships between co-existing low-grade metabasite phases — generally taken as a pointer to equilibrated systems. This technique was applied by Bevins & Robinson (1993) to show, in metabasites from Wales, that the change from prehnite–pumpellyite to prehnite–actinolite facies assemblages was not due to changes in pressure–temperature but to a change in the whole-rock MgO/(MgO + FeO) ratios. They showed that the changeover from the prehnite–pumpellyite to prehnite–actinolite facies took place at a whole-rock MgO/(MgO + FeO) ratio of *c.* 0.54. This was an empirical demonstration of facies overlap as predicted from the petrogenetic grid of Frey *et al.* (1991).

Geothermobarometric methods have also been newly used to determine temperature and pressure of low-grade metabasites and have given results that are geologically realistic (Powell *et al.*, 1993; Digel & Gordon, 1995). The positive results arising from petrochemical projections suggests that such thermobarometric approaches will be more widely applied in future as thermodynamic data for low-

grade minerals become better constrained. The effect of CO_2 in the fluid phase has long been thought to be influential in restricting the development of low-grade calcsilicate phases such as pumpellyite and prehnite. Recent thermodynamic modelling by Digel and Ghent (1994) has demonstrated that mole fractions of CO_2 as low as 0.002 do in fact result in suppression of these diagnostic phases in low-grade metabasites.

1.3 Very low-grade metamorphism in a global setting

Low-temperature metamorphic rocks are found over large areas of the Earth's continental and oceanic crust and so it is to be expected that they develop in a variety of tectonic settings. The three generally accepted broad global metamorphic associations of the Earth's crust, in terms of subduction, continental orogenic and arc/ridge settings, are shown on a *P–T* diagram in Fig. 1.1. In these three associations the low-temperature field forms an obvious continuum into the higher grade settings and thus an integral part of the metamorphic progression in any of the associations.

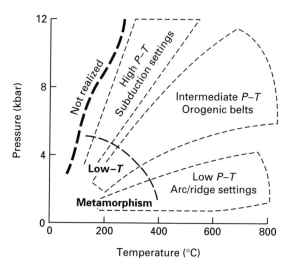

Fig. 1.1 Pressure–temperature (*P–T*) diagram showing schematic view of the three metamorphic facies series, Low-*P*, Intermediate-*P* and High-*P*, taken generally as distinctive of arc/ridge, orogenic belts and subduction settings respectively.

The most clear association is in respect of ocean crust with a progression from the hydrothermal metamorphism at ocean ridges into low-temperature metamorphism associated with the early stages of subduction. As shown by Alt in Chapter 6, the ocean ridge metamorphic style is for a step-like increase in grade with depth, and for prehnite and pumpellyite to be of rare occurrence. Metabasites of oceanic crust origin obducted in orogenic belts, however, show a contrasting style with a continuous facies profile and regular occurrence of prehnite and pumpellyite. As has been documented by Schiffman and Staudigel (1994), some of this discrepancy can be explained by metamorphism associated with accretionary processes. Some differences, however, can be linked to the genesis of the ocean crust. A more continuous metamorphic facies progression with more abundant prehnite and pumpellyite has been reported for ocean crust sequences in ocean island settings such as Iceland and La Palma (Schiffman & Fridleifsson, 1991; Schiffman & Staudigel, 1995). These contrasts in parageneses must relate to differences in the geodynamic and thermal structures between the two settings and thus there is potential to elucidate the origin of metabasite complexes found in many accretionary terranes (Schiffman & Staudigel, 1994).

Rocks metamorphosed at subgreenschist levels are also found in many continental associations and include the style of metamorphism defined as 'burial' by Coombs (1960). In continental settings, although it is usually not possible to constrain metamorphic conditions as tightly as at higher grade (because of fine-grain size, disequilibrium, etc.), it is often possible to integrate better the metamorphism with the original stratigraphy and structure so that derivation of a tectono-metamorphic history can often be undertaken with some confidence. In many generalized and schematic sections of orogenic metamorphism (e.g. Fig. 1.2), low-grade metamorphism is traditionally shown as occurring in the high-level part of the metamorphic profile. Accordingly, it should be preserved on the periphery of the orogenic belts because of erosion of the high levels of the orogenic core through uplift. Although low-grade metamorphism is found on the margins of orogenic belts, it is developed in a wider variety of settings than in such schematic representations. For example, Robinson (1987) recognized a broad distinction between subgreenschist facies metamorphism as a continuum with orogenic metamorphism linked to convergent settings such as in the Alps (Frey, 1986), and settings of an extensional character as in marginal basins where there is a truncated zeolite-lowest greenschist facies series. In convergent settings, low-grade metamorphism is indeed widely developed in the periphery of many orogenic belts such as the Alps and the Caledonides of Scandinavia.

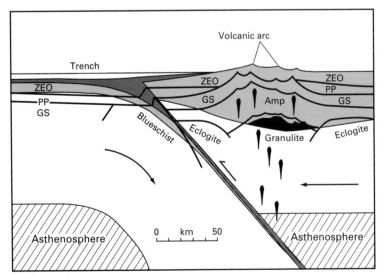

Fig. 1.2 Schematic view of metamorphic facies distribution in subduction and orogenic settings with arc development. After Spear (1993). GS = greenschist; ZEO = zeolite; PP = Prh–Pmp.

As detailed in Chapter 3, however, the apparent continuum from subgreenschist into higher grades of metamorphism in orogenic processes seems to be more apparent than real, with it being better related to processes associated with foreland thrusting than with the continental thickening in the orogenic core.

Subgreenschist metamorphism is widespread in the foreland regions of the Scandinavian Caledonides and there is an apparent progressive westward increase in grade towards the higher grade orogenic core that appears to concord with the above traditional model style (Fig. 1.2). Here, the low-grade metamorphic style recognized by studies of illite crystallinity investigations bears a close relationship to the thrusting style. Bevins *et al.* (1986) and Rice *et al.* (1989) recognized an inverted metamorphic pattern in the Finnmark region that they attributed to overthrusting of nappes carrying hotter, higher grade rocks from the orogenic core. In contrast, for the central Swedish Caledonides, Warr *et al.* (1996) linked the low-grade metamorphism to the variations in critical wedge geometry controlling the extent of burial of the rocks during the thrusting. Whether one or both interpretations of the metamorphic style apply, they both require that the metamorphism in these regions is a discrete and contrasting part to that of the higher grade zones of the Scandinavian Caledonides, rather than in the traditional style as portrayed in Fig. 1.2. Metamorphism linked to the foreland thrusting should be younger than for the traditional model and so there is potential to resolve between the contrasting models by radiometric dating. As detailed by Clauer and Chaudhuri in Chapter 7, however, dating of low-grade rocks is still subject to many problems, but there is potential to solve problems such as in the Scandinavian region with careful application through dating of size fractions in pelitic rocks.

Subgreenschist metamorphism also has a distinctive style that can be related to processes operating in the accretionary wedge setting of the convergent zone. As shown in Chapter 3, the illite crystallinity approach has sufficient resolution such that differences in grade between juxtaposed faulted units can be recognized in the Southern Uplands of Scotland. Here, grade increases from stratigraphically older to younger sequences in successive tracts of imbricated strata but in a syntectonic style rather than the posttectonic style associated with inverted metamorphic gradients. Such associations can be demonstrated by careful and detailed integration between metamorphic, structural and stratigraphic studies, such as have been undertaken in the Southern Uplands of Scotland (Merriman & Roberts, 1995). The very distinctive syntectonic metamorphic style documented using illite crystallinity for the Southern Uplands accretionary wedge, contrasts sharply with that proposed from a theoretical mechanical and thermal model of low-grade metamorphism developed in an accretionary wedge at a subduction margin (Barr *et al.*, 1991). This model has frictional heating along the decollement surface as the major source of heating without which little or no metamorphism would develop. The pattern of thermal development in the accretionary wedge was then used to derive the distribution of model metamorphic facies from a petrogenetic grid developed for the CMASH system, representative of low-grade metabasites. This gave a regional scale, progressive increase from unmetamorphosed and zeolite facies rocks through prehnite–pumpellyite to greenschist facies. The model was matched to the regional patterns of metamorphism developed in the accretionary wedge represented by the island of Taiwan, that shows a regional scale progressive metamorphism of this character (Fig. 1.3). This pattern contrasts strongly with that of the Southern Uplands where there is a discontinuous pattern of low-grade metamorphism with regular breaks in grade. The pattern of metamorphism in Taiwan, however, was deduced from 'occasional volcaniclastic rocks of basaltic to andesitic composition' (Barr *et al.*, 1991). Metabasites were chosen as it was argued that they gave rise to more diverse mineral assemblages that were stable over restricted ranges of temperature and pressure and thus more suited to designation of variation in grade than the pelitic rocks (Barr *et al.*, 1991). As documented here in Chapters 4 and 5, however, metabasite rocks can show marked variations in assemblage due to causes other than variations in pressure and temperature and may not be the best rock type with which to test the model. Following the success of the illite crystallinity approach in the Southern Uplands, there is potential to apply the same technique to pelites of the Taiwanese sequence to establish whether or not the metamorphic pattern suggested by the modelling can be confirmed. This

(a)

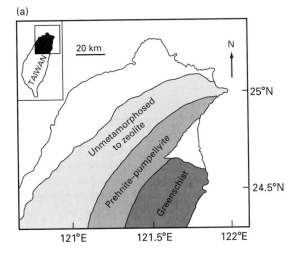

(b)

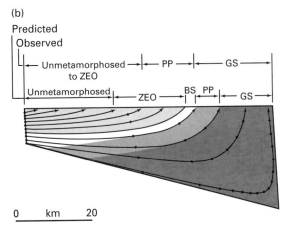

Fig. 1.3 Cainozoic metamorphism in northern Taiwan. (a) Map showing metamorphic facies distribution; (b) modelled metamorphic facies distribution for northern Taiwan, based on CMASH-FeO-Fe$_2$O$_3$ system at aH$_2$O = 0.5. Observed and predicted metamorphic patterns are shown for surface outcrop at top of diagram. (From Barr *et al.*, 1991.)

would be a valuable study as it should confirm or otherwise, whether the model of Barr *et al.* (1991) presents a realistic representation of the thermal structure in accretionary complexes.

Burial metamorphism is not a setting that can be easily displayed within schematic *P*–*T* diagrams such as Fig. 1.1, as it is normally regarded as a nonorogenic process. For example, Spear (1993) considers that burial metamorphism develops in sedimentary basins and that there are no associ-

ated thermal perturbations or deviatoric stress. He also identifies it as an anorogenic process occurring in plate interiors in large and undisturbed sedimentary basins, such as the Lower Palaeozoic, Michigan Basin or the present day basin of the Gulf of Mexico. While we agree that burial metamorphism can develop in anorogenic settings such as intracontinental basins and is characterized by an absence of deviatoric stress, we consider that the other parameters defined by Spear are not accurate attributes of many burial metamorphic terranes. The type burial metamorphism of South Island, New Zealand (Coombs, 1961), nor that extensively developed throughout large regions of the Andean mountain chain, were developed at active convergent margins and not in anorogenic settings.

The classical burial metamorphic region of South Island, New Zealand is recognized largely on structural and sedimentological evidence as an accretionary complex. Such a setting does not accord with anorogenic metamorphism and although there have been many studies of the metamorphism in volcaniclastic rocks, few data are available for metapelites and as yet no overall metamorphic model has been proposed. Conceptually, we envisage burial metamorphism as a process that operates in settings of differing genetic origin where rock sequences have accumulated to any great depth, and as detailed by Robinson (1987) many of these settings will be of an extensional basin character. Robinson (1993) recognized burial-type metamorphism as occurring in extensional basins of pre- and postorogenic character, in which the thermal flux was of differing origin. In the pretectonic setting, deep mantle uprise followed by thermal relaxation was the source of an enhanced heat flow, with thermal gradients not remaining static but decaying with time in response to the thermal relaxation. Such scenarios could develop in extensional basins formed at convergent margins such as back-arc and marginal basins. This model has been applied to the metamorphism of the Lower Palaeozoic sequences in the Welsh Basin (Bevins & Robinson, 1988), recognized as a back-arc basin. The development of a strong fabric in many areas of the basin is, however, considered by some to be definitive evidence of a syntectonic metamorphism and the pros and cons of the two models are detailed further in Chapters 3 and 5.

Low-grade burial-style metamorphism has also been recognized in postorogenic settings such as in the molasse basins of Devonian age in Norway and Scotland (Séranne & Séguret, 1987; Hillier & Marshall, 1992). In the Norwegian examples Séranne and Séguret (1987) recognized a transitional prehnite–pumpellyite to greenschist facies metamorphism, recorded by minerals in the sandy matrix of molasse conglomerates and interbedded sandstones, that was synkinematic with extension. Hillier and Marshall (1992) documented, in Devonian lacustrine basins, a burial metamorphism with temperatures reaching *c*. 250°C, using organic maturation methods and clay mineral assemblages. Robinson (1993) linked the origin of the metamorphism recorded in these settings to a two-dimensional numerical model of thermal structures in orogenic belts developed by Gaudemer *et al.* (1988). In this model, enhanced heat flow develops in wide (>*c*. 500 km) as opposed to narrow (<*c*. 500 km) orogenic belts because of lower heat loss and erosion rates. The Caledonian belt being wide would generate enhanced heat flux causing thermal weakening of the lithosphere resulting in extension recorded in extensional collapse basins (Gaudemer *et al.*, 1988), in which the infill would undergo metamorphism as a result of the high heat flow (Robinson, 1993). Accordingly, there is potential for this style of metamorphism to be recognized in other wide orogenic belts by examination of posttectonic molasse sequences.

Overall therefore, it appears that metamorphism in response to burial can be expected to occur in a variety of settings from extensional to accretionary, each having its distinctive features by which it can be recognized. The ability to integrate between metamorphic style, structure and original stratigraphy of the sequences is a powerful means to understand the interaction between low-grade metamorphism and the variety of settings in which it is developed. Much of the debate about its status appears related to its survival as a burial effect in relation to the intensity of any subsequent metamorphic/deformation overprint associated with basin closure and inversion.

1.4 Does low-temperature metamorphism matter?

As indicated at the beginning of this chapter, rocks affected by low temperature metamorphism are found over large areas of the Earth's outer continental and oceanic crust, suggesting that the processes involved in their formation are varied and diverse. For example, low-temperature metamorphism is a widespread process associated with crustal growth. From eruption at oceanic ridges to accretion onto continental margins, the alteration and recrystallization of ocean floor volcanic rocks are controlled by low temperature metamorphic processes. In parallel with volcanic rocks, low-temperature processes convert muddy sediments in oceanic trenches to shales and slates in accretionary prisms and these steadily add new crust to active continental margins. These processes have widespread implications for many major Earth systems as well as being of direct economic and environmental impact.

For example, the hydration of ocean crust during hydrothermal metamorphism at spreading centres as detailed in Chapter 6, and subsequent alteration of these metabasites as well as pelitic rocks in accretionary settings is of great influence during later stages of subduction. The recrystallization at low temperature in such settings, involving major fluid/rock interaction, will be a major influence on elemental budgets, especially of H_2O and other crustal volatiles during later subduction. The subsequent fate of such components during higher grade subduction metamorphism, whether released into the overlying mantle wedge or carried deeper into the mantle, will be of great significance as a primary control on the generation of arc volcanism, or for recycling of crustal materials into the mantle.

The processes of metamorphism, mass transfer and fluid rock interaction, operating at low temperature, are of relevance to many areas in terms of economic importance and/or environmental impact. One example is that low permeability mudrocks, such as shales and mudstones, are a cornerstone in current strategies for the underground disposal of radioactive wastes (e.g. Arch *et al.*, 1996; Horseman & Volckaert, 1996). In any case where a repository is hosted in mudrock formations, these rocks will form the final geological barrier against migration of radionuclides by sorption onto clay minerals. To a large degree, the barrier properties will depend on the diagenetic and very low-grade metamorphic history of the mudrock host, and particularly on the state of clay mineral reaction progress discussed in Chapter 2. Because mudrocks have relatively low thermal conductivity, the waste package is required to have low thermal power at the time of disposal in order to minimize alteration in the vicinity of radwaste repositories (Horseman, 1997). However, the potential effects on mudrocks of any thermal flux linked to a repository can be tested using natural analogues. As described in Chapter 3, many larger igneous intrusions into mudrocks have outer low-temperature cryptic aureoles and these local thermal effects are thought likely to be similar to that expected around repositories and thus can provide natural analogues for assessing the likely performance of mudrock-hosted repositories (Miller *et al.*, 1994). These natural analogue studies (Kemp & Rochelle, 1998) have shown that the retrogressive 'back-reactions' such as described from cryptic aureoles in Chapter 3, have an important part to play in 'self-sealing' any desiccation cracks that are likely to develop in mudrocks close to radwaste packages.

It is widely recognized that the smectite-to-illite reaction in pelitic rocks is closely associated with the initiation of hydrocarbon generation and its subsequent migration (e.g. Pollastro, 1993). There are, however, other features of this progression as detailed in Chapter 2 that can be of equal potential importance for petroleum exploration. Microfabric development linked to the transformation of smectite to illite and textural changes as a consequence of burial depth (Freed & Peacor, 1989) are associated with an improvement in the reservoir sealing properties of caps rocks. As described in Chapter 2, the new bedding-parallel microfabric that develops as the smectite-to-illite reaction progresses to completion also significantly reduces porosity, causing expulsion of fluids and gases including any hydrocarbons present. Such fundamental changes in lithification, previously regarded as simply 'compaction', also affect shale sonic velocity, one of the well established techniques used by oil exploration geologists for estimating depth of burial.

1.5 Mineral abbreviations

The use of mineral abbreviations in this book follows the scheme given by Bucher and Frey (1994).

2

Very low-grade metapelites: mineralogy, microfabrics and measuring reaction progress

R. J. Merriman and D. R. Peacor

2.1 Metapelitic rocks

2.1.1 Introduction

Pelitic rocks, including mudstone, calcareous mudstone (marl) and shale, and equivalent very low-grade metapelites, slate and phyllite, are the most abundant lithologies in basinal sedimentary sequences, where they typically form 45–55% of the succession (Tucker, 1981). Pelitic and metapelitic rocks are composed chiefly of phyllosilicate minerals. Such minerals include fine-grained clay minerals which may have a detrital or an authigenic sedimentary origin, detrital mica and chlorite flakes of silt and coarser sizes, clay minerals developed from burial diagenetic and very low-grade metamorphic reactions, and phyllosilicates produced by strain-related crystallization in tectonic microfabrics. Although clay minerals are strictly those found in clay size fractions, i.e. <2 μm-equivalent spherical diameter, this review will emphasize the importance of crystal thickness in the c^* direction rather than size in the crystallographic a–b plane. Hereafter, 'clay mineral' and 'phyllosilicate' may be used interchangeably as general terms for sheet silicates without implying any grain size difference. As discussed below, clay minerals of diagenetic origin form thin crystallites, commonly of a few tens of angstroms thick whereas phyllosilicates in tectonic microfabrics may exceed several thousand angstroms in thickness.

In this century, shales and mudstones have acquired a particular economic importance because of their role in the genesis of hydrocarbon fluids and their ability to form low permeability cap rock seals on hydrocarbon reservoirs. More recently, these pelites have become the focus of international research programmes to assess their potential as host rocks for radioactive and toxic waste repositories.

Slate has a long history of use since at least Roman times, providing not only roofing but also damp-proof courses, coping stones, sills, cisterns and laboratory worktops. Usage of roofing slate increased vastly as a result of the industrial revolution and the need for cheap housing. Because slate could be split into thin sheets, as little as 3–4 mm thick, and retain relatively high structural strength, housebuilders needed fewer and thinner timber supports for slate roofs compared with traditional tiles. Another advantage of slate proved to be its resistance to acids, for during the corrosive onslaught of 19th century industrial pollution many a slate roof survived better than the supporting stone and brickwork.

Although slate was one of the first rocks investigated with the petrological microscope (Sorby, 1853), the nature of the phyllosilicate minerals in slate and other pelitic rocks could not be revealed until the advent of X-ray diffraction (XRD) techniques and their application to clay minerals in the 1920s and onwards (Brindley & Brown, 1980; Moore & Reynolds, 1989). Systematic XRD studies of clay mineralogy in relation to sedimentary sequences began in the 1950s and much of the early work was linked to hydrocarbon exploration. As a result, several accounts of the progressive downhole illitization of smectite·in Tertiary sequences in the Gulf Coast, USA, appeared at approximately the same time (Burst, 1959; Powers, 1959; Weaver, 1959). Subsequently, the smectite-to-illite reaction was confirmed as a fundamental diagenetic process in pelitic sequences through the classic XRD studies of Hower *et al.* (1976). Regional studies of clay mineralogy in relation to very low-grade metamorphism led to the development of XRD-based clay mineral crystallinity techniques for indexing the transition from diagenesis to metamorphism in the 1960s. The application of these and other techniques to very low-grade terranes, described in

Chapter 3, has established the continuous nature of diagenetic and metamorphic processes in very low-grade metapelites.

One of the disadvantages of the XRD technique is that samples must be disaggregated to remove detrital minerals before investigating the authigenic clay minerals. The disruption of pelitic microtextures, particularly the disarticulation of delicate clay crystals and the preparation of separated material for XRD analysis, has generated controversy on the nature of layer coherency in smectite, I/S and illite, and the status of 'fundamental particles' (Nadeau, 1985). The application of the transmission electron microscope (TEM) to studies of pelites and metapelites allows clay minerals to be imaged *in situ* using intact ion-milled samples which preserve microtextures of clay grains and the microfabrics developed by diagenetic and very low-grade metamorphic processes (Peacor, 1992a). Moreover, TEM provides an independent method of measuring clay mineral crystallite size, for comparison with XRD-measured crystallinity, and analytical electron microscopy (AEM) can directly analyze individual clay crystals.

Over the past decade or more, the application of integrated TEM and XRD studies has provided evidence that clay minerals in pelitic and metapelitic rocks undergo a series of progressive mineralogical, compositional and microtextural changes at low temperatures, generally below 300°C. The main purpose of this chapter is to characterize these changes in terms of prograde mineral reactions and the development of microtextures and microfabrics.

2.1.2 Metapelitic zones and lithology

The anchizone (Kübler, 1967a,b, 1968) is widely recognized as a transitional zone of very low-grade metamorphism between diagenesis and low-grade (greenschist facies) metamorphism in metapelitic sequences (Kisch, 1983; Frey, 1987a). Definition of the anchizone using the Kübler index (KI in $\Delta°2\theta$) of illite crystallinity allows the prograde limit of diagenesis and the initiation of the epizone to be defined also (Kisch, 1990, 1991a). These three metapelitic zones and the subdivision of the anchizone are shown in Fig. 2.1, together with typical lithologies and their characteristic microfabrics. The zonal distribution of the main phyllosilicate reaction series and some indicators of progress in the smectite–illite–

muscovite series are also shown in Fig. 2.1, and are discussed in section 2.2.

The *late diagenetic zone* is broadly equivalent to the initial stage of deep epigenesis — 'zone of altered clay matrix' — of Kossovskaya and Shutov (1961, 1970), and also Kübler's (1967a, 1968) 'diagenetic zone'. Kisch (1983) preferred the term 'burial diagenesis', which he used synonomously with 'burial metamorphism' as used by Coombs (1961). Since sedimentary burial, however, is integral to diagenetic processes, burial is probably a non-essential descriptor. Here, late diagenesis is distinguished from early diagenesis (Fig. 2.1) largely on the basis of the smectite-to-illite reaction (section 2.2.2) and the typically postcompactional, bedding-parallel microfabric generated as the reaction progresses to completion (section 2.3.2). In contrast, early diagenesis in pelitic rocks leads to modification of the depositional fabric, through compaction and dewatering (e.g. O'Brien & Slatt, 1990; Bennett *et al.*, 1991), and the authigenic modification of clay mineral assemblages, generally below 100°C. These modifications are not considered further here.

Typical pelitic lithologies of the late diagenetic zone are claystone, mudstone and shale. The classification of pelitic lithologies and macrofabrics was comprehensively reviewed by O'Brien and Slatt (1990). Here the term mudstone is used for an indurated mud composed of approximately equal amounts of silt and clay that lacks fissility; a claystone is similar but contains more than two-thirds clay. The term shale is used for a fissile mudstone or claystone that splits easily into thin sheets along planes approximately parallel to bedding. At outcrop claystones and mudstones commonly show a tendency to spall into centimetre-size blocks along polygonal shrinkage cracks. Well-developed shrinkage polygons generally reflect the presence of smectite or mixed-layer illite/smectite. Late diagenetic shales may readily split into centimetre-size flakes and develop a litter of these flakes in degraded outcrops. Scanning electron microscopy (SEM) and TEM images reveal postcompactional microfabrics in the clay matrix of both mudstones and shales, and these have an overall bedding-parallel orientation (section 2.3.2). Scaly mudstones and scaly argillites are commonly found in accretionary sequences. These fabrics are generated by reorientation

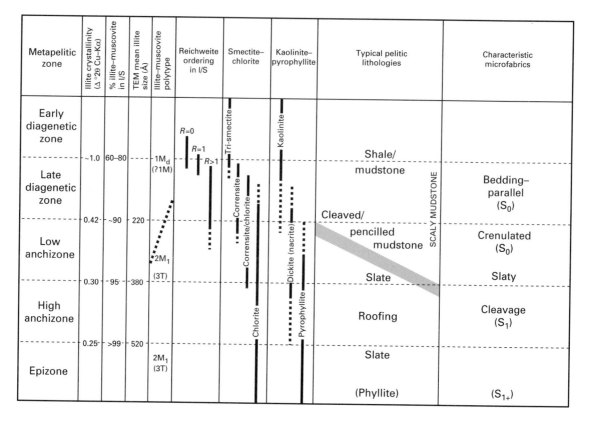

Fig. 2.1 Metapelitic zones showing associated lithologies and microfabrics. Illite crystallinity and other indicators of progress in the smectite–I/S–illite–muscovite series of minerals are used to index the zones. Distribution of trioctahedral phyllosilicates based on Árkai (1991), Hillier (1993b), Inoue and Utada (1991), and Jiang and Peacor (1994a,b). Distribution of kaolinite–pyrophyllite minerals after Frey (1987b), Ehrenberg et al. (1993), Spötl et al. (1993), Ruiz Cruz and Andreo (1996) and Wang et al. (1996).

and disruption of existing minerals on slip surfaces and do not develop pervasive microfabrics.

The changes in lithology that characterize the transition from the late diagenetic zone to the lower grade part of the anchizone — *the low anchizone* — were documented by Kisch (1991a). Intersection pencil structures and spaced cleavage are associated with a 'primary crenulation' in shales, mudstones and siltstones. At outcrop and in cuttings such lithologies typically weather to a litter of centimetre-thick pencils or flat tiles with a shape and size which is controlled by the bedding-cleavage relationship (Durney & Kisch, 1994). Characteristically, these

metapelites show the bedding-parallel microfabric crenulated by slaty cleavage microfabric (section 2.3.2). Although the appearance of pencil structure is a useful field indicator of incipient regional metamorphism, in some terranes bedding and cleavage fabrics may be subparallel over wide areas, particularly in some thrust belts. In such terranes slaty cleavage microfabric develops by enhancement of the bedding-parallel microfabric.

A subdivision of the anchizone into the *low anchizone* and the *high anchizone* at a KI value of 0.30 $\Delta°2\theta$ is sometimes made on the basis of regional metapelitic data (see Chapter 3) and field observations (Fig. 2.1). Because of the nonlinear nature of the Kübler index, the midpoint of the anchizone is taken at 0.30 $\Delta°2\theta$, as opposed to 0.34 $\Delta°2\theta$, and approximates to the midpoint in terms of the equivalent TEM mean crystallite size (see Fig. 2.19). Field observations suggest that this midpoint coincides with distinctive changes in lithology. Metapelites with a penetrative slaty cleavage, equivalent to the 'strong cleavage fissility' illustrated by Durney and Kisch (1994, Fig. 11e,f), progressively dominate lithologies of the high anchizone in many terranes (e.g. Merriman & Roberts, 1985; Ishii, 1988; Kisch, 1991b). Associated sandstones and greywackes may also develop a rough cleavage fabric. Most of the high quality roofing slates in Wales were exploited from areas of the high anchizone and epizone (e.g. Roberts & Merriman, 1985; Roberts *et al.*, 1991). Additionally, in some thermal aureoles the midanchizone represents the outer limit of concentric patterns of cryptic contact metamorphism (see section 3.4).

The transition from the high anchizone to the epizone is lithologically less distinctive than the low–high anchizone transition. Slates and phyllites are the characteristic lithologies of the epizone, and both rock types are notably more durable in terms of weathering properties than mudstones and shales. Some phyllites, with characteristically lustrous cleavage surfaces, may show more than one cleavage microfabric in the form of a second crenulation cleavage.

2.1.3 Metastable equilibrium and clay mineral reaction progress

Chemical equilibrium dominates the framework of metamorphic petrology through the concept of metamorphic facies. This derives from the observation that, for a given bulk rock composition, mineral assemblages change in simple, repetitive ways with changing pressure or temperature. Such regular predictable changes, as exemplified by classic Barrovian zones, were subsequently validated by a wealth of experimental data. One of the key elements of interpretation of such data is experimental reversibility, a necessary condition for true chemical equilibrium in laboratory experiments. Such conditions appear to be applicable to rocks which formed at or above temperatures corresponding to the greenschist facies, where relatively coarse-grained metapelites commonly display at least an approximation to chemical and textural equilibrium. However, these conditions do not extend to lower grade metapelites.

Pelites and metapelites formed at lower grades also commonly display regular, predictable patterns of change in mineral assemblages and textures, e.g. among dioctahedral clay minerals the sequence

involving smectite → mixed-layer illite/smectite → illite → muscovite is observed universally as a function of grade. These changes are coupled with a regular pattern of increasing crystal size and crystal perfection as commonly measured by illite crystallinity, i.e. as reflected in the width of the 001 peak of illite recorded by XRD. Such regular patterns have therefore commonly been inferred to be diagnostic of equilibrium conditions, and therefore to have states which are predictable functions of temperature. As such, there have been many attempts to characterize clay mineral related 'geothermometers' which, theoretically, should permit accurate determinations of temperature of formation on the basis of carefully studied phase relations under conditions of late diagenesis and very low-grade metamorphism (see section 2.5).

Lippmann (1981, 1982) emphasized, however, that within the temperature range normally accorded to very low-grade metamorphism, the ranges of solid solution of minerals increase with decreasing grade. Such a relationship is contrary to normal stable equilibrium conditions. Lippmann therefore hypothesized that such minerals do not represent chemical equilibrium. In support of this concept, Jiang *et al.* (1990b) and Jiang and Peacor (1994b) presented direct evidence that illite is a metastable phase because it has a composition which, to a first approximation, falls within the solvus between pyrophyllite and muscovite. Essene and Peacor (1995) have expanded on this concept, based in part on TEM observations of microtextures which are incompatible with stable equilibrium.

Peacor (1992a) noted that, with increasing grade, TEM data confirmed that crystal size increases (the principal cause of decreasing KIs of illite crystallinity), crystal defect densities decrease as shown by decreasing proportions of layer terminations in clay minerals, compositions become less variable, and microtextural relations become less complex as mixed-layering decreases. Overall there is a general trend to decreasing disorder. This trend is consistent with a series of prograde transitions in which pelites pass through a sequence of metastable states as they approach stable chemical equilibrium in the lithological sequence mudstone and shale, slate, phyllite and schist. This sequence of changes is one involving increasing homogeneity of composition of a given mineral, decreasing defect density (including decrease in mixed-layering) of individual crystals, increasing crystal size, and decrease in the number of minerals (phases). The end-product of this sequence is coarsely crystalline schist comprising a small number of homogeneous, relatively defect-free minerals which obey the Gibbs phase rule, to a first approximation, i.e. the higher-grade rocks approach a state of stable equilibrium.

Such a sequence can be viewed as one in which the transitions are compatible with the Ostwald Step Rule (Ostwald, 1897, 1900; Morse & Casey, 1988), i.e. any transition in mineral assemblage or texture is one in which the resultant free energy of the system is decreased, and therefore the system is nearer to that of stable chemical equilibrium. In pelites where the clay assemblages include minerals such as dioctahedral smectite, increasing temperature (grade) promotes reactions which ultimately give rise to muscovite. However, because the illite structure is more like that of smectite than that of muscovite, the rate of reaction of smectite-to-illite dominates as a first step, resulting in formation of a metastable mineral, illite. Likewise, crystal size increases in a gradual way as exemplified by the transformation of fine-grained shales to relatively coarse-grained slate and phyllite, the principal cause of changes in illite crystallinity.

The underlying cause of formation of transitional, metastable states, as opposed to a state of stable chemical equilibrium, is in reaction rates. Unlike relations at higher grades where the concepts of facies and stable equilibrium are applicable, reactions in pelites are sluggish. Morse and Casey (1988) have elegantly and simply emphasized that the basic reason for the occurrence of metastable phyllosilicates is the existence of a continuous series of intermediate structural/chemical states between clay minerals which form metastably at the lowest temperatures (e.g. smectite) and those which form under conditions of stable equilibrium at high temperatures. For example, the structures and compositions of the various mixed-layer illite/smectite clays and illite are intermediate to those of smectite and muscovite. Where smectite reacts to form a prograde member of the sequence there is a specific rate of reaction to each of the intermediate products. Although muscovite is the stable phase, the rate of reaction to form intermediate products is much faster than that to form muscovite, because their structures and compositions are more like those of reactant smectite. They therefore form first. In a second reaction, such intermediate products may transform to a dioctahedral phase still closer in composition to muscovite. Thus reactions occur in a series of steps in which products are in systems of lower free energy, i.e. consistent with the Ostwald Step Rule.

While there is considerable evidence from regional metapelitic sequences that reactions follow the Ostwald Step Rule, phyllosilicates formed in environments where reaction rates are enhanced, e.g. hydrothermal systems, tend to be the stable end-product in a given reaction series rather than an intermediate product. As discussed by Morse and Casey (1988), where reactions can follow different but parallel reaction pathways, the character of the products is controlled by the relative reaction rates, and each possible pathway contributes to the overall reaction rate.

Although very low-grade metapelites generally display metastable equilibria they also demonstrate a high degree of mechanical and structural stability because of the irreversible nature of prograde assemblages at surface P–T conditions. Hence the remarkable durability of some roofing slates. This explains also why experimentalists are generally unable to produce equilibrium reactions which are rigorously verified with experimental reversals. Nevertheless, mineral transitions in pelites as a function of metamorphic grade do generally follow well constrained sequences. Because reaction rates are constrained by kinetic factors, the reactants at a given step do not generally represent stable equilibrium, at least at grades below the greenschist facies, and cannot therefore be accurately and predictably associated with a specific temperature. On the other hand, the specific state of a given system can be ascribed to a given stage of the range of reactions, i.e. it can be characterized as corresponding to a state of *reaction progress*. This is generally the case because, with rare exception (see below), pelites are commonly metamorphosed in a more or less continuous prograde manner which, with increasing temperature, results in a unidirectional approach to the equilibrium state. In the case of the dioctahedral sequence smectite → illite → muscovite, the reaction progress would be specified by the compositional, textural and crystal structural relations of a given sample. Those states are generally defined by data obtained by XRD or TEM, as described in section 2.4.

2.2 Mineralogical relations

2.2.1 Phyllosilicate reaction series

As noted above, prograde sequences of minerals have been observed to occur, with few exceptions, in regularly repeating sequences as a function of grade (Fig. 2.1). There are two principal series, involving dioctahedral and trioctahedral clays, respectively:

1 dioctahedral 2 : 1 clays
   smectite → mixed-layer illite/smectite (I/S) → illite → muscovite
2 trioctahedral 2 : 1 clays
   smectite → mixed-layer chlorite/smectite (C/S) → chlorite.

The occurrence of one or the other sequence is related to bulk pelitic composition, e.g. basaltic or mafic rocks give rise primarily to the trioctahedral sequence as discussed in Chapter 4. The dioctahedral sequence usually is dominant in pelitic sediments, in part because of the presence of sialic or acidic crustal, plutonic or volcanic components, but it is normally accompanied by the separate triocta-hedral sequence. Where sediments have a high ferromagnesian component, e.g. where they are dominated by a more mafic volcanic component, trioctahedral minerals may even be dominant relative to dioctahedral minerals. That is, both trioctahedral and dioctahedral phases coexist, with their relative proportions varying according to bulk rock composition; Jiang and Peacor (1994a) have suggested that the relative proportions can be used as evidence of the tectonic setting of original sedimentation. The two common sequences are described below, and less common sequences described in subsequent sections. A summary of the main reaction series is shown in Fig. 2.1.

2.2.2 Smectite–I/S–illite–muscovite

The sequence involving dioctahedral clay minerals is, with increasing temperature, or grade: smectite → mixed-layer I/S → illite → muscovite. The full sequence from earliest diagenesis to epizonal (greenschist facies) metamorphism has not been observed in a single continuous sequence of meta-

pelitic zones. Where large portions of the sequence are observed (see Chapter 3), they are generally one of two types:

1 Early to late diagenesis or lowest anchizone in basinal sequences where pelites have been subjected only to sedimentary burial with associated over-burden stress, but without tectonic stress. Smectite, either of detrital origin or derived from volcanic glass, undergoes a transformation to illite at depths (temperatures) in the order of 3000 m (90°C). The illite so derived occurs in thin packets (thickness approximately 100 Å) with a bedding-parallel orientation (section 2.3.2). The classic setting for this range of the sequence is the Gulf Coast of the United States (Hower et al., 1976), where samples have been obtained from drill-core to depths of c. 6000 m. Samples from significantly greater depths, and thus of higher grade, have not been obtained from cores. They are therefore available only in rocks which have been uplifted and generally affected by tectonic stress.

2 Late diagenesis through the anchizone to epizone. Thin packets of bedding-parallel, authigenic illite are gradually replaced by larger crystals of muscovite several hundreds of angstroms in thickness oriented parallel to slaty cleavage microfabric. These musco-vite crystals continue to evolve into larger, mature crystals typical of schists in response to increased temperature coupled with tectonic stress.

The diagenetic and low-grade metamorphic ranges of the prograde sequence have been documented in many localities, and involve processes which, in detail, are subject to intense debate. In the limited space available here, we give a simple overview of the generally accepted mineralogical relations, followed by a brief description of relations as observed in three well characterized settings: (a) Gulf Coast sediments; (b) mudstones and slates of the Welsh Basin; (c) metapelitic rocks of the Helvetic Alps. In terms of integrated TEM and XRD tech-niques, the first two sequences are perhaps the two best studied, well defined examples which cover the full range of diagenetic and very low-grade metamorphic conditions in extensional basins. The third represents a classic fold-and-thrust belt where the transition from diagenesis through epizonal metamorphism has been characterized by detailed XRD clay mineralogy and TEM studies of white mica. Although not strictly metamorphic, diagenetic relations are described here because they are part of a continuous prograde sequence extending to the greenschist facies, including phases which are transitional to those of the anchizone, and are also represented by retrograde overprints on higher grade systems.

2.2.2.1 Mineralogical relations

Although the smectite structure is known to be of the 2:1 type, it has not been characterized in detail because crystals sufficiently large and well-enough ordered for conventional structure analysis have not been found, and probably do not exist. The end-member 2:1 dioctahedral phyllosilicate is pyrophyllite, the interlayer sheet of which contains no cations, H-bonding serving to bind adjacent 2:1 layers. Substitutions principally of Al^{3+} for Si^{4+} in tetrahedral sites, and of Mg^{2+} and Fe^{2+} for Al^{3+} in octahedral sites produce a 'net negative charge' which is balanced by interlayer cations, with the extent of substitutions becoming progressively larger in the sequence smectite $\rightarrow$ illite $\rightarrow$ muscovite. Ransom and Helgeson (1993) and Jiang et al. (1992a) collated available compositional data, and noted that each of the minerals displays a characteristic range of composition in terms of net negative charge, and therefore of charge-balancing interlayer cations. Interlayer cations are dominated by Ca, Na or K, with much recent data implying that smectite in marine sediments has K as the dominant interlayer cation (e.g. Masuda et al., 1996).

The relatively small net negative charge of smectite gives rise to minimal strength of bonding of interlayer cations to adjacent 2:1 layers. This in turn leads to unique attributes of smectite consisting of: (1) expansion or contraction of d(001) as water molecules that coordinate to interlayer cations are gained or lost in response to changes in partial pressure of water; (2) expansion of the interlayer with absorption of polar molecules such as ethylene glycol, glycerol or n-alkylammonium ion (e.g. Vali & Koster, 1986); (3) rapid exchange of interlayer cations with cations in pore fluids even at low temperatures.

Clay minerals intermediate to smectite and illite are referred to as 'mixed-layer I/S' (Reynolds & Hower, 1970). The nature and extent of mixed layering is controversial. For example, Środoń and Eberl (1984) and Środoń et al. (1992) concluded from collated data that a continuous series of mixed layering occurs. Other studies suggest that a discontinuous sequence of mixed-layer minerals is more common; for example Ransom and Helgeson (1993) with respect to composition, and Dong et al. (1997b) with respect to TEM observations of layer sequences. However, Dong et al. (1997b) emphasized that both ordered discontinuous and disordered continuous sequences may occur depending on formation conditions. The mineral rectorite is referred to as R1 I/S, indicating that where there are equal numbers of illite-like and smectite-like interlayers, there is an alternation in an ordered fashion of layers...ISISIS... (see Moore & Reynolds, 1989, for a review of the nomenclature and mineralogical relations). Similarly R2 (...IISIIS...) and R3 (...IIISIIIS...) ordered members of the sequence are known.

Much XRD data imply that the transformation of smectite to illite proceeds through a more or less continuous sequence (e.g. Hower et al., 1976), that is consistent with I/S being composed of illite- and smectite-fundamental particles (e.g. Nadeau et al., 1984b,c). Other data (e.g. NRM data) indicate that R1 I/S is a unique phase with Al/Si distributions in 2:1 layers being symmetric across interlayers, and not across octahedral sheets as in smectite and illite (Jakobsen et al., 1995). This implies that I/S consists of McEwan crystallites which are coherent sequences of smectite-like and illite-like layers, but not smectite or illite layers sensu stricto. Dong et al. (1997b) reviewed such relations, and described TEM observations supporting this view.

Illite has a net negative charge in the order of −0.75 per formula unit (pfu). Al contents of approximately 0.5 pfu are typical, with Fe^{2+} and Mg (phengitic) components being in the order of 0.1 pfu. The dominant interlayer cation is K^{1+}, but a solid solution series exists to brammallite, for which Na^{1+} is dominant. Ammonium may be a significant interlayer cation species, but its presence in small quantities often goes undetected due to the difficulty of analyzing for N or NH_4. Because of the much larger net negative charge which results in increased occupancy of interlayer cation sites as compared with smectite, illite is nonexpandable; interlayer cations are referred to as being 'fixed', i.e. the capacity for exchange is much diminished (e.g. Środoń et al., 1992).

Muscovite has a net negative charge approaching the ideal value of 1.0 pfu. It is the thermodynamically stable 2 : 1 dioctahedral phyllosilicate, with nearly complete solvi separating it from pyrophyllite and paragonite in the respective binary systems. Solid solution of Na and interlayer vacancies are thus extremely limited where equilibrium is attained (e.g. Guidotti, 1984; Jiang *et al.*, 1990b; Roux & Hovis, 1996). Although very low-grade, metastable phases may exhibit extensive solid solution in the case of the muscovite–paragonite system (Li *et al.*, 1994b), these cannot be readily distinguished from intimate mixtures of Na- and K-rich mica domains (Livi *et al.*, 1997). Phengite is the variety where net negative charge is in large part determined via substitution of Fe^{2+} and Mg in octahedral sites, with the amount of Si in tetrahedral sites exceeding 3 pfu, with less than 1.0 Al pfu. The amount of phengite solid solution has been related to pressure (Velde, 1965; see section 2.4.1.2). Guidotti (1984) gave an extensive review of aspects of the composition, structure and phase relations of muscovite and other micas in metamorphic rocks.

Fundamental particles. The concept of 'fundamental particles' and its corollary 'interparticle diffraction' was introduced in a series of papers by Nadeau and colleagues (Nadeau *et al.*, 1984a,b,c; Nadeau, 1985; Nadeau & Bain, 1986) and summarized by Nadeau (1998). Samples of smectite and R1 I/S, which were separated for preparation of XRD samples, gave particles 10-Å and 20-Å thick. They were referred to as elementary smectite and illite particles, respectively. Particles from samples having a higher proportion of illite layers had thicknesses which were multiples of 10 Å, and were referred to as 'fundamental illite particles'. They were assumed to be related internally by coherency between layers because they gave single-crystal SAED hk0 patterns. Cleavage was inferred to have occurred only on the expandable smectite interlayers, which were incoherent (turbostratic). The collection of separated fundamental illite particles was therefore theorized to duplicate the collection of coherent illite units in stacks of layers in original rocks, i.e. the stacks of layers in original rocks were thought to consist of sequences of coherently related illite layers separated by incoherently related smectite interlayers. When reconstituted as a sample for XRD analysis, the illite elementary particles were inferred to become separated by smectite-like, incoherent expandable interlayers, adjacent particles thus giving rise to 'interparticle diffraction'. The key element in these concepts is that, except for the sequence in which illite fundamental particles occurred, the XRD sample which is comprised of a sequence of illite fundamental particles behaves as would the original sequence of layers in original rocks.

Aspects of fundamental particle theory have been questioned on the basis of TEM data (e.g. Dong & Peacor, 1996; Peacor, 1998), which have demonstrated coherency across smectite (and illite) interlayers and cleavage within illite packets. Peacor (1998) has also emphasized that, although Nadeau and colleagues defined a working test for 'fundamental particles' involving single-crystal SAED hk0 patterns, such tests are virtually never used, separated particles being generally assumed to be 'fundamental particles' without adequate evidence. That is, separation of particles during sample preparation may create crystallites whose relation to original layer sequences in rocks is only indirect, in part different to that assumed by fundamental particle theory, but with relations which may be different for different rock types according to the severity of the separation process. On the other hand, Reynolds (1992) has demonstrated that, at least for his samples, there were no differences in XRD patterns of clays as obtained both from original rocks and separates. More research is necessary in order to define the relations between interlayers and layer sequences in original rocks and in those separates prepared for XRD analysis.

2.2.2.2 Gulf Coast burial diagenetic sequence

The sequence of clay minerals developed as a function of depth is well displayed in core samples to approximately 6000 m depth in the Tertiary mudstones and shales in the Gulf Coast of the United States. This sequence has been extensively studied, but the paper by Hower *et al.* (1976) is one of a series that brought definition to the mineralogical relations. The specific processes involved in the transformation are still controversial, various models ranging from dissolution and neocrystallization through solid-state replacement of individual layers. There is a vast literature on the subject, but a sampling of representative papers which discuss mechanisms include those by Inoue *et al.* (1990), Amouric and Olives (1991), Lanson and Champion (1991), Lindgreen and Hansen (1991), Eberl (1993), Huang *et al.* (1993), Whitney and Velde (1993), Huggett (1995), Elliott and Matisoff (1996), Nieto *et al.* (1996) and Dong *et al.* (1997b).

The now classic curve of Hower *et al.* (1976) showing the proportion of illite layers in I/S as a function of depth, implies that smectite at shallow depths consists of Reichweite ordering values (R) of $R = 0$ (disordered) mixed-layer I/S with approximately 25% illite layers. Starting at depths of approximately 3000 m, the proportion of illite layers as measured by XRD increases markedly and continuously as a function of increasing depth, reaching a value of about 80% at a depth of 3700 m (Fig. 2.1). Such 'illite' is really mixed-layer I/S with a small proportion of smectite, *sensu stricto*. Little additional change occurs up to depths approaching 6000 m.

TEM data show that at approximately 80% illite in I/S, marking the beginning of the late diagenetic zone, the illite consists of packets of layers 50–100 Å in thickness and having an orientation of the crystallographic *a–b* plane parallel or subparallel to bedding (e.g. Ahn & Peacor, 1986). Ho *et al.* (in review) have used an XRD method to show that there is little preferred orientation at depths more shallow than that of the I/S transition, but that there is a dramatic increase to well-defined preferred orientation concomitant with the transition, and no further change with increasing depth. Although the illite composition is restricted to a well-defined range, the illite is defect-rich, heterogeneous in composition, and complexly intergrown with layers or packets of layers of chlorite, berthierine (tri-octahedral, serpentine-type structure) and kaolinite. With increasing depth, the most distinctive change is a small increase in average packet thickness (e.g. Eberl, 1993).

2.2.2.3 Welsh Basin metapelitic sequence

Very low-grade metapelites from the Welsh Basin (see section 3.3.2.1) have been extensively studied by XRD and TEM using samples that represent a continuous range of grade from the late diagenetic zone through the epizone (e.g. Merriman & Roberts,

1985; Merriman *et al.*, 1990, 1992; Li *et al.*, 1994b). The sequence of change in texture, structure and composition of white micas is well defined and exhibited in part in TEM images shown in section 2.4.2.

Late diagenetic pelites are similar to those from the deep portions of the Gulf Coast sequence, with crystals of illite having a thickness range of 50 to 200 Å. Such thin packets are oriented parallel or subparallel to bedding, and commonly occur in complex stacks of separate packets. The layer stacking sequence is that of a disordered $1M_d$-like sequence, and in some cases, the illite has a large interlayer Na component, with a range extending over much of the solvus between muscovite and paragonite, but with compositions which are, as with all illite, deficient in interlayer cations and rich in tetrahedral Si, as compared with ideal muscovite or paragonite. Crystals typically show strain features, in response to tectonic stress. With increasing grade, thicker crystals occur in the bedding-parallel orientation having compositions tending toward end-member paragonite and muscovite, and with $2M_1$-dominant polytypism.

With the development of incipient slaty cleavage microfabric, strain-free micas are observed to attain a narrow range of orientations parallel to subparallel with the cleavage direction. In areas of open folding, the contrast between bedding-parallel and cleavage-parallel orientations tends to be sharp with few intermediate orientations (Ho *et al.*, 1996; van der Pluijm *et al.*, 1998), similar to the shale-to-slate transition at Lehigh Gap, Pennsylvania (Ho *et al.*, 1995). Such observations imply a reorientation mechanism consisting of dissolution and neo-crystallization, with dissolution in part related to the degree of tectonic stress-induced deformation. As consistent with that relation, cleavage-parallel packets are thicker, have well-defined subhedral shapes, consist of well-ordered $2M_1$ polytypes, and have compositions near end-member muscovite and paragonite.

Samples with incipient cleavage typically display coexisting bedding-parallel and cleavage-parallel modes of white mica. With increasing grade, the proportion of bedding-parallel mica decreases, as the proportion and crystal thickness of cleavage-parallel crystals increases. The end product is one in which relatively thick packets of well-defined, thermodynamically stable, defect-poor mica crystals (and trioctahedral chlorite and other minerals; see below) attain orientations parallel to cleavage, as a replacement of metastable, defect-rich illite which previously existed in the bedding-parallel orientation.

2.2.2.4 Helvetic Alps

The transition from diagenesis to metamorphism has been extensively studied in the external fold-and-thrust belt of the Helvetic Alps of eastern Switzerland (section 3.3.4.1). Using a combination of illite crystallinity and computer modelling of XRD data, Wang *et al.* (1996) characterized reaction progress in terms of four zones. Zone 1 is defined by the presence of $R = 0$ I/S with 20–25% illite, i.e. the early diagenetic zone (Fig. 2.1), but also contains $R = 1$ I/S with 50–85% illite and $R > 1$ I/S with $> 85\%$ I/S. Zone 2 is defined by the presence of $R = 1$ I/S with 50–85% illite and the absence of $R = 0$ I/S, but may also contain $R > 1$ I/S with $> 85\%$ I/S, i.e. it corresponds to the beginning of the late diagenetic zone (Fig. 2.1). Zone 3 is defined by the presence of $R > 1$ I/S with 85–90% illite and the absence of $R = 1$ I/S, but may also contain $R > 1$ I/S with $> 90\%$ I/S, i.e. it corresponds to the late diagenetic zone. Zone 4 is defined by the presence of $R > 1$ I/S with $> 90\%$ illite and the absence of all other ordering types, i.e. it corresponds to the beginning of the anchizone.

Liassic black shales occur over a wide area of Central Switzerland that extends across the transition from diagenesis through the epizone. The textural and compositional evolution of white micas representing this transition were characterized by Livi *et al.* (1997) using TEM and electron microprobe techniques. They showed that reaction progress is towards larger, more defect-free grains that approach end-member compositions. For example, diagenetic illite and I/S show a high degree of heterogeneity in terms of Si and Mg + Fe contents, whereas epizonal K-micas have near to end-member muscovite compositions and show a considerable reduction in phengitic component compared with anchizonal K-micas. Na-mica first appears in the late diagenetic zone or low anchizone from an illitic or smectitic precusor and forms nanometre-scale mixed Na-K micas + brammallite. These white micas evolve through the anchizone to form discrete

paragonite + muscovite + margarite intergrowths in epizonal metapelites.

2.2.3 Smectite–corrensite–chlorite

The sequence of changes for trioctahedral Mg, Fe-rich clay minerals is analogous to that for smectite–I/S–illite–muscovite. The specific variety of smectite is commonly the trioctahedral species saponite (Mg-dominant in octahedral sites), although there is a sequence between saponite (expandable; low net-negative charge) and vermiculite (higher net-negative charge, limited expandability). With increasing grade saponite generally is replaced by corrensite, which is ordered, 1:1, mixed-layer chlorite/smectite (C/S) or vermiculite/chlorite. Corrensite is in turn replaced by trioctahedral chlorite in the late diagenetic zone or low anchizone (Fig. 2.1).

The sequence of trioctahedral phases has been observed in a wide variety of environments, including:

1 Hydrothermal alteration of volcaniclastic rocks (Inoue *et al.*, 1984; Inoue, 1987; Inoue & Utada, 1991).

2 Prograde metamorphism of pelitic rocks from diagenesis through the epizone (Hillier, 1993b; Jiang & Peacor, 1994a,b; Dalla Torre *et al.*, 1996b; Schmidt & Livi, 1998; Schmidt *et al.*, 1999).

3 Precipitation from and replacement via hydrothermal solutions in pelitic, marine sediments at the Juan de Fuca spreading ridge (Buatier *et al.*, 1995).

4 Hydrothermal and regional metamorphism of primary Mg-silicates and glass in basic igneous rocks, including ophiolites and MORB (Kristmannsdóttir, 1979; Bettison & Schiffman, 1988; Shau *et al.*, 1990; Bettison-Varga *et al.*, 1991; Schiffman & Fridleifsson, 1991; Shau & Peacor, 1992; Alt, 1993, 1995; Robinson *et al.*, 1993; Schiffman, 1995; Schiffman & Staudigel, 1995; Bettison-Varga & MacKinnon, 1997; Schmidt & Robinson, 1997; see Chapter 4 for description).

5 Hydrothermal alteration of mafic minerals of amphibolites and gneisses (Beaufort & Meunier, 1994).

2.2.3.1 General mineralogical relations

The saponite–corrensite–chlorite sequence has been described in two different ways:

1 As mixed-layer C/S, in which intermediate members are viewed as comprising layers of chlorite and smectite. Corrensite, which is ideally *R*1 C/S, is viewed as having ordered, alternating layers which individually are equivalent to the respective end-members chlorite and smectite. This is based primarily on XRD data implying that there is a continuous sequence of mixed-layer phases from pure smectite to pure chlorite, or at least over a significant portion of the range of apparent mixed-layering, but some TEM data also support this relation (Bettison-Varga *et al.*, 1991);

2 As a system in which corrensite is unique in structure and composition, with chlorite-like and smectite-like layers, i.e. the layers are similar to, but with unique structure and composition, relative to chlorite and smectite layers, *sensu stricto* (Reynolds, 1988). Reynolds (1988) and Shau *et al.* (1990) asserted that corrensite, like rectorite, has a structure in which the proportion of Al^{iv} substituting for Si is symmetrical across interlayers rather than across octahedral sheets, thus requiring that interlayer charge in alternate interlayers be different, serving as a driving force for ordering of alternate high- and low-charge interlayers. Mixed-layer phases are thus viewed as mixed-layer smectite/corrensite, S/Co or corrensite/chlorite, Co/C. Much of the evidence for such relations has been derived from TEM observations (e.g. Shau *et al.*, 1990; Shau & Peacor, 1992; Jiang & Peacor, 1994c) but also from XRD data of sequences with variable grade (e.g. Schmidt & Robinson, 1997). Roberson (1988) has emphasized the difficulty in distinguishing between the two kinds of mixed-layering by XRD. Several authors have noted that both kinds of sequences are possible depending on degree of approach to equilibrium (e.g. Shau *et al.*, 1990), ordered mixed-layer sequences corresponding to the more stable assemblage.

2.2.3.2 *Sequence in metapelites*

The sequence of trioctahedral phyllosilicates occurring in metapelitic rocks of the Gaspé Peninsula, Quebec (see section 3.3.4.2) is typical of the zonal range from late diagenesis through the epizone. The trioctahedral clay minerals are directly associated with a smaller proportion of dioctahedral clays which range from illite with some mixed-layer

smectite through muscovite, as determined by crystallinity indices (Islam *et al.*, 1982).

In the late diagenetic zone the principal trioctahedral mineral is corrensite, presumably derived in part from mafic volcanic detritus where smectite originally formed as an intermediate phase. Much corrensite, however, was observed to be derived from detrital biotite which was replaced directly by complex interlayered phyllosilicates consisting primarily of corrensite, but also with smectite and chlorite layers (Jiang & Peacor, 1994a,b). As noted by Li *et al.* (1994a), much chlorite occurring in low-grade metamorphic rocks, including apparent detrital grains, has an origin in replacement of detrital biotite. Separate packets of corrensite averaging about 200 Å in thickness dominated the authigenic clays in the late diagenetic zone, but these contained some isolated chlorite layers, and were interstratified with separate packets of chlorite and mixed-layer Co/C. Similar TEM observations of corrensite in the late diagenetic zone of the Southern Uplands accretionary terrane (see section 3.3.3.1) were recorded by Merriman *et al.* (1995a).

Gaspé metapelites from the low anchizone have less corrensite and more chlorite, the chlorite packets having a more subhedral morphology. Average packet size has increased to approximately 500 Å. High anchizone rocks contain no corrensite, but the abundant chlorite grains contain some berthierine layers. Chlorite grains have a thickness of approximately 800 Å, and occur as separate packets which commonly have euhedral outlines, often forming thick stacks with packets of muscovite. The only trioctahedral mineral in epizonal rocks is chlorite, which occurs in packets commonly with euhedral outlines, with thicknesses averaging approximately 2000 Å, low defect densities and homogeneous composition, in sharp contrast with chlorite at lower grades.

The sequence of trioctahedral phyllosilicates found in a deeply buried but largely undeformed basinal sedimentary sequence has been documented by Hillier (1993b). These minerals are abundant in Devonian lacustrine mudrocks of the Orcadian Basin, northern Scotland, and represent the transition from the late diagenetic zone to the anchizone. Corrensite first appears at temperatures of approximately 120°C (based on vitrinite reflectance data), and formed either by reaction of Mg-rich trioctahedral smectite, or by reaction between Mg-rich carbonates and dioctahedral clay minerals. Corrensite persists to temperatures of approximately 260°C, i.e. through the low anchizone (Fig. 2.1), as a mixed-layer Co/C mineral intermediate between corrensite and chlorite. Only chlorite (+illite) is found in areas of highest maturity, equivalent to the high anchizone. Changes in composition associated with reaction progress largely relate to increases in tetrahedral Al from corrensite to chlorite, but there is no change in the Fe:Mg ratio. Only the IIb chlorite polytype is present throughout, and this appears to be the stable polytype in late diagenetic through epizonal metapelites (Walker, 1989).

2.2.4 Kaolinite–pyrophyllite

Kaolin minerals, including kaolinite, halloysite, dickite and nacrite, are the most commonly occurring 1:1 sheet silicates in pelitic and metapelitic rocks, although they are dominant only in unusual lithologies such as ballclays, fireclays and tonsteins. Kaolinite and both the 7-Å and 10-Å forms of halloysite are single layer 1:1 structures, whereas dickite and nacrite are double layer polytypes. Halloysite occurs mainly in soils, weathered and hydrothermally altered rocks, where it may also form by the hydration and rolling up of kaolinite crystals (Singh, 1996; Singh & Gilkes, 1992), and is not considered further here.

XRD data based on non-basal reflections are commonly used to distinguish kaolinite from its polytypes dickite and nacrite (Bailey, 1980). Although this is straightforward in the case of tonsteins and fireclays, polytype determination is more difficult in shales or slates where kaolin minerals are minor constituents. As a result, the less common polytypes dickite and nacrite have probably been reported as kaolinite in some pelitic sequences (Frey, 1987a). Polytypism in the kaolin group is readily determined by selected area electron diffraction (SAED) techniques (e.g. Buatier et al., 1997), and the increasing use of TEM studies is likely to clarify the distribution of kaolin polytypes in metapelites.

Kaolinite is widely recorded in diagenetic pelitic sequences (Kisch, 1983, and references therein). Clay mineral studies of deeply buried Carboniferous sequences in the Arkoma Basin (Arkansas–Oklahoma) indicate that kaolinite is present to

thermal maturities of $R_r = 1.9–2.1\%$ (Spötl et al., 1993), equivalent to the beginning of the low anchizone (Fig. 3.4). However, the transformation of kaolinite to dickite in the matrix of sandstones has been reported with increasing depth beginning at temperatures of approximately 120°C (Ehrenberg et al., 1993), i.e. in the late diagenetic zone (Fig. 2.1). In clastic, including conglomeratic, sequences in the Permo-Triassic of the Betic Cordilleras, Spain, dickite replaces kaolinite where associated pelitic rocks show illite crystallinity (KI) values in the range 0.25–0.53 $\Delta°2\theta$, i.e. ranging from the late diagenetic zone to the beginning of the epizone (Ruiz Cruz & Moreno Real, 1993; Ruiz Cruz & Andreo, 1996). In metapelitic rocks, strain also appears to influence the kaolinite–dickite–nacrite transformation. For example, Shutov et al. (1970) recorded nacrite replacing dickite in deformed veins, and Buatier et al. (1997) showed that dickite formed as a synkinematic mineral in black marls within thrust-fault zones.

In aluminous, Fe-poor pelitic rocks, kaolinite reacts in the presence of quartz to form pyrophyllite: 1 kaolinite + 2 quartz = 1 pyrophyllite + 1 H_2O.

This has been mapped as a reaction isograd in parts of the Swiss Alps (Frey, 1987b), and in terms of metapelitic zones has been located approximately at the late diagenetic-low anchizone transition (Wang et al., 1996). Based on vitrinite reflectance and fluid inclusion data, Frey (1987b) estimated the metamorphic conditions of the kaolinite–pyrophyllite isograd to be 240–260°C and 2.1 kbar at a water activity of 0.6–0.8 (see section 3.3.4.1). In normal K-bearing pelites kaolinite reacts to form illite or chlorite where Mg and Fe are available (e.g. Velde, 1968; Dunoyer de Segonzac, 1970; Boles & Franks, 1979). This reaction removes kaolinite from such pelites prior to the late diagenetic-low anchizone transition (Kisch, 1983), but more precise constraints in terms of illite crystallinity or other indicators of reaction progress are lacking.

2.2.5 Berthierine

Berthierine, an Fe, Al-rich mineral with a serpentine-like 1:1 structure, commonly occurs randomly interlayered with chlorite at all grades, and uncommonly as separate packets in association with chlorite (e.g. Jiang et al., 1992b, and references therein; Schmidt et al., 1999). Where it occurs at low grades, commonly as one to a few contiguous layers within chlorite packets, it has generally been interpreted to have formed as a metastable phase relative to chlorite (e.g. Jahren & Aagaard, 1989). The metastable berthierine layers are generally inferred to transform to chlorite with increasing grade. Indeed, Dalla Torre *et al.* (1996b) ascribed the origin of chlorite occurring at higher grades in the Franciscan Complex to transformation of both berthierine and smectite, the latter occurring as mixed-layer Co/C at lower grades. The formulae of berthierine and trioctahedral chlorite have the same form $[M_6Si_4O_{10}(OH)_6$, where M is primarily Fe, Mg and Al], thus implying polymorphism and the possibility that each may be stable over some range of *P–T*, but there is little overlap of the wide range of chlorite compositions with those of berthierine. Berthierine is generally Fe-rich relative to chlorite, but in the Kidd Creek massive sulphide deposit, very Fe-rich berthierine coexists with chlorite of apparently identical composition implying true polymorphism, as also concluded by Abad-Ortega and Nieto (1995). There is thus some possibility that very Fe-rich berthierine may be stable relative to chlorite over some range of *P–T*. Some of the Kidd Creek berthierine was thus interpreted to have formed as a replacement product of chlorite under retrograde metamorphic conditions.

2.3 Metapelitic microfabrics

2.3.1 Introduction

The property that enables slate to be split into thin sheets, slaty cleavage, has been studied for over 150 years. Prior to Sorby's early use of the microscope to examine slates, several papers published between 1835 and 1846 discussed electricity and terrestrial magnetism as causes of slaty cleavage. Subsequently, the influence of pressure was recognized by Sharpe (1847), who noted that flattening caused most of the distortion of fossils in slates, and later Sorby (1853) suggested that flaky minerals developed in a plane perpendicular to the direction of compression. Siddans (1972) gives a comprehensive review of early research into slaty cleavage. Despite Sorby's remarkable early observations, the detail of slaty cleavage microstructure proved too fine for resolution by optical microscopes, and was not fully understood until studied by electron microscopy over

100 years later. In the past 20 years application of SEM and particularly TEM has revealed the variety of microtextures and microfabrics that are found in pelitic and metapelitic rocks, and the interactive nature of deformation, strain and crystallization processes involved in their development (e.g. White & Knipe, 1978; Knipe, 1981; Bons, 1988).

Over the same time period, XRD techniques have also contributed to two important aspects of metapelitic fabric studies. With the increasing use of the illite crystallinity technique to characterize metapelitic terranes, Kisch (1989, 1991b) was able to investigate the correlation between degree of very low-grade metamorphism and the development of incipient slaty cleavage. Subsequently, using a combination of XRD intensity ratios measured on orientated rock slabs, Kisch (1994) measured phyllosilicate orientation in developing slaty cleavage fabrics. X-ray texture goniometry (Van der Pluijm *et al.*, 1994) has been used to study the crystallographic preferred orientation of phyllosilicates in relation to strain (e.g. Oertel, 1983; Oertel *et al.*, 1989), and the mechanical reorientation effects of slaty cleavage development (e.g. Ho *et al.*, 1995, 1996; Van der Pluijm *et al.*, 1998).

Here we do not propose to deal with the various aspects of tectonism that influence fabric formation in pelitic rocks, nor do we discuss the morphological classification of slaty cleavage, but will follow the scheme of Powell (1979), elaborated in Borradaile *et al.* (1982). The main purpose of this section is to describe the microtextures and microfabrics developed in the matrix clays of pelites and metapelites, and comment on their relationships with metapelitic reaction progress.

2.3.2 Microfabrics of the late diagenetic zone

One of the criteria proposed by Coombs (1961) as characteristic of burial metamorphism is the absence of a penetrative tectonic fabric, such as slaty cleavage. This is also the essential textural characteristic of the zone of 'initial epigenesis' proposed earlier by Kossovskaya and Shutov (1961, 1970), based on optical petrography of sandstones and pelitic and metapelitic rocks from the Russian and Siberian platforms (summarized in Kisch, 1983, Table 5.1; Frey, 1987a, Fig. 2.1). The same definition was adopted by Kisch (1983) for burial diagenesis which he broadly equated with burial metamorphism and the 'Diagenetic' Zone of Kübler (1967a, 1968), referred to here as the *late diagenetic zone* (Fig. 2.1). Although this restriction was initially proposed before the application of electron microscopy to pelitic microtextures, subsequent SEM and TEM studies have confirmed that microfabrics of tectonic origin are poorly developed or absent, whereas postcompactional, bedding-parallel microfabrics are a typical feature of mudstones and shales that have been deeply buried (Peacor, 1992a).

The development of a primary clay microfabric, i.e. the preferred orientation of platy clay minerals, in mudstone and shale is largely the result of three interactive processes (Bennett *et al.*, 1991).

1 Physicochemical processes, such as wave, current and gravity forces, and electrochemical mechanisms, influence the way clay plates are initially stacked together; face-to-edge (flocculated), face-to-face and edge-to-edge chains are typical particle stacking types.

2 Bio-organic, particularly bioturbation processes, commonly result in random clay fabrics similar to flocculated particles.

3 Burial diagenesis influences clay microfabric development in two ways:

(a) by postdepositional modification related to overburden stress and consolidation, including loss of water and gases (O'Brien & Slatt, 1990);

(b) by diagenetic mineral reactions that modify and cement the primary depositional microfabric, and result in an overall bedding-parallel alignment of the crystallographic *a–b* plane of clay minerals.

The development of a bedding-parallel microfabric in pelitic rocks appears to be associated with the smectite-to-illite reaction, and was first illustrated by the TEM lattice fringe images of Ahn and Peacor (1986) in a study of the samples originally characterized by Hower *et al.* (1976). Clay matrix microtextures in Tertiary shales from Gulf Coast wells show that, in the initial stages of the reaction, smectite-rich I/S typically show wavy or anastomosing, discontinuous layers with variable orientation (Ahn & Peacor, 1986, Fig. 10). Illite crystallites developed at the expense of smectite are straight and relatively defect-free. With increasing depth, illite crystallites thicken and with progressive elimination of smectite these eventually coalesce at *c.* 5500 m to form an approximately bedding-parallel fabric

(Fig. 2.2). The formation of this fabric results in a sharp increase in the preferred orientation of illite recorded by X-ray texture goniometry (Ho *et al.*, in review). Low-angle grain boundaries are common and some illite crystallites are slightly curved, both features reflecting the poor orientation of precursor smectite grains. Microtextures in compactional pressure shadows formed on either side of quartz silt grains confirm the postcompactional origin of the microfabric. Figure 2.3(a) shows the microfabric extending into the shadow with little evidence of compaction around the quartz grain. Since compactional pressure shadows generally form in the plane of compaction and this is approximately equivalent to bedding in non-tectonized pelites, the clay matrix fabric, i.e. crystallographic *a–b* planes, appears to be approximately bedding-parallel. In contrast, smectite-rich clays at shallow burial depths in the Gulf Coast shales commonly show a compactional microfabric of smectite and I/S wrapping silt grains (e.g. Lee *et al.*, 1985, Fig. 2). The poor degree of compaction of illite-rich clay around detrital quartz grains is clear evidence of the postcompactional, late diagenetic origin of the bedding-parallel microfabric, and demonstrates that it is not simply an inherited depositional parallelism of platy clay minerals.

Similar bedding-parallel fabrics have been recorded in lowest-grade mudstones and shales of metapelitic

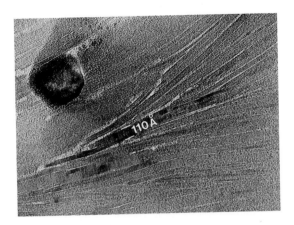

Fig. 2.2 Transmission electron microscope lattice fringe image showing bedding-parallel microfabric developed in thin illite crystallites. Tertiary Gulf Coast shale, depth 4745 m. (Photograph by H. Dong.)

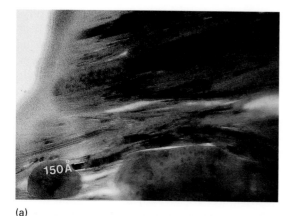

(a)

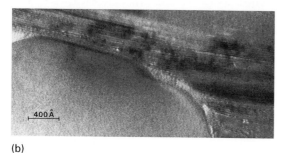

(b)

Fig. 2.3 Transmission electron microscope lattice fringe images. (a) Illite crystallite in bedding-parallel microfabric that has grown into a compactional 'shadow' between quartz grains. Tertiary Gulf Coast shale, depth 5747. Photograph by H. Dong. (b) Bedding-parallel microfabric in illite and chlorite showing lack of compaction against quartz grain. Lower Palaeozoic mudstone, Welsh Basin. (Merriman *et al.*, 1990; Fig. 4A, with permission.)

sequences in other geotectonic settings. In late diagenetic zone mudstones from the Welsh Basin (see section 3.3.2.1), illite and chlorite intergrowths form straight or slightly curved, defect-free crystallites in bedding-parallel orientation (Fig. 2.3b). TEM and XRD measurements indicate that the mean crystallite thickness of illite is typically of the late diagenetic zone (mean = 116 Å; Merriman *et al.*, 1990), although Jiang *et al.* (1990a) showed that ordered I/S has formed in this rock as a hydrothermal replacement of illite. Microtextural relationships with detrital quartz grains, however, indicate that these clay matrix microfabrics are essentially post-compactional (Merriman *et al.*, 1990, Fig. 3A), and originated during burial metamorphism (Roberts *et al.*, 1996). Similarly, Jiang *et al.* (1997) recorded

thin, subparallel chlorite and illite crystallites in the clay matrix fabrics of late diagenetic mudstones from the Gaspé Pennisula (see section 3.3.4.1). The illite is typically a defect-rich, $1M_d$ polytype with a volume-averaged crystallite thickness of 70 Å. Despite local random orientation of the microfabric, an overall preferred orientation parallel to bedding is noted by Jiang *et al.* (1997).

Bedding-parallel microfabrics are generally poorly developed in late diagenetic zone pelites from accretionary prism sequences. Accreted calcareous mudstones from the Barbados Ridge generally develop random fabrics and little or no fissility, particularly close to zones of décollement (Taylor *et al.*, 1991). Samples from a deep well in the Barbados accretionary prism show that a planar fabric develops below *c.* 1100 m, but clay minerals show typical compactional microfabrics around detrital and authigenic grains (Bryant *et al.*, 1991). Buatier *et al.* (1992) show that the initial stage of the smectite-to-illite transition is presently taking place at 500 m below the sea bed in the Barbados accretionary prism. Typical microfabrics, however, consist of poorly orientated, discontinuous, thin smectite crystallites with dispersed illite crystallites up to 150 Å thick. Although bedding lamination is well developed in late diagenetic zone mudstones in the accretionary prism sequences of the Scottish Southern Uplands (section 3.3.3.1), Merriman *et al.* (1995a) found that clay matrix microfabrics are poorly developed. TEM images show that phengitic white mica occurs as dispersed thin crystallites whereas chlorite and corrensite, the dominant phyllosilicates, commonly show kinking and bending due to a cryptic disruption of the microfabric associated with early stratal disruption.

Scaly fabrics in pelitic rocks are commonly associated with stratal disruption in accretionary sequences, particularly in shale-matrix mélange deposits and adjacent to zones of décollement. Such fabrics are characterized by anastomosing polished and slickensided fracture surfaces, but are seldom pervasive (Lundberg & Moore, 1986). Although SEM studies show that phyllosilicates within scaly fabrics are well oriented, they are the result of reorientation and disruption of existing minerals on slip surfaces with minimal recrystallization (Moore *et al.*, 1986). Microstructural studies of fault fabrics associated with scaly pelites suggest that

that zones of well-oriented phyllosilicates may be localized at fault walls, but authigenic illite within the fault zones show random fabrics (Knipe, 1986).

2.3.3 The anchizone and slaty cleavage development

In metapelitic terranes, the transition from the late diagenetic zone to the anchizone is associated with the regional development of incipient slaty cleavage. Kisch (1991b) noted the association between the late diagenetic zone–anchizone boundary and pelites showing pencil structure and primary (S_1) cleavage crenulating bedding fabrics (S_0) (Fig. 2.1). A number of SEM and TEM studies have recorded the microtextures associated with the transposition of a bedding-parallel microfabric to slaty cleavage microfabric. For example, Weber (1981, Figs 2 and 3) recorded microfolding without syntectonic recrystallization from the Rheinische Schiefergebirge, Germany, where the 'primary sedimentary fabric was strong and homogeneous'. In the southern Appalachians, Weaver (1984) made systematic SEM observations of mineralogy and fabric development in shales and slates. Bedding-parallel fabrics are recorded in diagenetic grade shales but as grade increases towards the late diagenetic zone–anchizone boundary a pencil cleavage is developed in the shales. Pencilling is accompanied by the development of a spaced cleavage causing rotation of detrital mica

flakes and crenulation of the bedding-parallel fabric (Weaver, 1984, Figs 8–10). Similar microfabrics were found in late diagenetic to low anchizonal mudstones forming an extensive Lower Palaeozoic subcrop on the Midlands Microcraton of England (Merriman et al., 1993). These rocks behaved as a passive cover sequence during Acadian tectonism, and although they developed a spaced cleavage, the bedding-parallel microfabric is well preserved as a crenulated lamination (Fig. 2.4).

The transformation of a bedding-parallel microfabric to slaty cleavage microfabric has been studied in detail by Li et al. (1994b), using a combination of SEM back-scattered electron (BSE) and TEM lattice fringe imaging, and AEM. They investigated a weakly cleaved Silurian mudstone of low anchizonal grade (KI = 0.35) from the central part of the Welsh Basin. The mudstone is rich in chlorite-mica stacks representing in situ alteration of detrital biotite and other ferromagnesian grains (Milodowski & Zalasiewicz, 1991; Li et al., 1994a), and these are commonly oriented with the crystallographic a–b stacking planes approximately bedding-parallel. By using the stacking planes in chlorite-mica stacks as a reference, two orientations of the a–b planes of white mica and chlorite intergrowths in the mudstone matrix were distinguished. Bedding-parallel intergrowths contain the thinnest white mica (50–200 Å) and chlorite crystallites, and these contain a high density of defects including kinks, bends and layer terminations

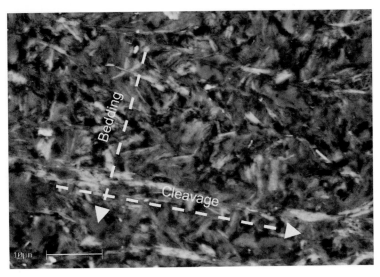

Fig. 2.4 Scanning electron microscope back-scattered electron image of crenulated bedding-parallel microfabric in Lower Palaeozoic shale. (Stockingford Shale Group; Merriman et al., 1993.) A spaced cleavage microfabric is developed as narrow, unpersistent domains of oriented white mica and chlorite. (Photomicrograph by J. M. Pearce.)

(Li *et al.*, 1994b, Fig. 3). White mica crystals in this orientation are mostly the $1M_d$ polytype and commonly have compositions intermediate between muscovite and paragonite, within the muscovite–paragonite solvus. Some relatively thick, defect-free white micas in the bedding-parallel orientation are the $2M_1$ polytype, but have similar intermediate compositions. Both types are thought to be derived from smectite during burial diagenesis, with thicker white micas forming by mimetic crystallization in the bedding-parallel orientation during deep burial which probably reached low anchizonal conditions (see section 3.3.2.1). The bedding-parallel white micas are commonly observed to be bent, kinked and strained adjacent to thicker white mica crystallites in cleavage-parallel orientation (Li *et al.*, 1994a, Fig. 2). Cleavage-parallel white micas have packet thicknesses of $200\,\text{Å}$ to $2\,\mu\text{m}$, and are strain-free, compositionally mature muscovite or paragonite crystals showing both $2M_1$ and $3T$ polytypism.

During the initial development of cleavage microfabric, mechanical rotation of mineral grains and pressure solution are the dominant processes. The nature of the interaction between these two processes and the microstructures created have attracted several TEM and X-ray texture studies. An early TEM study of slaty cleavage microstructure examined the variation in microfabric development around mesoscopic folds in Ordovician slates and sandstones at Rhosneigr, Anglesey, northern Wales (Knipe & White, 1977; White & Knipe, 1978; Knipe, 1979, 1981; Borradaile *et al.*, 1982, pp. 140–147). The slates are composed largely of white mica and chlorite (>70%), and are at anchizonal grade (B. Roberts, personal communication). Cleavage microfabric is initially developed by crenulation, with mechanical rotation causing bending, kinking and fracturing of the bedding-parallel phyllosilicates. By this means a spaced cleavage microfabric is formed by recrystallization along fractures, inter- and intracrystal slip on (001) and crystallization of phyllosilicates in dilational sites. The interactive mechanical and crystallization processes that characterize the initial stages of cleavage microfabric formation generate two microstructural domains (Knipe, 1981). Sites where the phyllosilicates are concentrated and crystallographic *a–b* planes are aligned parallel with the cleavage microfabric are termed P-domains; sites where phyllosilicates are oriented with crystal-

lographic *a–b* planes at high angles to P-domains, and where quartz grains are concentrated, are termed Q-domains (Fig. 2.5).

The influence of mechanical and pressure solution processes was contrasted in studies of the transition from mudstone to slate in the Ordovician Martinsburgh Formation at Lehigh Gap, Pennsylvania. Using TEM techniques, Lee *et al.* (1986) found that clay matrix microtextures in the mudstones show intergrowths of thin, slightly curved, subparallel illite (1M + 2M polytypes) and chlorite crystallites, with numerous edge dislocations and layer terminations. Transitional mudstones retain these microtextures but also show illite crystallites lacking defects. Two orientations of phyllosilicates were noted, parallel to bedding and parallel to cleavage, but very few intermediate orientations were observed. In the slates, illite and chlorite crystallites were found to be straight, relatively thick crystallites, free of defects, and showing straight semicoherent crystal boundaries. On the basis of the TEM microtextures in only the authigenic (matrix) clay minerals, Lee *et al.* (1986) concluded that dissolution and neocrystallization, i.e. pressure solution, dominated the cleavage fabric-forming process at Lehigh Gap. However, Ho *et al.*

Fig. 2.5 Low magnification transmission electron microscope image of microstructural domains. P-domains contain straight, mottled phengite crystallites oriented WNW–ESE, aligned parallel with slaty cleavage. Q-domains contain randomly oriented phengite crystallites.

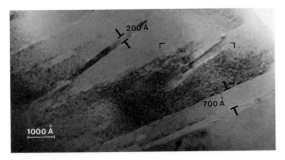

(a)

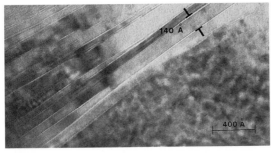

(b)

Fig. 2.6 (a) Transmission electron microscope lattice fringe image of thick white mica crystals in cleavage-parallel orientation enclosing packets of thin white mica crystallites. (b) Detail of framed area showing thick mottled mica overgrowing the basal crystallite of a stack of thin subparallel crystallites. (From Merriman *et al.*, 1990, with permission.)

(1995) used SEM and X-ray texture goniometry to study both the authigenic (matrix) clay minerals and detrital phyllosilicates. They found that mechanical rotation of detrital phyllosilicates in a bedding-parallel orientation dominated the process in the initial stages of the development of the transitional zone between mudstone and slate, whereas pressure solution was more important in the slate zone.

The transition from late diagenetic mudstone to anchizonal slate was characterized in terms of white mica crystal growth by Merriman *et al.* (1990) using both TEM and XRD measurements. Mean white mica crystallite thickness was found to increase rapidly with the transformation from mudstone to slate, whereas the proportion of crystal defects decreased. TEM microtextures suggest that in the cleavage-parallel P-domains white mica thickens by dissolution and recrystallization of thin illite crystallites inherited from rotated bedding-parallel

microfabrics. Figure 2.6 shows a thick white mica crystal in cleavage-parallel orientation overgrowing a stack of thin illite crystals characterized by low-angle grain boundaries and layer terminations (Merriman *et al.*, 1990). In this way the microfabric of typical late diagenetic pelites is annealed to form the thicker phyllosilicates of the cleavage microfabric. Similar microtextures were recorded in a well cleaved anchizonal metabentonite (Fig. 2.5), where 1M and 2M phengitic mica polytypes are rotated and thickened into the cleavage microfabric (Merriman *et al.*, 1995a).

The role of intracrystalline deformation of phyllosilicates is an important aspect of the TEM study made by Bons (1988). He reported on the development of anchizonal slaty cleavage in Cambro-Ordovician sequences in the Ori Dome, forming part of the Axial Zone of the Pyrenees. Bending, kinking and folding of phyllosilicates in the initial stages of cleavage microfabric formation is accompanied by intracrystalline deformation with the formation of dislocations, dislocation walls (subgrain boundaries) and new grain boundaries. Dislocations provide pathways for enhanced diffusion and as they move they create new pathways for pressure-solution processes. Bons (1988) noted the different response to intracrystalline deformation shown by chlorite and muscovite. Chlorite deformation is mainly by intracrystalline slip and dislocations are commonly aligned into dislocation walls forming subgrains; rotation of subgrains with further deformation can lead to high-angle grain boundaries. In contrast, muscovite rarely shows evidence of intracrystalline slip or polygonization and appears to deform by plastic and elastic processes causing rapid defect migration and crystal growth.

Differences in chlorite and phengite intracrystalline microtextures were also observed by Merriman *et al.* (1995b) in a TEM and XRD study of microfabric development in the Scottish Southern Uplands accretionary terrane. Over wide areas of the terrane slaty cleavage has an approximately bedding-parallel orientation so that anchizonal slates are commonly mistaken for shales. The slates, however, show a well-developed cleavage microfabric with many features that are indicative of strain-related crystal growth associated with accretionary tectonics and syntectonic metamorphism. Nanometric-scale folds are common (Fig. 2.7) and their geometries suggest

Fig. 2.7 Low magnification transmission electron microscope image of nanometric-scale fold in intergrown chlorite (chl) and phengitic white mica (phen) crystallites showing strain fields (kinks) resulting from dislocation by intracrystal slip on (001). (From Merriman *et al.*, 1995b, with permission.)

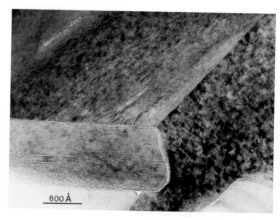

Fig. 2.8 Transmission electron microscope lattice fringe image of thick phengitic white mica crystals in epizonal slate.

that they formed by progressive intracrystalline and grain boundary slip parallel to (001). Fold-forming chlorite shows a high density of strain-related defects including stacking faults, layer terminations and kinks representing the strain field of dislocations associated with slip on (001). Where fold-hinge fracturing has occurred, however, the chlorite annealing the fractures is strain-free. Intergrown phengitic white mica rarely shows strain microtextures and appears to have a greater capacity to store strain energy, enabling it to recover from subgrain development by grain boundary migration, i.e. dislocation creep. As a result, Merriman *et al.* (1995b) found that white mica developed thicker crystallites than chlorite in response to the variable strain rates that characterized the assembly of the Southern Uplands accretionary prism.

2.3.4 The anchizone–epizone transition

In terms of microfabric formation, the transition from the high anchizone to the epizone is simply a continuation of the processes initiated with the late diagenetic zone-low anchizone transition. Several studies have suggested that, at this stage the crenulated fabrics inherited from lowest grade metapelites are progressively obliterated by pressure solution crystallization (e.g. Knipe, 1981; Weber, 1981; Lee *et al.*, 1986). Lateral extension of the cleav-

age P-domains appears to progress by dissolution and crystallization along the domain borders (Knipe & White, 1977), combined with grain and intra-grain boundary migration in the core of the domains (Bons, 1988; Merriman *et al.*, 1995b). White mica developed in epizonal cleavage microfabrics forms relatively thick, straight, defect-free crystallites (Fig. 2.8). Chlorite also forms thick, straight crystallites, but retains more defects than white mica and consequently tends to develop crystals with more subgrain boundaries and higher percentage lattice strain (Árkai *et al.*, 1996).

In summary, the formation of slaty cleavage microfabric in the anchizone and epizone is a response to the combined effects of thermal and strain energy in the early stages of regional metamorphism. Van der Pluijm *et al.* (1998) have pointed out that the contribution of either source to the total energy of the system is interchangeable and complementary, favouring mechanical grain kinking and rotation processes in low-energy environments, and chemically driven processes such as grain dissolution and neocrystallization in high-energy environments.

2.4 Measuring reaction progress

2.4.1 X-ray diffraction techniques

2.4.1.1 Illite crystallinity

The most widely used method of determining grade

in metapelitic sequences is the XRD-based illite crystallinity technique. It was introduced by Kübler (1967a,b, 1968) as a method of identifying the transitional 'anchimetamorphic zone' between diagenesis and the 'epimetamorphic zone' of low-grade metamorphism in metapelitic sequences. The subsequent definition of the anchizone using the Kübler index of illite crystallinity also defined the prograde limit of diagenesis and the initiation of the epizone (Kisch, 1990, 1991a), as shown in Fig. 2.1. The illite crystallinity technique measures changes in the shape of the first basal reflection of dioctahedral illite–muscovite at a spacing of approximately 10 Å. Although a number of methods have been developed for measuring the 10-Å peak profile (Frey, 1987a; Blenkinsop, 1988), the method first used by Kübler (1964, 1967a,b, 1968) has been adopted most widely. Kübler's method measures the width of the 10-Å peak at half-height above the background, and latterly (see Kisch, 1980) the measurement is expressed as small changes in the Bragg angle, $\Delta°2\theta$ (Fig. 2.9). This measurement is referred to as the Kübler index (KI) and KI values are in $\Delta°2\theta$.

When the term 'illite crystallinity' was introduced, Kübler and others (e.g. Kisch, 1983) used it cautiously, often in quotation marks, largely because the physical reasons for the change in the 10-Å peak profile were poorly understood. 'Illite' was regarded as a crystallographically ill-defined clay mica, and 'crystallinity' was not fully acceptable as an appropriate term for the 'ordering' process that was believed to control the shape of the 10-Å peak (Kübler, 1967a). Indeed other factors were known to influence peak shape, particularly the

experimental and measuring conditions, and also interference from other phyllosilicates (Frey, 1987a, pp. 18–21). However, through the application of TEM techniques to samples with well characterized KI values, Merriman *et al.* (1990, 1995b) showed that the width of the 10-Å peak is largely controlled by the thickness of illite–muscovite crystallites. The TEM measurements confirmed relationships predicted by the Scherrer equation (see section 2.4.1.4) whereby the thickness of small crystals, i.e. *crystallites*, of illite–muscovite is inversely proportional to the 10-Å peak width. Hence, as illite–muscovite crystallites thicken in prograde metapelitic sequences, so the 10-Å peak profile becomes narrower and the value of the Kübler index decreases. Although lattice strain also influences the 10-Å peak width, percentage strains in illite–muscovite crystallites tend to be small and also decrease with increasing grade (Árkai *et al.*, 1996). Here the term *crystallinity* is regarded as a reasonably succinct description of the state of *crystallite thickness and lattice strain*, and is used throughout with this meaning.

As discussed in section 2.2.2, illite is the first phase to generate a 10-Å peak in the smectite–I/S–illite–muscovite reaction series, and evolves to more mature phengite, muscovite and other white micas in very low-grade metapelites. Prograde changes in illite crystallinity are therefore an inseparable part of the progress of interrelated reactions, such as the reduction in smectite interlayers, polytypic transformations and a decrease in compositional heterogeneity of members of the series. Although it is only one of several indicators of reaction progress in the series, it has the advantage of being a relatively simple measurement and is sensitive over a wide range of metapelitic conditions, from the beginning of the late diagenetic zone to the beginning of the epizone. Clearly, the 10-Å peak measurement may include illite in addition to other white micas such as phengite and muscovite, and for this reason illite crystallinity is not the most appropriate term. Nevertheless, *illite crystallinity* (IC) is now enshrined in the literature and is retained here as mineralogical 'shorthand' for the crystallite size (and strain) variations in dioctahedral micas developed in the smectite–I/S–illite–muscovite series, as indicated by XRD measurements of the 10-Å peak.

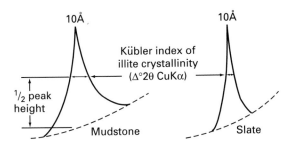

Fig. 2.9 The Kübler index of illite crystallinity measured on 10-Å X-ray diffraction peak profiles typical of a mudstone and a slate.

Sample preparation. It is essential to use only the

neocrystallized or authigenic clay mineral components of pelitic and metapelitic samples for IC measurements, and these must be separated from detrital micas prior to XRD analysis. The effects of some of the different separation methods used for IC studies have been reviewed by Kisch and Frey (1987), and recommended procedures for sample preparation were drawn up by Kisch (1991a) as a result of an international workshop on IC techniques. The main points of the recommendations are summarized below:

1 Crushing. The aim of this process is to reduce hand specimens to mm-sized granules suitable for more gentle disaggregation. Initial crushing should use a jaw-crusher, ensuring first that any loose coating (e.g. soil, algae, lichen) is first wire-brushed from the hand specimens. This is followed by use of a hammer-mill with a built-in sieve, or agate/porcelain pestle and mortar, with regular (c. 30 s) sieving between grinding. Disc mills and ball mills should be avoided, unless a pilot study can first establish what degree of crystallite size reduction, i.e. peak broadening, is caused by different grinding times in these mills.

2 Ultrasonic disaggregation and dispersion. This technique is routinely used to disperse clay minerals prior to gravity settling and separation of the <2-μm fraction. No specific recommendations were made due to insufficient data on this method. The overall conclusion, however, is that ultrasonic disaggregation probably has little deleterious effect on inherent crystallite size, provided that it is carried out in tanks or beakers, not with probes, and restricted to less than 30 min. The use of dispersing agents is not normally necessary with this technique, but some pyrite-rich black shales may require 1–2 ml of $0.1 \, mol \, l^{-1}$ Calgon (sodium hexametaphosphate) in order to prevent flocculation.

3 Size separation methods. No particular separation technique was recommended (e.g. settling or centrifugal), but each laboratory should routinely check the effectiveness of their method by SEM calibration or sedimentometer (e.g. Warr & Rice, 1994, p. 144). The <2-μm size fraction is recommended for IC measurements, and an earlier practice of removing the <0.1-μm fraction should be discontinued (see Merriman et al., 1990).

4 Oriented mounts of <2-μm fractions. Oriented mounts are used to enhance the basal reflections of white micas and other phyllosilicates during XRD analysis. Some laboratories sediment, i.e. pipette, the <2-μm material onto glass slides and others smear a slurry of the clay onto the slide. Robinson et al. (1990) found no consistent differences in KI values derived from the two techniques. The thickness of clay on the mount influences both the peak intensity and width (Krumm & Buggisch, 1991), but there is no consensus on the optimum thicknesses for minimizing peak width variations. Although Krumm and Buggisch (1991) suggested very thin mounts of $0.25 \, mg \, cm^{-2}$ of clay or less, Kisch (1991a) recommended avoiding thin mounts and using at least $3 \, mg \, cm^{-2}$ of clay based on his own results and those of Frey (1988). Warr and Rice (1994) also recommended clay thicknesses of $>3 \, mg \, cm^{-2}$, and Lezzerini et al. (1995) suggested $1–5 \, mg \, cm^{-2}$ and normalization of KI values.

5 Cation saturation and ethylene glycol solvation. Ca-saturation is recommended where smectite or I/S is suspected or likely to be present, particularly in late diagenetic zone pelites. This technique ensures that Ca rather than Na is the dominant interlayer cation in smectite and gives better separation of smectite from illite on XRD traces. Saturation with Sr was preferred by Eberl et al. (1987), and Warr and Rice (1994) found that Sr-saturation produced even-textured clay mounts with little evidence of shrinkage. Further separation of smectite and illite peaks is achieved by ethylene glycol (EG) saturation which expands the basal smectite peak to 17 Å. EG saturation is often used to sharpen very broad 10-Å peaks, but there is little experimental evidence on whether this is purely an effect of clay mineral expansion or also reflects some degree of particle rearrangement within the mount. Kisch (1991a) recommended giving both air-dried and EG-solvated KI values when this technique is used. Note that, unless stated otherwise, KI values are measured on air-dried samples.

X-ray diffractometer settings. IC measurements are made using Ni-filtered CuKα radiation and the normal generator conditions for Cu radiation which are generally 40 kV and 30 mA. For 10-Å peak measurements most studies scan the range 7–10°2θ at a slow scan rate of $1–0.5°2\theta \, min^{-1}$. The main recommendations made by Kisch (1991a) on X-ray diffractometer settings relate to the effects of the scan rate and time constant on peak broadening when diffraction patterns are recorded on a strip chart recorder. In a study of instrumental effects on IC data, Kisch (1990) showed that, for a given sample, the amount of peak broadening remains the same when the time constant (TC) used conforms to the conditions:

TC (s) ≤ 0.5 [receiving slit width ($°$)/scan rate ($° \text{ min}^{-1}$)] $\times 60$.

If these conditions are not met then increased peak broadening will result, particularly at high scan rates. Modern X-ray diffractometers are usually fitted with step-scanning motors, automatic divergence slits and a graphite monochromator. Moreover, diffraction data are digitized, thus avoiding problems associated with strip chart recorders, and analyzed using a computer program which will automatically subtract background before measuring peak width (e.g. Warr & Rice, 1994). The variety of software now in use for measuring diffractometer data strengthens the case for interlaboratory standardization (see below).

Errors and precision. An analysis of the precision of the IC technique was made by Robinson *et al.* (1990), who examined some of the sources of error that influence the 10-Å peak width. These included sample preparation techniques, machine variations using internal standards, intrasample and inter-sample variation. Errors were expressed as:

Mean error (%) = (standard deviation at 1σ/ mean) $\times 100$.

The smallest errors are associated with machine variation (<5%), whereas intrasample errors are up to 12% and intersample errors from multisampling at three sites are up to 14%. On the basis of their analysis Robinson *et al.* (1990) recommended that

the intervals used to contour IC data should not be less than 0.1 $\Delta°2\theta$ if the contouring is to have a high degree of confidence (>0.8).

Interlaboratory standardization. One of the long-running problems of IC measurement is correlation between laboratories. No matter how closely the recommendations of the IC working group are followed (Kisch, 1991a), variations due to sample preparation methods and machine conditions will arise which are specific to individual laboratories. In the past, different laboratories have calibrated their results to polished slate standards supplied by B. Kübler and H. J. Kisch, but this approach has several shortcomings. Firstly, the slate standards are useful for calibrating narrow peak widths, i.e. higher grade metapelites, but less useful for the broader peaks (>0.35 $\Delta°2\theta$) of lower grade pelites. Secondly, the standards are used as slabs inserted directly into the diffractometer and do not require any laboratory preparation, thus by-passing one of the sources of error identified by Robinson *et al.* (1990). Lastly, individual laboratories tended to use the slate standards to calculate their own anchizone limits instead of correcting their data to the scales associated with the standards (Blenkinsop, 1988; Kisch, 1990). The approach proposed by Warr and Rice (1994) attempted to remedy some of these problems with a set of four interlaboratory standards representing metapelites from the late diagenetic zone to the epizone, plus a muscovite flake. The crystallinity index standards (CIS) are supplied as <20-mm rock chips which can be processed in the same way as other samples prior to XRD analysis (e.g. Krumm *et al.*, 1994). These are analyzed using the laboratory's chosen machine conditions, and the mean of three repeat measurements plotted against the CIS values (Table 2.1). A regression analysis derived from the plot will allow a correction to be made as part of

Table 2.1 Crystallinity index standards. (From Warr & Rice, 1994, Table 3.)

| CIS sample number | SW1 | SW1 | SW2 | SW2 | SW4 | SW4 | SW6 | SW6 | MF1 |
Preparation	AD	GY	AD	GY	AD	GY	AD	GY	AD
Kübler index ($\Delta°2\theta$)	0.63	0.57	0.47	0.44	0.38	0.38	0.25	0.25	0.11
Error (%)	3.7	4.6	7.1	6.8	4.8	5.7	5.6	6.9	5.9
Intensity (cps)	533	484	627	631	853	814	585	585	124550
10-Å d value	10.146	10.103	10.101	10.072	10.071	10.082	10.082	10.080	9.945

a spread-sheet calculation, and typically will be in the form: $IC_{(CIS)} = 1.511564 \times IC_{(MYLAB)} - 0.02956$. The muscovite standard (MF1) is a single thick crystal showing no size broadening which is used to determine inherent machine broadening effects (see section 2.4.1.4). A set of CIS samples and further details can be obtained from Dr L. N. Warr, Geologisch-Paläontologisches Institut, Ruprecht-Karls-Universität, Im Neuenheimer Feld 234, 69120 Heidelberg, Germany.

2.4.1.2 K-white mica b cell dimension

Variation in the b unit cell dimension of K-white mica is widely used as a comparative geobarometer in low- and very low-grade metapelitic rocks. Although initially applied to greenschist facies terranes (e.g. Sassi, 1972; Sassi & Scolari, 1974; Fettes et al., 1976), XRD measurements of the b dimension were subsequently used to characterize pressure facies series in a range of very low-grade metapelitic terranes (e.g. Padan et al., 1982; Kemp et al., 1985; Robinson & Bevins, 1986). It has been customary to use the symbol b_0 for this spacing but since there appears to be no crystallographic rationale for this, the universally accepted symbol b is used here and throughout.

The a and b cell dimensions of muscovite are affected by two substitutions (Guidotti et al., 1989):

(1) $(Mg, Fe^{2+})^{vi} + Si^{iv} = Al^{vi} + Al^{iv}$;
(2) $(Fe^{3+})^{vi} = Al^{vi}$.

Substitution (1) is referred to variously as the 'phengite', 'celadonite' or 'Tschermak' substitution, or (1) and (2) may be combined and called the 'celadonite' substitution (Guidotti, 1984). The crystallographic changes caused by (1) and (2) have been quantified by Guidotti et al. (1989) using single crystal XRD studies combined with EMP analyses of natural and synthetic $2M_1$ muscovites. Although both the a and b dimensions decrease with increasing $Na/(Na + K)$ ratio, the effect of this ratio is not important when $Na/(Na + K) \leq 0.15$. In Na-poor muscovite Guidotti et al. (1989) found that b varies linearly with the contents of the octahedral and tetrahedral sheets in accordance with the regression equations:

(3) b (Å) $= 8.9931 + 0.0440 \Sigma (Mg + Fe^{2+} + Fe^{3+})$,

b (Å) $= 9.1490 - 0.0258 \Sigma (Al^{vi} + Al^{iv})$,
b (Å) $= 8.5966 + 0.0666 Si$.

The use of the phengite component of K-white mica for geobarometry is discussed in section 2.5.5. Although Velde (1965) suggested that changes in the phengite component are largely a response to pressure, it is not clear how such changes relate to progress in the smectite–I/S–illite–muscovite reaction series. For example, our published work has used ATEM data from a suite of late diagenetic to epizonal metapelites to study compositional changes in illite–muscovite with prograde crystallite thickening. Whereas increases in K and Al^{iv}, and decreases in Si accompany the progressive thickening of illite–muscovite crystallites, the Fe + Mg component showed no such systematic changes with crystallite thickness. Indeed, clay minerals which form at low temperatures may be metastable with enhanced solid-solution ranges which may be inherited from their precursors and not be representative of equilibrium as a function of pressure or other state variables. This suggests that the progressive development of illite/phengite at the expense of smectite, with associated changes in K, Al^{iv}, Si and increases in IC, may result in little change in the proportion of octahedral Fe + Mg. Since the phengite content of illite–muscovite is related to geotectonic setting (see section 3.3.1), this suggests that it may depend on basinal heat-flow characteristics and reaction kinetics in very low-grade metapelites (see section 2.5.5).

Measuring the K-white mica b cell dimension. Prior to publication of the above regression equations, it was common practice to measure the b dimension by X-ray diffractometer techniques using the dioctahedral mica (060) spacing and a suitable internal standard, usually the quartz (211) peak at 1.541 Å. In order to enhance the (060) reflection polished rock slabs were used, cut normal to the bedding/cleavage fabric. While this technique was probably justified in greenschist facies rocks, the effects of unreacted detrital micas and other minerals (Frey, 1987a, pp. 56–57) poses problems when slabs are used for measurements of very low-grade metapelites. Separated <2μm and 6–2μm size fractions (Padan et al., 1982), and side-packing sample holders can be used to overcome the problem. However,

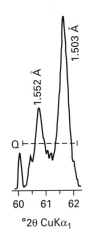
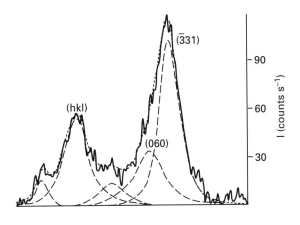

Fig. 2.10 Decomposed X-ray diffraction pattern showing resolution of two peaks $(060, \overline{3}31)$ contributing to the reflection at 1.503 Å used for measuring the K-white mica b cell dimension. (Reproduced from Huon et al., 1994, with permission.)

overlapping (060) and $(\overline{3}31)$ reflections from di-octahedral mica (Frey et al., 1983) still pose a problem. Huon et al. (1994) used peak deconvolution techniques to resolve the two reflections, but found that the $(\overline{3}31)$ reflection was in some cases stronger than (060) (Fig. 2.10). Although the separation between the two reflections resolved by this method is small (0.0021–0.0056 Å), this is multiplied sixfold by the conversion of the $d(060, \overline{3}31)$ spacing to the b value. Wang et al. (1996), however, tested the reliability of measurement of the $d(060, \overline{3}31)$ value against the calculated b dimension from a unit cell refinement using 20 reflections, and they found good linear correlation.

When it is possible to obtain high quality AEM data, the regression equations of Guidotti et al. (1989) can be used to calculate the b dimension (Merriman et al., 1995a,b). This approach offers several important advantages over the XRD method:

1 Detrital micas can be avoided;

2 Only data from authigenic micas, diagenetic or metamorphic in origin, are used to calculate the b dimensions;

3 Na-rich mica can be identified by AEM and excluded from the data set;

4 White micas in microstructures or microfabrics identified at the TEM level can be characterized.

Geotectonic setting. K-white mica b dimensions are usually determined for a suite of approximately 30–50 samples from an area of uniform grade and the data displayed as cumulative frequency curves. Using this method of display, Sassi and Scolari (1974)

were able to demonstrate that different geotectonic settings were commonly characterized by a well defined grouping of b values, generally with $1\sigma <$ 0.01Å. Subsequently Guidotti and Sassi (1986) collated published b dimension data for greenschist and blueschist facies rocks which were characterized in terms of the facies series of Miyashiro (1961). They also presented a 'qualitative plot' of b dimensions as a family of curves in P-T space using a grid constructed around the stability fields of the Al_2SiO_5 polymorphs and those of other mineral reactions. The inferred geothermal gradients associated with b values from Guidotti and Sassi (1986, Fig. 1) are shown in Table 2.2.

Guidotti and Sassi's b value curves do not extend into subgreenschist facies rocks, but a considerable amount of b dimension data from very low-grade metapelites have been published in the past decade. These generally corroborate the groupings found in the pressure series of greenschist and blueschist facies rocks, but with greater overlap than that indicated in Table 2.2. In terms of the three geotectonic settings of very low-grade metamorphism identified in Table

Table 2.2 K-white mica b dimension and pressure facies series (Guidotti & Sassi, 1986).

Facies series	b (Å)	Inferred geothermal gradient
Low pressure	<9.000	>35°C km⁻¹
Intermediate	9.000–9.040	25–35°C km⁻¹
High pressure	>9.040	<25°C km⁻¹

3.1, the highest *b* values are found in accretionary settings and are typically in the range 9.02–9.06 Å (see section 3.3.3 for details). For example, Cloos (1983) found that phengitic micas from shale and shale-matrix mélange in the Franciscan Complex have *b* values >9.030 Å. In a study of accretionary metamorphism in the Klamath Mountains, California, Goodge (1995) found that a range of metasedimentary rocks in the prehnite–actinolite, blueschist and greenschist facies were predominantly grouped in the range 9.02–9.06 Å. Similarly, *b* dimensions calculated from AEM analysis of white micas in metapelites in the Franciscan Complex are in the range 9.01–9.06 Å (unpublished results). In the accretionary terrane of the Scottish Southern Uplands, values are in the range 9.020–9.044 Å (Kemp *et al.*, 1985; Merriman *et al.*, 1995b). Examples from extensional basins include the Welsh Basin where Robinson and Bevins (1986) found the mean *b* value of 150 samples to be 8.995 Å (see section 3.3.2.1), and the English Lake District where Thomas (1986) reported *b* values in the range 8.96–9.00 Å from the Skiddaw Group (see section 3.3.2.3). Unpublished *b* values derived from AEM data from both areas show similar values and ranges. Values intermediate between those of accretionary and extensional settings characterize the external fold-and-thrust belt metamorphism of Alpine collisional settings (Table 3.1). In the Scandinavian Caledonides, Padan *et al.* (1982) recorded *b* values in the range 9.030–9.037 Å; Warr *et al.* (1996) found values in the range 9.01–9.06 Å, and Rice *et al.* (1989) recorded similar or slightly higher *b* values. Values in the Variscan fold belt of NW Spain are in the range 9.015–9.041 Å (Gutiérrez-Alonso & Nieto, 1996); in the Sierras Austales fold-and-thrust belt, Argentina, Von Gosen *et al.* (1991) recorded *b* values in the range 9.00–9.04 Å. The wide ranges of *b* values that characterize some Alpine collisional settings probably reflects a polygenetic metamorphic history, as demonstrated by Bevins and Robinson (1995) in the Northern Sierra Nevada, California, and Sassi *et al.* (1995) in the Carnic Alps.

2.4.1.3 *Chlorite crystallinity*

Although chlorite, together with illite–muscovite, occurs abundantly in metapelites, crystallinity measurements are not as widely used to index

reaction progress in the trioctahedral phyllosilicates as they are in dioctahedral phyllosilicates. Prior to the work of Árkai (1991), several studies (see Frey, 1987a, and references therein) had established that the chlorite 7-Å half-height peak width (ChC_{002}) showed good positive correlation with illite crystallinity, although ChC_{002} tended to be smaller, i.e. showed improved crystallinity, compared with illite–muscovite. Similar results were reported by Duba and Williams-Jones (1983) using the chlorite 14-Å peak width, ChC_{001}.

Árkai (1991) made a regional study of metapelitic rocks in north-east Hungary (see section 3.3.4.4) to compare ChC_{002} and ChC_{001} values, and correlate these with illite crystallinity (KI), vitrinite reflectance, conodont alteration index (CAI) and metabasic mineral facies. ChC values for both peaks, air-dried and ethylene glycol-solvated, were measured using the same conditions and separation techniques as those used for illite crystallinity (KI) measurements. Slate standards were used to calibrate the KI data, and also the ChC data, to the metapelitic zonal limits of Kübler (1975). Relatively good ($r = 0.75–0.85$) positive linear correlation was found between the two chlorite peak width values, and between these and KI values derived from metapelites ranging from late diagenetic zone to epizone. Using this approach Árkai (1991, Fig. 5) defined the anchizonal limits as: $ChC_{001AD} = 0.31–0.43$ $\Delta°2\theta$ and $ChC_{002AD} = 0.26–0.33$ $\Delta°2\theta$. Further studies in this region by Árkai and Ghabrial (1997) showed that chlorite crystallinity could also be used to estimate grade in meta-igneous rocks which lacked indicative mineral assemblages.

In a TEM study comparing chlorite and white mica crystal growth in microfabrics, Merriman *et al.* (1995b) found good correlation between ChC_{001} and TEM-measured mean chlorite crystallite thickness based on 14-Å lattice fringes. Across a range of metapelitic zones chlorite showed variable correlation with white mica crystallite thicknesses: slightly thicker crystallites (but same number of 14-Å layers) compared with white mica in the late diagenetic zone; thinner in the anchizone and considerably thinner than white mica in the epizone. These differences were attributed by Merriman *et al.* (1995b) to the differing effects of strain on the crystal growth of the two minerals. In anchizonal and epizonal metapelites chlorite tends to retain more dislocations which reduce the overall crystallite size compared with white mica.

Several regional studies using chlorite crystallinity have found poor correlation with KI data. For example, Wang *et al.* (1996) found that ChC_{002} showed 'enhanced' crystallinity, i.e. higher apparent grades, compared with illite crystallinity. Moreover, in metapelitic sequences ranging from diagenetic to epizonal the overall correlation was poor. Similarly, Dalla Torre *et al.* (1996a) found no obvious correlation between ChC_{002} and KI values in a study of metashales from the Franciscan Complex. Here again, chlorite was found to be well crystallized, although only samples free of mixed-layer C/S were measured for ChC_{002} data.

In general, chlorite crystallinity appears to be a less reliable indicator of regional metapelitic reaction progress than illite crystallinity. In part this may be due to a tendency for the smectite-to-chlorite transition to follow parallel reaction pathways with differing amounts and types of intermediate products in response to different heat-flow rates. For example, fewer intermediate phases appear to characterize the tri-smectite-to-chlorite transition in geothermal systems (Robinson *et al.*, 1997a). This may explain the fact that in some sequences intermediate mixed-layer C/S or Co/C may persist at higher metapelitic grades than I/S (Fig. 2.1). In addition, chlorite crystal growth appears to be more variably affected by strain than illite–muscovite, and this can give poor correlation between the techniques in strongly tectonized sequences. Overall, the chlorite crystallinity technique appears to be best suited to basins with a simple heat-flow history and minimum tectonization. It also provides a method for correlating grade in meta-igneous rocks with that of metapelites (Árkai & Ghabrial, 1997).

2.4.1.4 Size and strain measuring techniques

The interaction between changes in crystallite size and the distribution of lattice strains is an integral part of reaction progress in the phyllosilicate minerals found in metapelites (Peacor, 1992a). Some lattice strains are induced by tectonic stress, for example by mechanical rotation of grains during slaty cleavage formation (section 2.3.2), and some are introduced during crystal growth, by polytypic transformation and by dissolution/crystallization processes. For

example, the reaction of smectite to form random I/S introduces a one-dimensional disorder in smectite crystallites parallel to c*; as illite crystallites progressively thicken, smectite layers are proportionally reduced and disorder decreases. Different lattice strains might be introduced if the newly formed illite is subsequently deformed in a slaty cleavage microfabric. A number of studies have made measurements of the size and strain components of phyllosilicates in very low-grade metapelites using XRD peak profile analysis. This technique analyses the profile of reflections produced by X-ray diffractometry in terms of the three main components of broadening (Klug & Alexander, 1974): (i) machine or instrumental effects; (ii) the effective crystallite size (N) or the mean size of coherent X-ray scattering domains, measured normal to the (hkl) planes giving rise to the peak; (iii) lattice strain or the various types of lattice imperfections and defects that give rise to non-periodic features of crystal structure.

Several methods have been used to calculate the effects of crystallite size and strain from XRD peak profiles, including single peak methods such as Scherrer, variance and Voigt, and the multipeak Warren–Averbach method. Some of the first XRD measurements of illite–muscovite crystallite size used the Scherrer equation to derive the effective crystallite size N from peak broadening β (Scherrer 1918, in Klug & Alexander, 1974):

$$\beta = K\lambda/Nd \cos \theta$$

where K is a constant generally taken to be 0.9 (but see Nieto & Sánchez-Navas, 1994; Drits *et al.*, 1997), λ is the radiation wavelength (1.5418 Å for CuKα, d is the interplanar spacing (≈ 10 Å), and θ is the Bragg angle in radians. Weber *et al.* (1976) used the Scherrer measurement to propose a modified illite crystallinity index based on domain size (e.g. Dunoyer de Segonzac & Bernoulli, 1976; Árkai & Tóth, 1983), and Kübler (1984) suggested replacing 'illite crystallinity' by the term 'Scherrer width' (*largeur de Scherrer*), but neither of these suggestions has been widely adopted. By use of the Scherrer measurement it is possible to convert the Kübler index, corrected for instrumental broadening with a suitable muscovite standard, to an effective crystallite size $N_{(10Å)}$. Plots derived from this type of conversion (see Fig. 2.19) show a non-linear correlation between $N_{(10Å)}$ and KI values and an exponential increase in crystallite size once epizonal conditions are achieved, marking the sensitivity limit of the IC technique (Merriman *et al.*, 1990). Scherrer measurements have also been used to correlate XRD-derived illite–muscovite crystallite thicknesses with those measured by TEM (e.g. Merriman *et al.*, 1990, 1995b; Nieto & Sánchez-Navas, 1994; Dalla Torre *et al.*, 1996a; Jiang *et al.*, 1997); these correlations are further discussed in section 2.4.3.

The use of the Scherrer equation has several shortcomings. For example, the equation is valid for estimating crystallite thicknesses only where the size distribution is very limited (Krumm, 1997). Also, the Scherrer 'constant' K has different values for different thickness distributions (Drits *et al.*, 1997). Furthermore, the size and strain components of peak broadening are not separately measured and the general assumption that the lattice strain contribution is negligible tends to underestimate crystallite size (Jiang *et al.*, 1997).

Some of the peak profile analysis methods which are used to obtain both the size and strain components are described by Árkai *et al.* (1996). The *variance method* of Wilson (1963) uses a plot of W/R against $1/R$, where W is the variance of the peak profile integrated over the range R. Mean crystallite size and lattice strain values are derived from the intercept and slope of the straight line fitted to points representing different ranges. Using this method without instrumental correction, Árkai and Tóth (1983) showed that a prograde increase in the mean crystallite thickness of illite–muscovite is accompanied by a decrease in lattice strain.

The *Voigt method* makes use of the different shape components of the peak profile which are attributable to size and strain; broadening caused by small crystallite size is described by a Cauchy function whereas a Gaussian function describes broadening due to lattice strain. A parabolic fit to the natural logarithms of the Fourier coefficient of the peak profile can be used to derive the integral width values for the size and strain components; the linear coefficient is proportional to the Cauchy function whereas the parabolic coefficient is related to the Gaussian contribution. These two integral widths can be used with the appropriate Scherrer equation to calculate crystallite size and strain (see Árkai *et al.*, 1997). The Voigt method was used by Árkai *et al.* (1996) to show that both illite–muscovite and chlorite crystallite thicknesses increase with prograde metapelitic conditions but lattice strain decreases. Percentage strain in chlorite, however, is generally greater than that in illite–muscovite, consistent with TEM evidence of more strain-related textures in chlorite.

The *Warren–Averbach* (W–A) method of peak profile analysis separates the broadening effects of crystallite size and lattice strain by using the Fourier coefficients of two or more orders of the same (hkl) reflection (Warren & Averbach, 1950). This method depends on the increase in strain-related broadening with diffraction angle, as exhibited by higher orders of the same reflection, in contrast with the lack of change related to size broadening in higher order reflections. Warr and Rice (1994) used the W–A method with various combinations of the 001, 002, 003 and 005 basal reflections to calculate the mean crystallite size of illite–muscovite in CIS samples (Table 2.1). Subsequently, Warr (1996) preferred the single peak method, a variation of the Voigt method, for calculating illite–muscovite crystallite size because it was more precise and six times faster than the W–A method. The W–A method has also been used to model the crystallite size distributions in metapelitic rocks, and such distributions typically show bell-shaped patterns skewed towards larger sizes, which have been interpreted as characteristic of Ostwald ripening (Eberl & Środoń, 1988; Eberl *et al.*, 1990). Eberl and co-workers have developed a modification of the Bertaut–Warren–Averbach (BWA) technique which they have applied to determination of crystal size distributions of illite and I/S crystallites, concluding that most distributions are lognormal (Drits *et al.*, 1997, 1998). Their technique is coded in the computer program called Mudmaster (Eberl *et al.*, 1996). They have also developed a method of treatment of I/S whereby polyvinylpyrrolidone (PVP-10) is used to first prevent swelling of the smectite component, and thereby to prevent interparticle diffraction, giving rise to measurement of independent particles which assumedly are comprised only of illite-like interlayers which they assume are fundamental particles (Eberl *et al.*, 1998). The crystal size distributions so measured with the BWA technique approximate one of two types, lognormal or asymptotic.

2.4.1.5 Computer modelling of X-ray diffraction patterns

Calculated or modelled XRD patterns are widely used at the research level for identifying clay mineral species, for determination of the state of order in sequences of mixed-layer minerals, and for producing sets of intensity patterns as an aid to quantification of clay mineral mixtures. A commercially available program, NEWMOD© (Reynolds, 1985), is routinely used in many XRD laboratories for this purpose. The program requires input to a menu of variables including chemical variables, structural parameters and crystallite size range, as well as machine conditions and X-ray beam geometry, in order to generate one-dimensional XRD patterns of clay minerals. Where some variables are not known, the program will model with default values, but a complete set of well defined variables will generate the best simulations of natural sample patterns. Because of the operating speed and memory of current computers, and improvements to the software, once some basic variables are established others can be refined by iterative processes.

In pelitic sequences, modelled XRD patterns are commonly used to measure reaction progress in terms of relative proportions of mixed-layer minerals, such as percentage illite in I/S, and the stacking sequence, i.e. the Reichweite ordering (R). Iterative methods are also used to model the shape of a specific XRD peak, e.g. the 10-Å peak, using the distribution and the mean size (thickness) of crystallite populations (Moore & Reynolds, 1989), also referred to as coherent scattering domains (e.g. Lanson & Besson, 1992).

Computer modelling techniques are also used to model the contribution of phases which generate the overlapping peaks that produce broad diffraction bands. For example, Lanson (1990) developed methods for the deconvolution or *decomposition* of XRD patterns that can fit up to six 'elementary' curves of phases that are suspected of contributing to a complex diffraction band. These methods have been applied to investigating smectite–I/S–illite reaction progress in diagenetic sequences from the Paris Basin (Lanson & Champion, 1991; Lanson & Besson, 1992; Lanson & Velde, 1992), and the Illinois Basin (Gharrabi & Velde, 1995). Decomposition techniques were also used by Huon *et al.* (1994) to identify K-white micas in very low-grade meta-limestones (see section 2.4.1.2).

2.4.2 Transmission electron microscope techniques

2.4.2.1 Overview of transmission electron microscope techniques

There are three kinds of transmission electron microscope (TEM):

1 The TEM which in its dedicated form is capable of providing near-atomic scale images. It provides data over the full range of the static-beam/sample interaction, typically to diameters of a few hundred angstroms;

2 The dedicated scanning transmission electron microscope (STEM), which as in an SEM, utilizes a focused beam which may be rastered over areas to provide averaged chemical data, secondary electron images (and other types of image, depending on the detectors of choice), and element maps where an EDS detector is present;

3 The 'STEM', where that term is used in a colloquial sense, consisting of a TEM for which the optical system can be altered to that where the beam may be focused and scanned. Such 'STEMs' may be used in either TEM or scanning mode and they comprise the kind of instrumental system which is most commonly used for geological research.

TEM techniques have developed to the point where they are now routinely used in characterization of clay minerals in rocks subject to diagenesis through low-grade metamorphism. They have a powerful advantage in that they simultaneously provide information in three primary ways:

1 Textural information can be obtained through TEM imaging over a broad range of magnifications up to near-atomic resolution;

2 Crystal structure characterization can be obtained through selected area electron diffraction (SAED), which in turn leads to imaging which includes 'lattice fringe imaging' in which 001 planes of phyllosilicates are commonly imaged, and 'structure imaging';

3 Quantitative chemical data can be obtained from the precise areas characterized by other means, using energy dispersive spectroscopic (EDS) techniques, with resolutions approaching 100 Å.

TEM data are obtained over very small sample volumes, however, and care must be taken that such observations are representative of the geological features being investigated. XRD, by contrast, provides data over relatively large volumes, but it is averaged over materials which may be heterogeneous. Where TEM observations are to be made, it is therefore advisable to work through a variety of methods from lowest resolution to TEM observations. A typical sequence would concern field work with mineralogical and/or chemical analysis of bulk samples in order to define representative trends. The mineralogical analysis generally emphasizes XRD data. Optical microscopic study of thin sections, followed by scanning electron microscopy (including BSE imaging) should be carried out on the same sections which will eventually serve as TEM mounts if ion-milled samples are to be used. Such observations permit typical areas of key samples to be chosen for TEM observations, the number of samples so investigated generally being small because of the time requirements. Only then should selected areas be ion-milled for TEM observations. Those observations should then only be acceptable if they are shown to be representative of the features defined by other techniques on bulk samples.

Sample preparation. There are three principal classes of TEM sample: ion-milled, diamond microtomed, and 'holey-C' powder mounts.

1 Ion-milling is generally preferred because small areas of thin sections are used as starting materials, giving rise to samples which retain the textural relations present in the rocks (e.g. Li *et al.*, 1994a,b). Where smectite layers occur, impregnation with a resin which prevents collapse of smectite (001) layers to 10 Å (thus causing ambiguity with respect to 10-Å illite layers) is necessary (Środoń *et al.*, 1992; Kim *et al.*, 1995).

2 Clay mineral 'separates' which are obtained through grinding and a variety of subsequent treatments may be immersed in epoxy, including Spurr or L. R. White resins to prevent layer collapse, and subsequently ion-milled or treated with a diamond microtome for direct production of electron-transparent samples. Such separates may be treated with *n*-alkylammonium hydrochloride to cause expansion of smectite-like interlayers, the degree of expansion reflecting layer charge (Vali & Hesse, 1990).

3 Alternatively, powdered material may be spread on the surface of commercial available mounts consisting of a grid of Cu wires (or Be in the case of mounts to be used for chemical analysis) overlain by a thin layer of 'carbon'. Such mounts almost invariably have (001) crystal layers normal to the beam, and therefore give rise only to (hk0) diffraction data, with no information on layer sequences. However, special mounts of separates which have been shadowed by Pt can be used to determine sample thickness (e.g. Nadeau, 1985).

Orientation of the samples prior to ion-milling or microtoming is essential where observations are required of 001 diffractions or 001 lattice fringes, both of which are fundamental to characterization of clay minerals. Such data require that (001) be normal to the section. That can be optimized in ion-milled samples of thin sections by orienting thin sections normal to bedding and/or slaty cleavage. Where clay separates have been spread on flat surfaces and impregnated, sections must be obtained with surfaces normal to layering.

2.4.2.2 Selected area electron diffraction (SAED)

This method derives its name from the fact that an aperture can be inserted into the electron-optical column to limit the area of the sample which is intercepted by the electron beam, as chosen by the operator while viewing the TEM image. A number of apertures of different diameter are available so that the portion of the sample which contributes to the electron diffraction pattern can be selected by the operator. A push of a button then alters the beam path so that the electron diffraction pattern of the area chosen is projected onto the viewing screen or camera. In practice, as one views the sample, it can be translated via *x–y* manual controls until some feature of interest is observed. The image may then consist of a number of crystals, one of which is of particular interest. An aperture is chosen which limits the field of view only to that crystal, and the diffraction mode is chosen. The diffraction pattern which is observed then corresponds only to the crystal selected in the area of the aperture.

The diffraction pattern is generally restricted to those reciprocal lattice points in the plane normal to the electron beam (the zero-order Laue-zone), i.e. if c* is parallel to the beam, only hk0 reflections may be observed. The sample holder may then be tilted so that other orientations may be viewed. Although the 'standard' sample holders permit samples to be tilted only about the single holder axis, other holders are available which also permit tilting about a second horizontal axis or rotation about an axis normal to the sample. A very large fraction of the reciprocal lattice of a given crystal can thus be explored.

Because flat clay mineral grains are preferentially aligned parallel to the surface of holey-C mounts, only hk0 reflections are generally available. The hk0 reflections are similar for all conventional phyllosilicates, so they do not aid in identification. Those reflections nevertheless provide important data, as they may be used to judge relative degree of coherency of layers. Single crystals in which all layers are coherently related give rise to hk0 diffraction patterns (a*–b* plane) which are hexanets.

Turbostratic stacking causes each layer to be randomly rotated about its normal (c*) with respect to adjacent layers, the collective diffraction pattern of all such layers consisting of circles, each being equivalent to a single reflection (or symmetrical set of reflections) from a single-crystal pattern (e.g. Nadeau *et al.*, 1984b). Partial coherency gives rise to spotty circles. Freed and Peacor (1992a) showed for Gulf Coast samples, that hk0 patterns of I/S become progressively less powder-like as the proportion of illite layers increases. Nadeau *et al.* (1984c) have defined fundamental particles as those which give single-crystal patterns, i.e. those in which all layers are coherently related.

Samples which have c* parallel to their surface (normal to the electron beam) give 001 diffractions which may be used in forming lattice fringe images (see below; Guthrie & Veblen, 1989a,b). As illustrated by Peacor (1992a), such diffraction patterns have reciprocal lattice vector c* and a vector normal to c* (e.g. Fig. 2.11). The vector which is normal to c* can be any vector in the a*–b* plane. Because clay minerals are pseudohexagonal, that includes a* and b*, and their pseudohexagonal equivalents. A clay mineral grain with (001) parallel to the beam (bedding or cleavage parallel to the beam, in general) may therefore give diffraction patterns such as a*–c*, b*–c*, only c*, or pseudohexagonal equivalents, depending on their specific orientation. All STEMs or TEMs are fitted with axes about which the sample may be rotated, so that if the c* axis is parallel to the

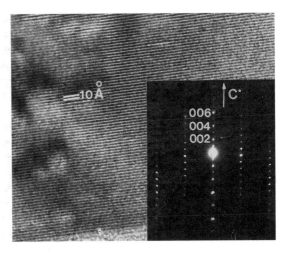

Fig. 2.11 Selected area electron diffraction (SAED) pattern of the 2M₁ polytype of muscovite from a bentonite of epizone grade, superimposed on 002 lattice fringes with a spacing of 10 Å. Starting at the origin (bright spot, middle), the vertical rows of reflections have indices 00l, 11l and 22l. The periodicity in all rows corresponds to d(001) = 20 Å. 00l reflections with l odd (e.g. 001, 003, 005) occur here only through a process of multiple diffraction due to the presence of 11l and 22l reflections, but they are generally absent in most SAED patterns of 2:1 phyllosilicates. (Photograph by H. Dong.)

instrument's rotation axis, the crystal may be rotated about c* and successive a*–c*, b*–c* and c* (only) patterns may be seen. If both a*–c* *and* b*–c* patterns are observed simultaneously, the sequence is in general at least partly turbostratic (incoherent layer interfaces).

2.4.2.3 Lattice fringe imaging

When an SAED pattern contains 001 reflections plus non-001 reflections, for example, an objective aperture may be inserted into the electron-optical column which blocks out all reflections except the 001 reflections which are allowed to pass through the aperture. The image of the sample will then be formed only with 001 reflections. However, the image produced from only the 001 reflections will only include information about the sample which is inherent in the reflections allowed to pass through the aperture and form the image. For example, if 001 reflections are used, only information relating to the z-atom coordinates may be obtained.

If the image is formed with 001 reflections only, the image will be the Fourier transform of those reflections. It will therefore consist only of a sequence of 'fringes' in the area of the image occupied by the crystal giving rise to the 001 reflections. The fundamental spacing of those fringes is $d(001)$. Such fringes therefore permit the values of $d(001)$ of individual layers to be directly observed and measured as shown in Fig. 2.12. The relation between lattice fringe contrast and clay mineral structure and composition has been defined by Guthrie and Veblen (1989a,b), who provided a quantitative basis for interpretation of lattice fringe images.

One of the principal problems with lattice fringe images concerns the propensity for smectite inter-layers to dehydrate in the vacuum of an ion mill or TEM, with d-values of specimens changing from typical values of 13–14 Å to 10 Å. Because illite 001 fringes have a spacing of 10 Å, ambiguity results in differentiating between illite and smectite layers, although pure illite and smectite may each have characteristic appearances (Ahn et al., 1986). The relations defined by Guthrie and Veblen (1989a,b) and exemplified in the images of Veblen et al. (1990) have defined the imaging conditions which permit such differentiaton in mixed-layer I/S even where smectite layers have collapsed. As noted above, treatment with a resin which prevents collapse of layers provides an additional means of identifying individual layers.

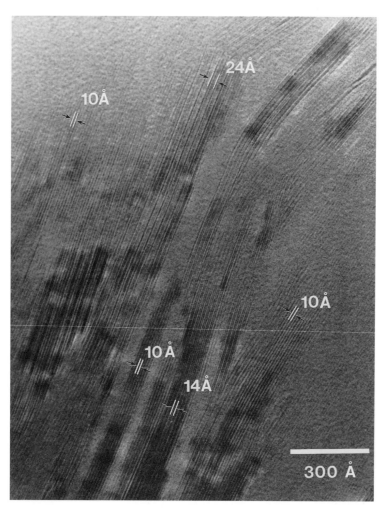

Fig. 2.12 Lattice-fringe image of a late diagenetic zone shale with fringes of various characteristic spacings. Packets of fringes correspond to the minerals illite (10-Å spacing), chlorite (14 Å) and corrensite (24 Å). Disordered layer sequences are indicated by some fringes of corrensite having a spacing ≈ 34 Å, corresponding to one chlorite and two smectite (saponite layers). Areas of the sample for which no fringes appear contain the same minerals, but crystallites are not oriented parallel to the beam, and so they do not diffract and cannot give rise to lattice fringes. (Photograph by Wei-Teh Jiang.)

2.4.2.4 Analytical electron microscopy (AEM)

With the addition of an X-ray detector to the column, characteristic X-rays emitted by the area of the sample interacting with the electron beam may be analyzed and the composition of that sample area determined (see Peacor, 1992b, for a more detailed discussion). The detectors used are invariably solid-state detectors, in part because they are sensitive to a wide range of the spectrum of emitted X-rays, and take up minimal space. The resulting energy-dispersive spectra (EDS) provide quantitative data on elements with atomic number, Z, greater than about 10 (Na). Detectors which utilize the recently developed B-composite or aluminized windows in place of conventional Be windows are capable of providing qualitative data for elements with atomic numbers as low as that of boron ($Z = 5$).

The thick samples used in the electron microprobe cause scattering of the electron beam within the sample, resulting in production of X-rays over a sample volume (area) which is large compared to the diameter of the incident beam. In the case of the TEM, however, the sample is so thin that little increase in cross-section of the beam occurs in the sample. The area of analysis is only slightly larger than that of the beam diameter. In practice, such thinness is achieved on the edges of ion-milled samples, where when viewed in TEM mode, samples are seen to be 'electron-transparent'. In TEMs, the beam diameter is in the order of tens of nanometres, and so resolution is an order of magnitude greater than that of an electron microprobe. The distinct advantage of a STEM in this regard, however, is that the focused beam may provide chemical analyses for areas of diameters approaching 100 Å in instruments for which the beam is produced by a Pt or LaB_6 filament. In instruments fitted with the newer field emission guns (FEG), areas of analysis may be much less than 100 Å. In an instrument fitted with scanning electronics, quantitative analyses may therefore be directly obtained from the specific grains for which SAED patterns and lattice-fringe images have also been obtained in TEM mode. It is a simple matter to then switch to scanning mode, where imaging can be obtained using secondary electron, back-scattered electron, or transmitted electron images, so that the areas identified in TEM mode can be directly analyzed in STEM mode.

A significant advantage of AEM relative to electron microprobe analysis is that where AEM analytical data are obtained from very thin areas, no matrix correction factor (ZAF) is necessary, i.e. X-ray intensity is directly proportional to element concentration. However, a disadvantage lies in the lack of knowledge of accurate sample thickness, so that the volume of excitation, to which the X-ray intensity is also proportional, is usually unknown. Cliff and Lorimer (1975) showed that the ratios of X-ray intensities of two elements are proportional to the ratios of their concentrations:

$$I(Y\mathrm{K}\alpha)/I(X\mathrm{K}\alpha) = \mathrm{k}\, c(Y)/c(X),$$

where $I(Y\mathrm{K}\alpha)$ is the intensity of $\mathrm{K}\alpha$ radiation from element Y, k is a proportionality constant, and $c(Y)$ is the atomic concentration of element Y. Since all significant clay minerals contain Si, in practice one uses samples of well-analyzed standards of silicates to obtain values of the constants, k, for ratios of known $c(\mathrm{Si})/c(X)$, where a different standard is used for each element X. The EDS analysis of an unknown is then processed by eliminating background and integrating intensities to obtain the ratios of the intensities of $\mathrm{SiK}\alpha$ to each of the other cations. The 'k-values' are then used to produce concentration ratios. A distinct advantage of AEM relative to electron microprobe analyses is that the standard k-values need be measured only once and then permanently stored, i.e. once determined, no additional standard data need be obtained at the time of analysis of an unknown.

The last step in analysis of data involves normalization of the analyses to a formula, the specific formula dependent on identification of the analyzed mineral via the full complement of observations (SAED, lattice fringes, EDS analysis, etc.). AEM analyses do **not** directly give rise to element weights per cent as do electron microprobe analyses. Weights per cent may be calculated from the normalized formulae, but they then include any errors inherent in the assumptions on which normalization was based. They should therefore generally not be presented without appropriate caveats, as most readers would otherwise associate with them the kinds of precision normally associated with weights per cent which are directly obtained by the more familiar electron microprobe analyses.

Some of the more common problems associated with AEM analyses include:

1 Diffusion of elements away from the area of beam/sample interaction, a problem which is most serious with alkalies and especially with K (Van der Pluijm *et al.*, 1988). Where grains are of sufficient size, this problem is minimized by rastering over as large an area as possible. K. J. T. Livi [pers. comm.] has noted that no net diffusion occurs where grains on holey-C grids are entirely included within the beam when operating in TEM mode.

2 Lack of knowledge of the oxidation state of Fe (and Mn), and therefore ambiguities in normalization of formulae to numbers of oxygen atoms.

3 Lack of analytical data for light elements, including N as present in NH_4, and therefore errors in normalization. These and other problems were discussed by Peacor (1992b) who recommended that these effects can be minimized by normalizing formulae to totals of octahedral and tetrahedral cations assuming either a dioctahedral or trioctahedral formula; that is at the expense of some error, however, to the degree that a given structure has octahedral occupancy which is not ideally dioctahedral or trioctahedral.

The end result of AEM analysis is a normalized formula that in ideal cases approaches those obtained through electron microprobe data in accuracy and in sensitivity. With respect to the latter, concentrations as small as 0.2 wt% can be measured for first-series transition elements.

The use of EDS spectra in a qualitative way when surveying samples is an especially powerful tool. Many investigators will by now be familiar with the power of combining back-scattered electron imaging on an SEM with qualitative EDS analysis. Whenever

a grain in, say, a thin section is observed with unusual contrast or texture, a rapidly obtained EDS spectrum usually suffices for identification. In the same way, EDS analysis can be used on the TEM, so that the operator rapidly learns to correlate specific contrast and texture features with specific minerals such as

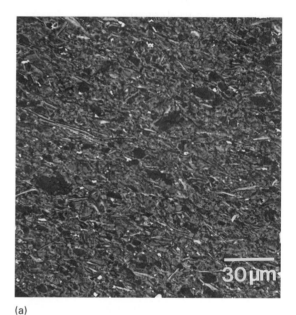

(a)

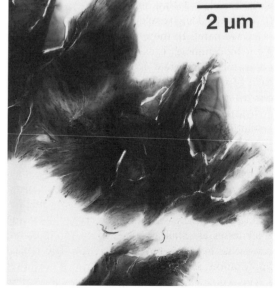

(b)

Fig. 2.13 (*Right*) Back-scattered electron (BSE) scanning electron microscopy image (a) showing the texture of a diagenetic grade shale from the Gaspé Peninsula, Quebec. All grains large enough to be resolved are detrital in origin, and correspond primarily to quartz (dark black), and altered biotite (light contrast, elongated parallel to bedding). Matrix phyllosilicates (primarily chlorite and illite) cannot be resolved in the BSE image, but they occur as packets in cross-section with variable contrast in the low-resolution transmission electron microscope image (b) where some homogeneous detrital grains of quartz are also shown. (Photograph by Wei-Teh Jiang.)

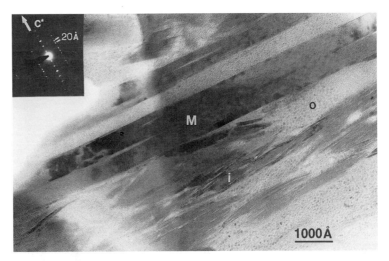

Fig. 2.14 Intermediate-resolution transmission electron microscope image showing a large detrital grain of muscovite (M) within a matrix of organic kerogen (O) and illite (I). Detrital grains of muscovite are seldom observed to be as small as that shown. The illite occurs in small subparallel packets, whose diffraction pattern (not shown) corresponds to $1M_d$ polytypism. The inset diffraction pattern is of the muscovite, and corresponds to the well-defined $2M_1$ polytype. (From Hover et al., 1996, with permission.)

chlorite or muscovite. It is also possible to obtain scanning images using the EDS spectrum, so that so-called 'dot maps' are obtained with contrast proportional to the concentration of a given element. For example, a map of the distribution of K readily defines the distribution of mica relative to chlorite.

2.4.2.5 TEM characteristics of dioctahedral clay minerals as a function of grade

The commonly observed prograde sequence of dioctahedral, 2:1 clay minerals is: smectite → $R1$ I/S → illite [bedding-parallel orientation] → muscovite [cleavage parallel]; more than one member of the sequence commonly coexist, complicated also by the occurrence of separate detrital grains. Fortunately each of the species has a distinct combination of characteristics which permit their identification and distinction. Figure 2.13(a) is a relatively low-resolution back-scattered electron SEM image showing typical grain-size distributions for clays of two types: (1) large, angular detrital grains which have sizes in the micrometre range; (2) a fine-grained matrix of a mixture of grains too small to be resolved by BSE imaging. Such matrix minerals are generally authigenic in origin, with individual crystallites having thicknesses as small as 100 Å. They are therefore resolvable by TEM, but not by SEM. The typical appearances of detrital and matrix clays, the latter in the sequence of increasing grade, are illustrated in the following descriptions:

1 Detrital mica (Fig. 2.14). Detrital dioctahedral clay grains are almost inevitably of mature muscovite derived from relatively high-grade rocks, as opposed to illite or smectite. Grains are seldom thinner than thousands of angstroms, there usually being a very sharp division between the large detrital grains and smaller authigenic grains. SAED patterns show sharp non-(001) reflections typical of well-ordered, coherent layers, which define $2M_1$ or, less frequently, 3T polytypism. Lattice fringe images show packets of sharp, straight layers of constant 10-Å spacing with homogeneous contrast, the boundaries of which commonly exceed the imaged area. Layer terminations are rare. As a result of beam damage, there is a typical 'mottled' contrast which may increase in intensity with increasing beam exposure. Compositions approach those of ideal interlayer vacancy-free mica.

2 Smectite (Fig. 2.15). Lattice fringe images of smectite have fringes which are 'wavy and anas-tomosing' and of variable contrast, as a result of variable orientation from layer to layer and along layers. Layer terminations (edge dislocations) are common. Spacings in materials which have dehydrated and collapsed in the vacuum of the TEM are approximately 10 Å, in which case individual layers of smectite cannot be differentiated from illite. However, where smectite has been treated with an appropriate expanding agent (e.g. L. R. White resin), spacings are in the order of 12–13 Å. SAED patterns have diffuse 001 reflections, with only low

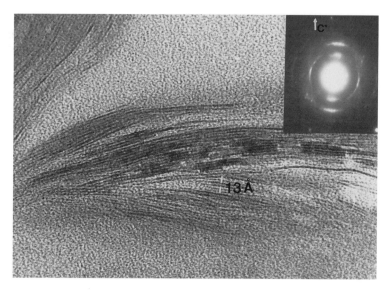

Fig. 2.15 Lattice-fringe image of smectite from an early diagenetic shale from the Gulf Coast, USA. The typical fringes are curved, show variable spacing averaging 13Å, show common layer terminations, and have variable contrast along layers caused by varying orientation relative to the electron beam. Areas where no layers occur also contain smectite, but are not in the correct orientation to give rise to lattice fringes. The inset diffraction pattern shows only 001 and 002 reflections of the 00l series, and they are broadened normal to c* due to variable layer orientation, and diffuse parallel to c* due to variation in layer spacing. Sample treated with L. R. White resin to prevent collapse of smectite interlayers. (Photograph by H. Dong.)

orders being detectable. Non-001 reflections are commonly so weak, diffuse (with streaking parallel to c*), and poorly defined as to be unresolvable. When present they are non-periodic, with both 0kl and h01 reflections being observed, these features being a function of predominantly turbostratic stacking (Dong & Peacor, 1996). Layer terminations are common. AEM analyses display spectra with small Al peaks, commonly with some Mg and Fe.

3 *R*1 I/S (Fig. 2.16). *R*1 I/S has the general appearance and characteristics of smectite, but lattice fringes are generally less wavy and more linear, and SAED patterns show higher orders of 001 reflections. Non-001 reflections are more intense, but still are non-periodic and poorly defined. The key to identification is in using specific focusing conditions which produce alternating dark and light contrast in fringes as a reflection of differences in alternate

Fig. 2.16 Lattice-fringe image of clay minerals in a late diagenetic zone K-bentonite, from the Welsh Basin. Sequence of fringes of illite/smectite (I/S) show variable spacing and contrast, with some fringes being uniformly darker or lighter than adjacent fringes. A series of fringes corresponding to *R*1 I/S on the right, middle, has periodicity 2 × 10 = 20Å as defined by dark fringes. Sequences of fringes having uniform 10-Å spacing and contrast correspond to packets of illite (I). The inset selected area electron diffraction pattern shows typical diffuse, non-periodic non-00l reflections. (From Dong *et al.*, 1995, with permission.)

interlayers (Guthrie & Veblen, 1989a,b; Veblen *et al.*, 1990). Where samples are impregnated with a suitable agent, the smectite-like interlayers do not collapse, and the periodicity is commonly observed to be 21 Å. AEM analyses show more Al and K than smectite, but are generally indistinguishable from those of illite (see below).

Illite—bedding parallel illite (Fig. 2.17). This illite, which is commonly directly formed via reactions in which smectite was a reactant, occurs as fairly well-defined subparallel packets with mean thickness of approximately 70–100 Å (see section 2.4.3). The subparallel packets generally inherit the orientation of smectite layers, with packets being parallel to bedding on average. Two types of orientation are

seen at contacts with larger grains of detrital minerals such as quartz: (a) illite may replace the smectite that forms compactional wraps around detrital grains and is aligned parallel to the grain surface; (b) in less compacted microtextural domains, such as the 'shadows' associated with compaction around rigid grains, authigenic illite forms bedding-parallel packets (see section 2.3.2). Contrast of fringes is relatively constant, reflecting the more-nearly parallel relative orientation of layers, and layer terminations are uncommon. SAED patterns show relatively sharp 001 reflections out to high orders, but non-001 reflections are still diffuse, poorly defined, and largely nonperiodic. Although reflections are not continuously diffuse as in the ideal $1M_d$ polytype, such illite is referred to as being $1M_d$ (Dong &

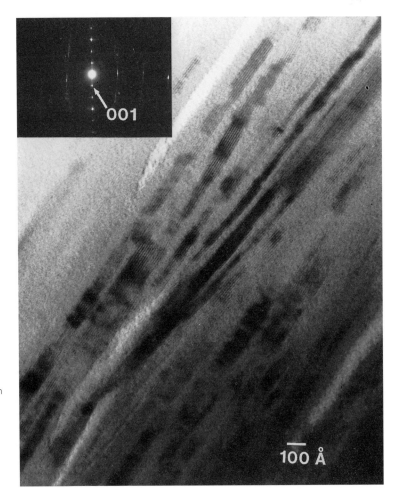

Fig. 2.17 Lattice-fringe image of thin packets of late diagenetic illite from the Gaspé Peninsula, Quebec. Packets of straight fringes with 10-Å spacing are oriented parallel to bedding and are separated by thin areas occupied by illite whose layers are not in correct orientation to give rise to 00l selected area electron diffraction (SAED) reflections. The inset SAED pattern shows a typical well-defined set of 00l reflections, but with diffuse and non-periodic non-00l reflections. (Photograph by Wei-Teh Jiang.)

Peacor, 1996). Both h0l and 0kl reflections may occur in the same patterns as a reflection of some turbostratic stacking, but the degree of coherency is increased relative to $R1$ I/S or smectite, and stacking is predominantly of the $2M_1$ type (Dong & Peacor, 1996). AEM analyses indicate Al contents intermediate to those of smectite and muscovite.

Illite–muscovite — cleavage parallel (Fig. 2.18). Such material occurs in subparallel packets with well-defined boundaries, with the average packet orientation being parallel to cleavage, largely as a result of dissolution of bedding-parallel illite and neocrystallization of larger crystals normal to maximum prevailing tectonic stress. Crystals are typically in the order of several hundreds of angstroms in thickness and give sharp reflections with periodic 0k1 or h01 reflections consistent usually with $2M_1$ polytypism, but with 3T polytypism being frequently observed. Lattice fringes are straight, of constant orientation and therefore of constant contrast, and layer terminations are rare. AEM analyses show that in the anchizone such micas range from phengite to muscovite or paragonite in composition, whereas compositions of epizonal micas invariably correspond to nearly ideal muscovite or paragonite, and rarely margarite (e.g. Livi *et al.*, 1997a,b).

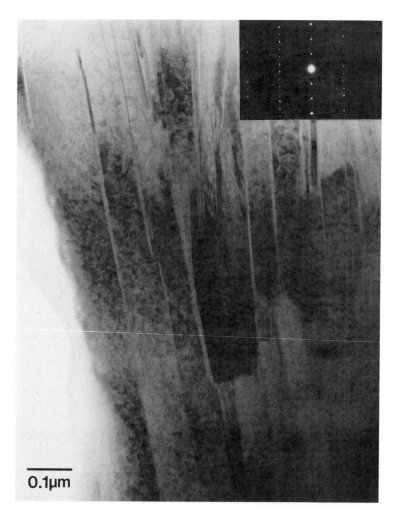

0.1μm

Fig. 2.18 Intermediate-resolution transmission electron microscope (TEM) image of epizonal grade packets of muscovite, oriented parallel to the direction of cleavage. Packets in the order of 1000 Å in thickness display the mottled texture typical of micas. The inset selected area electron diffraction (SAED) pattern has sharp 0kl reflections with 20-Å periodicity (relative to the 10 Å of the 00l reflections) typical of $2M_1$ polytypism. (Photograph by Wei-Teh Jiang.)

2.4.2.6 Crystallite size distributions

Crystallite sizes have been directly determined by TEM using both powdered 'separates' and ion-milled rock samples, whereas XRD has been extensively used to indirectly determine crystallite thicknesses using <2-μm separations. Two different techniques have been used for direct measurements of powdered samples, TEM on Pt-shadowed samples, and atomic force microscopy (AFM). The Pt-shadowing technique, perfected by Nadeau *et al.* (1984b), involves spreading powdered samples on a flat substrate such as a mica cleavage flake, and evaporating Pt onto it, the Pt source being set a known angle to the surface. The sample then becomes coated with a thin layer of Pt except for a 'shadow' on the side of a grain away from the source, the width of the shadow being proportional to the height of the grain. The sample is placed in the TEM, and images show sharp contrast in areas coated with Pt and those which are not, allowing the shadow width, and thus the height of a given grain to be directly determined. The width and length of grains can also be directly determined. This technique has been widely used to determine particle size distributions (e.g. Nadeau & Bain, 1986). Eberl *et al.* (1990) pointed out that such distributions generally approximate log-normal distributions. Alternatively, grain shapes and thicknesses can be directly determined using AFM (e.g. Lindgreen *et al.*, 1992).

The sizes of individual crystals can be directly measured from TEM images, as shown by Merriman *et al.* (1990, 1995b). They directly measured the thicknesses of individual packets in large numbers of TEM images of different areas of ion-milled portions of the same samples. Individual packets were measured by counting the number of regularly spaced fringes parallel to c* between discontinuities, such as grain boundaries with other phases, stacking faults and layer terminations caused by dislocations. Each defect-free sequence of layers can be regarded as a crystallite, or a coherent scattering domain in terms of its contribution to X-ray diffraction. The advantage of this technique is that the grain sizes have been unaffected by the separation technique, i.e. they have not been cleaved into smaller particles, an effect which is assumedly more significant where easily-cleaved smectite interlayers are present. Jiang *et al.* (1997) and Li [pers. comm.] have compared measurements of grain thicknesses in ion-milled and powdered samples, and found decreases in thickness of factors up to six. Li *et al.* (1998) noted that grinding can produce log-normal distributions where none existed in undisturbed samples. These relations imply that only the crystal size distributions determined from samples which have not been powdered produce functions which correctly reflect actual rock distributions.

Where crystal size distributions are to be compared with mean crystal sizes (thicknesses) as determined by XRD techniques such as the Warren–Averbach method (see section 2.4.1.4), it is important that the distributions be made with the same definitions of 'size'. The contribution of a given grain to the XRD profile is proportional to grain volume. Rather than counting a given grain once, it must be divided into a number of equal-size areas so that the number of such areas times the thickness is proportional to the volume of a given grain (Jiang *et al.*, 1997).

Observation of strain by TEM, for comparison with XRD-determined values, is much more difficult. Bons (1988) has observed complex defects in chlorite, but such measurements are much too time-consuming because large numbers of grains, each of which must be individually studied, are required for statistically meaningful results. Jiang *et al.* (1997) have approximated strain by the proportions of easily identified layer terminations, a form of edge dislocation, and compared the results with values determined by analysis of XRD patterns. The results showed that strain decreased with increasing metamorphic grade, as a function of the generally observed trend of increasing perfection of crystals with increasing grade (Peacor, 1992a; Árkai *et al.*, 1996).

2.4.3 Correlation of X-ray diffraction and transmission electron microscope measurements

Several studies have used TEM measuring techniques on ion-milled samples to explore correlation between crystallite thickness and size calculations derived from XRD measurements. Correlation studies are based on converting XRD peak width measurements to an effective crystallite size, also referred to as coherent scattering domain (CSD) size, using one of the methods described in section 2.4.1.4. To date, studies of this type have mainly focused on correlating TEM measurements of illite–muscovite crystallite thickness with illite crystallinity (KI) measurements, although both Merriman et al. (1995b) and Árkai et al. (1996) also attempted to correlate chlorite measurements. A number of such studies have converted the KI limits of the anchizone to equivalent crystallite sizes (Table 2.3).

In one of the early studies of this type, Merriman et al. (1990) used the Scherrer equation to convert KI measurements to CSD sizes and a synthetic fluorophlogopite mica to correct for instrument broadening effects. No correction was made for the effects of strain, which was assumed to be significant only for the late diagenetic zone sample. The mode of the TEM-measured white mica thickness populations showed the best correlation with CSD sizes for the late diagenetic and anchizonal metapelites, but the epizonal sample showed little correlation with either the mode or the mean. However, the presence of Na-mica in the samples studied by Merriman et al. (1990) probably caused broadening of the XRD peaks.

In a similar study, Nieto and Sánchez-Navas (1994) used the Wilson variance method to subtract contributions due to strain, and muscovite flake standard MF1 (Warr & Rice, 1994) to determine the instrumental contribution to peak broadening. They found that CSD sizes derived by this method showed good correlation with the mode of the TEM-measured white mica thickness populations. Based on this study, Nieto and Sánchez-Navas (1994) proposed that the KI limits of the anchizone, i.e. 0.42 and 0.25 Δ°2θ, are equivalent to 300 Å and 900 Å in terms of crystallite size (Table 2.3).

Both chlorite and white micas were used to correlate TEM and XRD size measurements by

Merriman et al. (1995b), but only samples containing Na-poor phengitic K-micas were measured. KI and ChC_{001AD} measurements were converted to CSD sizes using the Scherrer equation, and muscovite and clinochlore flakes used to correct for instrument broadening effects. No corrections were made for the effects of strain. Merriman et al. (1995b) found good correlation ($r^2 = 0.91$) between the mean of TEM thickness measurements for phengite (10 Å) and chlorite (14 Å) and the CSD sizes, whereas the mode of these populations showed poor correlation.

Árkai et al. (1996) calculated CSD size and percentage lattice strain for illite–muscovite and chlorite by four different methods and compared these with TEM crystallite thickness measurements. Significant differences were found between the four methods (Scherrer, Voigt, variance and Warren–Averbach), and between results derived from using different standards for instrument broadening effects. The best correlations were found between the mean of the TEM-measured thickness populations and the Scherrer calculated CDSs for both KI and ChC_{001AD} measurements. The Voigt method gave the next best correlation, and the median of the TEM thicknesses showed better correlation with Scherrer and Voigt CDSs for the diagenetic sample measured by Árkai et al. (1996).

A different approach to TEM measurements was adopted by Jiang et al. (1997) who ion-milled both intact metapelite samples and clay separates bonded in resin from the same samples. In making their TEM measurements, Jiang et al. (1997) assumed that a given crystallite contributes coherently scattered X-rays in proportion to its volume. Where the thickness of a crystallite varied along its length, it was divided into domains of constant thickness

Table 2.3 Illite–muscovite (10 Å) crystallite thicknesses for the KI limits of the anchizone.

	Kübler index (Δ°2θ)	1	2	3	4	5
Late diagenetic–anchizone transition	0.42	230 Å	300 Å	230 Å	230 Å	220 Å
Anchizone–epizone transition	0.25	480 Å	900 Å	520 Å	480 Å	520 Å

1, Merriman et al. (1990), Scherrer method; 2, Nieto & Sánchez-Navas (1994), variance method; 3, Warr & Rice (1994), Warren–Averbach method; 4, Jiang et al. (1997), Voigt method; 5, Fig. 2.19, TEM mean thickness.

and the average of several thickness measurements was shown to be proportional to the crystallite volume. These thickness measurements were correlated with Voigt analyses of 10-Å XRD peak profiles from a suite of prograde metapelites representing the late diagenetic to epizone transition. With the exception of the late diagenetic sample, reasonable correlation was found between TEM-measured thicknesses of the clay separates and CSD sizes determined by the Voigt method and calculated from the Scherrer equation. However, average crystallite thicknesses in the intact rock samples are up to five times larger than those measured in the clay separates. Jiang *et al.* (1997) proposed volume-averaged thicknesses of 230 and 480 Å for the limits of the anchizone (Table 2.3). Similar differences in illite–muscovite crystallite size were found by Li *et al.* (1998), who compared TEM-measured mean thicknesses with those obtained by XRD 10-Å peak profile analysis. Much smaller differences were found when TEM size measurements were made on the ion-milled clay separates used for XRD analysis. Because of the errors involved in deriving crystallite size distributions from clay separates, Li *et al.* (1998) recommended that only distributions obtained from intact rock samples, e.g. ion-milled samples, should be used as indicators of diagenetic and metamorphic crystal growth conditions.

TEM crystallite thicknesses from ion-milled metapelites representing a range of geotectonic settings are plotted against illite crystallinity (KI) in Fig. 2.19. Both the mean and the median of the TEM-measured crystallite thickness populations show an inverse relationship with the XRD-measured KI; the mode of TEM thickness populations (not plotted) does not show this type of relationship. Calculated crystallite thicknesses (CSDs) derived from the Scherrer equation using the KI values are plotted for comparison. It is clear that both measured and calculated crystallite sizes have similar distributions and this suggests that, despite its shortcomings (see section 2.4.1.4), the Scherrer calculations reflect changes in average size in terms of contributions from populations of different thicknesses. Lowest grade, immature pelites with large KI values have a narrow range of crystallite thicknesses and this is reflected in similarities between TEM mean and median values, which may also coincide with the population mode. As grade increases and crystallite

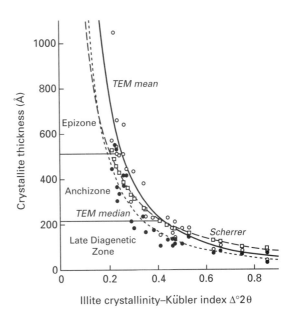

Fig. 2.19 Transmission electron microscope (TEM) measured illite–muscovite crystallite thickness along c* plotted against the Kübler index. Mean (○) and median (●) values of the TEM-measured populations, and thicknesses calculated from the Scherrer equation (□) are shown for comparison. The limits of the anchizone (KI 0.25 and 0.42) are shown as TEM mean thicknesses. (Based on data from Árkai *et al.* (1996), Merriman *et al.* (1995a,b) and unpublished data with contributions from S. R. Hirons, S. J. Kemp and B. Roberts.)

size distributions spread out, the TEM mean and median values diverge. Up to the beginning of the anchizone, TEM mean values show better agreement with the calculated values, whereas TEM median values agree better with calculated values in the anchizone and epizone. However, the TEM mean values for the anchizone, 220 and 520 Å in terms of the KI limits, show the best agreement with those determined by other studies shown in Table 2.3.

2.4.4 Retrogression

A wealth of data has clearly demonstrated that the majority of low-grade sequences represent prograde reactions, with the clay minerals being a valid measure of reaction progress and grade. Studies discussed later, however, show that care must be exercised in associating a given metapelitic grade with a specific clay mineral assemblage, or even with one interval of a complete sequence. Moreover, such studies suggest that only where textural relations have been imaged and when observed relations clearly define the mode of formation of a given clay mineral can that mineral be directly associated with a given origin with absolute certainty.

A sequence of clay minerals observed by Price and McDowell (1993) from the Precambrian Freda Formation in Wisconsin appeared to be a typical prograde sequence as defined by XRD data from drill core samples. Initially, the smectite identified in shallow samples was inferred to have been replaced by illite in deeper samples, although the transition occurred over a narrow depth range. As a function of depth, therefore, reaction progress appeared to closely parallel the classic prograde Gulf Coast sequence as defined by Hower *et al.* (1976). Price and McDowell (1993), however, noted that drill core samples from the Freda Formation in Michigan did not display the same sequence. They suggested that the smectite of shallow samples might not be the precursor of illite, but rather was derived by reaction of fluids with illite which had formed during an earlier, prograde event.

In a TEM study of the same samples investigated by Price and McDowell (1993), Zhao *et al.* (1998) observed, however, that the illite occurring at great depths had characteristics typical of anchizonal metamorphism (mean packet size, composition, deformation textures, etc.), not illite-rich I/S typical of that formed by transformation directly from smectite. The $R1$ I/S of shallow samples was found to occur in packets identical to those of illite of greater depth, i.e. the $R1$ I/S had replaced packets of illite which formed at higher grades (Fig. 2.20). Furthermore, detrital muscovite in deeper samples displayed few alteration features, whereas muscovite in shallow samples was seen to have been directly replaced by smectite and I/S. This study suggests that conclusions based only on mineralogical relations (prograde diagenesis in this case) determined by methods such as XRD should, in selected cases, be regarded as tentative. Where there is reason for doubt, textural relations must be defined by direct observation by TEM and/or SEM in order to determine reaction processes.

Nieto *et al.* (1994) observed replacement of triocta-hedral chlorite by dioctahedral smectite on a regional basis. Nieto [pers. comm.] has also documented large-scale replacement of chlorite and muscovite

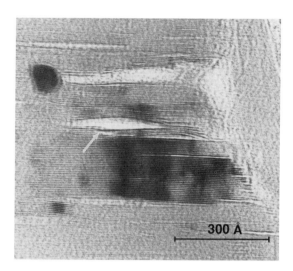

Fig. 2.20 Lattice-fringe image of detrital muscovite from the Freda formation, Wisconsin, USA. The straight fringes with mottled, dark contrast are 10-Å fringes corresponding to relict, detrital muscovite. The arrow points to a transition zone between fringes of muscovite on the left and smectite (curved fringes with variable spacing) on the right. Lens-shaped areas with no fringes are caused by contraction of smectite layers upon dehydration. (Photograph by Gengmei Zhao.)

by smectite in pelites of the San Vicente sediments, and of replacement of illite formed during low-grade metamorphism by I/S in pelites of the Basque-Cantabrian Basin, Spain. The term 'retrograde diagenesis' has been used for such reactions, akin to the term retrograde metamorphism, but implying conditions usually associated with the early or late diagenetic zones.

Although the reactions described by Nieto *et al.* (1994) and Zhao *et al.* (1998) occurred on a regional scale, similar reactions have been observed on a more localized scale. For example, Jiang *et al.* (1990a) found I/S occurring in mudstones from the Welsh Basin in an area where detailed isocryst mapping indicated that prograde metamorphism had proceeded to the low anchizone. TEM micro-textures showed that illite had reacted with fluids and was locally replaced by $R1$ I/S. The fluid activity was associated with a major fault zone which was active during and after the main regional metamorphic event in the Welsh Basin, and this was assumed to be the cause of local retrograde diagenesis.

In the cases described above, reactions characteristic of the early or late diagenetic zone, where smectite is normally found in prograde sequences, occurred in rocks which had previously undergone prograde metamorphism of anchizone or epizone grade. Such retrograde diagenesis as observed on a regional and local scale appears to be linked with tectonic events of low intensity during which fluid flow and fluid/rock ratios were enhanced.

2.5 Geothermometry and geobarometry

The field of metamorphic petrology, especially, has undergone a major transformation in the past few decades with the identification of a plethora of mineralogical/petrological relations which may be used to constrain or specify pressure or temperature of formation, i.e. geothermometers or geobarometers (see Essene, 1989, for a general review and references). A key to application of such relations was the recognition of the need to establish that conditions in both experimental systems and natural rocks were those of stable chemical equilibrium. With respect to the former, the standard test is that of experimental reversibility. Metamorphic petrologists have traditionally emphasized studies of rocks which form at greenschist facies conditions or above, conditions where coarse-grained rocks generally approach chemical and textural equilibrium, and for temperatures for which reaction rates are sufficiently rapid to permit experimental reversibility.

The discussion in section 2.1.3 emphasized that specific, well-recognized sequences of minerals and mineral textures occur with increasing grade of diagenesis and very low-grade metamorphism. It was the recognition of such sequences in higher grade metamorphic rocks which led to the development of geothermometers and geobarometers. The repetitive nature of those sequences in very low-grade metamorphic and diagenetic systems suggested that geobarometers and geothermometers should exist for rocks forming under such low-temperature conditions. A variety of systems, some involving clay minerals, have been defined. Relations which have been defined as quantitative or qualitative geothermometers include the smectite-to-illite reaction, illite crystallinity, chlorite composition, polytypism, vitrinite reflectance and conodont alteration index, whereas the phengite content of white mica has been related to pressure. Some of these systems are described briefly below.

In section 2.1.3 it was emphasized that many authors have concluded that clay minerals are generally metastable, and that specific temperatures or pressures cannot be ascribed to them (e.g. Essene & Peacor, 1995). This is also the case for organic materials. Nevertheless, a state of 'reaction progress' can be ascribed to any assemblage, using one of the so-called geothermometers or geobarometers identified above.

If minerals, especially clay minerals, tend to exist in states of metastability in low-grade rocks, is it possible to associate given pressures and temperatures with specific metamorphic grades? Essene and Peacor (1995) pointed out that although reactions are controlled by a host of factors affecting reaction kinetics (e.g. fluid/rock ratio, fluid composition, etc.), where such conditions are constrained to be approximately equal, there should be an approximate correlation between reaction progress and grade. That is, in regional metamorphic sequences where there is similarity of bulk rock compositions, sedimentary mineral assemblages, tectonic stress gradients, fluid/rock ratios and temperature, reaction progress as measured, for example by illite crystallinity, has been found to correlate well with metamorphic grade. Despite the cautions emphasized by Essene and Peacor (1995), it seems clear that when prograde sequences of some localities are compared, the geothermometers 'work', at least where such localities can be shown to have been constrained by similar geological variables. Such correlations must be treated with caution, however, as they do not generally represent true equilibrium, and disparate values of one or more variables such as tectonic stress, bulk rock composition, time, fluid composition or fluid/rock ratio may result in very different correlations of reaction progress and metamorphic grade (temperature) in other sequences. It must be kept in mind, however, that some authors contend that at least some relations involving clay minerals can be used to specify precise temperatures (e.g. Aja & Rosenberg, 1992). In any event, the likelihood of the existence of chemical equilibrium and therefore of the applicability of geothermometers and geobarometers increases with increasing grade and is commonly achieved under greenschist facies conditions.

It may also be argued that measurements of reaction progress in the range that is characteristic of very low-grade metamorphism span a limited temperature range (*c.* 100–300°C) for which there is a wide range of textures, mineral assemblages and mineral compositions. Hence, large changes in those collective variables are characteristic of a relatively narrow range of temperatures. These changes in clay minerals parallel those observed in organic materials over a similar temperature range as they react (mature) during sedimentary burial processes. However, since organic materials, particularly vitrinite, are mainly effected by temperature changes they are widely used as indicators of thermal maturity (e.g. Sweeney & Burnham, 1990). In some cases this has enabled clay mineral reaction progress to be thermally indexed, particularly the smectite-to-illite reaction, as discussed later. Other organic indicators, such as conodont alteration index, have also been correlated with illite crystallinity on a regional scale, as discussed in Chapter 3 (see section 3.2.3).

In the following sections, relations are described for some clay mineral systems which have been most widely used as geothermometers. Although their applicability as quantitative geothermometers is subject to question, the systems which are described below are almost universally accepted as indicators of reaction progress.

2.5.1 Illite–smectite reaction

The reaction in which smectite is a reactant and illite a product was recognized to occur over a predictable range of depth in mudstones of the Texas Gulf Coast, USA (Powers, 1959, 1967; Burst, 1969; Perry & Hower, 1970). The now classic paper by Hower *et al.* (1976) emphasized that the depth-related transformation occurred over the temperature range of approximately 60 to 90°C, as indicated by down-hole measurements of well temperatures. The same range has, at least to a first approximation, been verified in many other studies of this type. A closely similar temperature range, with detailed correlation between temperature and specific proportions of illite and smectite in mixed-layer I/S, was experimentally determined by Sass *et al.* (1987), Aja (1991), Aja *et al.* (1991a,b) and Aja and Rosenberg (1992), although May *et al.* (1986) have questioned the validity of any

experiment involving such fine-grained heterogeneous clay minerals.

The use of the illite–smectite reaction as a measurement of temperature was questioned by Jiang *et al.* (1990b), who showed that illite is metastable relative to the thermodynamically stable assemblage pyrophyllite–muscovite. Loucks (1992) presented data implying that the composition of illite does not approximate a mixture of pyrophyllite and muscovite, but Jiang *et al.* (1994) showed that these data were inaccurate. Essene and Peacor (1995) reviewed the many factors leading to conclusions which demonstrated that clay minerals in general, and illite and smectite in particular, are metastable. Aja and Rosenberg (1996) questioned those views, and Essene and Peacor (1997) subsequently reaffirmed their contention that most clay mineral associations are metastable.

Essene and Peacor (1995) noted that even if one assumes that illite and smectite are metastable in all occurrences, the illite–smectite reaction can nevertheless provide information regarding temperature of formation of illite in three ways:

1 identification of approximate maximum temperatures (e.g. Pollastro, 1990, 1993; Bish & Aronson, 1993), assuming that smectite cannot exist where temperature, as one factor affecting reaction kinetics, has reached a limiting value;

2 with careful experimental analysis of those factors, including temperature, which determine reaction rate (Huang *et al.*, 1993);

3 recognition that geological conditions constraining factors affecting reaction rates (e.g. temperature, fluid/rock ratio, concentrations of alkalis in solution, pH) are similar to those where temperature has been determined, such as Gulf Coast mudstones, as reviewed at length by Velde (1992). Essene and Peacor 1995) cautioned that temperatures so determined are approximate at best.

The errors inherent in the use of the illite–smectite reaction are best illustrated by the differences in degree of transformation in different types of sediments that are interbedded and have therefore been subjected to the same burial temperatures. This is especially true for bentonites where reactions are retarded relative to enclosing shales (e.g. Šucha *et al.*, 1993; Li *et al.*, 1997). One possible cause of the difference may be the presence of organic acids in shales due to maturation of organic material (e.g.

Small, 1993, 1994). Whatever the causes, however, the marked differences in reaction progress for smectite in sediments of the same age and which have been affected by the same temperature, is definitive evidence of the problems inherent in using the illite-to-smectite reaction as a geothermometer. The formation of illite in very young sediments near ambient temperatures, under conditions normally ascribed to smectite (e.g. Turner and Fishman, 1991) further illustrates the problems.

2.5.2 Illite and chlorite crystallinity

As discussed in section 2.4, illite and chlorite crystallinity are primarily a function of crystallite thickness with lattice strain and other imperfections being a secondary contributor. Illite and chlorite have crystallite thicknesses in the order of 100 Å where they have formed in late diagenetic pelites. Such small crystals have relatively large concentrations of defects as observed by TEM (e.g. Jiang *et al.*, 1997). The combination of large surface area relative to volume, and high defect concentrations, causes such crystallites to have larger free energies and to be metastable relative to larger, less imperfect crystals. Dissolution of some smaller crystals and crystallization of solutes on others causes the number of crystals to decrease and the crystal sizes to increase. This has been ascribed to the very specific process of Ostwald ripening (Eberl *et al.*, 1990). However, such a ripening process is defined for closed system conditions and at constant temperature, conditions not met in the geological systems under consideration. Other investigators (e.g. Dalla Torre *et al.*, 1996c; Merriman *et al.*, 1995b; Jiang *et al.*, 1997), have described the ripening process as being related to anisotropic, tectonically induced strain, and have described TEM-observed textural relations which document increase in crystal size and decrease in strain. The key factor here, however, is that increase in crystal size is a continuous process in which the mean value of the crystal size distribution increases with increasing grade, the resulting larger crystals being more stable (of lower free energy). Only where crystals have become macroscopically coarsened do they approach a condition of stable textural equilibrium.

A given value of illite or chlorite crystallinity is therefore a function of reaction progress, and rep-resents a state which is controlled by kinetic factors. Crystal sizes are governed by the relative rates of growth of crystals, pressure solution and recrystallization, which are in turn controlled in part by fluid activity, temperature and tectonically induced strain. Since temperature is only one factor controlling growth conditions it is unlikely that experimental conditions could be achieved where all other factors are constrained and changes in crystallinity could be directly indexed with temperature measurements.

A more pragmatic approach that involves correlation with organic maturity indicators is discussed in Chapter 3 (see Fig. 3.4). A number of studies have correlated illite crystallinity with organic maturity indicators.

2.5.3 Chlorite geothermometers

As reviewed by de Caritat *et al.* (1993), there are two types of chlorite geothermometer:

1 Based on variation in chemical composition as determined by thermodynamic analysis of specific reactions. These include relations described by Walshe (1986) for chlorite coexisting with quartz, and by Hutcheon (1990) for chlorite–carbonate assemblages.

2 Based on empirical calibrations of composition variation as a function of known increase in temperature. Cathelineau and Nieva (1985) proposed a thermometer based on Al content of chlorite as determined for chlorite in cores from the Los Azufres geothermal system, Mexico. These relations were subsequently modified by Cathelineau (1988) and by Kranidiotis and MacLean (1987) and Jowett (1991). Aagaard and Jahren (1992) correlated increasing temperature with an increase in Al content of chlorite coexisting with illite.

De Caritat *et al.* (1993) applied several of the geothermometers to their own set of chemical data and compared the results, noting that the correlation was poor and that none of the thermometers gave reliable temperatures over the studied range. Shau *et al.* (1990) and Jiang *et al.* (1994c) showed through direct TEM observations that chemical analyses of fine-grained chlorite which forms in the very low-grade metamorphic environment are usually contaminated, in part by the presence of a mixed-layer smectite (corrensite) component. Such chemical data are therefore not representative of a thermodynamically stable phase, such stability being a requirement of a geothermometer. Similar observations have been made by Schmidt *et al.* (1999). The critical analytical factor in analyses subject to such contamination is imbalance in tetrahedral and octahedral Al, a factor which also affected most of the analyses of de Caritat *et al.* (1993). The imbalance in Al is greater for increased mixed-layering, which is more likely at low temperatures. In addition, TEM data show that lower temperatures are consistent with smaller crystals, for which analytical contamination is more likely. The Al content in so-called 'chlorite' is thus regularly, but non-quantitatively, related to grade.

Hillier and Velde (1991) compiled data for the variation in composition of chlorite with temperature for several prograde sequences, finding the usual regular decrease in octahedral vacancies and increase in tetrahedral Al (i.e. trend to equality in numbers of octahedrally and tetrahedrally coordinated Al). The range in composition for samples, however, is in the same order of magnitude as the total variation for all samples at all grades. They also emphasized the problem of analytical contamination, especially at lower grades, metastability of low-grade phases, and influence of bulk rock composition. Zane *et al.* (1996) correlated thousands of analyses of chlorite with mode of occurrence and concluded that composition trends are strongly correlated with bulk rock composition, and cannot be used as a geothermometer. Such cautions are especially significant in that reaction progress may be significantly different, e.g. in pelites and metabasites.

It is clear that in any given prograde sequence there is a regular progression in the composition of chloritic (not necessarily chlorite, *sensu strictu*) material as *measured*, say, by electron microprobe analysis, even though such compositions are commonly contaminated with phases other than chlorite. Such variation cannot be a rigorous 'geothermometer', as reviewed by Essene and Peacor (1995). To a first approximation, however, and where bulk rock composition and geological parameters such as thermal gradient, tectonic stress, etc., are equivalent, reaction progress will occur to similar degrees as a function of temperature. That is, a given chlorite thermometer appears to 'work'. For a given locality, there may even be excellent correlation with temperature. Such relations should be applied to

other localities or rock types only with extreme caution insofar as reaction progress rather than absolute temperature is measured. It can also be argued that, for chlorite and other low-grade 'thermometers', they all 'work' well in many circumstances. As noted above, however, that is in part because the range of temperature characteristic of very low-grade metamorphism is relatively small, and the overall changes in texture and mineral assemblage large. Even estimates of temperature based on the simplest observations on hand specimens or thin sections are seldom very far off the mark.

Lastly, we note that Essene and Peacor (1995) concluded that chlorite thermometers based on theoretical thermodynamic analysis were also flawed, in part because activity–composition relations are not accurately known. As a result of these collective relations, all previously proposed chlorite geothermometers must be considered to be inaccurate methods of measuring temperature, as opposed to being useful indicators of reaction progress, and they should be applied with extreme caution. Much more work is necessary to establish accurate thermodynamic parameters and determine accurate compositions in homogeneous phases shown to have formed under conditions of stable equilibrium before chlorite geothermometers can be used with confidence on a strictly quantitative basis.

2.5.4 Polytypism of chlorite and white mica

Subsequent to the pioneering work of Yoder and Eugster (1955), a great quantity of research has demonstrated an apparent prograde sequence of polytypism for white micas from $1M_d$ to 1M to $2M_1$ (e.g. Velde & Hower, 1963; Maxwell & Hower, 1967; Hoffman & Hower, 1979). There has been some implication of a direct relationship to temperature, e.g. Walker (1993) noted that temperatures corresponding to transitions from one polytype to another in Salton Sea sediments correlated well with temperatures experimentally determined by Velde (1965). However, Grubb *et al.* (1991) noted that no 1M polytype had ever been detected in SAED patterns of diagenetic white mica; likewise, pure $1M_d$ illite had never been observed. Dong and Peacor (1996) confirmed the latter observations and noted that I/S, illite and smectite in shales from the

Texas Gulf Coast consisted of layer sequences which were partially coherent and partially incoherent (turbostratic). Furthermore, where layer relations were coherent, stacking sequences did not correspond to ordered 1M or $2M_1$ sequences; rather sequences were largely random, with local ordering, but primarily of $2M_1$ polytype. They noted that XRD patterns of such heterogeneous material should not be interpreted as being due to diffraction from independent 1M or $2M_1$ units, but as a complex function of partially and locally ordered units, in contrast with the conclusions of many XRD studies.

Bailey and Brown (1962) suggested that chlorite polytypism might vary as a function of temperature, type I polytypes being restricted to lower temperature regimes. Walker (1993) and de Caritat et al. (1993) have critically reviewed the occurrences of chlorite polytypes. They concluded that a well-defined correlation of polytypism with temperature does not exist. Schmidt and Livi (1998), on the other hand, presented TEM data which showed that polytypes Ibb and Iibb occur in diagenetic and anchizonal samples, whereas only Iibb occurs in epizonal samples, although in this case the chlorite occurs in metagreywackes rather than metapelites. They noted, however, that there is a general increase in stacking order with increasing grade, as consistent with a general trend toward equilibrium.

It is well known that polytypes differ by vanishingly small free-energy increments, and that only trace amounts of elements in solid solution may affect formation conditions of a given polytype relative to others. The ubiquitous occurrence of muscovite as the $2M_1$ polytype suggests that it is the stable form. The occurrence of $1M_d$ white mica as the disordered form which precipitates at the lowest temperatures of formation is likewise understandable from the point of view of the general high entropy of phases formed at very low temperatures, i.e. stacking faults in coherent layer sequences and the occurrence of incoherent layer boundaries are defects whose density normally increases with decreasing temperature of formation. Such high-energy, metastable phases should change toward the stable $2M_1$ form. Dong and Peacor (1996) pointed out that there is no crystal-chemical rationale for the occurrence of 1M white mica, as consistent with its rarity in very low-grade metapelitic sequences. Nevertheless its occurrence in hydrothermally derived clays (e.g. Lonker & FitzGerald, 1990) suggests that it may form under as yet unspecified conditions.

The wealth of data on variations in white mica and chlorite polytypism as a function of temperature are generally consistent with predictable trends. The data, however, imply that there is no predictable, accurate correlation with temperature and that polytypic sequences should not be used other than as indicators of reaction progress (Fig. 2.1).

2.5.5 Phengite geobarometer

Experimental data of Velde (1965) indicate that the magnitude of the phengite component of mica is primarily a function of pressure. He obtained experimental data relevant to the following two reactions, the balanced formulae being those of Essene (1989), with the two balanced formulae for each reaction corresponding to different end-members:

Phengite = K-feldspar + chlorite + quartz + water

$$3K_2Mg_2Al_2Si_8O_{20}(OH)_4 = 6KAlSi_3O_8 + Mg_6Si_4O_{10}(OH)_8 + 2SiO_2 + 2H_2O,$$

or

$$5K_2MgAl_3Si_7AlO_{20}(OH)_4 = Mg_5Al_2Si_3O_{10}(OH)_8 + 2K_2Al_6Si_6O_{20}(OH)_4 + 6KAlSi_3O_8 + 2SiO_2 + 2H_2O.$$

phengite = K-feldspar + phlogopite + quartz + water

$$3K_2Mg_2Al_2Si_8O_{20}(OH)_4 = 4KAlSi_3O_8 + K_2Mg_6Si_6Al_2O_{20}(OH)_4 + 6SiO_2 + 4H_2O,$$

or

$$6K_2MgAl_3Si_7AlO_{20}(OH)_4 = K_2Mg_6Si_6Al_2O_{20}(OH)_4 + 3K_2Al_4Si_6Al_2O_{20}(OH)_4 = 4KAlSi_3O_8 + 6SiO_2 + 4H_2O.$$

Experimental data for the latter reaction were provided by Massonne and Schreyer (1987), and subsequently reinterpreted by Bucher-Nurminen (1987).

The phengite component may be determined either directly by chemical analysis, or by inference from the cell dimension b, as noted in section 2.4.1.2. Use of the above or other reactions to determine pressure should, at first sight, depend critically on three factors (Essene, 1989):

1 Experimental data for Fe-bearing phases and for reversed reactions of stable phases, criteria which are not currently met.

2 Accurate models of activities of solid solution components which, for complex phases such as chlorite and phengite, can now only be approximated.

3 Most importantly, verification that in rocks to which the barometer is applied, the buffering phases are present under equilibrium conditions, e.g. that K-feldspar is present. Essene (1989) pointed out that this is generally not the case. In the absence of K-feldspar, phengite barometry provides only minimum limits of pressure.

Most of the applications of the phengite geobarometer (see section 2.4.1.2) have been carried out without regard to one or more of the above factors, especially the condition of a buffering assemblage. Such applications assume that the phengite component of muscovite is primarily a function only of pressure, and not of bulk rock composition. Velde (1967) noted that the Si content of muscovite appeared to be independent of bulk rock composition in a limited group of samples, but the data of Brown (1968) implied the opposite. A review of relevant data is also provided by Guidotti (1984).

Authigenic clay minerals formed during early diagenesis may have solid solutions which are much more extensive than for higher temperatures, and related to kinetic factors rather than equilibrium. As reactions proceed at higher grades, reactant clay minerals such as illite may inherit such components, giving rise to phengite components which are unrelated to pressure. This suggests that the phengite component may in part be a measure of reaction progress rather than pressure, i.e. it may be dependent on reaction kinetics in very low-grade, subgreenschist facies rocks. Where heat flow is low, as in accretionary settings, phengite contents are typically high, perhaps because there is insufficient thermal energy to cause equilibration of the phengite component of neoformed illite. By contrast, the phengite component of illite–muscovite is typically small in extensional basins, because early basinal heat flow tends to be relatively high. These white micas commonly develop in bedding-parallel microfabrics during the late diagenetic stages of reaction progress, and have evolved, phengite-poor compositions (e.g. Li *et al.*, 1994b). Subsequent strain-enhanced crystal growth that may result from deformation and basin inversion does not result in significant changes in the composition of the white mica. In using

the phengite barometer, it is therefore essential to determine that reactions have progressed to a point where chemical equilibrium has been achieved.

The large number of studies described in section 2.4.1.2 imply that the phengite content of muscovite seems to function well as a qualitative indicator of broadly defined pressure differences, despite the concerns expressed by Essene (1989) and others. It also seems clear, however, that a truly quantitative geobarometer must be firmly based on a resolution of those concerns.

2.6 Overview of conditions of very low-grade metamorphism

The observations reviewed above indicate that clay minerals in pelitic and metapelitic rocks exist in a series of metastable intermediate structural and chemical states which form in response to the low temperatures and sluggish reaction rates that characterize very low-grade metamorphism. Despite the low temperatures ($<300°C$), thermal energy is widely recognized as the dominant influence on the progress of reactions in clay minerals (e.g. Frey, 1987a; Huang *et al.*, 1993; Pollastro, 1993). However, there is also ample evidence that strain energy associated with overburden and tectonic stresses also makes an important contribution to reaction progress in clay mineral assemblages. The interaction between these two energy sources is considered later, and their influence on regional metapelitic patterns developed in different geotectonic settings is discussed in Chapter 3.

There is little doubt that reactions in clay mineral systems are sensitive indicators of changes in thermal conditions in basinal sequences and very low-grade metamorphic terranes (Frey, 1987a; Peacor, 1992a). Several studies (see Eslinger & Glasmann, 1993, and references therein) have shown that certain well characterized clay mineral transformations can be correlated with organic indicators of thermal maturity, such as vitrinite reflectance and hydrocarbon generation zones. For example, Pollastro (1993) provided abundant evidence that progress in the smectite-to-illite reaction can be correlated with increased temperatures related to sediment burial and that reactions occur over the same temperature interval as those controlling hydrocarbon generation. Other studies have gone further and used clay mineral reactions as geothermometers to estimate maximum temperatures and geothermal gradients in basinal sequences (e.g. de Caritat *et al.*, 1993; Price & McDowell, 1993). However, the uncritical use of such reactions for geothermometry has been challenged by Essene and Peacor (1995, 1997) who pointed out that clay minerals do not represent equilibrium assemblages. Unlike the situation at higher grades, there are no reversible clay mineral reactions in very low-grade metapelitic rocks nor parameters akin to vitrinite reflectance that can be directly and consistently converted to accurate (palaeo)temperatures.

At present the best approach to thermal indexing of clay mineral reaction progress is through correlation with other basin maturity indicators, the most reliable of which appears to be vitrinite reflectance (Sweeney & Burnham, 1990). Using this approach (see section 3.2.3), the transition from the early diagenetic zone to the late diagenetic zone is indexed at *c.* 100°C, and the late diagenetic zone–anchizone transition occurs at *c.* 200°C (see Fig. 3.4.). These temperature estimates are based largely on correlations derived from basins with normal heat flow characteristics, i.e. geothermal gradients in the range 25–30°C km⁻¹. However, the kinetic response of clay mineral reactions in basinal sequences can be significantly different from that of vitrinite. Several recent studies in basins with abnormally high or low geothermal gradients have suggested that clay mineral reactions may be more sensitive to geological heating rates than organic materials. For example, Velde and Lanson (1993) found different rates of reaction progress in terms of percentage of illite in I/S and vitrinite reflectance R_r when they compared the Paris Basin with the Salton Sea geothermal system. Both the clay mineral and the vitrinite reactions show maximum response to a geothermal event spanning 10^4–10^5 years in the Salton Sea system. In contrast, vitrinite in the Paris Basin has matured in response to a suspected intrusive event, whereas clay minerals have not because the duration of heating was insufficient. Elsewhere, Hillier *et al.* (1995) found that low-heat-flow basins tend to show the smectite-to-illite reaction to be in advance of vitrinite reflectance whereas the reverse was observed in some high-heat-flow systems. Hence correlation between organic and clay mineral reaction progress is dependent

on thermal history, particularly on basinal heat flow characteristics and geological rates of heating.

Temperature changes in sedimentary sequences are brought about by burial beneath overburden rocks resulting from sedimentation processes or tectonism, or both. As depth of burial increases, overburden stress generated by vertical loading is the main cause of compaction, leading to pore reduction and fluid migration. In late diagenetic zone pelites these processes are interactive both with the progress of clay mineral reactions, which causes loss of interlayer water, and with the growth of bedding-parallel microfabric which reduces porosity. Differential compactional stresses, generated for example between competent sandy/silty laminae and less competent clay laminae in shales, or even by compactional bending of clay minerals around rigid sand/silt grains, provides strain energy for clay mineral reactions. Strain-induced defect migration (dislocation creep), possibly as screw dislocations (Baronnet, 1992), most likely contributes to the polytypic transformation from the $1M_d$ illite polytype of smectite-rich I/S to dominant $2M_1$ polytypism in illite-rich I/S, as documented by Dong and Peacor (1996). The increased dominance of $2M_1$ illite directly influences crystal thickening (e.g. Dong *et al.*, 1997b) and hence illite crystallinity, and in turn both contribute to bedding-parallel microfabric development as the smectite-to-illite reaction progresses to completion. Such changes account for the sharp increase in preferred orientation associated with transformation of smectite to illite observed by Ho *et al.* (in review).

With increasing depth of burial, pore fluid pressure in typical pelites is likely to increase in response to several factors. These include compactional dewatering, organic maturation and hydrocarbon generation, dehydration caused by clay mineral reactions, pore reduction caused by diagenetic clay mineral growth and aquathermal pressuring caused by temperature increase. In most cases the fluids produced by these processes slowly escape, and as they migrate they mobilize and facilitate the cation exchanges between reactants and products of clay mineral reactions. Overpressuring can develop in shales and mudstones where the pore fluid pressure at a given depth exceeds the hydrostatic pressure gradient (Osborne & Swarbrick, 1997). It can be generated during diagenesis by any combination of the above factors, but is likely to be effective only when the system is well sealed, the sealing having been ascribed to the change in texture occurring with the transformation of smectite to illite (Freed & Peacor, 1992b). Whereas there is growing evidence that overpressuring retards organic reaction progress, as measured by vitrinite reflectance (Dalla Torre *et al.*, 1997; Carr, 1998), the effects on the smectite-to-illite reaction are still uncertain. A close relationship between overpressuring and the onset of the smectite-to-illite reaction in the Gulf Coast shales has been described by Bruce (1984), but it is not clear to what extent the reaction has been inhibited. However, Colten-Bradley (1987) suggested that dehydration of smectite may cause overpressuring, and this in turn inhibits further dehydration. It is likely that once overpressured systems return to hydrostatic pressures, as a result of basin inversion for example, any retardation of reaction progress would be restored, particularly if tectonic stress is also involved.

In basinal settings where burial diagenetic and very low-grade metamorphic reactions are free of tectonic stress and affected only by increasing overburden stress and temperature, it appears that little change in texture and mineralogy occurs subsequent to the completion of the smectite-to-illite reaction, as illustrated for example by the Gulf Coast shales. Although there is some evidence for minor increase in thickness of illite (e.g. Eberl, 1993), clay minerals forming a bedding-parallel microfabric are little changed to the greatest depths and hence highest temperatures from which samples have been obtained in the Gulf Coast sequence. Samples characteristic of higher grade metapelites have largely been obtained elsewhere from sequences which have been subjected to tectonic stress, and thus to a different system of variables. Present knowledge suggests therefore that a window may exist between burial-related processes and processes driven by tectonism within which reaction progress shows little change.

When pelitic rocks are deformed in response to tectonic stress, slaty cleavage develops as a result of interactive mechanical deformation and thermally driven crystallization processes (e.g. Knipe, 1981; Van der Pluijm *et al.*, 1998). Several studies have found that the overall effect of tectonic stress on clay mineral reaction progress is to accelerate crystal growth (e.g. Merriman *et al.*, 1990, 1995b; Árkai *et al.*, 1996; Jiang *et al.*, 1997). Indeed Árkai *et al.* (1996) found that lattice strain and crystal growth are interrelated, and as illite–muscovite and chlorite crystals grew thicker in slaty cleavage microfabrics the amount of residual lattice strain in both minerals was reduced. The increased rate of crystal thickening of clay minerals in cleavage microfabric requires large activation energies for rapid progress in terms of Ostwaldian steps (Essene & Peacor, 1995). A possible source is elastic and plastic strain energy which is initially stored in deformed phyllosilicates and released during dislocation creep, causing migration of defects into grain boundaries and growth of thicker, defect-poor crystals (e.g. Bons, 1988; Merriman *et al.*, 1995b).

All the processes described above are collectively consistent with a drive towards a condition of stable equilibrium. That is, fine-grained pelitic systems are comprised of a complex assemblage of heterogeneous organic and inorganic phases, each of which is individually metastable (e.g. smectite or illite), and which collectively are entirely incompatible with requirements of the Gibbs phase rule. All changes, whether towards increasing crystal size as measured by illite crystallinity, decrease in defect densities produced both during initial crystallization or by stress, or increasing compositional homogeneity in clay minerals, proceed towards a simpler system in which a small number of homogeneous phases coexist in stable chemical and textural equilibrium, those conditions being achieved in the epizone.

2.7 Future research

In the past decade the focused use of TEM studies on metapelitic sequences previously characterized by reconnaissance XRD surveys of grade has been a key factor in advancing knowledge of metapelitic rocks. This approach, including wherever possible the integration of structural and stratigraphical information with patterns of very low-grade metamorphism, is a prerequisite for future work, enabling samples for TEM study to be carefully selected so as to be representative of a specific aspect or sequence of metamorphic conditions. Whereas TEM permits the observer to make direct observations of samples which have not lost their physical integrity, the observations require equipment not available to most workers. In addition, data acquisition and interpretation are time consuming and difficult, especially for factors such as crystallite size distributions where large numbers of observations must be made in order to obtain accurate representations of bulk samples. On the other hand, XRD is a technique which is readily available, easily and rapidly giving results which are representative of much larger volumes. Almost all XRD data are obtained from samples which are 'separated' in some way, and the effect of such separation on the states of clay minerals in original rocks is still not well established. The degree to which sample separation affects crystal size distributions, mean sizes, defect states and the coherency of interlayers are examples of factors which are still under debate. Only when such factors have been determined can some parameters measured by XRD be interpreted as representative of the states of clay minerals in original rocks.

In some ways, the relation between powder XRD data and TEM data of clay minerals is also analogous to that between powder XRD and single-crystal XRD data of minerals such as feldspars, many years ago. Once single crystal research established a direct relation between structure parameters and some characteristic of XRD patterns, the XRD patterns could be routinely and confidently used to measure parameters such as composition and structural state. Direct observation of compositional and structural parameters of clay minerals by TEM plays a similar role relative to XRD patterns, and it is essential that much more data be obtained which compares the origins of parameters such as illite crystallinity and defect states as determined by both XRD and TEM.

On a regional scale, studies which integrate structural and stratigraphic data along with both XRD and TEM data are certain to produce a clearer picture of reaction progress in all of the clay mineral series considered above, but particularly for the kaolinite–pyrophyllite series for which details are currently sparse. The rate of a specific reaction may be very different in different rock types, one determining variable being texture as relates to porosity and to closed versus open system behaviour, and in turn to fluid/rock ratio. Much more work is needed on prograde sequences which involve a variety of coexisting rock types. On a more localized scale, the interrelationship between prograde and retrograde (back-reaction) transitions in clay mineral

assemblages within thermal aureoles is a promising area for future research. In such settings the effects of tectonic stress are minimal, enabling studies to focus on reactions affected mainly by thermal energy and relatively high fluid/rock ratios.

It is now clear that geothermometers and geobarometers which utilize characteristics of clay minerals in very low-grade metamorphic environments generally give inaccurate absolute temperatures and pressures, but are useful indicators of reaction progress. Nevertheless, some approximation to absolute temperatures can often be obtained where care is taken and where temperatures are known from other sources for comparison, i.e. some geothermometers are said to 'work', at least in specific cases. Given that it is generally accepted that clay minerals are metastable in very low-grade systems, it is possible to focus on those factors which determine rates of reaction, to isolate determining parameters in example systems for which temperatures are well known, and determine the degree to which thermometers actually do work. For example, the Cathelineau and Nieva 'thermometer' should clearly not be applied to just any other system subject to variables other than those in the type section studied by them, but it may be possible to define those systems in which it does work, and to restrict application to such systems. Much more work of this nature is necessary before appropriate interpretations can be made of absolute temperatures determined with any so-called thermometer.

Although it is clear that reaction progress is affected by several factors, the contribution made by prograde increases in temperature is probably paramount, and is also of greatest interest in terms of correlation with mineral facies in very low-grade metabasic rocks and organic maturity indicators. In the light of the studies reviewed above, it seems probable that in very low-grade metapelites the rate of temperature change with burial depth (geothermal gradient) or time, strongly influences reaction progress in clay minerals. This may be another consequence of the Ostwald Step Rule whereby reactions can follow different but parallel reaction pathways, with the character of the products being controlled by the relative reaction rates. Future evaluation of the effects of heating rates on the main clay mineral reaction series and their different reaction pathways in natural systems is crucial to a better understanding of reaction kinetics in metapelitic rocks. Clay minerals promise to be at least as sensitive to thermal conditions in sedimentary basins as established organic maturity indicators, and may prove to be better indicators of a range of basinal heat flow conditions.

3 Patterns of very low-grade metamorphism in metapelitic rocks

R. J. Merriman and M. Frey

3.1 Introduction

In the decade or more following the reviews of Kisch (1983) and Frey (1987a), studies of very low-grade metamorphism have flourished. These reviews and subsequently a series of international conferences organized by IGCP 294 (Very Low-Grade Metamorphism), promoted activity in a wide variety of metapelitic terranes and encouraged the use of more advanced analytical techniques to complement and qualify established methods. As discussed in Chapter 2, these studies have considerably advanced our understanding of the continuum of prograde processes that occur in clay minerals during the transformation from diagenesis to metamorphism. In addition, the automation of X-ray diffraction (XRD) analysis has vastly increased the data generated from metapelitic terranes. The result has been a greater interest in the interpretation of the patterns produced by the spatial distribution of metapelitic data, especially when combined with structural and stratigraphical information.

Early studies of patterns of very low-grade metamorphism were largely concerned with textural zonation in relation to relative depth of burial. This was the essence of the depth controlled stages of regional epigenesis and metagenesis recognized by a number of Russian geologists, particularly Kossovskaya et al. (1957) and Kossovskaya and Shutov (1961, 1970). Much of this early work was not widely accessible because it was published in Russian journals, and became better known largely through the reviews of Kisch (1983) and Frey (1987a). Hence the seminal work of Coombs (1961) on burial metamorphism probably influenced, and no doubt inspired, many early studies of very low-grade metamorphic patterns. Coombs (1961) recognized that diagenetic and very low-grade metamorphic processes are a continuum resulting from the burial

of sedimentary and volcanigenic strata. Recognition of the progressive nature of diagenesis and very low-grade metamorphism by Coombs and others, including Burst (1959), Weaver (1959, 1960), Perry and Hower (1970) and Hower et al. (1976), helped to dismantle the conceptual barriers between the sedimentary petrologist/clay mineralogist and metamorphic petrologist.

The concept of the anchizone (Kübler 1967a,b, 1968), a transitional zone of incipient metamorphism between diagenesis and low-grade (greenschist facies) metamorphism, is fundamental to pattern recognition in metapelitic terranes. Perhaps an equally important consequence of Kübler's work was to bring the XRD techniques of the clay mineralogist to systematic regional studies of very low-grade metamorphism. Kübler's technique of measuring changes in the XRD peak profile of illite–muscovite, the most common series of clay minerals in many very low-grade metapelites, provided a specific index of the anchizone in terms of illite crystallinity. By so defining the anchizone in terms of the Kübler index (KI in $\Delta°2\theta$) of illite crystallinity, both the prograde limit of diagenesis and the initiation of the epizone could also be defined (Kisch, 1990, 1991a). Consequently, three zones are most commonly used to characterize patterns in very low-grade metapelitic sequences: *late diagenetic zone* (KI > 0.42); *anchizone* (KI < 0.42 > 0.25); *epizone* (KI < 0.25).

The techniques and concepts discussed in Chapter 2 have enabled these zones to be better characterized and understood. It is now evident that the Kübler index and the sequence of metapelitic zones it delineates is related to reaction progress in the smectite–I/S–illite–muscovite series (see section 2.4.1). Although the Kübler index specifically measures increases in the thickness of illite–muscovite crystallites, these crystallographic changes are linked with progressive changes in chemical and mineralogical heterogeneity, polytypism and lattice strain. Parallel changes in the smectite–corrensite–chlorite series occur through the metapelitic zones in the same prograde fashion. Changes in microfabric and lithology accompany reaction progress through the metapelitic zones, as mudstone and shale evolve to form slate. Thus the metapelitic zones provide a general indication of the state of reaction progress in assemblages of clay minerals, as well as a guide to the type of pelitic lithology that can be expected (see Fig. 2.1).

In Chapter 3 we discuss how relationships between depth of burial, geothermal gradient, stratigraphy and structure govern the patterns of metapelitic zones in very low-grade terranes. We review studies of very low-grade terranes and have selected these to illustrate some typical metapelitic patterns which have been well characterized by a variety of techniques. Our review suggests that several specific patterns can be recognized and related to geotectonic setting (see Table 3.1). With the benefit of a better understanding of clay mineral transformations we have also attempted to correlate reaction progress in clay minerals with those in organic matter, as a means of integrating studies of basin maturity (see Fig. 3.4).

3.2 Sampling and data interpretation

3.2.1 Field sampling

This section provides guidance on the systematic sampling and field recording techniques employed in the study of metapelitic terranes. The laboratory treatment of the samples and the analytical techniques used are described in Chapter 2.

Pelitic lithologies with typically more than 50% clay and silt, including claystone, mudstone, shale, slate, phyllite and schist, are the most suitable for studies of metapelitic grade. Calcareous pelites, i.e. marls and metamarls, have also been widely used in some terranes. Experience suggests that muddy fine sandstones are also acceptable so long as the clay mineral component exceeds approximately 40%. The way pelitic rocks break or split, either naturally in the weathered outcrop or when hammered, can be a useful field guide to suitable rock types and their grade. Diagenetic mudstones tend to break readily into centimetre size blocks when hammered,

and in outcrop have a tendency to spall into similar-sized blocks along polygonal shrinkage joints. Shrinkage polygons may reflect the presence of an expanding mineral (smectite or smectite in a mixed-layer mineral). Shales (i.e. fissile mudstones) in the zone of diagenesis have similar characteristics, but readily split into centimetre size flakes, and degraded outcrops typically produce a litter or scree of these small flakes. Where a spaced or fracture cleavage is developed in the late diagenetic zone or lower anchizone, claystones and mudstones tend to split into pencils or flat tiles with a shape and size which is controlled by the bedding-cleavage relationship (Durney & Kisch, 1994). In temperate or alpine climates, mature outcrop and road cuttings in this type of pelite are characterized by a scree or litter of elongated fragments. As slaty cleavage becomes better developed, metapelites tend to be more resistant to outcrop degradation and extensive scree is less common. Pelites of this type will often require splitting with a hammer and chisel. Good quality roofing slates will resist splitting even with a chisel, and are even less easily broken across the splitting direction. These characteristics are typical of high anchizonal and epizonal metapelites; in natural outcrops they are associated with restricted screes and mature cuttings which seldom degrade to a litter of fragments. Similar weathering characteristics generally hold for phyllites and schists in temperate climates, but shape and splitting properties vary with the thickness and regularity of cleavage/schistosity planes.

Fine grained tuffs and metatuffs deserve special mention. K-bentonites are altered very fine-grained tuffs largely composed of mixed layer illite/smectite (I/S) (e.g. Środoń et al., 1986; Huff et al., 1988), but despite the predominance of authigenic illitic minerals they are not ideal for metapelitic grade studies. Because of their high I/S content many K-bentonites in late diagenetic and lower anchizonal sequences tend to have lower crystallinities (higher KIs) than those found in associated mudstones. For example, numerous K-bentonites in the Moffat Shale Group at Dob's Linn, Scottish Southern Uplands terrane (Merriman & Roberts, 1990), consistently give KIs in the range 0.8–1.2 $\Delta°2\theta$, whereas mudstones in the same sequence have KIs in the range 0.4–0.6 $\Delta°2\theta$. Sluggish reaction progress in bentonites may be due to the inhibiting effect of early diagenetic replacement of volcanic glass by K-rich zeolites and

K-deficient smectite (e.g. Masuda *et al.*, 1996), the low organic content of bentonites (Li *et al.*, 1997), and more rapid fixation of K in I/S minerals in shale at low temperatures (Šucha *et al.*, 1993). Differences in the percentage expandability of I/S in K-bentonites and shales converge at temperatures above 150°C in deeply buried sequences (Šucha *et al.*, 1993). Where the regional grade exceeds the middle anchizone (KI < 0.30 Δ°2θ) and slaty cleavage is well developed, K-bentonites tend to prograde to fully illitized *metabentonites* (<5% smectite interlayers) with KIs similar to associated metapelites (Roberts & Merriman, 1990; Merriman *et al.*, 1995a). Since they lack detrital white micas, well-cleaved metabentonites are ideal for dating slaty cleavage development (Dong *et al.*, 1997a; see also Chapter 7).

Limestones and sandstones, including greywackes are not suitable for metapelitic grade studies, largely because they do not contain sufficient clay. Moreover, these lithologies can either inhibit clay mineral reactions, through early cementation (e.g. limestones and some sandstones), or promote reactions through porosity and pore fluid movement, which are not comparable with reaction conditions in pelitic rocks. Some carbonate-cemented mudrocks, particularly carbonate nodules, have been used to compare pre- and postcementation clay mineral assemblages (C. V. Jeans, personal communication). Highly pyritic black shales and mudstones can cause problems with flocculation during separation of clay fractions.

Fresh rock should be sampled whenever possible. Unweathered mudrocks tend to show a range of grey colours from dark-grey or black, to green- or blue-grey, to pale greys. Weathering in temperate climates modifies these colours to shades of brown, reddish-brown, yellow or khaki, usually characterized by the presence of Fe-oxyhydroxides such as goethite or lepidocrocite in clay fractions. Alteration of chlorite to vermiculite and illite to I/S or kaolinite may result from severe weathering in temperate climates. Tropical weathering may result in the complete replacement of the metapelitic clay mineral assemblage by smectite or kaolin-group minerals (e.g. Vicente *et al.*, 1997), and in extreme cases by Al-oxides or Al-hydroxides in bauxitic clays. Note that shales and slates associated with typical redbeds, i.e. hematized clays, commonly show no alteration to clay mineral assemblages despite an intense red or purple colour. The degree and nature of weathering

in pelitic rock terranes may determine the sampling strategy adopted. For instance, in deeply weathered tropical terranes it may be necessary to carry out a pilot XRD study to determine the nature of alteration at a variety of sites, including natural outcrop and river sections, road cuttings, quarries and boreholes (if available) to establish which type of site has best preserved metapelitic mineral assemblages.

Large samples are not necessary for the separation techniques described in section 2.4.1.1. Approximately 200 g will provide enough rock for several < 2-μm mounts, and ideally should include some small blocks around 3 × 3 × 6 cm for a thin section if additional optical, scanning electron microscopy (SEM) or transmission electron microscope (TEM) work should be required. Geological and lithological information relevant to the metamorphic history of the sampled site should be recorded in the field. This includes proximity of igneous rocks, particularly intrusives which may have overprinted regional metamorphic patterns; major faults and particularly shear zones which may have retrogressed or enhanced the regional pattern; and evidence of hydrothermal alteration or mineralization. A record of the lithology and the nature of tectonic fabrics is a crucial part of a metapelitic study (see Durney & Kisch, 1994). As noted above, lithology and the development of slaty cleavage fabrics can be used as indicators of metapelitic grade. In the laboratory this information provides a check against the fidelity of KI measurements and general mineralogy. For example, it is unusual for a slate with a well developed cleavage fabric to possess a large KI value (>0.50 Δ°2θ) and contain minor smectite, and might indicate that sample numbers have been interchanged. All field records can eventually be brought together in a spreadsheet format to which laboratory data can be added, and this maintained as the database for the metapelite study.

Determining what is a representative sample can present problems in the field. Where there is much local variation in lithology, some studies recommend taking up to five pelite samples from each site to establish intersample variation (e.g. Hunziker *et al.*, 1986). Other studies have multisampled at sites that are representative of the major metapelitic zones recognized in a region (e.g. Roberts *et al.*, 1991). In this case five samples were taken in an outcrop area of approximately 4 m². The aim of multisampling is to determine the errors and precision involved, particularly if data are used to generate contoured maps (Robinson *et al.*, 1990). Sample errors and the precision of the illite crystallinity technique are discussed in section 2.4.1.1.

The total number of samples collected from a study area will depend on three main factors:

1 the size and nature of the terrane, particularly the degree of exposure, weathering, accessibility and lithological variability;

2 knowledge of the geology, especially the regional and local structure, and stratigraphic succession;

3 the resources available for the study to be carried out.

Restrictions imposed by any of these factors will govern the number of samples collected. In turn, this will determine whether the study represents a metamorphic traverse or a reconnaissance, or a more comprehensive survey capable of generating a metamorphic map, as discussed below.

3.2.2 Methods of displaying regional metapelitic data

The earliest displays of patterns in metapelites tended to be part of more comprehensive correlation schemes, such as that proposed by Kisch (1983) which incorporated the stages of regional epigenesis and metagenesis of Kossovskaya *et al.* (1957) and Kossovskaya and Shutov (1961, 1970). With the systematic application of XRD techniques to regional studies of metapelitic rocks, borehole and stratigraphic profile plots of clay mineral variation in relation to depth became a popular way of displaying data. In one of the early examples of such studies, the distribution of clay minerals in Triassic and Liassic clays, shales and slates was used by Frey (1970) to demonstrate the transitional nature of the diagenetic zone–anchizone boundary in the Swiss Alps. The regional pattern was characterized using a combination of the stratigraphical distribution of clay minerals, feldspar, carbonates and quartz in sampled profiles, with the distribution of key minerals through the metapelitic zones defined by Kübler's illite crystallinity index. Similar techniques were subsequently used and developed in other studies (e.g. Kisch, 1983; Weaver, 1984; Merriman & Roberts, 1985; Roberts *et al.*, 1991). Diagrams generated from this type of study commonly show minerals in < 2-μm

separations as ranges through metapelitic zones (Fig. 3.1). This method of displaying data is useful for illustrating prograde sequences of co-existing phases, and the metastable overlap of illite polytypes which is characteristic of reaction progress in many clay mineral assemblages.

In a study of the coal rank associated with very low-grade metamorphism, Kisch (1974) modified the zonal scheme of Kossovskaya and Shutov (1961, 1970) to show clay mineral distribution in relation to relative depth of burial, from early diagenesis through 'burial metamorphism' to 'regional' metamorphism. Further modifications to this scheme were made by Kisch (1983), adding the metapelitic illite crystallinity zones of Kübler (1967a,b, 1968) and the depth-related mineral facies recognized by Coombs (1960, 1961). Using a series of studies of metamorphism in the Alps, Frey (1986) correlated metapelitic zones (based on KI) with mineral assemblages in metabasic rocks, coal rank and fluid inclusion data. Although more time consuming, integrated studies of this type undoubtedly offer the best method of understanding patterns and *P–T* conditions of metamorphism in very low-grade terranes. Diagrams generated by this type of study and reviews (e.g. Kisch, 1987, Fig. 7.3) represent the first comprehensive correlation of mineral and textural zones with progressive depth of burial, ranging from early diagenesis to early greenschist facies regional metamorphism. Another version of this type of chart (Fig. 3.4) is discussed in section 3.2.3.

Annotation of cross-sections with crystallinity data has been used to illustrate the relationship between tectonics and very low-grade metamorphism. Kübler (1967a) used this method to illustrate the increase in metapelitic grade associated with the development of slaty cleavage in the western Pyrenees. Cross-sections combining illite crystallinity, coal rank and fluid inclusion data were used by Frey *et al.* (1980) and Breitschmid (1982) to demonstrate inverted metamorphic patterns associated with nappe emplacement in the Lake Lucerne–Reuss Valley region of the Alps.

Relationships between stratigraphy, depth and metapelitic grade have been explored in regions with well established stratigraphical successions. Where there are sufficient measurements to produce statistically viable results for each formation, the KI range with mean values can be displayed as a series

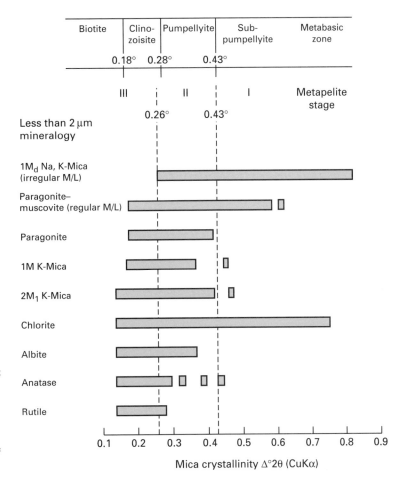

Fig. 3.1 Variation in the less than 2-μm mineralogy of metapelitic stages in the northern part of the Welsh Basin. Mica crystallinity uses the Kübler index (KI). Approximately equivalent metapelitic zones are: Stage I = late diagenetic zone; Stage II = anchizone; Stage III = epizone. Equivalent metabasite zones from the same region are also shown, with zonal divisions correlated with isocrysts on a metamorphic map. (From Merriman & Roberts, 1985, Fig. 7, with permission of the Mineralogical Society.)

of horizontal bars stacked in stratigraphical order (Fig. 3.2). This method has been used to reveal patterns of burial metamorphism (e.g. Davies *et al.*, 1997), a metamorphic hiatus (e.g. Merriman & Roberts, 1985) or high strain zones (Roberts *et al.*, 1991). A similar approach has been used with tectonostratigraphic successions, using ranges and mean KI values to characterize the patterns of very low-grade metamorphism in an accretionary terrane (Merriman & Roberts, 1995; Merriman *et al.*, 1995b) (see Fig. 3.10).

Maps showing the distribution of index minerals have not been as widely used to interpret patterns of very low-grade metamorphism as they have in some classic studies of higher grade terranes (e.g. Barrow, 1893, 1912; Kennedy, 1948). Generally they have been used to represent mineral zone boundaries based on appearance or disappearance of minerals,

rather than isograds based on index minerals resulting from specific reactions (e.g. Frey, 1974; Weaver, 1984). However, the kaolinite + quartz = pyrophyllite + H_2O isograd has been mapped in parts of the Alps (Frey, 1987b; Wang *et al.*, 1996) and in Taiwan (Yang *et al.*, 1994). Maps showing XRD measurements derived from specific phyllosilicates have been extensively used to characterize patterns of very low-grade metamorphism, particularly in the past decade. An early example of combining regional distribution of illite crystallinity measurements with structural geology is that of Kisch (1980). He was able to demonstrate that overthrust nappes of high-grade rocks in the Central Caledonides of western Sweden had resulted in anchizonal metamorphism in the underlying autochthon. Árkai (1983) plotted KI values, metabasic index minerals, vitrinite reflectance,

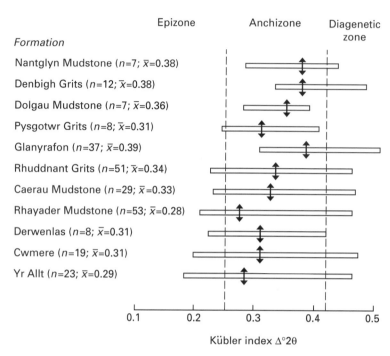

Fig. 3.2 Variations in the mean (arrows) and range (bars) of Kübler indices through the stratigraphical succession in the central part of the Welsh Basin. n = number of samples; $\bar{x}$ = mean KI values. (From Merriman *et al.*, 1992, Fig. 3. Reproduced by permission of the Director, British Geological Survey, © NERC.)

graphite ordering and K-white mica b cell dimension on a regional geological map to characterize the Alpine metamorphism of NE Hungary. As part of a regional survey of the Southern Appalachians, USA, Weaver (1984) constructed a map of mineral zones, KI and the first appearance of slaty cleavage to assess the separate effects of temperature and tectonics. Metamorphic maps combining the distribution of metapelitic zones with metabasite mineral assemblages were used by Merriman and Roberts (1985) and Bevins and Robinson (1988) to interpret the patterns of very low-grade metamorphism in the Welsh Basin.

Contoured metamorphic maps. Contouring of illite crystallinity data has been used to generate metamorphic maps in terranes where an evenly spaced distribution of metapelite samples is possible. The technique was introduced by Roberts and Merriman (1985) and has been used in a number of regional studies (e.g. Fortey, 1989; Warr & Robinson, 1990; Awan & Woodcock, 1991; Hesse & Dalton, 1991; Roberts *et al.*, 1991). Although a close and even spacing of samples generally increases the statistical validity of contours, the resources available for the

metapelite survey usually dictate the number and distribution of samples collected. Typical recorded intervals range from 1 to 1.5 km between samples (Roberts *et al.*, 1990) or 1 sample per 2.5 km² (Merriman & Roberts, 1995), to approximately eight samples per 100 km² (Hesse & Dalton, 1991). An appraisal of the precision of the contouring technique by Robinson *et al.* (1990) showed that the spacing of isocrysts (contours of equal crystallinity KI) should be at least 0.1 $\Delta°2\theta$ apart. Following from this work isocrysts tended to be limited to those defining the anchizone (0.42 and 0.25 $\Delta°2\theta$) and the middle of the anchizone (0.30 $\Delta°2\theta$), and less commonly subdividing the diagenetic zone and epizone (e.g. British Geological Survey, 1992a,b, 1993a,b; Fortey *et al.*, 1993; Merriman *et al.*, 1993; Roberts *et al.*, 1996).

Manual and computer techniques are used to produce contoured maps. Manual contouring uses data points plotted on an outline geological or structural map and interpolates the position of an isocryst by measuring and proportioning between adjacent values. With this technique the effects of geological discontinuities, such as unconformities or faults, or discordant patterns across unconformities and around intrusions, can be accommodated as the map evolves (see Fig. 3.11). Computer contouring uses interactive surface modelling (ISM), or similar software, to produce the contoured map. Generally, algorithms are used to calculate a surface consisting of a grid of several thousand nodes (depending on size and scale) for the entire area of study. In the final iteration each grid node is assigned a KI value as a distance-weighted average of the four nearest datapoints, and the designated contours drawn from this grid (Roberts *et al.*, 1990). Krigged data have also been used with GMAP and MACGRIDZO programs to produce contoured crystallinity maps (Warr *et al.*, 1996). Maps generated by computer usually lack any geological framework and before producing a final version those elements of the structural and/or igneous geology which are likely to influence the pattern of metamorphism can be superimposed. At this stage, most computer generated contoured patterns require manual modification, for example to reflect the concentric patterns associated with postmetamorphic intrusive aureoles, or to adjust contours for the effects of postmetamorphic faulting (Fig. 3.3).

Patterns of very low-grade metamorphism revealed by contoured maps have been used to elucidate a variety of basinal, tectonic, magmatic and metamorphic events. For example, Roberts and Merriman (1985) found that an early burial metamorphic event characterized the strata below the sub-Ordovician unconformity in North Wales. In the English Lake District, Fortey *et al.* (1993) identified a burial metamorphic episode, which they related to arc development that predated the late Caledonian deformation and associated high-anchizone to epizonal metamorphism. Beneath the mainly Mesozoic cover of the English Midlands, Merriman *et al.* (1993) showed that the platform formed by the Midlands Microcraton is characterized by widely spaced contours in late diagenetic strata, but northeastwards these change to more closely spaced contours in higher grade rocks of a concealed late Caledonian fold belt. Adjacent to late Caledonian granitic plutons in the Scottish Southern Uplands terrane, overprinted patterns due to thermal effects are consistently more extensive than aureoles delineated by visible hornfelsing (e.g. British Geological Survey, 1992a, 1993d). Displacement of isocryst patterns has been used to estimate the latest movements on major strike-slip faults (e.g. Roberts *et al.*,

1991). Metamorphic cross-sections can be constructed from contoured maps as a means of reconstructing structural and metamorphic history. Using this technique, Roberts *et al.* (1996) restored sections across the Welsh Basin to explore basin structures at the time of early burial metamorphism. Warr *et al.* (1996) used a combination of contoured KI data and metamorphic cross-sections to model the tectothermal evolution of the Caledonian foreland thrust belt of central Sweden. Further examples of the interpretation of contoured maps are discussed below.

3.2.3 Metapelites and basin maturity

When sedimentary basins fill and subside organic materials contained in sedimentary sequences undergo a series of irreversible reactions in response to burial. The potential to generate oil, gas and coal by such reactions is an important aspect of *basin maturity*. Estimating the timing of thermal maturity and the maximum depth of burial associated with basin development, including any subsequent uplift, are crucial factors in assessing hydrocarbon potential and prospectivity. Basin maturity studies rely largely on organic indicators, such as vitrinite reflectance, and modelling of organic maturation is a highly developed discipline capable of providing kinetic models of basin evolution (e.g. Waples, 1994).

During sedimentary burial, clay minerals contained in mudstones and shales undergo diagenetic and very low-grade metamorphic reactions, detailed in Chapter 2, which are equivalent to those of organic materials. Reaction progress in both clay mineral and organic materials is irreversible under normal diagenetic and anchizonal conditions, so that uplifted sequences generally retain indices and fabrics indicative of maximum maturity and burial. Systematic correlation of mineral and organic maturity indicators was initiated more than three decades ago (see section 3.2.2), as a means of integrating the basin maturity studies of the mineralogist and the petroleum geologist. Although each approach has its own advantages, in general they are complementary or at least compensate where one or other cannot be applied. For example, although organic materials are more reactive than clays to low temperature changes, they are a scarce component of many sedimentary sequences. In contrast, clay minerals are almost invariably present in basinal sediments and may be

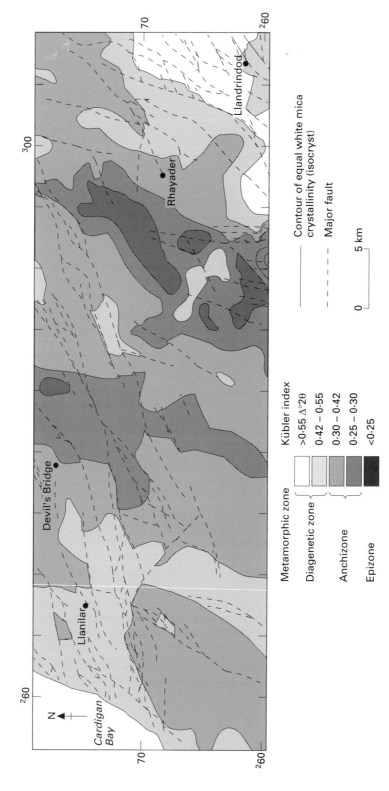

Fig. 3.3 Computer-generated map of Kübler indices from the central part of the Welsh Basin. Based on 663 data points and approximately one pelite sample per 1.5 km². Contours are manually modified to show the likely effects of postmetamorphic faulting on the pattern of burial metamorphism. (From Merriman *et al.*, 1992, Fig. 2, reproduced by permission of the Director, British Geological Survey, © NERC; see also Roberts *et al.*, 1996.)

the only means of assessing maturity in some barren red-bed sequences, and also in basins developed before the evolution of terrestrial plants. One of the problems of correlating mineral and organic materials is that they may react differently to the physical conditions of sedimentary burial. In some basins for example, a higher than normal geothermal gradient appears to accelerate organic maturity where the same conditions have had no equivalent effect on illite crystallinity (e.g. Bevins *et al.*, 1996). Elsewhere rapid burial of sedimentary sequences or overpressuring can inhibit organic maturation by preventing escape of volatiles (e.g. Dalla Torre *et al.*, 1997; Carr, 1998). Strain caused by deformation can enhance vitrinite maturity (e.g. Teichmüller, 1987), and also promote illite crystal growth (see section 2.3.2). Among other things, the correlation scheme presented in Fig. 3.4 will help to establish the type of basinal or tectonic conditions where mineral and organic materials can be expected to react differently.

The most widely used indicator of thermal maturity in sedimentary rocks is the vitrinite reflectance technique. It measures changes in the optical properties of plant and woody debris caused by progressive loss of volatiles and incipient graphitization during maturation. Vitrinite reflectance (random reflectivity, R_r or maximum reflectivity, R_{max}) is particularly useful for estimating maximum temperatures from dispersed organic particles in sedimentary rocks, and is also used to indicate maturation or *rank* of coal (e.g. Teichmüller, 1987). Using vitrinite reflectance values, Horsfield and Rullkötter (1994) divided the progressive maturation of sedimentary organic matter into three stages in relation to oil and gas generation. During diagenesis ($R_r < 0.5\%$) kerogen is formed, the major precursor of petroleum. The succeeding stage of catagenesis ($0.5\% < R_r < 2.0\%$) is the main stage of oil formation, with cracking of oil to produce hydrocarbon gases (wet gas) occurring in late catagenesis. Only dry gas, mostly methane, is produced in the stage of metagenesis ($2.0\% < R_r < 4.0\%$). These stages have been correlated with coal rank, clay mineral reaction progress and metapelitic zones in Fig. 3.4, adapted from the correlation scheme of Merriman and Kemp (1996) and Fig. 2.1.

Because of the greater availability of data, only reaction progress of the smectite–I/S–illite–muscovite series is used to index clay mineral and metapelitic

zones on the left side of Fig. 3.4. A series of XRD traces show typical clay mineral assemblages in <2-µm fractions of metapelites, prograding with depth from top to bottom in the diagram. Depth estimates for the zonal boundaries are based on 'normal' geothermal gradients of 25–30°C km^{-1}. The Early Diagenetic Zone is characterized by the formation of authigenic smectite and/or kaolinite, although very fine-grained illite, illite/smectite and chlorite of detrital origin may also be present in some <2-µm fractions. The Early Diagenetic Zone is correlated with the Diagenetic Stage of organic maturation, characterized by kerogen formation from the selective preservation of resistant cellular materials with variable proportions of the residues of less resistant amorphous organic materials. Beds rich in plant remains, e.g. peat, form low rank lignite and sub-bituminous coals at the Diagenetic Stage. The smectite-to-illite transition begins low in the Early Diagenetic Zone, at temperatures in the range 70–90°C (Freed & Peacor, 1989).

The transition from the Early Diagenetic Zone to the Late Diagenetic Zone marks the transformation to 60–80% illite in I/S, and is correlated with the transition from diagenesis to catagenesis and the beginning of oil generation from kerogen, the opening of the *oil window*. In many sedimentary basins these important changes take place at burial depths of 3–4 km and temperatures of about 100°C (Pollastro, 1993). In the Late Diagenetic Zone, illite begins to form discrete crystals thick enough to be measured by TEM, and as the reaction proceeds to illite >85% of I/S the 10-Å peak profile is usually sufficiently well defined to be measured using the Kübler index. The base of the oil window at $R_r = 1.3\%$ marks the beginning of gas generation (Horsfield & Rullkötter, 1994), and is associated with further reduction in smectite content of I/S as illite crystals continue to thicken with deeper burial. Coal rank increases through the zone of catagenesis, from sub-bituminous to bituminous with a progressive loss of volatile matter from 45 to 15%. Note that the mineralogical transformations that characterize the Late Diagenetic Zone are approximately equivalent to those of burial diagenesis (Kisch, 1983) and burial metamorphism (Coombs, 1961). Characteristically, the partial to complete recrystallization of clay matrix minerals is on a regional scale but without the development of a tectonic fabric (see section 2.3.2).

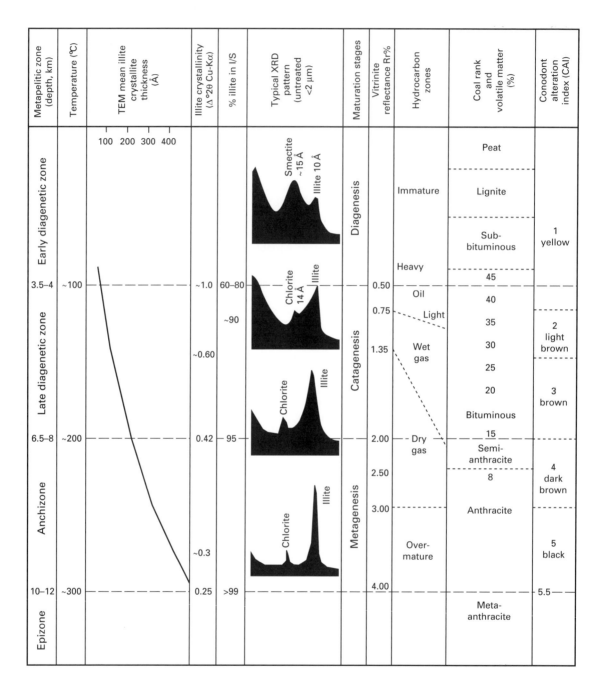

Fig. 3.4 Basin maturity chart showing correlation of reaction progress in the smectite–I/S–illite–muscovite series of clay minerals with the maturation stages of Horsfield and Rullkötter (1994) and some organic maturity indices. (Modified from Merriman & Kemp, 1996, with personal communication from N. MacCormack).

The beginning of the anchizone and very low-grade metamorphism is indexed at KI = 0.42 Δ°2θ or a mean illite crystallite thickness of 220 Å (see Fig. 2.19) and CAI > 4 (Garcia-Lopez *et al.*, 1997). The minimum temperature for the onset of the anchizone is 200°C (Kisch 1987; see also Mullis *et al.*, 1993).

This is correlated with the onset of metagenesis at $R_r = 2.0\%$, associated with dry gas generation from kerogen and high rank semi-anthracite coals. Within the anchizone, smectite is virtually eliminated as the rate of illite crystallite thickening is accelerated by tectonic fabric development in metapelites. A weak tectonic fabric may first develop in the late diagenetic zone (Kisch, 1991b), when CAI values exceed 3 (Garcia-Lopez *et al.*, 1997). By the middle of the anchizone (KI = 0.3 $\Delta°2\theta$) a penetrative slaty cleavage commonly characterizes metapelitic rocks, and coals are anthracite. Dry gas generation has generally ceased by $R_r = 3.0\%$ and overmaturity characterizes the floor of metagenesis at $R_r = 4.0\%$.

The transition to the epizone and low-grade metamorphism is indexed at CAI >5.5 (Garcia-Lopez *et al.*, 1997) and KI = 0.25 $\Delta°2\theta$. Mean illite crystallite thickness is typically 520 Å and illite compositions are those of mature phengite or muscovite micas. Vitrinite reflectance values of $R_r > 4.0\%$ are associated with meta-anthracite coals prograding to semigraphite in the high-grade epizonal rocks (Teichmüller, 1987). The transition from the anchizone to the epizone (subgreenschist to greenschist) occurs at approximately 300°C (Bucher & Frey, 1994).

3.2.4 Pattern recognition

Application of the various methods of interpreting regional metapelitic terranes outlined above has led to the recognition of several specific patterns of very low-grade metamorphism (Fig. 3.5). A *burial metamorphic pattern* is characterized by an increase in metapelitic grade with stratigraphic age (Fig. 3.5a), and is generally assumed to develop as a series of regionally extensive ground-parallel surfaces defined by metapelitic zones. Such patterns are equivalent to regional metamorphic mineral zones developed in metabasic rocks (e.g. Neuhoff *et al.*, 1997; see Chapter 5). Burial patterns have been recognized in settings where the stratigraphical succession is well defined and the regional geology can be related to the pattern of metamorphism. Several criteria have been used to recognize a burial pattern in metapelitic sequences:

1 a basin-wide correlation between grade and age of strata such that grade generally increases into older strata (e.g. Robinson & Bevins, 1986);

2 combined structural and metamorphic maps show highest grades in anticlinal areas and lowest grades in synclinal areas (e.g. Hesse & Dalton, 1991);

3 combined structural and metamorphic cross-sections show an approximate parallelism between isocryst surfaces and formational boundaries, with grade generally decreasing in the direction of younging (e.g. Roberts *et al.*, 1996);

4 where sufficient data are available, plots of ranges and mean values of KI show a down-section increase in metapelitic grade (e.g. Fig. 3.2; Roberts *et al.*, 1991).

A burial pattern develops *in situ* and represents the earliest response to depth-controlled diagenetic and very low-grade metamorphic reaction progress caused by basin subsidence. A uninterrupted sequence of metapelitic zones is characteristic of a discrete episode of burial and low-grade metamorphism. Interrupted or discordant regional patterns, described below, are indicative of basin inversion and post-metamorphic faulting. The field gradient strongly influences the regional development of a burial pattern. Where gradients are low the depth interval over which critical changes occur may be so large that a pattern may be difficult to establish. For example, in a basinal setting where burial has occurred under a geothermal gradient of 20°C km^{-1}, the anchizone alone may extend over 5 km of stratigraphic thickness. In such a setting the subsidence required to generate a pattern which includes the diagenetic zone, anchizone and the epizone might exceed 15 km (Fig. 3.4); such thicknesses are beyond those expected from theoretical analysis of basin formation. Hence a burial pattern is more likely to be recognized in basins where the field gradient was high, typically >35°C km^{-1}, and metapelitic zones are most closely spaced (e.g. Roberts *et al.*, 1996).

A different type of burial pattern may be developed in accretionary terranes. When sequentially younger wedges of sediment are buried and underplated in an accretionary prism, younger strata buried under older strata may develop a pattern of grade increasing from older to younger rocks. An *accretionary burial pattern* comprises a sequence of continuous metapelitic zones developed *in situ* during synchronous under-thrusting, burial and very low-grade metamorphism (Fig. 3.5b). Unlike normal burial metamorphism, an accretionary burial pattern typically shows grade increasing into stratigraphically younger tracts of strata which were progressively buried beneath an

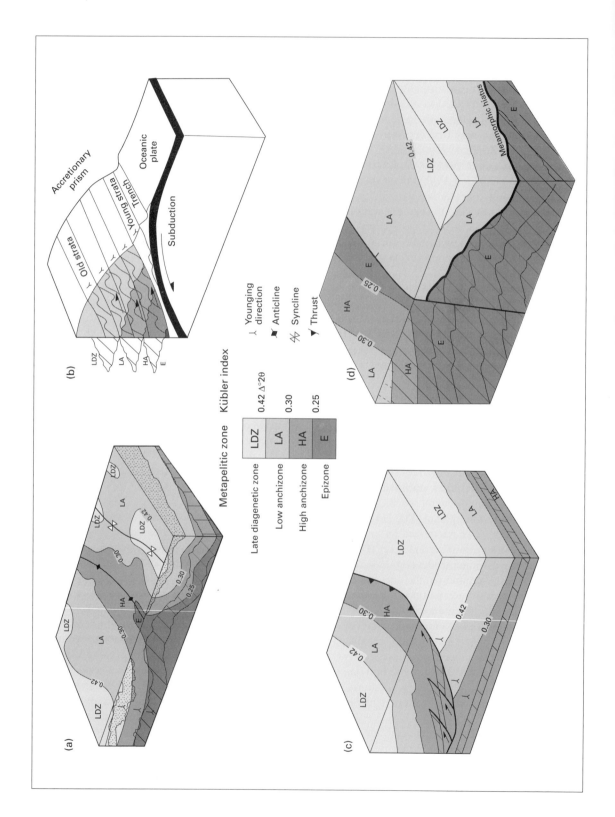

Metapelitic zone   Kübler index

	LDZ	0.42 Δ°2θ
Late diagenetic zone		
Low anchizone	LA	0.30
High anchizone	HA	0.25
Epizone	E	

⊥ Younging direction

◣ Anticline

⚹ Syncline

▼ Thrust

overburden of older, imbricated strata forming a thrust stack. Such patterns show two significant differences from inverted metamorphic patterns, discussed below. First, accretionary burial patterns do not show an (apparent) inversion of the field gradient because the pattern is essentially developed by burial of inverted stratigraphical sequences. Second, the patterns develop syntectonically as the accretionary prism is assembled, unlike inverted metamorphic patterns which are intrinsically post-tectonic. Accretionary burial patterns can only be recognized in terranes with a well characterized tectonostratigraphy. The main characteristics of this type of burial pattern are: (a) plots of ranges and mean KI values on tectonostratigraphical diagrams show grade increasing into sequentially younger tracts of strata (e.g. Fig. 3.10); (b) metamorphic maps show that metapelitic zones and isocrysts are generally conformable with the tract bounding faults which imbricate the succession, and that grade generally increases into younger tracts of strata (e.g. Merriman & Roberts, 1995).

An *inverted metamorphic pattern* is characterized by grade increasing upwards in relation to structure causing an apparent inversion of the metamorphic field gradient (Jamieson *et al.*, 1996). Such patterns do not develop *in situ* and are generally the result of tectonic assembly by thrust faults juxtaposing the pattern of higher grade or more deeply buried strata, over lower grade or less deeply buried strata (Fig. 3.5c). The characteristics of inversion generally require that a pattern of metamorphism existed prior to thrusting, hence inversion patterns are intrinsically postmetamorphic, even though local modification to the pattern may be syn- or post-tectonic. The main criteria used to recognize an inverted metamorphic pattern in metapelitic sequences are:

1 combined structural and metamorphic cross-sections showing transported metamorphic patterns (e.g. Frey *et al.*, 1980; Frey, 1988);

2 metamorphic maps showing lower grades in rocks forming the footwall or autochthonous sequences

and higher grades in the hanging wall or alloch-thonous sequences (e.g. Warr *et al.*, 1996);

3 a general lack of conformity or a metamorphic hiatus between metapelitic zones above and below a thrust surface.

A *metamorphic hiatus* is indicated wherever the continuity of metapelitic zones is interrupted by the absence of one or more zone from a sequence so that, for example, the lower anchizone is in contact with the epizone (Fig. 3.5d). On a regional scale a metamorphic hiatus may be associated with a major unconformity and record two separate metamorphic episodes. Where a metamorphic hiatus is suspected at an unconformity, KI data should be contoured separately in the two areas bordering the mapped unconformity. Discordant patterns separated by an unconformity indicate two metamorphic events, even where a major hiatus is absent (Roberts & Merriman, 1985). A metamorphic hiatus may also result from thrust faulting removing one or more metapelitic zones, for example during nappe emplacement, and is commonly associated with inverted metamorphic patterns. Normal faulting can also create a metamorphic hiatus by juxtaposing down-faulted low-grade metapelites against higher grade metapelites (Figs 3.5d and 3.20).

3.3 Regional patterns of very low-grade metamorphism

3.3.1 Geotectonic setting

Two contrasting geotectonic settings are responsible for the pressure–temperature–time (*P–T–t*) paths that characterize the transition from diagenesis to metamorphism (Robinson, 1987). Collisional settings are the most widely recognized and are developed in regions of crustal thickening at convergent plate margins. Extensional settings are associated with the development of major crustal rift systems, including marginal and back-arc basins (Fig. 3.6).

In a review of metamorphic belts, Maruyama *et al.* (1996) recognized two types of collisional setting. A-type collision involves subduction of continental margin assemblages beneath another continent; B-type collision involves subduction of oceanic crustal rocks beneath an active continental margin. A-type or Alpinotype (Bally, 1975) subduction generates very low and low-grade metamorphic rocks in settings

---

Fig. 3.5 (*Opposite*) Patterns of very low-grade metamorphism. (a) Burial metamorphic pattern in folded strata; (b) accretionary burial pattern; (c) inverted metamorphic pattern; (d) metamorphic hiatus. See text for explanations.

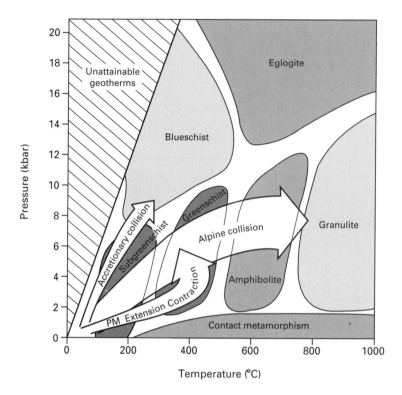

Fig. 3.6 Typical pressure–temperature–time (*P–T–t*) paths for the main geotectonic settings of very low-grade metamorphism. The approximate position of metamorphic facies is shown on the *P–T* grid of Bucher and Frey (1994, Fig. 4.2). *P–T–t* paths are clockwise for both types of collisional settings and counter-clockwise for extensional settings. PM = passive margin setting.

referred to here as *Alpine collisional settings* (Table 3.1). Such settings tend to be long-lived zones of convergence where the transition from diagenesis to very low-grade metamorphism is usually part of a continuum that follows an essentially clockwise *P–T–t* path. This path most commonly passes into intermediate pressure-type Barrovian sequences (Fig. 3.6), less commonly into high pressure-type blueschist facies rocks (Maruyama *et al.*, 1996). The *P–T–t* path followed in Alpine collisional settings reflects initial rapid basin or trench infilling and burial along a moderately steep *P/T* gradient. Subsequent phases of deformation, folding and crustal thickening are commonly associated with peak metamorphic temperatures, so that with time a clockwise decrease in the *P/T* gradient takes place. Finally, uplift and cooling caused by changing buoyancy in the subducted lithosphere beneath Alpine-type orogenic belts leads to a further decrease in the *P/T* gradient (Fig. 3.6). Whereas the internal zones of mature Alpine-type orogenic belts develop a range of greenschist to amphibolite facies rocks, very low and low-grade rocks survive in the external fold-and-

thrust belts where allochthonous sequences are in contact with autochthonous cover sequences. In these belts typical K-white mica *b* cell dimensions (9.00–9.03 Å) are characteristic of the intermediate pressure facies series of Guidotti and Sassi (1986), but values can vary widely.

Very low and low-grade rocks are also found in *accretionary collisional settings* generated by B-type subduction of oceanic crust. In such settings sedimentary strata are sheared off the descending oceanic plate, shortened, rotated and imbricated to form an accretionary prism on the outboard side of the overriding continental crust. Rapid burial results in a steep *P–T–t* path through the subgreenschist to blueschist facies fields (Fig. 3.6), but uplift may follow closely and the retrograde path usually involves little thermal overprinting (Maruyama *et al.*, 1996). Accretionary burial patterns may be developed in metapelitic sequences, but are detectable only where the tectonostratigraphy is well characterized. Typical K-white mica *b* cell dimensions are in the range 9.02–9.05 Å and are characteristic of the intermediate and high-pressure facies series (Guidotti & Sassi, 1986).

Table 3.1 Some characteristics of the main geotectonic settings of very low-grade metamorphism.

	Extensional	Alpine collisional	Accretionary
Typical basinal lithologies	Mudstone/shale ≥ sandstone, pyroclastic rocks > limestone	Flysch sandstone and shale, shallow water limestone	Turbidite sandstone + siltstone > mudstone/shale > chert, pillow basalt limestone, Mn-nodules, mélange
Tectonometamorphic timing	(a) Deposition and low-P burial metamorphism followed by basin inversion and higher P overprint (b) Passive margin without tectonic overprint	Long-lived contractional setting; polyphase metamorphism with thermal overprinting common	Deposition, burial metamorphism and deformation a single event; little thermal overprinting
P–T–t path	Anticlockwise	Clockwise	Clockwise
Structural style	Variable slaty cleavage development; discrete slate belts in zones of highest strain	External fold-and-thrust belts show intense slaty cleavage in allochthon and weak cleavage in autochthonous cover sequences	Imbricated strata, steeply inclined; variable slaty cleavage development; cleavage and bedding subparallel in underplated slate belts
Typical metapelitic patterns	Burial patterns in late diagenetic-low anchizonal strata; high anchizone-epizone in slate belts associated with inversion tectonics	Late diagenetic-anchizone in autochthon high anchizone-epizone in allochthon; inverted metamorphic patterns associated with thrusting	Anchizone in underplated slate belts; epizone in high strain zones; accretionary burial patterns
K-white mica b dimension	8.98–9.01 Å	9.00–9.03 Å	9.02–9.05 Å

Metapelitic rocks in *extensional settings* develop a restricted range of facies, typically from diagenetic to low greenschist (Fig. 3.6). Robinson (1987) suggested that the restricted range of metamorphism is the result of a limited thermal budget related to the commonly transient nature of extensional basins, and their relatively shallow crustal setting. Unlike collisional settings, cooling in extensional basins is the result of subsidence caused by thermal relaxation in the mantle. Hence a typical *P–T–t* path is counterclockwise (Fig. 3.6), with the oldest strata in the basinal sequence showing the highest grade and *P/T* gradient increasing with time and sedimentary infilling; the final phase of contraction and uplift causes a further anticlockwise rotation of the *P/T* gradient. The term *diastathermal* metamorphism was introduced by Robinson (1987) and Robinson and Bevins (1989) to describe the development of diagenetic to epizonal/low greenschist facies rocks caused by early heating of the basin fill in an extensional setting. Burial patterns are characteristically developed in metapelitic sequences, but may be overprinted by strain-enhanced crystallization associated with the tectonics of basin contraction and inversion. K-white mica *b* cell dimensions are typically small, 8.98–9.01 Å, and characteristic of the low-pressure facies series (Guidotti & Sassi, 1986).

The main characteristics of these three settings, summarized in Table 3.1, form the broad framework for the review that follows. Such settings are not immutable and evolve in response to the geodynamics of lithospheric plate activity. A back-arc or marginal basin might open, develop a sedimentary fill with a pattern of diastathermal metamorphism and subsequently contract, deform and invert. The intensity of the overprinting associated with inversion will determine whether or not the characteristics of diastathermal metamorphism are preserved. Similarly, in some accretionary settings evidence of the transition from diagenesis to metamorphism might be preserved only as fault blocks resulting from rapid uplift. Rocks formed in both settings might eventually be amalgamated into typical Cordilleran or Alpine terranes. Some of the patterns of very low-grade metamorphism generated in these settings are described below.

3.3.2. Extensional settings

3.3.2.1 The Welsh Basin

The Welsh Basin is the type example of a diastathermal setting, where relatively high heat flow generated by extension has produced a burial pattern of very low-grade metamorphism (Fig. 3.7). Although the basin contracted and inverted during a late Caledonian orogenic event, deformation has not erased the burial pattern. The numerous studies which have helped to characterize very low-grade metamorphism in the Welsh Basin have also led to debate on the relative importance of the extensional and contractional phases of metamorphism, and the role of thermal and strain energies in advancing metapelitic reactions.

Despite its relatively limited extent, the Welsh Basin has played a prominent role in advancing the understanding of very low-grade metamorphic processes, perhaps surpassed only by the Swiss Alps and South Island, New Zealand in terms of the impact of studies generated therein. Since the late 1970s more than 30 papers, maps and memoirs have been published on various aspects of very low-grade metamorphism in the Welsh Basin. In scale these range from basin-wide studies of metapelitic and metabasic grade to TEM studies of nanotextures in phyllosilicates. Several innovative developments have derived from these studies including the technique of contouring illite crystallinity data, methods for recognizing burial metamorphic patterns and the concept of diastathermal metamorphism. This work has helped to establish the Welsh Basin as the type example of an extensional setting and is here accorded a lengthier description than some other examples of this setting.

The Lower Palaeozoic Welsh Basin (British Geological Survey, 1994) comprises a 15-km thick succession of Cambrian to Silurian mudrock-dominated marine sediments and volcanic rocks. It developed on Precambrian basement forming part of the Eastern Avalonian microcontinent (Thorpe *et al.*, 1984). Extensive marginal basin volcanism occurred in the upper Ordovician, erupting rhyolitic and basaltic lavas and tuffs of calc-alkaline to alkaline composition (Howells *et al.*, 1991). Volcanism ceased in the late Ordovician, and was followed by Silurian infilling of the basin in a series of fault-controlled sub-basins (Woodcock, 1984). Basin inversion began towards the end of the Silurian, with red-bed deposits indicating initial uplift in Ludlow time (415–420 Ma), followed by deformation and inversion culminating in the Lower Devonian (*c.* 400 Ma) as the Acadian tectonometamorphic event (Soper *et al.*, 1987). Folds generated by the deformation have an axial-planar slaty cleavage with predominantly NE–SW (Caledonian) trend. The eastern margin of the basin is formed by the Midland microcraton, a platform separating the Welsh Basin from the concealed Anglian Basin which also contains Lower Palaeozoic strata (Pharaoh *et al.*, 1987; 1991; Woodcock & Pharaoh, 1993). The northern and northeastern extent of the basin is terminated by Triassic basins. Some of the contemporaneous basin-bounding Precambrian rocks are found in the northwest, on Anglesey and Llŷn, and similar rocks probably host the Mesozoic basins of the Irish Sea along the western termination of the Welsh Basin (Fig. 3.7). The Variscan Front cuts the southern margin of the basin (British Geological Survey, 1996).

Regional studies of metapelitic grade based on illite crystallinity (Merriman & Roberts, 1985; Roberts & Merriman, 1985; Robinson & Bevins, 1986) established the overall pattern of metamorphism across the basin (Fig. 3.7). These followed earlier studies which recognized a range of zeolite, prehnite–pumpellyite and greenschist facies assemblages in metabasic rocks (Bevins, 1978; Roberts, 1981; Bevins & Rowbotham, 1983). Although the restricted distribution of metabasic rocks prevented their use for basin-wide study of patterns, in places the application of both techniques resulted in correlation between metapelitic zones and mineral facies up to the biotite zone of the greenschist facies (Fig. 3.1). Across the basin grade ranges from late diagenetic along the eastern margins to epizonal in the centre, with grade broadly increasing with the age of strata from mid-Silurian in the east to Cambrian in the northwest. This led Robinson and Bevins (1986) to suggest that depth of burial was the dominant control on the pattern of metamorphism. In the northern

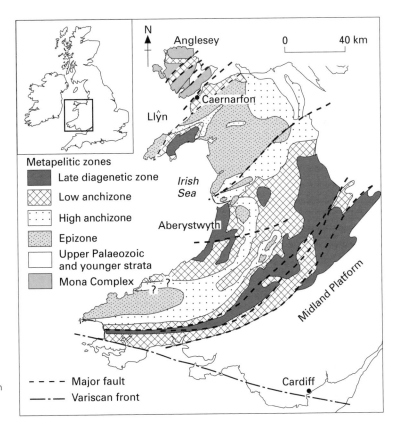

Fig. 3.7 Metamorphic map of the Lower Palaeozoic Welsh Basin. (Based on Robinson and Bevins (1986), with modifications from Roberts and Merriman (1985), Awan and Woodcock (1991) and Roberts *et al.* (1991, 1996).)

part of the basin, Roberts and Merriman (1985) recognized a pre-Ordovician burial metamorphism but overall found that the metapelitic pattern is more closely related to tectonism, with highest grades occurring in areas where a penetrative slaty cleavage is well developed. They suggested that this was due to strain-related crystal growth of white mica in slaty cleavage microfabric.

Detailed surveys of metapelitic grade in central Wales (Fig. 3.3) provided further evidence of both burial metamorphism and the influence of strain. For example, Roberts et al. (1991) and Merriman et al. (1992) found that grade generally increased from the late diagenetic zone to the epizone with depth of burial in successions 4–9-km thick. However, the successions also show zones of high strain related to contrasting lithology and ductility, and these can be recognized both from isocryst patterns and plots of depth against illite crystallinity (Fig. 3.2). The geological mapping of the central part of the basin (British Geological Survey, 1993b,c), revealed the approximate parallelism of isocrysts and formation boundaries, and this was used to infer a depth-controlled metamorphism by Roberts et al. (1996). On this basis they restored metamorphic cross-sections to the ground-parallel pattern that might be expected of burial metamorphism prior to folding and basin inversion. This revealed a strong parallelism between strata and isocryst surfaces in some late diagenetic to low anchizonal areas. Elsewhere, the restored sections showed that reverse movement on earlier extensional faults and growth of precursor folds had already begun when isocryst surfaces had been imposed. This suggested to Roberts et al. (1996) that metamorphism was 'accelerated or even initiated' by the early compression of the basin. Recent $^{40}Ar–^{39}Ar$ age dating of slaty cleavage, using metabentonites and slates from the northern and southern parts of the basin, suggest that an early metamorphic event at 414–421 Ma (mid-Ludlow) was followed by uplift and cooling in the Lower Devonian at around 390–400 Ma (Dong et al., 1997a).

The detailed XRD analysis which supported these regional studies revealed that systematic changes in < 2-μm mineralogy accompany increasing metapelitic grade and slaty cleavage development (Fig. 3.1). Late diagenetic pelites, characterized by the absence or very weak development of slaty cleavage, contain $1M_d$ illite with generally < 15% smectite interlayers,

homogeneous Na, K-mica (Jiang & Peacor, 1993) and chlorite. Slaty cleavage development ranges from weak to very well developed in anchizonal metapelites, and as this fabric intensified, discrete $2M_1$ K-mica and paragonite developed at the expense of illite and Na, K-mica, with minor albite and anatase. Epizonal slates commonly contain only discrete $2M_1$ K-mica and paragonite, chlorite and minor albite, anatase or rutile.

TEM studies in the Welsh Basin also revealed the importance of both burial and strain in the development of metapelitic microfabrics and white mica crystallite thickness. Some of the first ever TEM studies of slaty cleavage microstructure were those of Knipe (1979, 1981) using anchizonal slates from Anglesey (see section 2.3.3). He recognized two microtextural domains: P-domains are rich in phyllosilicates and represent the oriented slaty cleavage lamellae; Q-domains contain less well orientated minerals, mostly quartz and chlorite, which crystallized in the dilational domains between cleavage lamellae. TEM was used by Merriman et al. (1990) to study the transition from mudstone to slate in a prograde sequence of lower Ordovician metapelites from North Wales. The thickness of white mica/illite crystallites was found to increase as grade increased from the late diagenetic zone to the epizone and could be correlated with XRD-measured illite crystallinity (KI). A significant increase in crystallite thickness occurs in the anchizone and appears to increase rapidly as slaty cleavage microfabric develops. Li et al. (1994a,b) used TEM and AEM to identify two types of white mica in a lower anchizonal mudstone from central Wales: a bedding-parallel population is characterized by thin crystallites (50–200 Å) of $1M_d$ and $2M_1$ polytypes, some with typical K-poor and Si-rich immature illite compositions; a cleavage-parallel population consists of thicker (100 Å–2 μm), K- and Na-micas of mostly $2M_1$ and 3T polytypes. The studies of Li et al. (1994b) provided the first TEM evidence of microfabrics associated with the pattern of burial metamorphism in the Welsh Basin.

The diastathermal model of metamorphism proposed by Bevins and Robinson (1988) generated an ongoing debate on the metamorphic history of the Welsh Basin (Roberts et al., 1989). At the core of the debate is the relative importance of the extensional and contractional phases. Of the several lines of evidence advanced by Bevins and Robinson (1988) in support of early diastathermal metamorphism, the most robust are: (a) the pattern of burial metamorphism; (b) relatively high geothermal gradients in the range 36–52°C km^{-1} (Bevins & Merriman, 1988a; Bottrell *et al.*, 1990); (c) K-white mica *b* cell dimensions (mean = 8.995 Å; Robinson & Bevins, 1986) indicative of a low-pressure geotectonic setting. On the other hand, Roberts *et al.* (1989) argued that the metamorphic peak and final equilibrium was not reached until contraction and deformation, and this is supported by ^{40}Ar–^{39}Ar dating of the slaty cleavage (Dong *et al.*, 1997a). Perhaps the exceptional feature of the Welsh Basin is that mineralogical and textural evidence for both an extensional and contractional metamorphism is preserved. No doubt the Welsh Basin will continue to attract studies of very low-grade metamorphism and generate further debate.

3.3.2.2 South-west England Basin

South-west England formed part of an extensive and dynamic rift system during Devonian and early Carboniferous time. As sub-basins filled and developed they were inverted and thrust northwards to form a fold-and-thrust belt. Two patterns of very low-grade metamorphism related to burial and thrust stacking are preserved, and these are locally overprinted by contact aureoles of Variscan granite plutons.

The Upper Palaeozoic basin of south-west England forms part of the external Variscides of western Europe. Development began in the early Devonian, probably as a series of marginal sub-basins lying to the south of an Avalon terrane which was still in the process of 'soft' collision (docking) with Laurentia to the north (Soper & Woodcock, 1990). The basin fill is variable, but predominantly comprises deep water mudrocks and turbiditic sandstones, outer shelf limestones and interbedded alkali and tholeiitic basalt lavas and tuffs. Tectonic evolution of the basin was asymmetric, with a passive margin on its northern side characterized by extensional trapdoor type basins, and an active margin to the south. The active margin, representing the northern edge of the Variscan front, generated a series of major thrusts which advanced northwards in space and time, closing and inverting the basins developed along the passive margin (Holder & Leveridge, 1994;

Leveridge *et al.*, 1998). Two episodes of regional deformation are recognized. Closure of the more southerly Gramscatho Basin and inversion of the central Trevone Basin in the late Devonian to early Carboniferous is associated with the emplacement of a family of major thrust nappes. The early deformation is characterized by anchizonal to epizonal grades and a slaty cleavage associated with tight to isoclinal folds. A second phase of deformation in the Upper Carboniferous resulted in out of sequence thrusting in the Trevone Basin, and inversion of the Culm Basin, a foreland basin lying to the north of the major nappes. Emplacement of a chain of granite plutons with associated mineralization occurred in the late Carboniferous–early Permian.

A regional synthesis based on over 25 studies of very low-grade metamorphism in the basin was made by Warr *et al.* (1991, and references therein). Metapelitic grade ranges from late diagenetic to epizonal, the younger Upper Carboniferous rocks generally showing lower grades (late diagenetic–lower anchizone) than the Lower Carboniferous to Lower Devonian strata (Fig. 3.8). Slaty cleavage development is spatially related to metapelitic grade, with penetrative fabrics tending to be associated with anchizonal to epizonal grades whereas late diagenetic to lower anchizonal rocks typically show a weak, nonpenetrative cleavage fabric. K-white mica *b* cell dimensions indicate both low-pressure and intermediate-pressure facies series.

In parts of the basin the regional pattern of metamorphism shows a very general relationship between metapelitic grade and the stratigraphic succession indicative of burial (Warr *et al.*, 1991, Figs 1 and 2). There is some evidence that a burial pattern was acquired prior to folding and thrusting. For example, in the central Trevone Basin isocryst patterns are folded by a regional south-facing syncline (Pamplin, 1990), suggesting the pattern predated the D1 backthrusting fold phase. Elsewhere, postburial inversion patterns have resulted from deformation and thrust emplacement, as in the Tintagel high-strain zone (Warr, 1991). Along the active southern margin overthrusting of high-greenschist facies metapelites and metabasites (Start Complex) also represent inverted metamorphic sequences. Although emplacement of late granitic plutons has not extensively overprinted the regional pattern, a regional association of epizonal grades with

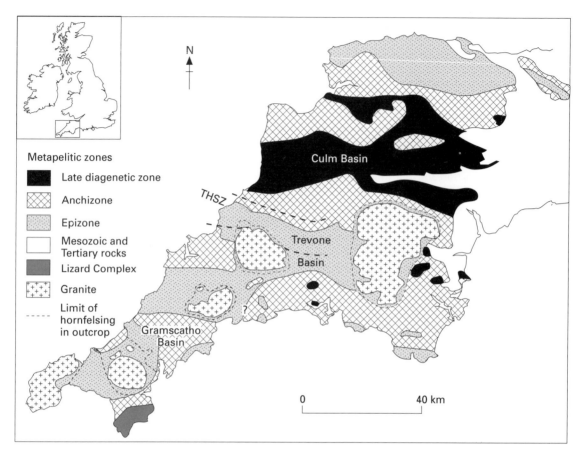

Fig. 3.8 Metamorphic map of the Palaeozoic rocks of south-west England. The epizone may include higher grade rocks within the limits of hornfelsing. (Based on Warr *et al.* (1991), with modifications from Warr (1995) and Leveridge *et al.* (1998).) THSZ = Tintagel high-strain zone.

the subsurface extent of the batholith (Fig. 3.8) suggests that a more extensive pattern of thermal overprinting is concealed in the subcrop.

The tectonothermal model proposed by Warr *et al.* (1991) recognized two patterns of very low-grade metamorphism in northern and southern areas of the region. In the north, K-white mica *b* cell dimensions are indicative of a low-pressure facies series, and together with vitrinite reflectance data are consistent with a relatively high heat flow. This evidence and an overall burial pattern of metamorphism together suggest a diastathermal setting. The southern area lacks a consistent relationship

between grade and the stratigraphical succession, while K-white mica *b* dimensions tend towards more intermediate facies series values. Consequently Warr *et al.* (1991) suggested that regional metamorphism in the south was related to thrust stacking and crustal thickening. Hence the south-west England Basin appears to be an example of an extensional basin that contracted to form a fold-and-thrust belt.

3.3.2.3 *English Lake District and Isle of Man*

In northern England, two contrasting settings of very low-grade metamorphism are juxtaposed. Ordovician arc and back-arc sequences were metamorphosed in an extensional setting, with high basinal heat flow locally elevated by Caledonian intrusions. South of the arc terrane, crustal downwarping created a foreland basin with atypically low heat flow within which Silurian strata were deeply buried before being uplifted and deformed in the late Caledonian. This is a region where the effects of high and low geothermal gradients on metapelitic patterns can be compared, and has attracted studies of illite crystallinity, K-white mica b cell dimensions and microfabrics, including K–Ar and Ar–Ar age dating of slaty cleavage.

The Lower Palaeozoic inliers of the English Lake District and the adjacent Isle of Man (Fig. 3.9) represent sedimentary basins that developed on Avalonian crust lying to the south of Iapetus, a former oceanic basin. Basinal sequences, which range in age from early Ordovician (Tremadoc) to late Silurian (Pridoli), have been regionally metamorphosed to very low and low grades, initially in response to extensional processes and associated volcanism, and subsequently as a result of basin contraction and inversion associated with the closure of Iapetus. In the Lake District the succession comprises three major groups of strata. The Skiddaw Group is a >3.5-km thick sequence of Lower Ordovician (Tremadoc–Llanvirn) mudrocks and turbidite sandstones (Cooper

& Molyneux, 1990) forming the major outcrop in the Northern Fells Belt and smaller inliers in the Central Fells Belt (Fig. 3.9). This group is similar in age range and lithology to the Manx Group occupying much of the Isle of Man, some 80 km to the west (Molyneux, 1979). In the Lake District the Skiddaw Group is succeeded by the Borrowdale Volcanic Group, a >6-km thick sequence of andesitic and dacitic tuffs and lavas of mostly mid-Ordovician (Caradoc) age occupying the Central Fells Belt (Petterson et al., 1992). The volcanic rocks are overlain by the Windermere Supergroup, a foreland basin sequence of thin late Ordovician shelf deposits succeeded by Lower Silurian hemipelagites and thick late Silurian turbidites (Kneller et al., 1993). The Windermere Supergroup crops out in the Southern Fells Belt and further south-eastwards in the Craven Inliers. Calc-alkaline granites associated with the Caradoc volcanism intrude the Skiddaw and Borrowdale Volcanic Groups, and a granitic batholith may be concealed at a shallow depth (c. 4 km) beneath both groups (Lee, 1986). On the Isle of Man, the Manx Group is intruded by two exposed plutons and other concealed granites (Roberts et al., 1990). These three terranes, representing parts of a volcanic arc, a back-arc or inter-arc basin and a foreland basin, were juxtaposed by the closure of Iapetus during progressive collision with the southern margin of Laurentia (Soper et al., 1987; Barnes et al., 1989).

Fig. 3.9 Metamorphic map of the Lower Palaeozoic rocks of the English Lake District and Isle of Man. Based on maps and accounts published by Fortey (1989), Fortey et al. (1993, 1998), Hirons (1997), Roberts et al. (1990) and Thomas (1986). Note that greenschist facies rocks in thermal aureoles may be included in the epizone, and that subgreenschist facies volcanic rocks in the Central Fells Belt are shown as anchizonal.

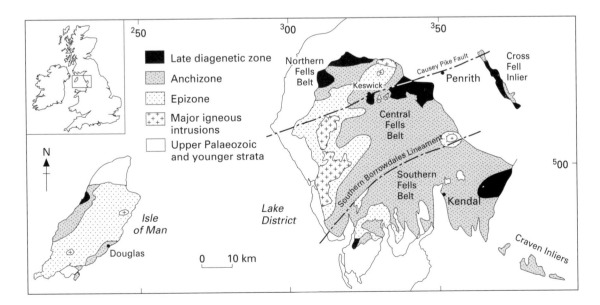

Detailed studies of metapelites, including contoured KI data, indicate that metamorphic grade in the Skiddaw Group ranges from late diagenetic to epizonal (Fortey, 1989; Fortey et al., 1993). Late diagenetic mudstones mainly consist of illite with chlorite, minor K, Na-mica, illite/smectite, kaolinite or pyrophyllite and quartz. These mudstones commonly contain chlorite-mica stacks and show a strong bedding-parallel fabric (Fortey et al., 1998). Anchizonal and epizonal slates consist largely of $2M_1$ K-white mica and chlorite, with paragonite and K, Na-mica, and minor pyrophyllite, albite, rutile and quartz. Slaty cleavage fabrics are strongly developed in metapelites of the high anchizone and epizone, where S_1 may appear as closely-spaced pressure-solution seams accompanied by one or more later, weakly developed fabrics. In contrast, a weak S_1 cleavage fabric is developed in some late diagenetic or low anchizonal metapelites, where it cuts the bedding-parallel fabric (Fortey et al., 1993). K-white mica b cell dimensions are generally in the range 8.96–9.00 Å (Thomas, 1986), indicative of a low-pressure geotectonic setting. According to Fortey et al. (1993), the overall pattern of metamorphism in the Skiddaw Group suggests a three-stage metamorphic history. Early burial metamorphism, characterized by late diagenetic to low anchizonal grades, occurred under a higher than normal field gradient ($c. 35°C\,km^{-1}$). This event was probably related to extensional high heat flow, with associated volcanism and granitic batholith emplacement. Subsequent tectonic thickening accompanied anchizonal to epizonal metamorphism and the development of S_1 slaty cleavage, and suggests closure of the extensional basin. Finally, post-S_1 uplift is associated with southerly-directed thrusting and further folding which modified the pattern, particularly in areas of high strain.

A metapelitic survey of the Manx Group, Isle of Man, used closely spaced samples (1–1.5 per km^2), and compared white mica crystallinity measurements using the Kübler index with those derived from the Weber index (Roberts et al., 1990). Two contoured isocryst maps, histograms and mineral range diagrams, similar to Fig. 3.1, were used to correlate the two techniques. Metapelitic grades range from late diagenetic to epizonal, and the <2-μm mineralogies characteristic of different grades are closely similar to those found in the Skiddaw Group. Although grade over much of the central and eastern

part of the Isle of Man is related to contact effects from granite intrusions, Roberts et al. (1990) also detected an earlier regional metamorphism which was largely anchizonal. K-white mica b cell dimensions (B. Roberts, personal communication) show a similar range to those from the Skiddaw Group and again suggest a low-pressure geotectonic setting.

No systematic studies of metapelitic grade have been carried out in the Borrowdale Volcanic Group because of the general scarcity of suitable lithologies. Although Thomas (1986) recorded a series of K-white mica b cell measurements, the wide scatter of values was attributed to the compositional variability of the tuffaceous lithologies studied. Petrographical studies of secondary mineral assemblages, outside of thermal aureoles, record rare prehnite and occasional pumpellyite, consistent with subgreenschist facies metamorphism (Allen et al., 1987; Fortey et al., 1998).

A reconnaissance study of white mica crystallinity by Thomas (1986) established that anchizonal metamorphism is widespread in the Windermere Supergroup. A more detailed survey of metapelitic grade of the entire basin, currently in progress (Hirons, 1997), suggests that grade ranges from high anchizone in the west, through mid-anchizone in the Kendal area to low anchizone and late diagenetic grades in the east. Some recent studies of specific areas have also been carried out as part of the geological survey of the Lake District. For example, Fortey et al. (1998) recorded a range from lower anchizone to epizone in the Ambleside area, with highest metapelitic grades associated with the steep limb of a regional (Westmorland) monocline. Differences in grade are considered to reflect variation in strain associated with the rapid subsidence, closure and Acadian uplift of the foreland basin (Kneller & Bell, 1993). An intermediate to high-pressure geotectonic setting is indicated by a series of K-white mica b cell measurements, including those of Thomas (1986; $\bar{x} = 9.035$ Å) and Hirons (in Smellie et al., 1996; $\bar{x} = 9.037$ Å). Such a setting is consistent with the proposed development of the foreland basin as a crustal downwarp on the leading edge of the Avalonian microcontinent during the late stages of collision with Laurentia (Kneller et al., 1993). Despite relative low heat flow in the basin, high anchizonal metapelites were extensively developed as a result of locally intense deformation and the growth of strain-related fabrics (Hirons, 1997). K–Ar and Ar–Ar age dating of slaty cleavage in the Windermere Supergroup of the Craven Inliers (Fig. 3.9) shows that it formed between 418 ± 3 and 397 ± 7 Ma, with the younger end of the age range indicating that the probable age of Acadian uplift is late Lower Devonian (Merriman *et al.*, 1995a).

3.3.3 Accretionary settings

3.3.3.1 Southern Uplands of Scotland

The Southern Uplands of Scotland is probably the most intensively studied accretionary terrane world-

wide, with a history of research spanning more than 150 years that involved some of the pioneers of geology. It was here that Lapworth established the importance of graptolite biostratigraphy in 1846, and this allowed Peach and Horne to rapidly map the entire terrane and publish a regional memoir nearly 100 years ago (see Stone, 1995). Despite good biostratigraphical control, the significance of relationships between the stratigraphy and the structure, i.e. the tectonostratigraphy, was not appreciated until way-up techniques were applied to working out younging directions across the terrane. Once the tectonostratigraphical pattern was revealed, McKerrow *et al.* (1977) and Leggett *et al.* (1979) proposed that the terrane represents an accretionary prism. The unusual pattern of accretionary burial metamorphism, described in section 3.2.4, could only be established by integrating metapelitic data with the tectonostratigraphy as shown in Fig. 3.10.

The Southern Uplands terrane is essentially an imbricate thrust belt of Lower Palaeozoic turbiditic sandstones and shales. Outcrop within the terrane

Fig. 3.10 Variations in the mean and range of Kübler indices (KI) plotted on a tectonostratigraphical diagram for the Rhins of Galloway, SW Scotland (from Stone, 1995, Fig. 32, reproduced by permission of the Director, British Geological Survey, © NERC). The diagram represents a generalized cross-strike section from oldest strata (NW) to youngest strata (SE). The age of each fault-bounded tract of turbidite sandstone is indicated in the left-hand column. Metapelitic grade is indicated in the right-hand column.

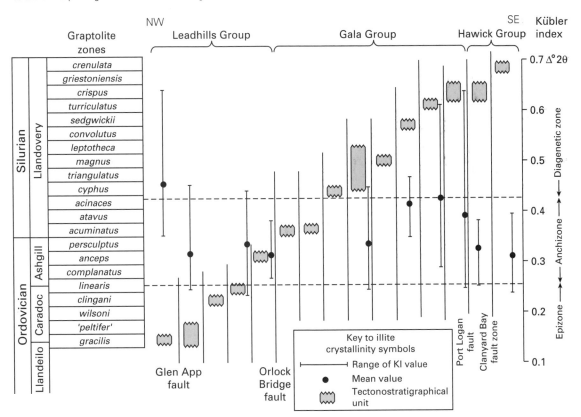

is controlled by strike-parallel faults, initiated as early thrusts, which divide strata into a series of NE–SW trending tracts or tectonostratigraphical units (Fig. 3.10). Within each fault-bounded tract, beds are steeply dipping to vertical and predominantly young towards the north-west. However, graptolite faunas in the shales indicate that, despite the north-westerly younging direction within tracts, overall the tectonostratigraphy becomes sequentially younger towards the south-east (Fig. 3.10). McKerrow *et al.* (1977) and Leggett *et al.* (1979) argued that this tectonostratigraphical pattern developed in an accretionary prism generated by closure of an ancient ocean (Iapetus) separating the active southern margin of Laurentia from Avalonian crust. The prism built up above a north-west-dipping subduction zone as slices of sedimentary cover were sheared off the descending plate, underplated, shortened and rotated to form an imbricate stack. Thus successively younger slices of strata were added to the south-easterly-prograding stack while retaining their north-westerly younging direction.

The earliest studies of very low-grade metamorphism in the Southern Uplands established that the prehnite–pumpellyite facies is widely developed in volcaniclastic greywackes and in less commonly occurring metabasic volcanic rocks (Oliver & Leggett, 1980). Reconnaissance studies of metapelitic grade showed that anchizonal rocks are the most common, and that K-white mica *b* cell dimensions are consistent with medium to high-pressure facies series conditions (Oliver *et al.*, 1984; Kemp *et al.*, 1985). More detailed surveys of metapelitic grade began in the late 1980s as part of the geological resurvey of the Southern Uplands, and used the improved tectonostratigraphy established by the British Geological Survey (e.g. Stone, 1995) for sampling and data interpretation. Metapelitic surveys, which use a sampling density of approximately 1 pelite per $2.5\,km^2$, have continued to be linked to the geological survey and published as metamorphic maps in the marginalia of 1:50000 geological maps (British Geological Survey, 1992a,b; 1993a,d). Metapelitic surveys currently cover just over half of the terrane.

Although early studies suggested that the overall pattern of metamorphism in the Southern Uplands increased from the south-east towards the north-west (e.g. Kemp *et al.*, 1985), more detailed metapelitic

surveys indicated considerable variations in metamorphic trends, and include patterns which are characteristic of accretionary burial, i.e. a trend of grade increasing from older into younger strata (Merriman & Roberts, 1995). A typical contoured metapelitic map resulting from a study of the south-western part of the terrane is shown in Fig. 3.11. Across much of the area isocrysts are generally subparallel to the strike-parallel faults which are commonly the tract boundaries. Metapelitic zone boundaries are commonly terminated by faults and grade may increase or decrease abruptly from tract to tract, indicating a close relationship between grade and the imbrication of the succession. In relation to tectonostratigraphy the distribution of KI data shows that grade increases south-eastwards through the Leadhills Group, then decreases further south-eastwards into younger tracts of the Gala Group, but then increases again into the younger Hawick Group (Fig. 3.10). Hence the northern and southern parts of the terrane display patterns typical of accretionary burial. Concentric patterns of isocrysts typical of contact aureoles are found around granitic intrusions, but these are not displaced across a major sinistral strike-slip fault in the centre of the area (Anderson & Oliver, 1986), indicating postfault emplacement of the granite. Elsewhere in the south-west of the terrane similar patterns show grade increasing into younger strata and suggest that subduction-related metamorphic patterns were generated until at least mid-Llandovery time (Hirons *et al.*, 1997; Barnes, 1998).

The development of a tectonic fabric in the metapelitic rocks is closely associated with changes in grade and mineralogy in this part of the Southern Uplands. Lowest grade late diagenetic pelites are generally uncleaved mudstones and shales, and consist of illite ($2M_1$) and chlorite, with variable subordinate amounts of corrensite, illite/smectite, kaolinite and albite (Merriman *et al.*, 1995b). Anchizonal shales and slates commonly develop a bedding-parallel slaty cleavage, and consist of $2M_1$ phengite and chlorite, with minor albite and quartz; corrensite and paragonite (including K, Na-mica) are absent. In the high anchizone (KI 0.25–0.30) and the epizone, a continuous slaty cleavage is developed, both in the metapelites and associated greywackes. Highest grade epizonal slates contain $2M_1$ phengite and chlorite, with minor albite, quartz and rutile. TEM studies

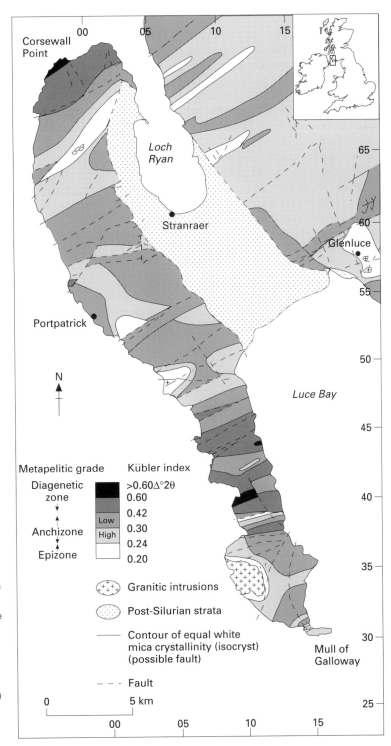

Fig. 3.11 Contoured metamorphic map of the Rhins of Galloway, SW Scotland, based on approximately one pelite sample per 2.5 km² (modified from Stone, 1995, Fig. 31, reproduced by permission of the Director, British Geological Survey, © NERC). The pattern of regional metamorphism shows metapelitic zones approximately subparallel with the (thrust) fault pattern. Around granitic intrusions concentric metapelitic patterns overprint the regional pattern.

of the white mica and chlorite indicate that the thickness of crystallites correlates with the XRD-measured crystallinity indices, and that crystallites thicken with increasing grade (Merriman *et al.*, 1995b). Differences in the prograde increase of white mica and chlorite crystallite thicknesses are attributed to the effects of strain. White mica appears to store strain energy more effectively than chlorite and this enables mica to recover from subgrain development and develop thicker, defect-free crystallites, particularly under anchizonal and epizonal conditions. AEM data show that the white mica is paragonite-poor phengite with b cell dimensions in the range 9.020–9.044 Å (Merriman *et al.*, 1995b), indicative of an intermediate to high-pressure series. Chlorite compositions range from diabantite with intergrown corrensite in the late diagenetic zone, to ripidolite in the epizone. Mineral compositions indicate that metamorphic temperatures range from 150 to 320°C, under an inferred field gradient of <25°C km^{-1}. These conditions suggest that accretionary burial ranged

from 6km in the late diagenetic zone to at least 13km in the epizone.

In the south-western part of the Southern Uplands terrane the overall pattern and conditions of metamorphism are consistent with synchronous burial and metamorphism of strata in an imbricate thrust stack (Merriman & Roberts, 1995, 1996). A tectono-metamorphic model is shown in Fig. 3.12, and has been adapted from Sample and Moore (1987). Less deeply buried diagenetic zone strata, which form an imbricate thrust fan in the upper part of the stack, represent the sediments scraped off the toe of the advancing accretionary prism. More deeply buried strata form a thrust duplex developed beneath the diagenetic zone strata and represent underplated rocks. At both levels the structural geometry of imbrication is similar, and both contain evidence of soft sediment deformation, stratal disruption and cleavage development. However, strata in the more deeply buried underplated duplex are generally at anchizonal grade and show more

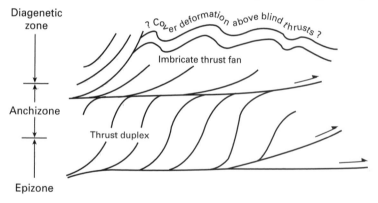

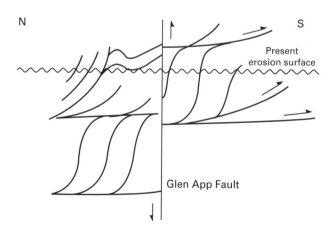

Fig. 3.12 The development of a pattern of accretionary burial metamorphism within an imbricate thrust stack, SW Scotland. (From Stone, 1995, Fig. 33, reproduced by permission of the Director, British Geological Survey, © NERC.)

intense development of slaty cleavage. The age of strata accreted into the upper and lower levels varies according to processes operating at the toe of the prism (e.g. Moore *et al.*, 1991). Consequently, at different stages of accretion older strata at diagenetic grade may form the overburden on younger strata which acquire higher grades, giving rise to a pattern of accretionary burial metamorphism. As the thrust stack was assembled, strata in the thickened rear of the stack were progressively rotated towards the vertical, at the same time introducing a tectonic dip to the metapelitic zones developed during the earlier stages of burial. At the present level of erosion in the Southern Uplands terrane these relationships have been preserved by subsequent reactivation of thrusts as normal faults (Fig. 3.12).

Patterns of accretionary burial can be expected in other subduction complexes but only where there is sufficiently detailed tectonostratigraphy and metapelitic data to allow recognition of diagnostic metamorphic trends. Accretionary burial patterns do not develop where very low-grade metamorphism and imbrication of basinal sequences are not synchronous, as for example in typical fold-and-thrust belts. Where a stratigraphical sequence develops a normal pattern of burial metamorphism prior to contraction and thrusting of the basin fill, the inherited pattern of older rocks showing higher grades than younger rocks is generally retained in very low-grade metapelitic sequences, whatever the structural geometry of the fold-and-thrust belt.

3.3.3.2 *Kodiak Accretionary Complex*

The Kodiak Accretionary Complex is part of the eastern Aleutian arc-trench system, active since the early Jurassic. Components of the accretionary prism include Jurassic blueschist and greenschist terranes, Upper Cretaceous mélange deposits and abundant turbidites, Palaeocene turbidites and volcanic rocks, and Eocene–Oligocene turbidites. Neogene basins overlie some of the accreted Eocene–Oligocene turbidites, while Quaternary sediments are being accreted to the prism at the shelf edge to the modern trench (Moore *et al.*, 1991). About 70% of the Kodiak Accretionary Complex comprises the Kodiak Formation, a slate and greywacke belt of Maastrichtian age which was underplated into the prism in the late Cretaceous over an estimated 12-

Ma timespan (Sample & Moore, 1987). Much of the slate belt consists of coherent, thrust-bounded packets or tracts which are landward-dipping, and about 20% consists of zones of disrupted sandstones with a scaly argillite matrix. These disrupted beds are associated with both pre-accretion deformation and strike-slip faulting which postdates slaty cleavage development.

From observations on the nearly continuous sea cliff exposures on Kodiak and adjacent islands, Sample and Moore (1987) proposed a sequence of structural events which includes early soft-sediment deformation, tectonic stratal disruption followed by thrust faulting and slaty cleavage development. Slaty cleavage is typically steeper than both the bedding and thrust faults, and varies from slaty to phyllitic in the central belt where it locally transposes the bedding fabric. White mica, quartz, carbonate and pyrite are the main metamorphic minerals identified, but prehnite and pumpellyite occur in scattered outcrops. White mica crystallinity measurements were made on 2–6-μm fractions from 'argillite' samples, using Weber's (1972a,b) illite crystallinity technique (Hb_{rel}), together with vitrinite reflectance determinations (Sample & Moore, 1987). The Hb_{rel} values range from 110 to 280, approximately equivalent to the late diagenetic zone through to the epizone, with a mean value (162) in the high anchizone, using the correlation with KI values proposed by Blenkinsop (1988). As discussed in Chapter 2, the use of Weber's technique and other methods of measuring illite crystallinity have been superseded by measurement of the Kübler index (KI) which has the advantage of being inversely related to crystallite thickness (see Fig. 2.19). Vitrinite reflectance values (R_r) are mostly in the range 3.5–4.0 within a mean of 3.73, indicating a burial temperature of 225°C according to Bostick *et al.* (1978). Both the crystallinity and vitrinite data show little evidence of regional cross-strike variation, although no tectonostratigraphy is presented for the 60 km+ transect through the Kodiak Formation (Fig. 3.13).

The structural and mineralogical data indicate that the Kodiak Formation developed by duplex underplating to the base of the accretionary prism (Sample & Moore, 1987, Fig. 14). Based on a geothermal gradient of 20°C km^{-1}, Sample and Moore (1987) estimated a burial depth of at least 10 km, consistent with the mainly anchizonal metapelitic zone in the

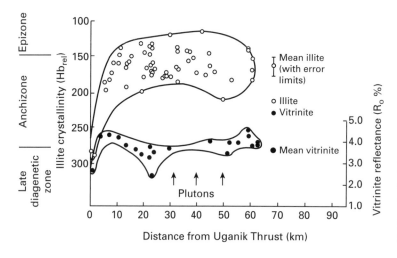

Fig. 3.13 Variation in illite crystallinity (Hb_{rel}) and vitrinite reflectance (R_o%) in a cross strike traverse through the Kodiak Formation, Alaska (from Sample & Moore, 1987, Fig. 14, with permission). Metapelitic zonal limits converted from Hb_{rel} values using Blenkinsop (1988).

slate belt. The upper part of the prism consists of the strata sheared off and imbricated at the toe of the wedge, and this provided a thick overburden which preserved the coherency of the underplated strata. There are strong similarities between the Kodiak Accretionary Complex and the Southern Uplands Terrane (section 3.3.3.1) in terms of tectonic style, metapelitic grade, estimated temperatures and inferred depth of burial. With improvements in the tectonostratigraphy and more metapelitic and organic maturity data, this terrane promises to provide considerable insights into the development of metamorphic patterns in accretionary prisms.

3.3.3.3 Franciscan Complex

For many geologists the Franciscan Complex is probably the world's best known blueschist belt because of the large number of studies that have been published from the complex and their influence on our knowledge of subduction-related processes. Many of these studies are of metabasites occurring as blocks of eclogite, amphibolite and blueschist in mélange deposits. It is only in the past 10–15 years that the shale matrix of mélange deposits and associated metapelitic rocks have attracted the studies they deserve as the most abundant rock type in the complex. Some of these studies suggest anomalous relationships between illite crystallinity and tectonic fabric development, and also infer different physical conditions for some metapelites compared with amphibole-bearing lithologies.

The Franciscan Complex is an assemblage of accretionary terranes extending from southern Oregon to the Coast Ranges of California on the Pacific coast of North America. The dominant rock types are greywacke and shale, and shale-matrix mélange. Large blocks and slabs of mafic volcanic rocks, serpentine, radiolarian chert and limestone commonly occur in mélange deposits, but are volumetrically minor constituents. Three major belts, Coastal, Central and Eastern, are recognized on the basis of age, lithology, structural style and metamorphism, and these are further subdivided into terranes (Jayko et al., 1986). Easterly-dipping subduction of the Franciscan Complex began in the late Jurassic with the accretion and blueschist facies metamorphism of the Eastern Belt (Wakabayashi, 1992). Further accretion of most of the oceanic terranes in the Franciscan occurred in the Cretaceous, and accretion of the Coastal Belt began in the latest Cretaceous–Paleocene and continues at present (Wakabayashi & Unruh, 1995).

Despite their relative abundance, studies of shale and slate are few compared with those describing the high-grade blocks of eclogite, amphibolite and blueschist in the Franciscan Complex. One of the earliest studies was that of Cloos (1983), who compared the shale forming mélange matrix, metashales in high-pressure/low-temperature sequences from the Diablo Range, and low-pressure/low-temperature shales from the Great Valley sequence. The metashales from the mélange matrix and the Diablo Range have similar mineral assemblages consisting of: phengitic white mica + chlorite + albite + titanite + trace amounts of lawsonite, pumpellyite or calcite. Illite crystallinity data (measured in mm) from the mélange matrix and the Diablo Range metashales indicate anchizonal grades, and K-white mica b cell dimensions are generally >9.030 Å.

Underwood *et al.* (1995) studied the thermal maturity of two subterranes along the San Gregorio–San Simeon–Hosgri fault zone, one of the important components of the San Andreas fault system (Fig. 3.14). Bedded greywackes in the Point Sur subterrane locally contain laumontite, and shales (some with 'phacoidal' cleavage) give illite crystallinity (KI) values in the range 0.60–0.80, typical of the late diagenetic zone. Mean random vitrinite reflectance (R_r) values range from 0.8% to 1.4% and indicate palaeotemperatures of 125–180°C. In the southern part of the subterrane R_r increases to 2.9% and KI increases slightly to 0.45. Anomalous high values of R_r (4.8–5.0%) and KI (0.26–0.30) are associated with mélange matrix emplaced as faulted-squeezed diapirs. To the south-east in the Lucia subterrane, R_r values increase to 3.4–4.9% and KIs to 0.18–0.40. Although regional grade is generally higher in the Lucia subterrane (blueschist and prehnite–pumpellyite facies), the higher temperatures indicated by the vitrinite reflectance data suggest a hydrothermal overprinting centred on the Los Burros gold district (Underwood *et al.*, 1995).

XRD, TEM, EMP, fluid inclusion and vitrinite reflectance data were determined from a series of traverses through the Franciscan Complex of the Diablo Range by Dalla Torre *et al.* (1996a). The rocks studied (Fig. 3.14) include metashales forming interbeds in greywacke sandstones, mudstones and shales forming the matrix of various mélange units and metagreywackes. Many of the metashales display a schistosity parallel to bedding, and some show a crenulation cleavage. $2M_1$ K-white mica, chlorite, berthierine, corrensite, mixed-layer chlorite/corrensite, paragonite, kaolinite, lawsonite and minor amphibole and jadeitic pyroxene were identified in the samples. Illite crystallinity values (KI) range from 0.28 to 0.64 $\Delta°2\theta$, i.e. late diagenetic zone to high anchizone, whereas vitrinite reflectance (mean R_r) values are in the range 1.6–2.5%. Detailed TEM and AEM investigations of K-white micas indicate that composition changes from phengitic cores to muscovitic rims in the a–c, b–c, or intermediate

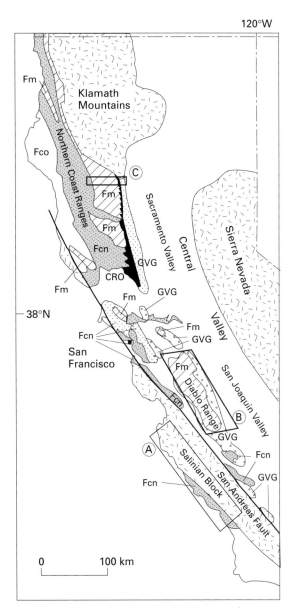

Fig. 3.14 Geological sketch map of the Franciscan Complex, California, showing study areas discussed in the text (modified from Wakabayashi & Unruh, 1995, Fig. 1). A, Underwood *et al.* (1995); B, Dalla Torre *et al.* (1996a,b,c); C, Merriman *et al.* (unpublished data). GVG, Great Valley Group; CRO, Coast Range ophiolite; Fm, blueschist or blueschist–greenschist facies Franciscan rocks; Fcn, Franciscan central belt; Fco, Franciscan coastal belt.

planes. Prograde core-to-rim changes in composition are associated with a reduction in crystal defects and an increase in muscovite at the expense of phengitic K-white mica. The KI data from the Diablo Range are regarded by Dalla Torre *et al.* (1996a) as reflecting a mixture of numerous small phengite coherent X-ray scattering domains and subordinate but larger muscovite domains, and they caution against using KI data for indicating peak metamorphic conditions. Dalla Torre *et al.* (1996c) regarded the vitrinite reflectance and illite crystallinity data as retarded relative to physical conditions deduced from amphibole-bearing shale, which indicate a transition from blueschist to greenschist facies conditions. In a separate study of chlorite textures and compositions, Dalla Torre *et al.* (1996b) suggested that rocks from the Diablo Range experienced a high-pressure/low-temperature metamorphism (>8 kbar, <230°C) followed by lower pressure overprint (5–8 kbar, 230–350°C). Chlorite crystallization is considered to have occurred during the lower-pressure overprint from precursor berthierine and smectite.

TEM and XRD techniques were used (Merriman *et al.*, unpublished data) to study a suite of metapelites from the Mendocino National Forest in the Eastern Franciscan belt (Fig. 3.14). The suite includes schist, slate, shale and mudstone lithologies representing two late Jurassic–early Cretaceous terranes metamorphosed under blueschist facies conditions. The eastern-most Picket Peak terrane largely comprises crenulated mica schist and schistose metagreywacke whereas the structurally underlying Yolla Bolly terrane contains abundant metagreywackes with shale/slate interbeds, and a major mélange unit (Jayko *et al.*, 1986). Metapelitic grade in the mudrock samples generally increases as tectonic fabrics in associated greywackes become more pronounced (Jayko *et al.*, 1986, Table 3), with grade ranging from the late diagenetic zone to the high anchizone (KI 0.26–1.10 Δ°2θ). As previously observed by Dalla Torre *et al.* (1996a), the unpublished studies found that grade is generally lower than expected in view of the tectonic fabrics developed in the metapelitic lithologies; for example, typical epizonal lithologies such as schists and phyllites show anchizonal KI values, and some slates show late diagenetic zone KIs. Although the presence of paragonite, intermediate Na,K-mica and I/S in many samples may have caused 10-Å peak broadening, TEM thickness measurements of white mica crystallites show reasonable correlation with KI values (see Fig. 2.19). Corrensite occurs extensively in mudstones and slates, and

lawsonite is intergrown with the matrix phyllosilicates of some metapelites and is also common in associated greywackes. K-white mica *b* cell parameters are generally large (range 9.01–9.06 Å; mean 9.03 Å), consistent with the very low-field gradients estimated for the terranes (<15°C km^{-1}; Jayko *et al.*, 1986).

3.3.3.4 South Island, New Zealand

The concept of burial metamorphism was developed as a result of a series of regional studies of the collage of late Palaeozoic and Mesozoic accretionary terranes, previously called the New Zealand geosyncline, forming the Eastern Province of South Island, New Zealand (Fig. 3.15). In the southern part of the Eastern Province, a down-section sequence of zeolites and other hydrous Ca–Al-silicates define a series of depth-related zones in volcanigenic greywackes, siltstones and tuffs 9–10-km thick (Coombs, 1954; Coombs *et al.*, 1959). The Eastern Province largely comprises subduction-generated arc, forearc and

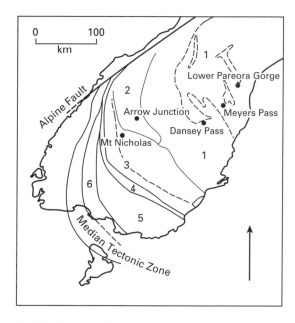

Fig. 3.15 Tectonostratigraphic terranes of Eastern Province, South Island, New Zealand (from Coombs *et al.*, 1996, Fig. 1). The dashed lines indicate limits of Haast Schist overprint (mostly Pmp–Act facies and higher). Eastern Province Terranes: 1, Torlesse; 2, Aspiring; 3, Caples; 4, Dun Mountain-Maitai; 5, Murihiku; 6, Brook Street.

trench terranes that extended along the eastern margin of the Gondwana plate. A major suture, the Median Tectonic Zone, divides the Eastern Province from the Western Province which consists of Palaeozoic metasedimentary and metavolcanic rocks intruded by mid-Palaeozoic to Cretaceous plutons (Coombs & Cox, 1991). The New Zealand Alpine Fault has displaced the Median Tectonic Zone some 460 km by dextral movement mostly during the Cenozoic. Currently the Pacific is overriding the Atlantic plate along the New Zealand Alpine Fault, exposing a narrow belt of amphibolite facies schists along the eastern side of the fault which are possibly derived from depths of 25 km (Cooper, 1980).

An excursion guide by Coombs and Cox (1991) provides brief descriptions of the major terranes in the Eastern Province and summarizes the characteristics of very low and low-grade regional mineral facies. The Torlesse Terrane comprises two subterranes separated by the Esk Head Mélange. The younger Pahau subterrane is late Jurassic to early Cretaceous in age, whereas the Rakaia subterrane is Permian to late Triassic. Both consist of turbidite sandstones with subordinate shales and minor pillow basalts associated with manganiferous and radiolarian cherts. Together they form a complex accretionary prism with thick tracts of strata which generally become sequentially younger towards the north and north-east (Coombs & Cox, 1991, Fig. 1). Grade ranges from zeolite facies to greenschist, and higher grades in the Haast Schists. The Aspiring Terrane is situated between the Torlesse and Caples terranes, and is possibly predominantly oceanic in origin. It comprises mainly pelitic schists and metabasites, with minor metacherts, marble and serpentinite. To the west and south of the Torlesse and Aspiring terranes, the Caples Terrane consists of a heterogeneous marine flysch, with metavolcanic and pelitic rocks. Grade ranges from prehnite–pumpellyite through pumpellyite–actinolite to greenschist, with local development of lawsonite–albite–chlorite assemblages. The Dun Mountain–Maitai Terrane comprises the Dun Mountain ophiolite belt with mélange zones, and thick mostly epiclastic sandstones and siltstones, including thin ash-fall tuffs, of the Maitai belt. Grade ranges from heulandite- and smectite-bearing siltstones at Kaka Point in the southeast of the Terrane, to lawsonite–albite–chlorite assemblages in vertical, strongly cleaved sandstones and slates

in the north-west. The Murihiku Terrane lies to the south-west of the Dun Mountain–Maitai Terrane, with which it is partly in faulted contact. A major asymmetrical syncline, the Southland Syncline, is the dominant structure in the Murihiku Terrane, and exposes some 9 km of Triassic to Jurassic volcanogenic sandstones and siltstone, including many vitric tuffs. Grade is entirely within the zeolite facies. The Brook Street Terrane forms a wedge between the Murihiku Terrane and the Median Tectonic Zone (Coombs & Cox, 1991, Fig. 1), and is essentially part of an accreted volcanic arc. Grade ranges from zeolite to prehnite–pumpellyite facies, with local evidence of high thermal gradient alteration.

Despite the wealth of data derived from studies of very low and low-grade metamorphic mineral assemblages in mostly volcaniclastic rocks, studies of metapelitic assemblages are few and systematic surveys of metapelitic patterns in the Eastern Province are sadly lacking. A reconnaissance study by Kisch (1981) reported illite crystallinity and vitrinite reflectance values from Southland and Otago, and another by Warr (1996) compared mineral facies data with illite and chlorite crystallinity. The zeolite facies rocks in the Dun Mountain–Maitai Terrane at Kaka Point gave 10-Å peaks too broad for valid measurements, reflecting a high content of mixed-layer illite/smectite and illite/chlorite in <2-μm fractions (Kisch, 1981). TEM studies of the Kaka Point siltstones confirmed the presence of illite–chlorite, including $R = 1$ ordered sequences, in part derived from dissolution of smectite (Ahn *et al.*, 1988). Subsequent TEM and AEM studies of the heulandite-rich bentonites show that smectite-rich $R = 0$ I/S is the dominant clay mineral (Li *et al.*, 1997). The low proportion of illite in the I/S and the presence of disordered Fe-rich berthierine suggests that diagenetic transformations in the bentonites are retarded compared with clays in associated siltstones. For example, mainly late diagenetic zone illite crystallinity (KI) values are reported from a suite of siltstone samples representative of the Kaka Point sequences by Jeans *et al.* (1997). Kisch (1981) also obtained vitrinite reflectance values ($R_{r\,oil}$) of 0.44, 0.69 and 1.02% from Kaka Point, consistent with the transition from early to late diagenetic conditions (Fig. 3.4), and in broad agreement with a maturity temperature of 80°C derived from a conodont colour alteration index obtained from the same area (Paull *et al.*, 1996).

In the Torlesse Terrane, both Kisch (1981) and Warr (1996) recorded late diagenetic KI values from mudstones associated with zeolite facies rocks in the Lake Benmore area (see Coombs & Cox, 1991; Stop 1.3), whereas in the Waimate Gorge epizonal values (KI 0.26, 0.28) are associated with prehnite–pumpellyite facies metagreywackes (Coombs & Cox, 1991; Stop 1.1). At Longslip Creek (Coombs & Cox, 1991; Stop 1.5), lineated slates associated with pumpellyite–actinolite facies metagreywackes gave epizonal values (KI 0.24, 0.27), while TEM studies show thick white mica crystallites (mean 939 Å) with highly phengitic compositions and calculated *b* cell dimensions (see section 2.4.1.2) in the range 9.033–9.036 Å (R. J. Merriman, unpublished data). In the Caples Terrane, siltstones associated with prehnite–pumpellyite facies metagreywackes from Balclutha Quarry (Coombs & Cox, 1991; Stop 5.3) gave KI values transitional between the late diagenetic zone and the anchizone (Kisch, 1981). In the Maitai Group at Key Summit (Coombs & Cox, 1991; Stop 3.1), blue-grey silty slates associated with lawsonite–albite–chlorite facies sandstones are epizonal (KI 0.23), and TEM measurements indicate thick white mica crystallites (mean 797 Å; see Fig. 2.8) with highly phengitic compositions and calculated *b* cell dimensions in the range 9.036–9.055 Å (Kisch, 1981; R. J. Merriman, unpublished data). The study by Warr (1996) showed that in the Eastern Province the anchizone, as defined by Kübler (see Kisch, 1990), approximately equates with the prehnite–pumpellyite and the pumpellyite–actinolite facies, the epizone with the greenschist facies, and the late diagenetic zone with the zeolite facies.

3.3.4 Collisional settings

3.3.4.1 Glarus Alps

The Glarus Alps, comprising the eastern part of the Helvetic Alps in Switzerland, is a classical area for the study of very low-grade metamorphism. Since the mid-1960s more than 25 papers have been published on various aspects of this subject from the Glarus Alps. Several pioneering results have derived from these studies including the first published illite crystallinity data from the Alps (Frey, 1969a)

and the recognition (by XRD) of the widespread occurrence of K, Na-mica in metapelites, originally described as a mixed-layer paragonite/phengite (Frey, 1969b).

The Glarus Alps, including their eastern extension to the Rhine River, cover an area of 35 by 45 km in north-east Switzerland (Fig. 3.16). Tectonically the area belongs to the Helvetic zone along the northern margin of the Alps. To the north, nappes of the Helvetic zone override Tertiary clastic sequences of the adjoining Molasse basin, a foredeep which formed during the later stages of the Alpine collision. To the south and east, the Helvetic nappes are overlain by the Penninic nappe system. The Mesozoic sedimentary rocks making up the largest part of the Helvetic zone comprise essentially a carbonate shelf sequence of the northern European margin. In the course of the Alpine collision the Penninic nappes were thrust onto the Helvetic zone. During this process, the rocks of the Helvetic zone were buried, deformed and metamorphosed up to lower greenschist facies conditions. Hunziker *et al.* (1986) dated the main phase of metamorphism in the Glarus Alps at 35–30 Ma, based on concordant K–Ar, ^{40}Ar–^{39}Ar and Rb–Sr illite ages, while a second age group of 25–20 Ma was attributed to movements along the Glarus thrust (see below).

In the region discussed here, the Helvetic zone is subdivided by the Glarus thrust into the Helvetic nappes (above) and the Infrahelvetic complex (below) along a displacement of up to 40 km. The Helvetic nappes represent classic detachment tectonics, whereas the Infrahelvetic complex is a thick-skinned fold-and-thrust belt, cored by the Aar massif basement uplift. Four deformation phases are distinguished in the Glarus Alps (Schmid, 1975; Milnes & Pfiffner, 1977). During the first deformation phase (= Pizol phase) initial emplacement of exotic strip sheets (Sardona and Blattengrat units) and detachment of the North-Helvetic flysch took place. The second deformation phase (= Cavistrau phase) caused large-scale recumbent folds in a small area of the Infrahelvetic complex located in the southern part of the Glarus Alps. Most of the pelitic and marly rocks in the Glarus Alps show a more or less pronounced slaty cleavage which was generated during the third deformation phase (= Calanda phase). This is the main phase of ductile deformation with thrusting and accompanying folding in the Helvetic nappes

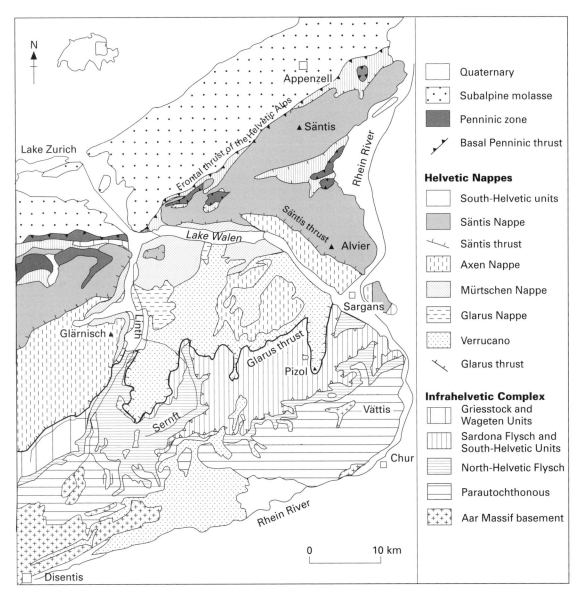

Fig. 3.16 Tectonic map of the Helvetic Alps of north-east Switzerland.

and the Infrahelvetic complex. The fourth phase of deformation (= Ruchi phase) is responsible for a crenulation cleavage related to movement on the Glarus thrust and created an inverted metamorphic pattern (see p. 94). The stratigraphic sequence of the Helvetic zone includes Permian conglomerates, Triassic dolomites, various thick limestones of Jurassic and Cretaceous age and Tertiary sandstones inter-bedded with shaly and marly horizons. Several Penninic klippes, resting on the basal Penninic thrust, are emplaced above the South-Helvetic units and the Helvetic Säntis nappe (Fig. 3.16). The South-Helvetic units comprise mainly limestones and marls of Upper Cretaceous to Eocene age, whereas the Penninic klippes consist mainly of Cretaceous North-Penninic flysch. For more detailed information on the geological evolution of the Helvetic zone, the reader is referred to Trümpy (1980), Pfiffner (1986) and Pfiffner *et al.* (1990).

The very low-grade metamorphism of the Helvetic zone of eastern Switzerland has been studied by many authors, using mainly the illite crystallinity technique and index minerals, supplemented by vitrinite reflectance and fluid inclusion data. Most recently, the results of HRTEM and AEM studies have also become available. Stable isotope data from the Glarus Alps are discussed in Chapter 8.

Approximately 2000 KI data have been produced from the Glarus Alps and adjacent areas, but many of these data are still unpublished. Detailed KI data have been published for the North-Helvetic flysch around the Linth and Sernft valleys (Rahn *et al.*, 1995, Fig. 3) and for the Helvetic zone between Appenzell and Chur (Wang *et al.*, 1996, Fig. 2); see Fig. 3.16 for location. In both these studies the approximate boundaries of the anchizone were mapped (Fig. 3.17), but contouring of illite crystallinity data (see section 3.2.2) has not been used in the Glarus Alps or anywhere in the Helvetic Alps. Contouring of data is difficult in regions of pronounced topographic relief, particularly where rapid lithological variations in the stratigraphical sequences are combined with a complex tectonic history which includes polyphase deformation (see above). In spite of these difficulties, patterns of distribution of KI data can be recognized and used to interpret the tectonometamorphic history of the region. The patterns show: (i) that within a single tectonic unit KI values decrease, hence grade increases, from the external part towards the internal part, i.e. from the north-west towards the south-east; (ii) within the stack of Helvetic nappes, KI values decrease downwards. This has been demonstrated, for example, for the area of the Glärnisch massif by Frey *et al.* (1973, Fig. 6) where KI data were collected from four different nappes over a vertical distance of 2000 m, corresponding to a structural relief of 3000 m (Fig. 3.16). Here the transition from the late diagenetic zone to the anchizone crosscuts nappe boundaries and is inclined 25–30° towards the north. This suggests that the pattern of incipient metamorphism was established after the nappes were assembled into a stack; (iii) An inverted metamorphic pattern is present between the epimetamorphic Verrucano nappe, the lowest Helvetic nappe of the Glarus Alps, and midanchizonal flysch units of the Infrahelvetic complex, separated by the Glarus thrust (Frey, 1988;

see also Árkai *et al.*, 1997). The amount of post-metamorphic thrusting along the Glarus thrust has been estimated to be 5–10 km based on the northward displacement of a specific isocryst in the Helvetic nappes with respect to the same isocryst in the Infrahelvetic complex (Rahn *et al.*, 1995; Wang *et al.*, 1996).

Various index minerals including kaolinite, pyrophyllite, K,Na-mica and paragonite have been identified in metapelites and metamarls of the Glarus Alps based on detailed XRD analysis. Kaolinite occurs in the uppermost Helvetic Säntis nappe whilst pyrophyllite is found in lower Helvetic nappes, and an isograd pertaining to the reaction Kln + Qtz = Prl + H_2O has been mapped (Frey, 1987b; Wang *et al.*, 1996, Fig. 6). This isograd crosses the Säntis thrust and, in terms of illite crystallinity data, is located approximately at the transition from the late diagenetic zone to the anchizone (Fig. 3.17). K, Na-mica and paragonite are widely distributed in the Glarus Alps and first appear at the beginning of the anchizone.

TEM and EMPA techniques were used by Livi *et al.* (1997) to study a suite of Liassic slates from the Glarus Alps. In addition to the index minerals mentioned above, brammalite, a Na-mica with a deficiency in interlayer cations, and the aluminous chlorite mineral sudoite were detected as neoformed minerals in the anchizone. Brammalite may be regarded as a precursor of paragonite and may also be mixed with K-rich mica on the nanometric scale. The paragonite-forming reaction must involve the reaction of a Na-source (feldspar or fluid) + illite/smectite + possibly a high-aluminium phase (kaolinite, pyrophyllite, aluminous chlorite).

Besides metapelites and metamarls, two other lithologies have provided useful information on metamorphic grade in the Glarus Alps:

1 In glauconite-bearing limestones of Cretaceous and Tertiary age, Frey *et al.* (1973) and Wang *et al.* (1996, Fig. 7) mapped two isograds. The stilpnomelane isograd occurs in the low anchizone (Fig. 3.17) and corresponds to the reaction Glt + Qtz ± Chl = Stp + Kfs + H_2O + O_2. The biotite isograd occurs close to the anchizone/epizone boundary and corresponds to the reaction Stp + Chl + Kfs = Bt + Qtz + H_2O (Brown, 1975, page 269).

2 In the Lower Oligocene Taveyanne greywacke of the Glarus Alps, a range of zeolite, prehnite–

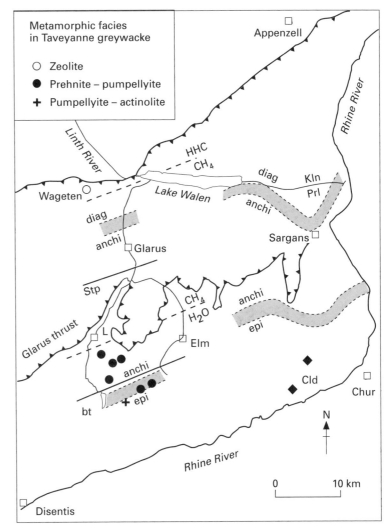

Fig. 3.17 Metamorphic map based on a synthesis of data from the Glarus Alps and adjacent areas. The late diagenetic zone/anchizone and anchizone/epizone boundaries are shown as broad bands. Pyrophyllite (Prl), stilpnomelane (Stp) and biotite (Bt) isograds are indicated as continuous lines. HHC/CH$_4$ and CH$_4$/H$_2$O limits based on fluid inclusion data, are shown as dashed lines. Localities where the Taveyanne greywackes contain diagnostic metamorphic mineral assemblages are indicated. Two chloritoid (Cld) occurrences west of Chur are also shown. Inset Fig. 3.16 shows location of map.

pumpellyite and pumpellyite–actinolite assemblages has been recognized (Rahn *et al.*, 1994, and references therein). The restricted distribution of metagreywacke prevented their use for a region-wide study of very low-grade metamorphic patterns, but in places resulted in correlation between metapelitic zones and mineral facies (Rahn *et al.*, 1994, 1995). Zeolite facies rocks are present in the Wageten slice at the northern Alpine border (Fig. 3.17) where KI data from adjacent marly shales indicate late diagenetic conditions, associated with a single vitrinite reflectance value of 1.8% R_{max}. Prehnite–pumpellyite facies rocks are widely distributed in Taveyanne metagreywackes of the North-Helvetic flysch, e.g.

south-east of Linthal (Fig. 3.17). Here, KI data indicate that middle to high anchizone conditions are associated with vitrinite reflectance values in the range 5.4–>8% R_{max}. Pumpellyite–actinolite facies assemblages were found in a single outcrop at Crap Ner, south of prehnite–pumpellyite facies occurrences. At this locality, a high anchizonal KI value and a vitrinite reflectance value of 8.6% R_{max} are recorded.

Fluid inclusion data have been particularly valuable in the investigation of very low-grade metamorphism of the Helvetic Alps. Mullis (1979, 1987) proposed a threefold zonation based on microthermometry in quartz veins, i.e. a fluid zone of higher hydrocarbons

(HHC) at $T < 200°C$, a methane zone at $T = 200$–$270°C$ and a water zone at $T > 270°C$. Results for the Linth and Sernft river area were compiled by Rahn *et al.* (1995, Fig. 4). The boundary between the HHC zone and the CH_4 zone is located in the Upper Helvetic nappes near the western end of Lake Walen (Fig. 3.17), with the zeolite facies locality of the Wageten slice belonging to the HHC zone but very close to the HHC/CH_4 boundary. The boundary between the CH_4 zone and the H_2O zone is located in the North-Helvetic flysch of the Infrahelvetic complex. This boundary was mapped near the villages of Linthal and Matt (Figs 3.16 and 3.17), and thus is oriented in a ENE–WSW direction, corresponding to the iso-reflection line of *c.* 6% R_{max}, and is parallel to the border of the Helvetic nappes, i.e. the Alpine border in this area. This suggests that the prehnite–pumpellyite facies occurrences south-east of Linthal belong, with one exception, to the H_2O zone.

The physical conditions responsible for the very low-grade metamorphism of the Glarus Alps have been derived by a variety of methods including phase equilibria, calcite–dolomite, chlorite, coal rank and stable isotope thermometry, fluid inclusion thermobarometry and K-white mica *b* cell dimensions. Metamorphic conditions at the pyrophyllite isograd were estimated from vitrinite reflectance and fluid inclusion data to be 240–260°C and 2.1 kbar at a water activity of 0.6–0.8 (Frey, 1987b). Based on new fluid inclusion data, however, a more recent temperature estimate for the onset of the anchizone is at *c.* 230°C (Mullis *et al.*, 1993). The fluid composition at the pyrophyllite isograd would then have to be either very CH_4- or CO_2-rich ($X_{CH_4} = 0.95$ or $X_{CO_2} = 0.999$; Livi *et al.*, 1997). *P–T* data for three localities or areas will be discussed.

1 The Wageten slice at the northern Alpine border may be considered as a Helvetic nappe unit situated above the Glarus thrust plane. Temperature values have been reported for two Taveyanne metagreywacke samples by Rahn *et al.* (1994) using chlorite thermometry (225 ± 27°C and 224 ± 37°C) and fluid inclusion thermometry (179 ± 19°C and 178 ± 19°C). A pressure of 1.3–1.5 kbar was derived from temperature conditions of 170–190°C, assuming a geothermal gradient of 30°C km^{-1}.

2 The Guggenegg locality belongs to the Axen nappe, the second uppermost Helvetic nappe in the Glarus Alps. Temperature estimates are available for one Liassic slate (Livi *et al.*, 1997a, sample MF647). For the assemblage Prl–Sud–Cal–Dol–Qtz–Gr phase equilibrium calculations indicate 300–340°C at an assumed pressure of 2 kbar. Temperature estimates from calcite–dolomite pairs range from 313 to 351°C, depending on the geothermometer.

3 The North-Helvetic flysch belongs to the Infrahelvetic complex. *P–T* data for eight Taveyanne metagreywacke samples from an area of 15 km^2 south-east of Linthal were provided by Rahn *et al.* (1994). Temperatures derived from chlorite Al^{iv} contents range from 275 ± 25 to 308 ± 32°C. Fluid inclusion data yielded a homogenization temperature of 247 ± 20°C for one sample and formation temperatures > 270°C for seven samples. Fluid inclusion homogenization temperatures can be derived from isochore constructions to estimate pressures with the help of the above temperature values. One sample was suitable for this purpose. For a temperature of 270°C, deduced fluid pressures of 2–2.4 kbar suggest an overburden of 8–9.6 km. Assuming a rock density of 2.5 g cm^{-3} and $P_{fluid} = P_{lith}$, the resulting geothermal gradient is 27–32°C km^{-1}. For 310°C the fluid pressure ranges between 2.4 and 3 kbar, yielding an overburden thickness of 10–12 km and a geothermal gradient of 24–29°C km^{-1}. These pressure estimates are consistent with K-white mica *b* cell dimensions of 9.0133 ± 0.0065 Å from a profile close to Elm (Frey, 1988), indicative of a medium pressure type and typical of an alpine collisional geotectonic setting (Table 3.1).

A synthesis of the data presented above, and those of other studies, suggests that the Glarus Alps evolved through a series of tectonometamorphic events. Sedimentation ended in the Lower Eocene (Helvetic nappes) or Early Oligocene (Infrahelvetic complex). This was rapidly followed by crustal shortening and deformation associated with the development of a southward-dipping subduction zone (Milnes & Pfiffner, 1977), and generated a regional pattern of metamorphic grade increasing from north to south both in the Helvetic nappes and in the Infrahelvetic complex. Radiometric dating indicates that the main phase of orogenic metamorphism occurred between 35 and 30 Ma (Hunziker *et al.*, 1986), during the Calanda phase of movements. In the Helvetic nappes, both the transition from the late diagenetic zone to the anchizone and the pyrophyllite isograd crosscut nappe boundaries, indicating that nappe transport mostly took place before regional metamorphism. In the Parautochthonous components of the Infrahelvetic complex, stilpnomelane and chloritoid neoblasts overgrow the schistosity generated by the Calanda movement phase, indicating that the metamorphism postdates the main deformation. Following the main internal deformation within the Helvetic nappes, they were transported passively northwards along the Glarus thrust giving rise to inverted metamorphic patterns. Local metamorphic events attributed to movements on the Glarus thrust have been radiometrically dated at between 25 and 20 Ma, suggesting that they are associated with the Ruchi phase of movement. Late movements on the thrust, however, continued until at least 9 Ma ago, as indicated by apatite fission track data (Rahn *et al.*, 1997).

3.3.4.2 Gaspé Peninsula

The Gaspé Peninsula forms the northern extension of the Appalachians into Quebec, eastern Canada, where two major tectonometamorphic events have generated low and very low-grade metamorphic sequences in Lower Palaeozoic rocks. A metapelitic study, probably the largest single survey of grade in terms of area, shows that a regional burial pattern is well preserved across much of the Peninsula.

During the mid-Ordovician Taconian collision, continental margin basins were thrust westwards over autochthonous Lower Palaeozoic shelf sequences on the Grenville basement of the Canadian Shield. The imbricated fold-and-thrust belt comprises the External Domain of the Taconian orogen (Pickering, 1987; St Julien & Hubert, 1975). According to Pinet and Tremblay (1995), the collision involved an island-arc terrane obducted over the continental margin, although the location of the (Iapetus) suture between the two terranes is problematical (Church *et al.*, 1996). The Acadian event in the late Lower Devonian inverted Middle Ordovician to Lower Devonian basinal sequences which had accumulated across the eroded fold-and-thrust belt, although in some cases basinal sequences were coeval with undeformed sequences in the Taconian belt. Acadian tectonics are characterized by major open folds and block faults. Sequences involved in the Acadian event rest with angular unconformity on the Taconian Belt in

the north-eastern part of the peninsular, but are in faulted contact in the west (Hesse & Dalton, 1991).

A metapelitic survey of the Gaspé Peninsula determined illite crystallinity from 2200 samples covering some 27 000 km^2, representing an approximate sampling density of 8 per 100 km^2 (Hesse & Dalton, 1991). White mica KI data, composition (from XRD 060 peak spacing), polytypes and chlorite composition (from XRD 001 and 060 peak spacings), are compared with vitrinite reflectance (R_r%) data from localized occurrences of asphaltic bitumen and conodont alteration indices (CAI). A contoured illite crystallinity map shows a general conformity between the major structural trends and isocrysts, with grade regionally increasing along strike from east to west. Grade ranges from late diagenetic (KI > 0.60) to epizonal, but late diagenetic and low anchizonal metapelites are predominant. The oldest rocks in the cores of anticlines are associated with the highest grades (KI ≤ 0.30), and both the metamorphic map and cross-sections are consistent with a depth-related pattern of burial metamorphism (Hesse & Dalton, 1991, Figs 8 and 9). Within the Acadian belt, normal, postmetamorphic faulting tends to preserve the burial pattern, i.e. down-faulted blocks juxtapose low grades against higher grades. Metamorphic inversion patterns, however, appear to be associated with thrusting in the northern parts of the Taconian belt.

3.3.4.3 Scandinavian Caledonides

The external foreland fold-and-thrust belt developed along the eastern margin of the Scandinavian Caledonides has attracted several studies of relationships between patterns of very low-grade metamorphism and thrust tectonics. These have involved illite and chlorite crystallinity, K-white mica *b* cell dimension and vitrinite reflectance measurements, combined with stratigraphic and structural mapping. Several of these studies have identified inverted metamorphic patterns, but widely differing thermal histories are invoked to explain the patterns.

In Scandinavia, a succession of nappes involving Neoproterozoic to Silurian strata were thrust eastwards and south-eastwards over the Baltic Shield during the Caledonian Orogeny (Fig. 3.18). Nappe emplacement occurred chiefly in the late Cambrian–

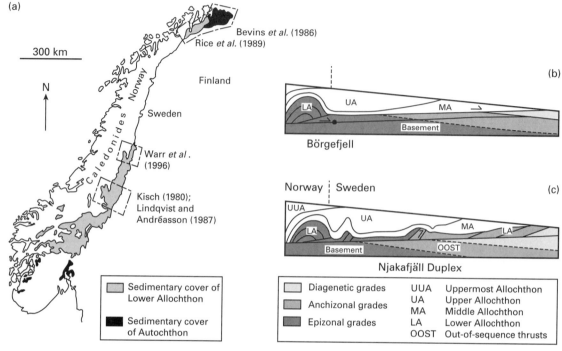

(a)

300 km

N

Finland

Sweden

Bevins *et al.* (1986)
Rice *et al.* (1989)

Warr *et al*.
(1996)

Kisch (1980);
Lindqvist and
Andréasson (1987)

| | Sedimentary cover of Lower Allochthon |
| | Sedimentary cover of Autochthon |

(b)

UA
LA
MA
Basement

Börgefjell

Norway ┆ Sweden   (c)

UUA
UA
LA   MA   LA
Basement   OOST

Njakafjäll Duplex

	Diagenetic grades	UUA	Uppermost Allochthon
	Anchizonal grades	UA	Upper Allochthon
		MA	Middle Allochthon
	Epizonal grades	LA	Lower Allochthon
		OOST	Out-of-sequence thrusts

Fig. 3.18 A. Location of studies of very low-grade metamorphism in the fold-and-thrust belts of the Scandinavian Caledonides (modified from Warr *et al.*, 1996, Fig. 1); B, tectonometamorphic model of the central Swedish fold-and-thrust belt showing very low-grade cover sequences of the Autochthon buried beneath the thrust wedge of Middle and Upper Allochthons. Postmetamorphic imbrication/accretion of the Lower Allochthon causes uplift around the Börgefjell area; C, migration of imbrication towards the toe of the thrust wedge and late out-of-sequence thrusting produced inverted metamorphic patterns. (Modified from Warr *et al.*, 1996, Fig. 13, with permission.)

early Ordovician (Finnmarkian) and middle to late Silurian (Scandian) phases of the orogeny. The succession of nappes comprises four major allochthons (Roberts & Gee, 1985): Uppermost Allochthon; Upper Allochthon; Middle Allochthon; Lower Allochthon. These allochthonous nappes were emplaced on a foreland consisting of crystalline basement overlain by Neoproterozoic to Early Palaeozoic cover sequences, together forming the Autochthon. Metapelitic studies have concentrated chiefly on the Autochthon and Lower Allochthon.

The earliest study was that of Kisch (1980) in central Jämtland, western Sweden (Fig. 3.18). He measured illite crystallinity and vitrinite reflectance from a suite of samples representing Lower Palaeozoic sequences in the upper two units of the Jämtland Supergroup. The sequences represent units of the Autochthon and the Lower Allochthon and include the Middle Cambrian to Upper Ordovician Tåsjön Group and the Silurian (Llandovery–Wenlock) Änge Group. Kisch (1980) found that grade generally increased from east to west, from the autochthon into the allochthon. Lowest grades occur in the autochthonous sequences where late diagenetic to low anchizonal metapelites are associated with vitrinite reflectance (R_{max}) values of 1.4–2.9%. Epizonal grades are associated with vitrinite reflectance values of 4.2–4.5% in the Lower Allochthon, extending to the west of the Offerdal Nappe of the Middle Allochthon. Although his study established the usefulness of illite crystallinity in the outer zones of thrust belts, Kisch (1980) was equivocal about the influence of thrusting on the pattern of very low-grade metamorphism and recognized that more detailed studies were needed.

One such study was that of Lindqvist and Andréasson (1987), who made detailed measurements of illite crystallinity and mica phengite content in the Röde Sandstone forming the Alsen Klippe, in the centre of the area studied by Kisch (1980). The Alsen Klippe preserves a section through the base of the Offerdal Nappe of the Middle Allochthon overlaying the Röde Sandstone of the Lower Allochthon. Fabrics observed in the sandstone are progressively transformed over a vertical distance of 80 m beneath the thrust plane, changing from a discontinuous irregular cleavage, to a spaced pressure solution cleavage which develops crenulations before it is eventually transposed into a mylonitic schist with biotite 'fishes' in the shear zone forming the base of the thrust. Lindqvist and Andréasson (1987) showed that up to the base of the mylonite, KI values of $2M_1$ K-white mica in <2-µm fractions are mainly within the low anchizone, but these change to those of the high anchizone and epizone in the mylonites in the base of the thrust. KI values (from <6-µm fractions) are generally lower in the shear zone along the base of the thrust, although this is attributed to the influence of biotite on 10-Å peak broadening. The inverted pattern was not considered to be consistent with heating from below, and downward heat flow from the nappe was also discounted by Lindqvist and Andréasson (1987). They concluded that local increases in temperature were unlikely and that prograde recrystallization was largely the result of storage and release of strain energy (i.e. annealing), aided by enhanced fluid transport within and adjacent to the thrust zone.

Different conclusions were reached by Bevins *et al.* (1986) and Rice *et al.* (1989) from a study of the Middle Allochthon, Lower Allochthon and Autochthon in Finnmark (Fig. 3.18a). Data from 203 illite crystallinity analyses and 75 K-white mica *b* dimensions were used to examine relationships between the emplacement of the Middle Allochthon and low-grade metamorphism in the Lower Allochthon and Autochthon during the Scandian phase of the Caledonian orogeny. Metamorphic grade in the Kalak Nappe Complex (Middle Allochthon) is inferred to have been at least in the middle greenschist facies when emplaced over the Gaissa Thrust Belt of the Lower Allochthon. Within the Gaissa Thrust Belt KI data indicate a metamorphic inversion pattern characterized by high anchizonal-low epizonal conditions near the roof thrust and late diagenetic-low anchizonal conditions in the vicinity of the floor thrust. K-white mica *b* dimensions increase

with metapelitic grade from low-pressure type near the floor to intermediate/high-pressure type near the roof of the Gaissa Thrust Belt. A metamorphic inversion pattern is also developed, generally at lower grades, between the Gaissa Thrust Belt and the underlying Autochthon, with highest grades (low anchizone) occurring close to the regional sole thrust. Rice *et al.* (1989) concluded that the inversion pattern in the Gaissa Thrust Belt and three related thrust stacks, all part of the Lower Allochthon, resulted from emplacement of the 'hot' Middle Allochthon. A downward heat flow is inferred by 'contact' metamorphism from the overthrust hot nappe. In the roof of the Gaissa Thrust Belt peak metamorphism and deformation of strata can be correlated with pre- to synthrusting, whereas these events in the floor area are considered to be syn- to postthrusting.

Following an investigation of the Vilhelmina area of central Sweden (Fig. 3.18a), Warr *et al.* (1996) discounted heating from hot nappe emplacement and shear heating as likely causes of low-grade metamorphism in this relatively flat region of the fold-and-thrust belt. The investigation combined a clay crystallinity survey with detailed structural mapping of Neoproterozoic to Ordovician cover sequences of the Autochthon and the Lower Allochthon. A computer contoured metamorphic map of the area shows that metapelitic grade increases from the late diagenetic zone in the undeformed Autochthon to the east, to epizonal conditions in the north-west associated with the internal thrust sheets of the Lower Allochthon. Warr *et al.* (1996) found that the sole thrust, along which the Allochthon was placed on the Autochthon, separates diagenetic grade pelites in the footwall from anchizonal metapelites in the hangingwall. Tectonic microfabrics are poorly developed or absent in the autochthonous pelites, but a well-developed continuous slaty cleavage fabric is found in the high anchizonal and epizonal rocks. No correlation, however, was found between strain determinations and KI data. Measurement of K-white mica *b* dimensions from 30 samples suggests an intermediate to high-pressure facies type consistent with low palaeogeothermal gradients and thrust-stacking.

Warr *et al.* (1996) proposed a four-stage tectono-metamorphic history for the Scandian phase of external fold-and-thrust belt development in central

Sweden. The initial stage involved emplacement of a wedge-shaped stack of thrusts consisting of the Middle, Upper and Uppermost Allochthons. These units are considered to have cooled prior to their emplacement over the external cover rocks which were progressively buried and deformed under very low-grade metamorphic conditions. Modelling of the external profile of the thrust wedge suggested that no inverted temperature gradients developed beneath the wedge, so that the regional pattern of isotherms was undisturbed and dipped parallel to the wedge surface. The second stage was characterized by postmetamorphic accretion of the very low-grade metasedimentary cover into the base of the thrust wedge to form the Lower Allochthon, and was associated with uplift in the Börgefjell area (Fig. 3.18b). Stage three involved further accretion into the toe of the wedge as the imbrication migrated towards the foreland. This resulted in emplacement of internally derived higher grade rocks over the lower grade strata of the external cover sequences,

producing typical inverted metamorphic patterns (Fig. 3.18c). The final stage, characterized by low strain out-of-sequence thrusting, also resulted in emplacement of higher grade over lower grade rocks.

3.3.4.4 Bükkium of north-east Hungary

Over the past 25 years more than 30 original papers and reviews have been published on Alpine and Variscan regional metamorphism of the Carpatho-Pannonian terranes of Hungary (Árkai, 1995). Many of these studies have investigated low and very low-grade metamorphism in the Bükkium of north-east Hungary, forming the innermost part of the western Carpathians (Fig. 3.19a). The Bükkium is part of the Pelso Superunit, bounded by the Western Carpathian Superunit to the north, and to the south by the Tisza Superunit comprising the Central Hungarian Autochthonous Unit and the South Hungarian Nappe Zone. Rocks of both the Pelso and Tisza Superunits provided a basement for

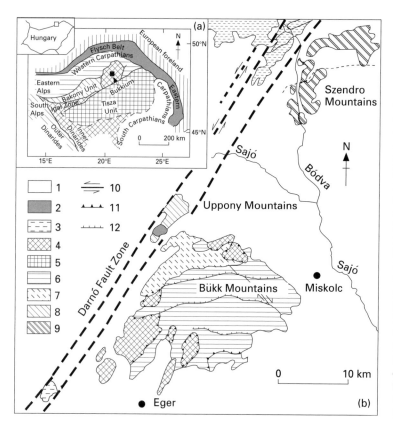

Fig. 3.19 (a) The Carpatho-Pannonian geotectonic framework of Hungary. (b) The Bükkium of north-east Hungary (after Árkai et al., 1996, with permission, Fig. 1). 1, Tertiary and Quaternary; 2, Gosau-type Senonian conglomerate (postmetamorphic); 3, South Gemer Unit (U. Permian-Mesozoic), diagenetic zone; 4–7, Bükk Mountains; 4, Szarvaskö-Mónosbél Nappe (mostly Jurassic), diagenetic and partly low anchizonal; 5, Kisfennsík (Little Plateau) Nappe, diagenetic; 6, Fennsík (Bükk Plateau) Parautochthon (Triassic and Jurassic), mainly anchizonal, locally epizonal; 7, Fennsík Parautochthon (M. Carboniferous-Permian), mostly anchizonal, partly epizonal; 8, Uppony Palaeozoic (U. Ordovician?-M.Carboniferous), transitional anchizone-epizone; 9, Szendrö Palaeozoic (M. Devonian-M. Carboniferous), epizonal; 10, strike-slip fault; 11, nappe boundary; 12, major thrusts within nappe.

the accumulation of Tertiary and Quaternary strata in the Pannonian Basin system which developed during Miocene subduction of the European plate beneath the Eastern Carpathians (Peresson & Decker, 1997). The Bükkium comprises three inliers forming upland regions within the Pannonian Basin system (Fig. 3.19b). In the Szendrö Mountains strata range in age from Lower Devonian (Emsian) to Middle Carboniferous (Bashkirian), and consist of metapelites and psammites with metalimestone intercalations (Árkai, 1983). The Uppony Mountains contain strata ranging in age from Upper Ordovician to Middle Carboniferous, and include metapelites, psammites and metalimestones, with minor amounts of metachert, metatuff and metabasic volcanic rock (Árkai *et al.*, 1981). By far the most extensive of the three inliers, the Bükk Mountains comprises the Fennsík Parautochthon, the Szarvaskö-Mónosbél Nappe and the Kisfennsík Nappe (Lelkes-Felvári *et al.*, 1996). The Fennsík Parautochthon consists of Middle Carboniferous to Upper Jurassic meta-sedimentary rocks including evaporites, radiolarian cherts, turbidites and two Triassic volcanic horizons. A Lower-Middle Jurassic magmatic complex with ophiolitic affinities characterizes the lower part of the Szarvaskö-Mónosbél Nappe, whereas the upper part consists of an Upper Jurassic olistolith. Upper Triassic limestones and intermediate metavolcanic rocks characterize the Kisfennsík Nappe.

In a series of metamorphic studies involving over 900 samples, mineral assemblage and textural data, illite and chlorite crystallinity (XRD and TEM data), K-white mica *b* cell dimensions, graphite ordering and vitrinite reflectance have been used to characterize regional grade in the Bükkium (Árkai, 1973, 1977, 1983, 1991, 1995; Árkai *et al.*, 1981; 1995). Illite crystallinity studies in the Szendrö Mountains indicate that metapelitic grade is mostly in the epizone, with greenschist facies assemblages of the chlorite zone and locally the biotite zone developed in associated metatuffs. Dispersed organic material is in the graphite-d_1 state (slightly disordered; Landis, 1971), consistent with greenschist facies conditions. K-white mica *b* cell dimensions are in the range 8.990–9.008 Å and characteristic of a low to intermediate pressure facies series, with no significant difference between the northern and southern units of the area (Árkai, 1983). Chlorite crystallinity values are low, and typical of the epizone (Árkai, 1991). In the Uppony Mountains, regional metamorphic grade is generally lower than in the Szendrö area, with average illite crystallinity values corresponding to the anchizone–epizone boundary (Árkai *et al.*, 1981). Chloritoid occurs in some metasandstones, and metacherts contain graphite-d_2, anthracite and metabituminite. K-white mica *b* cell dimensions are generally in the range 8.993–9.026 Å (mean = 8.996 Å) and characteristic of a low-pressure facies series. In the Bükk Mountains grade generally increases from the south to the north, from the late diagenetic zone to the epizone, although considerable variation is found in relation to the tectonostratigraphy (Árkai, 1983). K-white mica *b* cell dimensions are also variable, with those in the Fennsík Parautochthon indicating medium pressures, the Bükk Mesozoic rocks indicating low-pressure conditions and all others rocks indicating transitional low-to-medium-pressure conditions.

K–Ar dating of white mica from high anchizonal and epizonal rocks suggest that the regional metamorphism culminated between the Late Jurassic-Early Cretaceous (160–120 Ma) 'eo-Hellenic' phase of subduction of the Dinarides, and the mid-Cretaceous (100–95 Ma) Austrian phase of crustal thickening in the Inner Carpathians (Árkai *et al.*, 1995). Apatite fission track analysis indicates that the Bükk Mountains were exhumed less than 40 Ma ago after Tertiary burial beneath an overburden thickness of approximately 1 km.

3.3.5 High strain zones

In some very low-grade metamorphic terranes, high strain rates along faults and shear zones are characterized by ductile and ductile/brittle fabrics and the development of foliated rocks as greenschist facies conditions are locally approached (Barker, 1990). Although a number of studies have examined relationships between tectonic strain, slaty cleavage development and illite crystallinity, as reviewed by Frey (1987a, pp. 18–20), until recently few have specifically studied high strain zones. In the past 5 years, however, a number of studies have used clay crystallinity to examine the effects of shear zone tectonics and patterns of strain-related metamorphism developed in and around high strain zones. These studies suggest that two types of metapelitic pattern are developed in the vicinity of high strain zones.

High-level shear zones and thrust planes typically associated with thin-skinned thrusting may retain evidence of localized strain-induced reduction of phyllosilicate crystallite size. Kinked and mottled phyllosilicate crystals are typical of the polygonization microtextures retained in some high level shear zones. In contrast, where the shear zone was originally deeply buried, strain-related reduction in crystallite size appears to have been recovered and possibly enhanced by postshearing recrystallization. Examples of both types are described below.

The Moniaive shear zone. This is a major zone of ductile deformation in the Southern Uplands accretionary terrane of SW Scotland (section 3.3.3.1). Development of the zone was initiated in the early Wenlock (Silurian) above a basement discontinuity, and ended with granite intrusion at about 392 Ma (Stone *et al.*, 1997). The zone, up to 5 km wide, comprises several anastomosing subzones of high strain enclosing lenticular domains of relatively low strain (Phillips *et al.*, 1995). Although the shear zone has been traced some 90 km along strike, the widest and best characterized section lies between Moniaive and the intrusion south-west of New Galloway (Fig. 3.20). Strata within the zone are mostly steeply-dipping quartzose and pyroxenous greywackes with thin metapelitic interbeds, and belong to the Gala Group of Llandovery age (British Geological Survey, 1997; McMillan, 1998). Ductile deformation is largely taken up within the greywacke matrix where an intense shear-zone fabric is developed in phengitic white mica, chlorite and quartz. In the most highly deformed greywackes, quartz and feldspar clasts show evidence of cataclastic deformation, grain rotation and the development of mica beards and pressure solution seams. A mainly sinistral sense of shear is shown by asymmetrical beards, S-C fabric and extensional crenulation cleavage (Phillips *et al.*, 1995). Deformed pyroxenous greywackes contain actinolite–epidote–chlorite–phengite assemblages typical of the greenschist facies. Metapelitic lithologies possess an intense, locally lineated, slaty cleavage and associated phyllonites show S-C fabric.

Because of late extensional faulting, the metapelitic pattern developed across the shear zone is markedly asymmetrical (Fig. 3.20). Highest grade epizonal rocks are found in the centre of the zone and along its northern boundary formed by the Orlock Bridge Fault. North of the fault, grade decreases sharply to the low anchizone or late diagenetic zone, creating a metamorphic hiatus that represents a downthrow to the north of the Orlock Bridge Fault of at least 4 km, based on a field gradient of 25°C km^{-1} or less (Merriman *et al.*, 1995b). The greenschist facies/epizonal grades attained in the mostly intensely sheared rocks imply burial depths in excess of 12 km at the time of metamorphism. This is consistent with TEM evidence of thick phengite crystallites (mean thickness 663 Å) and the absence of strain microtexture in the mica. Although some chlorite crystallites retain strain microtextures, they also form thick crystallites with a TEM mean of 520 Å. Hence microtextures in the highest grade metapelites within the shear zone suggest recovery/recrystallization of the subgrain boundaries expected to develop at high strain rates. Grade gradually declines on the south-eastern side of the shear zone, although the metapelitic pattern associated with the zone is more extensive than foliation fabrics recorded in the greywackes. A maximum width of 7.5 km is indicated if the outer limit of the shear zone is taken at the mid-anchizonal isocryst (KI 0.30). Beyond this limit most strata are at late diagenetic to low anchizonal grades, typical of the regional pattern in this part of the Southern Uplands.

Glarus thrust. The relationship between crystal size and strain in phyllosilicates and the degree of tectonization associated with the Glarus thrust plane in the Swiss Helvetic Alps (section 3.3.4.1) has been investigated by Árkai *et al.* (1997). The study used a suite of slates and mylonites collected at measured intervals up to 170 m above and below the thrust plane. Epizonal grades characterize the overthrust Permian Verrucano strata of the hangingwall whereas the Tertiary Flysch forming the footwall is mainly mid-anchizonal (Frey, 1988). In the hangingwall, illite–muscovite crystallite size decreases and average lattice strain increases towards the thrust plane. Chlorite shows the opposite trend, i.e. crystallite size increases and lattice strain decreases as the thrust plane is approached. No such trends in illite–muscovite nor chlorite were observed in the footwall. Árkai *et al.* (1997) suggest that strain-induced reduction in illite–muscovite crystallite size, e.g. resulting from intracrystal slip and subgrain formation, was not recovered by postthrusting recrystallization

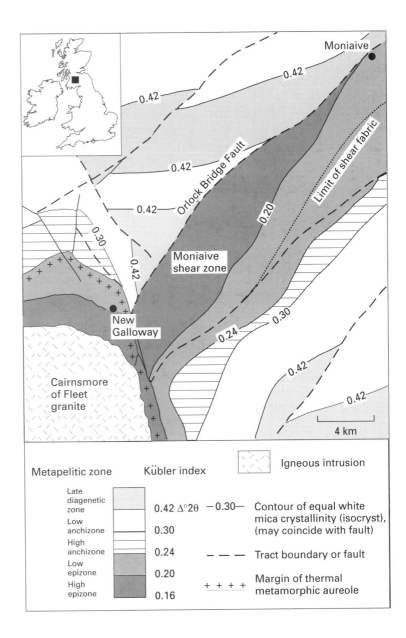

Fig. 3.20 The Moniaive shear zone, southern Scotland. (Modified from McMillan, 1998, reproduced by permission of the Director, British Geological Survey, © NERC.)

in the hangingwall. Unlike illite–muscovite, chlorite shows some evidence of recovery in the hangingwall which is inferred to result from fluid movement along and above the thrust plane.

Niarbyl, Isle of Man. The Niarbyl shear zone is a 100-m wide belt of intensely sheared pelites in the Lower Palaeozoic Manx Group of the Isle of Man (section 3.3.3.3). The predominantly sinistral

shearing postdated the regional cleavage and metamorphism. Phyllitic rocks in the zone are composed of well-orientated $2M_1$ K- and Na-rich micas with minor chlorite (Roberts *et al.*, 1990). Some of the phyllites have unusually high KI values, indicative of the late diagenetic zone. TEM images reveal that white mica packets originally 300–800 Å thick have been reduced to stacks of subgrains by strain-induced slip along (001) with associated dislocations and strain-fields (Merriman *et al.*, 1995b, Fig. 12). The common occurrences of kinked and mottled polygonization microtextures in phyllosilicate, and the anomalously high KI values suggest that the high structural level of the Niarbyl shear zone has not permitted postshearing recovery.

3.4 Low-temperature contact metamorphism

Patterns of contact metamorphism are generated in the vicinity of igneous intrusions, typically as aureoles that are broadly concentric or parallel with the margins of the intrusive body (Kerrick, 1991). Around larger bodies of plutonic rock the aureole usually consists of a series of zones of decreasing temperature, many of which correspond to the well characterized hornfels facies of contact metamorphism (Winkler, 1967; Bucher & Frey, 1994). Until quite recently the baked and faintly spotted rocks which represent the transition from diagenetic country rocks to the beginning of the albite–epidote–hornfels facies have received little attention. Many model reaction sequences of metapelitic aureoles are assumed to begin with the appearance of biotite, cordierite and/or andalusite (e.g. Pattison & Tracy, 1991). Over the past 10–15 years, however, the outer low-temperature zones (150–350°C) of several aureoles have been characterized by detailed XRD and, in a few cases, SEM and TEM studies, which have recorded phyllosilicate assemblages and their crystallinity. Some of these studies reveal that even around some small plutons (<3 km across) a cryptic aureole is developed in metapelites extending to 2–3 km beyond the outer limit of hornfelsing observed at outcrop. In terms of their metamorphic history, the aureoles which have been studied can be divided into those that predate and those that postdate regional metamorphism.

3.4.1 Aureoles predating regional metamorphism

In a comparative study of the effects of burial and contact metamorphism on the Cretaceous Mancos Shale, Nadeau and Reynolds (1981) detailed the changes in illite–smectite (I/S) adjacent to a Tertiary pluton near Cerrillos, New Mexico. Within a modelled 200°C isotherm and up to 2 km from the contact,

I/S in bentonites has $R = 2$ ordering whereas $R = 3$ I/S and chlorite are found in shales. At approximately 5 km from the contact random ($R = 0$) I/S characterizes both types of mudrock, whereas at 20 km the bentonites consist of smectite and the shales contain random I/S. Changes also occur in the proportion of illite layers in I/S with distance from the contact; 75–100% illite is found within the 300°C isotherm (<1 km), but between 100–150°C and about 5 km distance illite forms approximately 50% as the ordering changes to random I/S. Although illite crystallinity data were not used in this study, the changes in percentage of illite in I/S suggest that the transition from late diagenetic to anchizonal conditions would be encountered at about 1.5 km from the contact.

Similar mixed-layer mineral assemblages were reported in contact aureoles by Merriman and Roberts (1985) from Lower Palaeozoic mudrocks, but in this case contact metamorphism was followed by regional metamorphism. In the northern part of the Welsh Basin 90% of rectorite and pyrophyllite-bearing pelites are located close (<0.5 km) to outcrops of boss-like intrusions and sills. Associated minerals include illite, chlorite, corrensite, intermediate K, Na-mica and paragonite. TEM characterization of mineral assemblages in the margin of one basic sill identified intermediate K, Na-mica intergrown with paragonite and muscovite (Jiang & Peacor, 1993). This and another study of the intrusions (Merriman *et al.*, 1987; Bevins & Merriman, 1988a), indicate that the high-level intrusions and their aureoles have experienced varying degrees of hydrothermal alteration. White mica crystallinity (KI) data show that contact altered pelites containing $R = 1$ ordered mixed-layer clay minerals and pyrophyllite belong to the anchizone or late diagenetic zone. These metapelitic grades are consistent with thermal aureole temperatures of 200–300°C based on other occurrences of pyrophyllite and rectorite (Merriman & Roberts, 1985). Although they are bleached and baked close to intrusive contacts, the rocks are only mildly spotted and have failed to develop a penetrative slaty cleavage during the Acadian inversion of the Welsh Basin (see section 3.3.2.1). In contrast, nearby pelites which have experienced only regional metamorphism may develop a slaty cleavage and show higher metapelitic grades. The contact-altered rocks appear to have resisted the development of a penetrative cleavage because they were already baked and indurated by pre-tectonic intrusions at the time of regional deformation. The lack of a penetrative fabric in the contact-altered rocks has ensured the survival of metastable ordered mixed-layer minerals, which are rare elsewhere in the Welsh Basin (Roberts & Merriman, 1985).

3.4.2 Aureoles postdating regional metamorphism

Where a thermal aureole has developed after regional metamorphism, the pattern overprints and is generally discordant with pre-existing geological and metamorphic boundaries. In addition, the low-temperature metapelitic aureole is typically more extensive than mapable hornfelsing which is commonly recognized at the first appearance of biotite in the albite–epidote–hornfels facies. Concentric patterns in metapelites may develop a cryptic aureole extending up to 3 km beyond the outer limit of hornfelsing observed at outcrop, with an outer margin in the middle anchizone (British Geological Survey, 1992a, 1993a,d, 1997; Hirons *et al.*, 1997; Lintern & Floyd, 1997; McMillan *et al.*, 1998).

A typical overprinted aureole pattern is shown in Fig. 3.11. The well exposed pluton at the southern end of the Rhins of Galloway peninsula, up to 3 km across, has an inner concentric aureole of hornfelses which approximately corresponds with the limit of the epizone. Beyond this, high anchizonal, slightly spotted metapelites form a cryptic aureole extending up to 2 km from the contact, over-printing a broadly concentric pattern on the ENE-trending pattern of isocrysts generated by regional accretionary metamorphism (see section 3.3.3.1). In the centre of the peninsula, a boat-shaped aureole is developed around a poorly exposed intrusion at Cairngarroch and, a few kilometres to the east, a similar pattern is developed around a concealed intrusion. Both patterns extend across a major strike-slip fault but show no displacement, thus indicating postfault emplacement of the intrusions (Stone, 1995). In other regions the presence of concentric patterns without any surface expression of igneous rocks have been used to infer the presence of concealed intrusions (e.g. James, 1955; Roberts *et al.*, 1990).

Unlike aureoles that predate regional metamorphism, late aureoles that postdate and overprint the regional pattern rarely contain well-ordered mixed-layer clay minerals or pyrophyllite. In some late aureoles, however, anomalies occur in patterns that normally prograde towards the intrusive contact, and suggest local retrogression of metapelitic hornfelses. For example, anomalously low KI values, typically of late diagenetic zone mudstones, are found in some spotted (ex-andalusite) hornfelses in the aureole of the Criffle–Dalbeattie granite (Lintern & Floyd, 1997). Along a sampling traverse of approximately 1 km, KI values increased from 0.35 (low anchizone) to 0.64 (late diagenetic zone) in spotted hornfelses within 250 m of the contact (unpublished study). Initial results of SEM and TEM studies of these metapelites indicate that thin crystallites of illite are present in areas of incipient spotting replacing andalusite and cordierite and also some thicker mica crystals. The irregular distribution of such areas of retrogression possibly reflects the channellizing of fluids associated with thermal fracturing in the waning phase of contact metamorphism, most likely associated with localized hydrothermal cells (Hanson, 1995).

3.5 Regional controls on metapelitic patterns

The studies reviewed above suggest that burial has a fundamental influence on the patterns developed in very low-grade metapelitic sequences. Whether beneath an overburden generated by normal, basin-filling sedimentation or by tectonostratigraphical thickening in accretionary settings, the prograde sequence of metapelitic zones developed in many regions is largely a reflection of burial depth. Such patterns suggest that the overburden has a key role in determining both the thermal conditions and the compactional and tectonic stresses that contribute to reaction progress in metapelitic assemblages.

To achieve the temperatures of 100–300°C that characterize very low-grade sequences, burial beneath 4–12 km of overburden is required for a typical geothermal gradient of 25°C km^{-1}. There is, however, wide variation in the sedimentation rates that generate the overburden, and in the geothermal gradients found in sedimentary basins. In relation to the main geotectonic settings considered in section 3.3.1, forearc basins in accretionary settings are characterized by the highest sedimentation rates (1000–100 m/m.y.), whereas extensional basins experience much lower rates of sedimentation (100–10 m/m.y.), and typical foreland basin rates fall between these two (Deming, 1994). The lowest geothermal gradients are found in accretionary settings, as low as $10–15°C\,km^{-1}$, whereas extensional settings may experience the highest gradients of up to $50–60°C\,km^{-1}$. Interaction between sedimentation rates and the geothermal gradient controls the response of pelitic rocks to burial-related processes.

In accretionary settings, low heat flow and high sedimentation rates, which further depress the heat flow, commonly combine to generate extensive regions where, with local exceptions, metapelitic grade has not advanced beyond the high anchizone. Such conditions can also occur in some foreland basins where a combination of low heat flow and high sedimentation rates has also resulted in rapid burial and very sluggish reaction progress (see section 3.3.2.3). These conditions contrast sharply with extensional basins where heat flow is typically high in the early stages of basin development. Although sedimentation depresses the heat flow as the overburden accumulates, if sedimentation proceeds at a constant and intermediate or low rate, the thermal conductivity of the overburden will increase. Hence, the initially high heat flow of typical extensional basins may be reduced by sedimentation but heat flow may continue to be significantly higher than that in typical accretionary settings. In metapelitic sequences the overall result is that clay mineral reaction progress is more advanced and this is reflected in the wide range of metapelitic zones and especially the extensive areas of epizonal rocks commonly found in extensional settings.

In all the settings discussed above, strain energy associated with tectonic deformation has a major influence on the development of metapelitic patterns. This influence is clearly seen in the development of slaty cleavage microfabric which begins around the late diagenetic zone to low anchizone transition and becomes the dominant fabric in the high anchizone and epizone (see Fig. 2.1). In terms of clay mineral reaction progress, mineralogical and field evidence suggests that slaty cleavage begins to form once the smectite-to-illite reaction approaches completion at approximately 90% illite in I/S. This in turn suggests

that at least 5 km of overburden would be expected before a penetrative slaty cleavage microfabric could begin to develop in typical basinal sequences. Other nonpenetrative fabrics may develop in less mature and/or less confined pelites, and produce the scaly mudstones or scaly argillites found in disrupted mudstone sequences and mud diapirs.

In most metapelitic sequences the effects of thermal and strain energies are interactive and complementary (Van der Pluijm *et al.*, 1998). This is particularly well illustrated in some accretionary terranes where, despite the low heat flow charac-teristics, metapelitic reactions have typically advanced to middle or high anchizonal conditions. Because of the synburial deformation processes that characterize subduction zones, both compactional and tectonic stresses provide strain energy early and throughout the assembly of accretionary prisms. As discussed in Chapter 2, strain energy makes an important contribution to the process of crystal thickening, and also to related processes such as polytypic transformations and migration of crystal defects. As a result, tectonic fabrics may be relatively well devel-oped in late diagenetic pelites in some accretionary sequences. Moreover, the early availability of strain energy may explain the low temperature (30°C) recorded for the initiation of the smectite-to-illite reaction in the Barbados accretionary complex (Buatier *et al.*, 1992). High strain rates, however, can retrogress metapelites in certain settings (see section 3.3.5), and this may be a factor in the 'retarded' KI values reported from some accretionary settings (e.g. Dalla Torre *et al.*, 1996c).

In settings where burial and deformation were sequential events, basin history determines whether evidence of both events is preserved. For example, in some extensional settings reaction progress is initially influenced mainly by thermal effects related to burial, but overprinted by the effects of deformation and strain as the basin contracted and inverted. In moderately or weakly deformed basins of this type, highest grade metapelites showing the maximum effects of strain enhancement are generally found in the cores of regional anticlinal folds (e.g. Fig. 3.5a). Where the basin fill has been more severely deformed, the pattern of normal burial may be preserved in an external fold-and-thrust belt but locally inverted by late thrusts. In the most severely deformed basins the effects of strain completely overprint any evidence of a burial pattern.

The extent of low-temperature metapelitic patterns developed in contact aureoles is also controlled by basin history. Cryptic aureoles representing temperatures of < 350°C are more extensively developed in regions of low heat flow because, even at depths of 5–10 km, intrusions have been emplaced into country rocks where the ambient temperature is significantly below 350°C. In the Southern Uplands terrane for example, cryptic aureoles are commonly two to three times more extensive than the area delimited by mapable hornfelses. Conversely, cryptic aureoles associated with intrusions emplaced at depth in high heat flow basins tend to be limited because of the small overstep between magmatic heating and regional heat flow for temperatures below 350°C.

3.6 Conclusions

Regional very low-grade metapelitic rocks are found in three main types of geotectonic setting. In Alpine-type collisional settings they are preserved in external fold-and-thrust belts. In accretionary collisional settings older strata may form the overburden beneath which younger strata are underplated, resulting in accretionary burial patterns where grade increases into younger strata. In extensional basin settings a normal burial pattern is characteristic, where grade increases into older strata beneath an overburden of younger strata.

Depth of burial strongly influences regional patterns in very low-grade metapelitic rocks. The prograde sequence of metapelitic zones developed in many regions is the result of reaction progress in clay minerals promoted by the interaction of basin heat flow and strain associated with overburden and tectonic stresses. Recognition and study of the patterns generated during very low-grade metamorphism is possible because they are irreversible in relation to basin inversion tectonics. Indices of metapelitic reactions can be approximately correlated with organic maturity indicators and used for basin maturity studies. However, the effects of different heat flow systems and heating rates on clay mineral assemblages are presently poorly understood, and thermal indexation of reaction progress awaits further study. This is a particularly promising area for future research, with the possibility of developing applications for basin maturity modelling.

Cryptic low-temperature aureoles develop where regional very low-grade rocks are intruded by late plutons. These aureoles are most extensive in low heat flow regions where they may be two to three times wider than the observed limits of hornfelsing. Anomalously low-grade rocks may develop in parts of some aureoles as a result of retrogressive 'back reaction'. Low-temperature aureoles and back-reaction effects are another promising area for future research, as analogues for the barriers and host rocks used for the storage of heat-emitting radioactive waste.

Localized metapelitic patterns are generated around high strain zones. Around deeply buried high strain zones, annealing of lattice strains in phyllosilicates results in thicker crystallites and higher (epizonal) grades associated with the most highly sheared rocks. In some high-level shear zones and thrust planes, lattice strains are not fully recovered by postshearing recrystallization of phyllosilicates and grade may be locally retrogressed in the most highly sheared rocks.

4 Petrological methods for the study of very low-grade metabasites

P. Schiffman and H. W. Day

4.1 Introduction

Very low-grade metabasic rocks, from zeolitized basalts drilled from the ocean floor to greenschist found in orogenic belts, constitute major portions of the Earth's crust. Although their geological significance has been long appreciated, petrologists continue to struggle with attempting to quantify the pressure–temperature conditions under which these low-grade rocks are recrystallized.

There are myriad problems associated with the study of very low-grade rocks. First and foremost is that they are intrinsically fine-grained and 'messy' rocks. Thus the characterization of mineral parageneses typically requires more analytical 'firepower' than with their higher grade equivalents. Technological developments in the past few decades, such as the widespread availability of powerful personal computers and high-speed back-scattered electron (BSE) detectors on scanning electron microscopy (SEM) and electron microprobes, have greatly aided the study of fine-grained rocks. When these are employed in concert with rigorous methods for assessing the phase equilibria of higher grade mineral assemblages, they potentially afford petrologists a more routine ability to use mineral parageneses in quantitative studies of low-grade metamorphic terranes.

Although researchers now have adequate petrographic tools for studying fine-grained rocks, they will continue to provide a challenge when trying to extract quantitative information on their origin. Subgreenschist grade metabasites almost invariably exhibit petrographic evidence of incomplete recrystallization and therefore only partial equilibration of their metamorphic mineral assemblages. In addition, most low-grade rocks display evidence of complex igneous, sedimentary and/or hydrothermal histories, which precede any superimposed orogenic

or burial metamorphism. In low-grade metamorphic environments, this complexity is preserved and must be distinguished from any subsequent metamorphism.

The physical character, as opposed to strictly compositional properties of low-grade protoliths may significantly affect their recrystallization history. For example, within a given flow unit, fluid–rock interactions occurring under identical pressure–temperature conditions may produce markedly different mineral assemblages within a permeable scoriaceous flow top versus a less permeable massive flow interior. These sorts of features can even be observed on the scale of an individual thin section. Secondary mineral assemblages are strongly dependent on where they occur, i.e. within relict primary phases, within groundmass or matrix, or within vesicles and veins. Given these caveats, the systematic examination of the correspondence between protoliths and their secondary mineral assemblages has only recently been seriously investigated. Attempts at estimating intensive variables attending the metamorphic recrystallization of these rocks are in their infancy.

Metabasites, like all other low-grade metamorphic rock protoliths, are best studied using an integrated multianalytical approach, which might include electron microprobe, X-ray diffraction (XRD), transmission electron microscopy (TEM), stable isotope geochemistry and fluid inclusion analysis. Many of these techniques are discussed in detail elsewhere in this book. In this chapter, we present some basic methods for the systematic study of low-grade rocks using relatively standard petrological, petrographic and electron microprobe techniques. The electron microprobe continues to be an essential tool and electron imaging, typically using the modern BSE capabilities, provides necessary, but not necessarily sufficient, petrographic evidence for equilibrium amongst coexisting minerals. Compositional data provide the further evidence for establishing chemical equilibria amongst coexisting phases. Collectively, these data are essential for delineating systematic relations between mineral and bulk chemical composition, and ultimately for creating petrogenetic grids and estimating pressures and temperatures of formation.

4.2 Field study of very low-grade metabasites

Metamorphic studies in terranes containing low-grade rocks can be broadly subdivided into two categories: (1) ones that address questions about regional geology and tectonics; and (2) studies that address processes of fluid–rock interactions, e.g. in modern oceanic crust, ophiolites or associated ore deposits. Regardless of their specific goals, all studies inevitably start with basic questions such as: what are the diagnostic mineral assemblages in these rocks, and do these assemblages vary spatially within the study area? Since the diagnostic minerals and mineral assemblages only recrystallize within protoliths of appropriate bulk composition, it is essential to develop a strategy which optimizes the likelihood of obtaining useful samples.

Very low-grade metabasites encompass a wide range of protoliths, including lava flows of varying morphologies, hyaloclastic, pyroclastic or epiclastic tuffs and breccias, as well as various forms of hypabyssal intrusive rocks. Since many of these protoliths are fine grained, in the field it is often difficult to identify unequivocally the most mafic compositions that will contain the diagnostic sub-greenschist grade mineral assemblages. Many studies of very low-grade metamorphism are based in ancient volcanic arc and arc-ophiolite terranes, in which the protolith suite may exhibit a wide range in bulk compositions, essentially basalt to rhyolite. In such terranes, it is essential to identify and sample the basic rocks wherever possible. Even if protolith compositions were essentially basaltic, original magmatic differentiation processes (e.g. Bevins & Merriman, 1988b) and especially subsequent weathering or hydrothermal alteration may have profoundly modified these, enhancing compositional differences between, for example the exterior and interior of a given flow unit or individual pillow

basalt lobe (see discussion of these processes in Chapter 6). Tectonic processes such as extensional faulting at oceanic ridges, may also produce secondary porosity and markedly enhance porosity (Bettison-Varga et al., 1992). Dissolution during an early stage of hydrothermal alteration may also produce the same result (Harper, 1995).

In many basaltic protoliths, individual eruptive units may exhibit textural variations which control the physical properties (i.e. porosity and per-meability) of these rocks (Manning & Bird, 1995) and, consequently, their response to subsequent hydrothermal alteration or regional metamorphism. Collectively, the primary and superimposed physical and compositional heterogeneity of the protoliths may appreciably affect the nature of the metamorphic mineral assemblages which are ultimately developed (Rose, 1995). Dramatic differences in the primary permeability of basalt flows (i.e. high in scoriaceous flow tops and low in massive flow interiors) appear to control the degree to which these lithologies will recrystallize during subgreenschist grade metamor-phism (Schmidt & Robinson, 1997). Specifically, the impermeable portions of flows retain primary minerals (e.g. calcic plagioclase) and secondary minerals of lower grade than those which are regionally developed and in the more permeable portions of the same flow units (Jolly & Smith, 1972; Levi et al., 1982; Alt et al., 1986a; Bevins et al., 1991b; Schmidt, 1993).

Compositional heterogeneity can exist at the outcrop, hand specimen as well as thin section scale. The concept of 'metadomains' has been long recognized (Smith, 1968). Typically, compositional heterogeneity arises where original heterogeneity was enhanced by subsequent fluid–rock interactions. Thus, the scoriacious portions of a basalt flow, the fractured portions of a metabasite dyke or the pegmatitic portions of a gabbroic body may become calcium-rich metadomains due to filling of voids and/or replacement by calcsilicate minerals (e.g. Jolly & Smith, 1972; Rose & Bird, 1987; Bettison-Varga et al., 1992; Metcalfe et al., 1994). In seawater hydrothermal systems, glassy pillow margins and interpillow matrix (derived from spalled-off pillow margins) which experience relatively high integrated fluid–rock mass ratios will recrystallize to meta-domains composed predominantly of mafic layer silicates (Seyfried et al., 1978a).

In outcrop, metadomains, especially where they are composed predominantly of a single mineral, are readily recognized by their striking colour differences: black or dark green for smectite, chlorite and pumpellyite; pistachio green for epidote; white or pink for prehnite or zoisite. During very low-grade metamorphism of mafic rocks, primary minerals may also serve as catalysts for metadomain formation, and the composition of these minerals will dictate the secondary mineral assemblages which they will host. For example, olivine or orthopyroxene phenocrysts are pseudomorphed by Mg-smectite, serpentine and talc; plagioclase phenocrysts are replaced by Na- and Ca–Al silicates such as zeolites, albite, prehnite, pumpellyite and epidote; at greenschist grades, clinopyroxene phenocrysts become metadomains for actinolite (i.e. uralite).

Regional scale mapping of mineral zones in very low-grade metabasites is also possible. Specifically, zeolites developed within metamorphosed flood basalts (e.g. in Iceland (Walker, 1960a; Neuhoff *et al.*, 1997) and Minnesota (Schmidt, 1993)) may be sufficiently coarse-grained to allow for unequivocal hand specimen identification.

4.3 Primary features

Virtually all very low-grade metabasites, whether they occur as effusive, pyroclastic or epiclastic deposits, retain some relict igneous textures. These include (i) glass or fine-grained mesostasis, commonly palagonitized, (ii) primary (i.e. magmatic) minerals including plagioclase, pyroxenes (mainly clinopyroxene) and Fe–Ti oxides, and (iii) primary

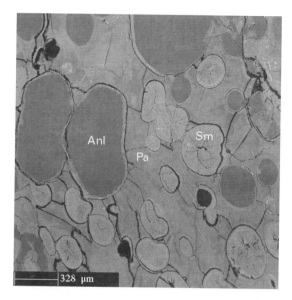

Fig. 4.1 Back-scattered electron image of palagonitized clast from hyaloclastic breccia, La Palma, Canary Islands. Palagonitization of basalt glass is accompanied by precipitation of secondary minerals (analcime and smectite in this example) which fill vesicles and destroy primary porosity. Anl = analcime; Pa = palagonitized glass; Sm = smectite.

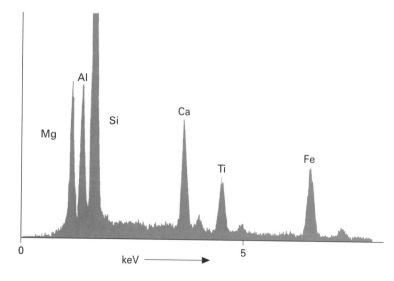

Fig. 4.2 Energy dispersive spectroscopic spectra between 0 and approximately 8 keV of palagonitized glass from sample shown in Fig. 4.1; analysis done at 15 keV. The composition of palagonite may be quite heterogeneous, even on the thin section scale, although it can generally be distinguished from mafic layer silicates which lack titanium.

porosity, mainly in the form of vesicles (and amygdules when filled with secondary minerals) or fractures (which become veins). Petrographic analysis of vesicle in fillings and crosscutting veins is a powerful tool for working out the paragenetic relationships amongst secondary minerals (e.g. Alt *et al.*, 1986a; Fridleifsson, 1984).

4.3.1 Glass and palagonite

Hyaloclastites are especially common in protoliths formed at significant water depths on the ocean floor. Under seafloor weathering and zeolite grade metamorphic conditions, primary glass transforms to palagonite, a hydrated alteration product (Peacock, 1926; Staudigel & Hart, 1983). The palagonitization process also produces secondary minerals (e.g. calcite, smectite and zeolites) which fill vesicles and interclast pore space (Hay & Iijima, 1968). The resultant material has a greatly reduced porosity with respect to the primary hyaloclastite (Fig. 4.1).

The composition of a palagonitized glass is complex (Fig. 4.2) and has a structure similar to the di- and tri-octahedral smectites that may crystallize within it (Eggleton & Keller, 1982; Zhou *et al.*, 1992). The compositional heterogeneity of palagonite is evident in the variable grey tones seen in BSE images (e.g. Fig. 4.3a) reflecting this material's variable states of hydration and cation content. The 'proto-smectite' property of palagonitized glass is reflected in a finite and measurable cation exchange capacity (CEC = 30–60 meq/100g), which is roughly half that of smectite (Schiffman & Southard, 1996). The CEC of palagonite and smectite in metabasites can be quantitatively measured *in situ*, on polished thin sections, using a CsCl-staining technique described by Hillier and Clayton (1992). The CEC can also be qualitatively assessed with X-ray dot maps (Fig. 4.3b).

At higher metamorphic grades, palagonitized glass recrystallizes into mafic layer silicates including smectite, corrensite and chlorite (Schiffman & Fridleifsson, 1991; Schiffman & Staudigel, 1995). In hydrothermal environments, these layer silicates may recrystallize in banded structures which mimic palagonitic textures.

4.3.2 Primary minerals

The transformation of anhydrous, primary minerals

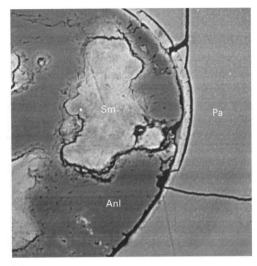

(a)

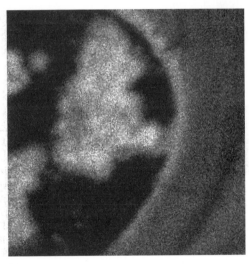

(b)

Fig. 4.3 Palagonitized, vesicular basalt after treatment with CsCl stain. (a) Back-scattered electron image, (b) Cs Lα X-ray dot map; abbreviations same as in Fig. 4.1. The intensity of the dots in (b) is a qualitative measure of the material's cation exchange capacity (CEC) for CsCl. The palagonitized basaltic glass in this sample has a CEC which is roughly half that of smectite. Analcime has no appreciable CEC as determined by this technique. Images are approximately 150 μm on edge.

into mainly hydrous, secondary minerals is a fundamental, but poorly quantified aspect of the recrystallization of mafic rocks. These mineral transformations proceed as irreversible, generally exothermic reactions that operate under conditions from very low-temperature (e.g. seafloor weathering) to greenschist metamorphic conditions. Olivine and orthopyroxene may be replaced by layer silicates and carbonates under very low-temperature conditions (Chapter 6), whereas clinopyroxene generally recrystallizes only in the lowermost greenschist facies (Fagan & Day, 1997). Plagioclase begins to recrystallize at intermediate (i.e. zeolite grade) temperatures. Since all primary minerals, as well as silicate glass, are metastable under very low-grade metamorphic conditions, their hydration and recrystallization are strongly affected by the physical properties of the protoliths, which control access of fluids to reaction sites.

Plagioclase. The alteration of primary plagioclase to albite or Na-zeolites is perhaps the most critical of all processes in very low-grade metabasites. The Ca and Al that are released help promote crystallization of calcite, mafic layer silicates, and calcsilicates such as prehnite, pumpellyite, epidote and titanite. These minerals form in pore spaces, veins and vesicles, or within the primary plagioclase itself. Albitization of primary plagioclase commonly proceeds with minimal disruption of primary textures, and delicate igneous quench textures may be preserved in basalts that have been completely albitized (Fig. 4.4).

Clinopyroxene. Uralitization (Fig. 4.5), the replacement of primary clinopyroxene by actinolite and lesser chlorite, is a process generally associated with the onset of greenschist conditions of metamorphism. For example, in Drillhole 15 from the Nesjavellir (Iceland) geothermal field, clinopyroxene remains unaltered in basalt samples recovered from temperatures approaching 300°C (Schiffman & Fridleifsson, 1991). In fossil hydrothermal systems, uralitization is first observed in samples which have epitode–albite–chlorite–quartz assemblages (e.g. Evarts & Schiffman, 1983; Alt *et al.*, 1986a). Although uralitization necessitates the hydration of clinopyroxene, the net uralitization reaction may entail dehydration because actinolite may form, at least in part, through reactions that require significant decomposition of chlorite (Fagan & Day, 1997).

Fe–Ti oxides. The stability of primary Fe–Ti oxides,

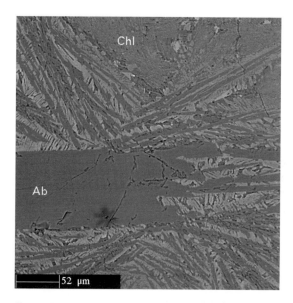

Fig. 4.4 Back-scattered electron image of quench texture in pillow basalt from ODP drill hole 858G; prominent calcic plagioclase has been albitized (Ab) and some groundmass patches have been recrystallized into chlorite (Chl), but dendritic clinopyroxene/plagioclase intergrowths in the groundmass are preserved.

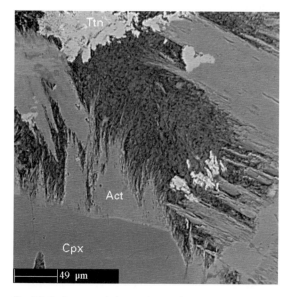

Fig. 4.5 Back-scattered electron image of partially uralitized clinopyroxene (Cpx) from mafic sills in ODP drill hole 857D. Mesostasis (centre) has been replaced by chlorite. Act = actinolite, Ttn = titanite.

principally titanomagnetite and ilmenite, during the very low-grade metamorphism of mafic rocks has been largely ignored by metamorphic petrologists, although it is of obvious concern to paleomagnetists, especially those investigating the magnetic properties of the oceanic crust (e.g. Kelso *et al.*, 1991; Gee *et al.*, 1993). Primary ilmenite and Ti-magnetite may be pseudomorphed by mixtures of low Ti-magnetite and titanite under low-grade, burial metamorphic conditions, and primary (exsolution) lamellar structures may be preserved as well; albitization is apparently an important factor in freeing up aluminium for recrystallization of titanite after Fe–Ti oxides (Kuniyoshi & Liou, 1976). Conversely, the hydrothermal alteration of metabasites under low pH, high fluid/rock conditions may result in leaching of Al and Si. Under these low pH conditions, primary Fe–Ti oxides may be pseudomorphed by 'blotchy' mixtures of low-Ti magnetite and TiO_2 (rutile or anatase), and the alteration phases do not appear to pseudomorph the primary lamellar structures.

4.3.3 Vesicles

Petrographic analysis of amygdules common to many metabasites is an integral part of the study of very low-grade metamorphism: critical index minerals and low variance mineral assemblages may occur exclusively in amygdules and not in other rock domains. Many studies have used the first occurrence of minerals in amygdules (and associated veinlets) to define isograds or mineral zones (e.g. Stoiber & Davidson, 1959; Walker, 1960a,b; Jolly & Smith, 1972; Schmidt, 1993; Schiffman & Staudigel, 1994). Recent phase equilibria studies (e.g. Springer *et al.*, 1992; Beiersdorfer & Day, 1995) have emphasized the need to identify low-variance mineral assemblages. Although these may occur within individual amygdules (e.g. as documented by Cho *et al.*, 1986, for the Kartmutsen metabasalts), low-variance assemblages in model systems require the co-existence of 'ubiquitous' phases, mainly quartz and albite. The latter are not commonly reported from amygdules.

In addition to using amygdules to identify the *first occurrences* of minerals within a sequence of metabasites, a few studies have employed amygdule *filling sequences*, in conjunction with cross-cutting vein relationships, to reconstruct prograde or retro-

grade thermal histories associated with burial or hydrothermal metamorphism (e.g. Levi, 1969; Fridleifsson, 1984). This technique has been used successfully for studying ocean floor metamorphism (e.g. Alt *et al.*, 1986a).

4.4 Secondary minerals

Secondary minerals are invariably both fine grained and xenoblastic, except where they occur in amygdules or veins. For these reasons, the unequivocal identification of secondary minerals in fine-grained rocks, when based exclusively on polarizing light microscope, may be unreliable. However, the increased availability of SEMs, which offer high speed BSE detection as well as energy dispersive spectrometry, has made electron-optic and X-ray spectrometric identification of these minerals much more routine. The following discussion focuses on ways in which these techniques can be effectively utilized to recognize key secondary minerals. The compositional variations of key minerals in very low-grade metabasites were recently reviewed by Beiersdorfer and Day (1995).

4.4.1 Mafic layer silicates

Layer silicates, including smectite, chlorite and products of their mixed layering, are perhaps the most abundant minerals in very low-grade metabasites. Since the smectite to chlorite transition is probably the most universally recorded mineral transformation in subgreenschist facies metabasites, it is essential to characterize the mafic layer silicate mineral parageneses in these samples. Standard optical petrographic methods, in which these layer silicates are differentiated on the basis of their absorption and interference colours, can be used effectively to broadly differentiate smectite from chlorite (e.g. as described by Bevins *et al.*, 1991b), but other techniques are required to more rigorously characterize interlayered chlorite/smectite. All workers should routinely use XRD techniques for the unequivocal *recognition* of mixed-layer chlorite/smectite. In particular, the basal spacing of chlorite/smectite will expand (to between 14 and 18Å) upon glycerol or ethylene glycol solvation and will collapse (to between 10 and 14Å) upon high-temperature heat treatment. Moore and Reynolds (1997) presented an excellent laboratory guide for XRD study of chlorite/smectite.

The unequivocal *identification* of chlorite/smectite minerals through XRD analysis of clay size fractions, however, may not be so straightforward. Although its status as a discrete phase is questioned (Reynolds, 1988), corrensite is generally accepted as a 50:50 mix of smectite- and chlorite-like layers, identified in powder patterns by its characteristic (glycol solvated) 32-Å superlattice reflection. Other forms of mixed-layer chlorite/smectite are more hotly contested. Many very low-grade metabasites contain layer silicates whose XRD properties (e.g. as indicated by peak migration curves; Reynolds, 1988) and microprobe compositions (as described below) imply that they contain >50% chlorite interlayers. But are these mixtures composed of discrete phases (i.e. corrensite + chlorite, Shau *et al.*, 1990; Shau & Peacor, 1992) or of smectite- and chlorite-like layers (Bettison-Varga *et al.*, 1991)?

Very low-grade metabasite samples commonly contain two or more mafic layer silicates (including both discrete and mixed-layer minerals). Even after rigorous deconvolution analysis of XRD powder patterns, the identification of these minerals may be ambiguous. Thus, for example, the XRD pattern of a two phase mixture of chlorite + mixed-layer chlorite/corrensite may be indistinguishable from that of a three phase mixture of chlorite, corrensite and regularly interstratified chlorite/smectite (Schmidt & Robinson, 1997), especially if one of the layer types in the mixed-layer phase is segregated in the overall mineral structure (Hillier, 1995).

Although not as definitive as XRD and HRTEM, BSE and X-ray microanalytical techniques offer relatively rapid, tentative identification of mafic layer silicates. In BSE mode, smectite and chlorite can be generally differentiated on the basis of their textures. Smectite, which will dehydrate extensively within the evacuated specimen chamber of the SEM/microprobe, tends to appear as packets of wavy crystallites separated by large open pores (Fig. 4.6a). Chlorite crystallites do not dehydrate and generally appear as denser, less porous aggregates (Fig. 4.6c,d). Corrensite generally exhibits an intermediate texture in BSE mode (Fig. 4.6b).

Smectite and corrensite can be tentatively differentiated from chlorite based on the presence of Ca in the latter as seen in energy dispersive spectroscopic (EDS) spectra (Fig. 4.7), although wavelength dispersive spectrometry (WDS) analysis is a more reliable method. Because the mafic layer silicates typically form in aggregates within groundmass or amygdules, even a fully focused electron beam will integrate the composition of tens to perhaps thousands of individual crystallites (as identified in HRTEM lattice fringe images, e.g. by Shau & Peacor, 1992; Shau *et al.*, 1990; Bettison-Varga *et al.*, 1991; Bettison & MacKinnon, 1997). The relative peak heights of Si Kα and Ca Kα that appear on EDS spectra can be used as an indication of the presence of smectite versus chlorite in these aggregates, although quantification of the SiO_2 content of mafic layer silicates seems to be the most reliable, non-XRD, non-HRTEM-based identification technique (Bettison & Schiffman, 1988). Using a 28 oxygen (or 56 anionic charge) basis for recalculating the structural formula of any trioctahedral smectite–chlorite mixture, values of Si exceeding 6.25/formula unit generally imply the presence of some smectite, either as interlayers or crystallites mechanically mixed with chlorite. As pointed out by Robinson *et al.* (1993), however, the presence of only minor, minute inclusions of either dioctahedral chlorite or non-layer silicate minerals, especially calcite, can make the Si cation content of discrete chlorite exceed the 6.25 value. Similarly, although the sum of (Na + K + Ca) correlates well with Si (e.g. Bettison & Schiffman, 1988; Robinson *et al.*, 1993), inclusions of non-layer silicate minerals may account for a substantial fraction of the reported interlayer cation contents of mixed-layer mafic silicates, particularly discrete chlorite (Essene & Peacor, 1995; de Caritat *et al.*, 1993).

Although the mechanics of the smectite to chlorite transition, and particularly the nature of mixed layering in this system, are still actively debated (e.g. most recently by Robinson *et al.*, 1993; Robinson & Bevins, 1994; Hillier, 1995; Schiffman & Staudigel, 1995; Walker & Murphy, 1995; Bettison-Varga & MacKinnon, 1997; Schmidt & Robinson, 1997), all workers agree that metabasites that contain dominantly smectite will eventually transform to those containing dominantly chlorite with increasing temperature. Recognizing the utility for further characterizing this important transformation, Bettison and Schiffman (1988) and later Schiffman and Fridleifsson (1991) proposed simple electron

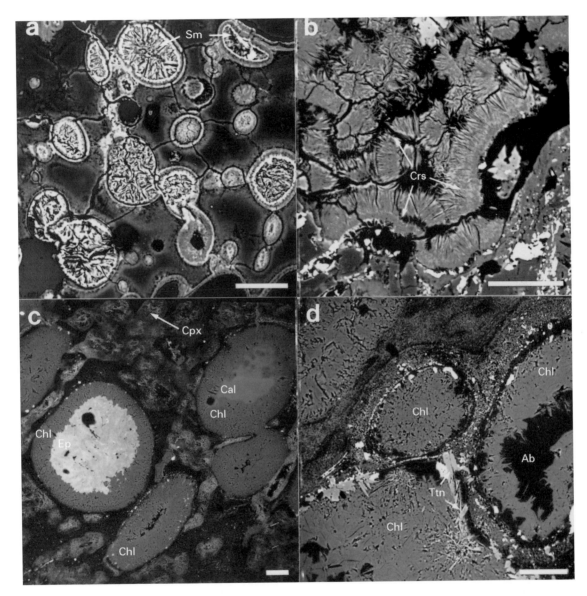

Fig. 4.6 Back-scattered electron micrographs of smectite, corrensite and chlorite in metabasites from the La Palma seamount series. (a) Smectite (Sm) filling vesicles in palagonitized hyaloclastite; (b) corrensite (Crs) lining vesicles in hyalocrystalline pillow basalt; (c,d) chlorite (Chl) filling and lining epidote (Ep), calcite (Cal) and albite (Ab) filled vesicles in fine-grained basalts; Ttn = titanite; Cpx = clinopyroxene. Scale bars are 50 μm. (From Schiffman & Staudigel, 1995.)

microprobe-based techniques for quantifying the percentage of chlorite in mixtures of chlorite/smectite. Subsequently, Bevins *et al.* (1991b) applied this method in conjunction with a chlorite geothermometer, based on tetrahedral versus octahedral Al content (Cathelineau & Nieva, 1985; Cathelineau, 1988), for assessing the grade of subgreenschist facies metabasites, and specifically for correlating the layer silicate and calcsilicate mineral parageneses. Corrensite- or chlorite/smectite have not been

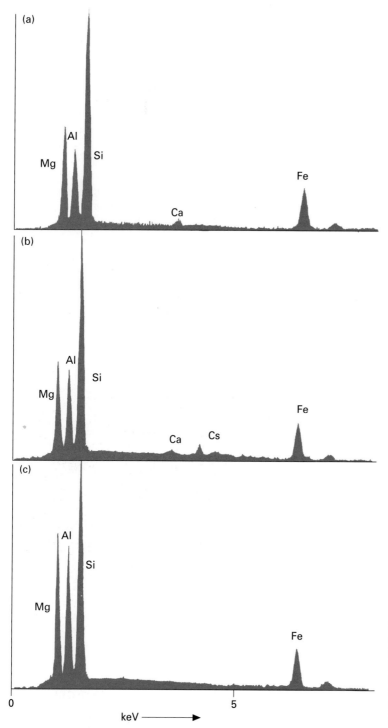

Fig. 4.7 Energy dispersive spectroscopic spectra of (a) smectite, (b) corrensite and (c) chlorite, between 0 and approximately 8 keV; analysis done at 15 keV. The corrensite was treated with a 1 mol l^{-1} CsCl solution resulting in cation (i.e. Cs for Ca) exchange (Schiffman & Southard, 1996).

reported from metabasite assemblages that contain epidote + actinolite, but they commonly occur in samples that also contain laumontite, prehnite and pumpellyite. The temperature interval for which these mixed-layer silicates exist in modern geothermal systems is approximately 150–270°C (Kristmannsdóttir, 1979; Schiffman & Fridleifsson, 1991). This is also the approximate temperature range over which the chlorite geothermometer indicates that mixed-layer silicates occur in very low-grade metabasites (Bevins *et al.*, 1991b; Schmidt & Robinson, 1997). The Cathelineau chlorite geothermometer has been criticized for both its empirical nature and because the reported Al variations in 'chlorite' are actually due to the presence of mechanically or structurally intergrown smectite or corrensite (de Caritat *et al.*, 1993; Essene & Peacor, 1995). Nonetheless, the method has proven to be a useful tool for comparing the relative grade of metabasites from a given study area, especially when used in conjunction with other techniques such as chlorite crystallinity (Arkai & Ghabrial, 1997), XRD deconvolution analysis (Schmidt & Robinson, 1997) and HRTEM (Bettison-Varga & MacKinnon, 1997).

4.4.2 Pumpellyite, prehnite and epidote

The identification of these key calcium–aluminium silicates is obviously crucial to any study of low-grade metabasites. If grain size is sufficient for using standard petrographic methods (e.g. occurrences in vesicles and veins), differentiation amongst these minerals can generally be made on the basis of diagnostic optical properties, particularly

1 absorption colours: prehnite = colourless, epidote = yellow to colourless (varies with Fe content), pumpellyite = green to colourless (varies with Fe content);

2 pleochroic scheme: pumpellyite exhibits preferential absorption for B = Y, which is the elongation direction;

3 interference colours: which are typically anomalous for pumpellyite;

4 extinction: prehnite commonly exhibits wavy or 'bow tie' extinction.

In many very low-grade metabasites, however, these minerals may occur only as fine-grained aggregates within primary plagioclase or recrystallized groundmass/matrix. Identification of these minerals

in such fine-grained aggregates is often impossible because (1) their refractive indices will covary with their Fe contents (Deer *et al.*, 1992), and (2) the aggregates may be mixtures of more than one mineral. Prehnite, pumpellyite and epidote can often be differentiated in BSE images (e.g. Fig. 4.8), but their grey scales in BSE images will also be strongly dependent on Fe/Al ratios for these minerals. Fine-grained, granular pumpellyite, in particular, may be extremely difficult to differentiate from epidote, titanite and mafic layer silicates (Evarts & Schiffman, 1983), Fine-grained, acicular pumpellyite can easily be misidentified as actinolite, chlorite or epidote. In these situations, EDS analysis is essential for at least the tentative identification (Fig. 4.9). Pumpellyite is readily distinguished from prehnite and epidote by its diagnostic small, but always present, Mg peak. Epidote and prehnite, however, are not readily distinguished from one another solely by EDS, and WDS analysis may be necessary even for qualitative differentiation.

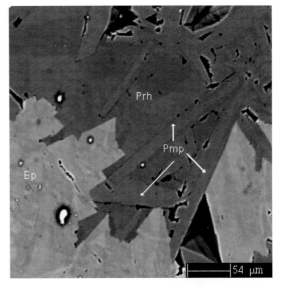

Fig. 4.8 Back-scattered electron image of pumpellyite, prehnite and epidote in metabasite from the La Palma seamount series. The grey level of these minerals appears to be strongly dependent on their relative iron contents: epidote (Ep) has approximately 13 wt% Fe_2O_3 as total iron; prehnite (Prh) has 4% and acicular pumpellyite grains (Pmp) have 6% (but also 2% MgO).

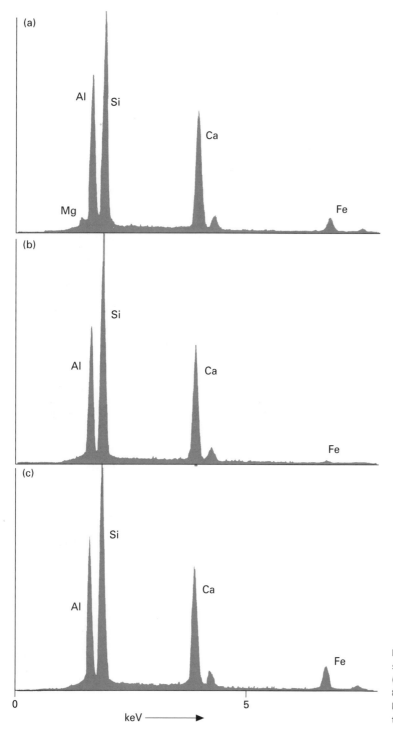

Fig. 4.9 Energy dispersive spectroscopic spectra of (a) pumpellyite, (b) prehnite and (c) epidote, between 0 and approximately 8 keV; analysis done at 15 keV. See Fig. 4.8 legend for origin and iron contents of these minerals.

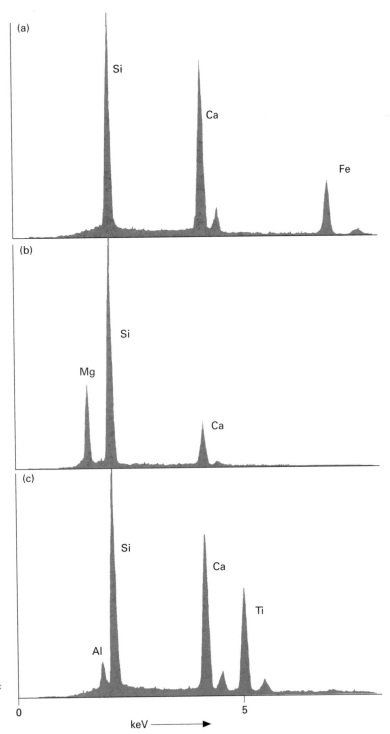

Fig. 4.10 Energy dispersive spectroscopic spectra of (a) andradite, (b) tremolite and (c) titanite, between 0 and approximately 8 keV; analysis done at 15 keV.

4.4.3 Other minerals

Most calcium silicates found in low-grade metabasites are readily identifiable by either BSE or EDS analysis. Andraditic garnet, coexisting with either epidote or pumpellyite, is not an uncommon amygdaloidal phase, especially in carbonate-bearing assemblages (Schiffman & Staudigel, 1995). Uralitic amphiboles, which may first form as 'beard'-like fringes on primary clinopyroxene (Fig. 4.5), have EDS spectra that readily distinguish them from other calcsilicates (Fig. 4.10).

Zeolites are probably the most difficult minerals to identify in low-grade metabasites because, with a few notable exceptions, their optical properties, especially relief and birefringence, are so similar (Deer *et al.*, 1992). In thin section it may often be possible to broadly subdivide unknown zeolites by their morphologies (e.g. as fibrous vs. equant) and then further subdivide fibrous zeolites based on their sign of elongation. Qualitatively, their BSE grey levels and their EDS spectra may not be very definitive. For example, using these techniques, certain Ca-rich zeolites may be difficult to distinguish from other Ca-rich zeolites or even from a calcic-rich plagioclase. But EDS is extremely useful for distinguishing Na vs. Ca-zeolites (e.g. wairakite from analcime in Fig. 4.11). And with some care, certain zeolites can be distinguished from feldspars on the basis of their EDS spectra. For example, analcime has a markedly higher Al/Si ratio with respect to albite (Fig. 4.11). Extensive solid-solution amongst zeolites may even make their identification with quantitative electron microprobe analysis a challenge. Fortunately, zeolites in very low-grade metabasites typically occur as relatively coarse, amygdule-filling grains, readily separated for identification by relatively unambiguous XRD analysis. For example, in a recent study of the low-grade metamorphism of Tertiary flood basalt sequences from East Greenland (Neuhoff *et al.*, 1997), a wide variety of amygdaloidal zeolites were identified through integrated optical, XRD and electron microprobe analyses.

4.5 Electron microprobe analysis of low-grade metabasites

Analysis of low-grade metamorphic minerals is complicated because many are fine grained, heat sensitive, or both. Samples may be difficult to polish due to the large hardness contrast between soft mats of layer silicates and more resistant calcsilicate grains. Hydrated minerals, particularly zeolites and some layer silicates, as well as palagonitized glasses, may be especially beam sensitive. Once proper analytical conditions that minimize beam damage are determined, suitable calibration and, especially, working standards must be chosen to maximize the accuracy of the mineral analyses. If minerals have variable oxygen contents, as in the case of pumpellyite and amphibole, it is also necessary to make some assumptions regarding ferrous/ferric iron contents. Finally, the precision and accuracy of analyses must be ascertained.

4.5.1 Analytical conditions

Optimum analytical conditions are ones that maximize count rates as well as peak/background ratios and also minimize systematic errors due to specimen damage, mainly from overheating. For the routine analysis of most low-grade metamorphic minerals, it is generally not necessary to generate X-rays whose energies greatly exceed 7keV (i.e. the approximate excitation potential for Fe Kα). Therefore, an accelerating potential of 15keV is generally sufficient. Counting times of 10s on both peak and background will generally provide relative counting errors well below 5% (e.g. counting errors for major elements in pumpellyite are presented in Table 4.1). As is discussed below, however, much more consideration should be given to beam current, as well as to the shape and size of the electron beam.

4.5.2 Analytical difficulties

The major problem with the analysis of low-grade metamorphic minerals is beam sensitivity. Overheating will result in dewatering of beam-sensitive hydrous minerals (e.g. zeolites and palagonitized glasses) or alkali migration (e.g. in analcime). These problems can be mitigated through the use of decreased beam currents and/or increased beam spot sizes for microanalysis. Both of these techniques have their drawbacks. Decreased beam currents correspond to decreased count rates which strongly limit analytical precision, most critically for minor

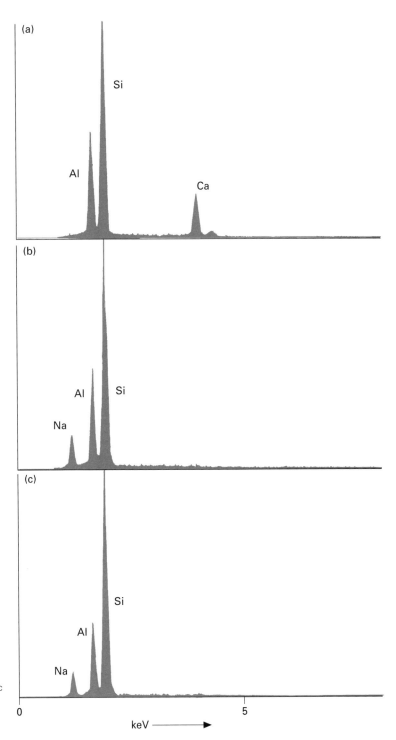

Fig. 4.11 Energy dispersive spectroscopic spectra of (a) wairakite, (b) analcime and (c) albite, between 0 and approximately 8 keV; analysis done at 15 keV.

Table 4.1 Chemical analysis of feldspar from a low-grade metabasalt, LaPalma (Sample LP127–3; P. Schiffman, unpublished).

$n = 10$	Mean	Standard deviation	RSD	RCE	$\frac{RSD}{RCE}$
SiO_2	68.01	0.50	0.007	0.009	0.8
Al_2O_3	19.56	0.43	0.022	0.016	1.4
CaO	0.48	0.21	0.440	0.381	1.1
Na_2O	11.24	0.23	0.020	0.029	0.7
K_2O	0.10	0.05	0.534	0.333	1.6
Total	99.38	0.47	0.005		

Cations per 8 oxygen

	Mean	Standard deviation	Standard error
Si	2.988	0.017	0.005
Al	1.013	0.020	0.006
ΣTet	4.001	0.008	0.002
Ca	0.023	0.010	0.003
Na	0.958	0.019	0.006
K	0.005	0.003	0.001
ΣAlk	0.986	0.022	0.007
Total	4.987	0.016	0.005
%An	2.3	1.0	0.3
%Ab	97.1	1.0	0.3
%Or	0.5	0.3	0.1
Total	100.0	0.0	

The analysis is the mean of 10 spots analyzed on a line profile across a single grain. Relative standard deviation (RSD) is the standard deviation divided by the mean value. The relative counting error (RCE) is the standard counting error of a typical background-corrected peak intensity divided by the background-corrected peak intensity (Bertin, 1978). Note that the totals reported are the mean and standard deviation of various sums, not the sum of the mean values. The percentages of end-member feldspars were calculated from the proportions of alkali and alkaline earth cations.

and trace elements. This precision can be recovered by using increased counting times (on peak and backgrounds), although this may in turn lead to specimen damage. Increased spot size may not be a practical solution when working with fine-grained crystals.

Most of the common hydrous calcsilicate minerals, including prehnite, pumpellyite and epidote, can be analyzed under 'routine' conditions for WDS (i.e.

10-nA beam current and a fully focused electron beam, approximately 1 μm in diameter). Zeolite minerals, however, are considerably more beam sensitive. As water is driven out of these minerals by overheating from prolonged exposure to the electron beam, the count rate of major cations (e.g. Si) will simultaneously increase (Reed, 1996). This behaviour can be readily verified in analyses performed with beam currents of 10 nA and a 1-μm spot diameter. Many zeolites (e.g. wairakite), however, will not dehydrate when analyzed with a rastered electron beam of identical current and effective beam size. This is because the heat of the electron beam is dissipated much more effectively under 60 000-times magnification rastered beam, which has the same effective diameter as the 1-μm fixed spot.

Palagonitized glass, found in many zeolite-grade metasites, is highly hydrated and also beam sensitive. Even under a rastered beam (6000×, effectively 10 μm in diameter) and at probe currents of 5 nA, Si Kα count rates increase by 5% after only 30 s of beam exposure (P. Schiffman, unpublished data). Both low beam currents (1 nA) and defocused beams (10 μm) may be necessary to ensure that no dehydration occurs during analyses of palagonitized glasses. By comparison, smectite may be much less susceptible to dehydration. Ferrell and Carpenter (1990) reported that dioctahedral, Na-montmorillonite analyzed with a 10-μm spot size was stable at 10 nA beam current for up to 90 s of exposure. Trioctahedral Ca-rich saponite, common to most very low-grade metasites, is stable under a fully focused beam at 10 nA for exposure periods up to 180 s (P. Schiffman, unpublished data).

Alkali migration in zeolites is more severe than in albite or microcline (cf. Ferrell & Carpenter, 1990) and is an analytical problem even at very low beam currents. At 1 nA and a fully focused electron beam, Na Kα count rates in analcime fall to 80% of their initial values after just 30 s; at 10 nA, beam diameters <20 μm result in significant Na loss after 60 s of counting time (P. Schiffman, unpublished data). Analysis of grains less than this diameter should be done with a lower beam current.

In summary, with a few exceptions, many of the hydrous minerals in metasites can be analyzed with a beam current of 10 nA and a small spot size, especially if the beam is rastered at a magnification equivalent to a fixed spot (e.g. 60 000 × = 1μm). Use of a rastered beam is also convenient if analytical points are being located during BSE imaging.

4.5.3 Standards

The ZAF formulations found in modern electron microprobe software packages ensure that the common metamorphic minerals found in low-grade metabasites can be accurately analyzed with common silicate and oxide standards. The most useful standards are synthetic crystalline materials with precisely prepared compositions or natural minerals that have been analyzed by multiple methods in many laboratories. The best or accepted analysis of such standards is considered to be the 'true' value for purposes of interlaboratory comparisons. Because good natural standards that have been analyzed in several laboratories are rare, most laboratories rely on a combination of synthetic standards and natural minerals for calibration.

Using appropriate beam conditions, most low-grade minerals can be analyzed with the same standards routinely used for higher grade minerals (Evarts & Schiffman, 1983; Bettison & Schiffman, 1988; Beiersdorfer & Day, 1995; Schmidt & Robinson, 1997). Zeolites can be analyzed using primary feldspar standards (e.g. Na, Si with albite; Al, Ca with anorthite; and K with orthoclase). Most other minerals common in metabasites (i.e. prehnite, pumpellyite, epidote, smectite, corrensite and chlorite) can be accurately analyzed using standards commonly used for amphibole. These include albite and omphacite (for Na), tremolite and periclase (for Mg), anorthite and corundum (for Al), quartz and tremolite (for Si), muscovite or orthoclase (for K), wollastonite or anorthite (for Ca), rutile (for Ti), rhodonite (for Mn) and hematite, fayalite or andradite (for Fe).

Secondary or 'working' standards should routinely be analyzed to monitor the day-to-day variation of analyses. Very few interlaboratory standards exist for the minerals found in very low-grade metabasites, and most laboratories create their own 'in-house' working standards. The compositions of natural minerals are commonly zoned and it is essential to document compositional ranges for such working standards.

4.5.4 Analytical uncertainties

Measurements should always be accompanied by a clear statement of analytical uncertainties. Any measurement has inherent systematic or random uncertainties and the importance of evaluating them cannot be overemphasized. For example, no statement about the significance of a similarity or difference between two measurements or groups of data should be made in absence of information about the size of uncertainties. Likewise, in the absence of an appropriate analysis of sources of error, it is impossible to determine whether trends exist in a suite of data, whether two variables are significantly correlated or, in general, if measurements are consistent or inconsistent with an hypothesis. Taylor (1982) and Bevington and Robinson (1992) have written excellent introductions to the basic concepts of error analysis that are appropriate for both novices and experienced analysts. Analyses without a statement of uncertainties are not useful data!

It is commonly not possible to publish complete sets of analytical data and it is standard practice to report 'representative analyses'. Unfortunately, many studies fail to report the population that is represented by an analysis. For example, a chemical analysis might be the result of analyzing a single spot with the electron microprobe or it might be the mean value of numerous such analyses. If the analysis is a mean value, how many spots were included? Does the mean represent analyses from a single mineral grain or several grains? Do the analyses represent a thin section or several? Without clear statements of the population that was sampled, the significance of data is greatly reduced and difficult to interpret broadly.

The goal of all analysts is to produce results that are both accurate and precise. That is, the results should be close to the 'true' value (accuracy) and repeated analyses should be reproducible within a narrow range (precision). The accuracy of an analytical method can be estimated by analysis of homogeneous standards (synthetic or natural) that are close to the composition of the desired unknown. However, most analytical uncertainties reported for microprobe analyses are measures of precision rather than accuracy. That is, the reported uncertainty is a measure of how well an analysis can be reproduced with repeated attempts. Most good laboratories have collections of homogeneous working standards that have been analyzed repeatedly by various analysts using methods that have been tested against other standards. If such working standards have compositions close to the unknowns being analyzed, the uncertainties associated with their repeated analysis are useful measures of the quality of analyses in the laboratory and should be reported.

4.5.4.1 Uncertainty in a single analysis

The uncertainty of a single point analysis, the estimation of minimum detection limits and the evaluation of the homogeneity of a group of spot analyses each depend on the error in determining X-ray intensities. The standard deviation in a single intensity determination, or standard counting error, is given by:

$$\sigma = \sqrt[2]{N}$$

where N is the number of counts accumulated. The standard counting error, appropriately modified for background corrections or other measurements required in the analytical procedure, is the most commonly used estimate of the uncertainty in the oxide analysis of a single spot (Bertin, 1978). The standard counting error is a true uncertainty because there is a 67% probability that an additional measurement will be within $\pm 1\sigma$ of the first.

The standard counting error is also useful for determining the minimum detection limits for an element. Although more elaborate definitions are available for very precise applications, a useful generalization is that a peak cannot be distinguished from the adjacent background unless its intensity exceeds three times the counting error in the measured background (Bertin, 1978).

Boyd *et al.* (1967) proposed that a sample should be considered inhomogeneous if the standard deviation of a group of analyses exceeded three times the uncertainty expected from counting error alone. Other methods have been proposed (Potts *et al.*, 1983; Mohr *et al.*, 1990), but this criterion is a simple and convenient way to detect chemical zoning.

It is unfortunate that standard counting errors are seldom reported in the literature. The calculation for background-corrected intensities is straightforward (e.g. Bertin, 1978) and, for most purposes, need only be carried out for a single representative analysis. Without a report of these uncertainties, it is impossible to make a meaningful comparison with other data, either from the same or other studies. Likewise, minimum detection limits are seldom reported, making it impossible to judge whether elements present in small amounts are analytically significant.

4.5.4.2 Uncertainty in multiple analyses

Chemical analyses are usually reported as the mean value of some modest number of analyses (n). Uncertainty is usually reported either as the *standard deviation* (σ) or as the *standard error of the mean* ($\bar{\sigma} = \sigma/\sqrt[2]{n}$). The two measures of uncertainty are useful for different purposes, but either can be calculated from the other provided that the number of measurements included in the mean value is recorded in the data tables.

The interpretation of the mean and the two measures of uncertainty depend on the absence of systematic error and the assumption that the errors are randomly distributed according to a gaussian distribution. The mean is the best estimate of the true composition only if there are no systematic errors in the analysis and only if the measurements are randomly distributed about the central value according to a gaussian function (the famous 'bell curve'). The standard deviation records the dispersion of the individual analyses about the mean value and indicates the uncertainty in a single measurement. For example, in a gaussian distribution, the probability that a single measurement is within $\pm 1\sigma$ of the mean is 67%. On the other hand, the standard deviation greatly overestimates the uncertainty in the mean value, which is given by the standard error of the mean.

Finally, it is essential to identify the sources of error that are included in the statement of uncertainty. Most commonly, uncertainties reported in the literature include the random error associated with the analytical method as well as the natural variation in the composition of the mineral being analyzed. Less commonly included are uncertainties in the standards being used, uncertainties in the calibration of the analytical method or uncertainties attributed to different analysts.

4.5.5 Criteria for a good analysis

Good analyses must be both precise and accurate. The accuracy of analytical methods in any laboratory can be estimated by analyzing primary standards. It is rarely possible, however, to rule out the existence of systematic errors in a particular analysis and we are usually confined to estimating the precision or reproducibility of a small number of analyses.

Feldspar. Most analysts judge the quality of analyses by whether the sum of all oxides analyzed is within a narrow range of 100% and whether the analysis is consistent with the expected stoichiometry of the mineral species analyzed. Both criteria are essential; an analysis summing to 100% is a necessary, but not sufficient criterion, for good analyses. Table 4.1 illustrates a chemical analysis of feldspar that is the mean of 10 analyses from a single grain. Although the standard deviations are significant only in the first decimal place, it is conventional to retain one or more digits than is formally justified if the numbers are to be used in further calculations or if the leading digit is 1 (Taylor, 1982). The analysis for total iron was omitted from this table because the average background-corrected intensity of the iron peak was less than three times the counting error in the measured background and, therefore, not distinguishable from it (Bertin, 1978). The total analysis lies well within the range of 98.5–101.5% that most analysts consider acceptable. Furthermore, there is no reason to suggest that the composition of the grain is heterogeneous because the relative standard deviations of the oxides are less than three times the uncertainty expected from counting error alone (RSD, RCE, Table 4.1) (Boyd *et al.*, 1967). Thus, on the basis of the oxide analysis alone, we judge that the mean analysis is acceptable and representative of the grain analyzed.

The calculated structural formula, however, suggests that the analysis is slightly non-stoichiometric (Table 4.1). The principles by which structural formulae are calculated from chemical analyses are well known and the calculation of feldspar is straightforward. Table 4.1 lists the mean value of the 10 calculated structural formulae as well as the standard deviation and standard error of the mean. Note that the standard errors imply that the mean values are significant in the third decimal place. Giaramita and Day (1990) have outlined the propagation of analytical uncertainties through structural formula calculations. The mean structural formula is not necessarily the same as the structural formula of the mean analysis, because the equation by which a structural formula is calculated is non-linear. However, the differences are insignificant in many cases (Giaramita & Day, 1990).

The best estimate of the total tetrahedral cations per eight oxygen is given by the mean and standard error (Table 4.1). Within the standard error of the mean, the total tetrahedral cations satisfy the requirements of stoichiometry. The total of alkali and alkaline earth cations, 0.986 ± 0.007, is two standard errors less than the required value of one and is therefore slightly, but significantly, deficient. A well crystallized, stoichiometric feldspar should also satisfy the criterion that:

$$Ca = Al\text{-}1 \text{ or } Ca - (Al\text{-}1) = 0.$$

Propagating the standard errors through this equation shows that the best estimate of this parameter is:

$$Ca - (Al\text{-}1) = (0.023 \pm 0.003) - (0.013 \pm 0.006)$$
$$= 0.010 \pm 0.007.$$

It is unlikely that this requirement of stoichiometry is violated, but more sophisticated tests would be required in order to beat this issue to death.

The slight deficiency in total alkali elements is probably a systematic error due to Na loss during the analysis. Although beam current was 2 nA and counting time was 10 s, the beam diameter was only 1 μm and some Na loss may have occurred. Nevertheless, restoration of 0.014 cations of Na to the formula unit would increase both Na and total cations so that the calculated per cent albite would increase only from 97.1 to 97.2%. We consider that this analysis of feldspar is sufficient for most petrological purposes.

Pumpellyite. The quality of analyses of other low-grade minerals, such as pumpellyite, is difficult to ascertain because the numbers of hydroxyl ions and trivalent ions are variable and cannot be measured directly in a microprobe study. The structural formula of pumpellyite contains 16 cations in four sites:

$$W_4X_2Y_4Z_6O_{20+x}(OH)_{8-x},$$

where W contains primarily Ca^{2+}, X contains both divalent (Fe^{2+}, Mn^{2+}, Mg^{2+}) and trivalent cations (Fe^{3+}, Al^{3+}), Y contains trivalent cations (Fe^{3+}, Al^{3+}) and Z contains Si^{4+} and Al^{3+} (Coombs *et al.*, 1976). The value of x is known to range from about 0.70 to 1.70, corresponding to 7.3–6.3 hydroxyl ions per formula unit, and more extreme values have been reported (Passaglia & Gottardi, 1973). The proportion of Fe^{3+} and Fe^{2+} is commonly adjusted so that all 16 cation sites are full, but depends also on the value of 'x'. As shown in Table 4.2, the value of 'x' adopted for a structural formula calculation affects not only the assumed number of oxygen and hydroxyl, but also the oxygen basis for the normalization and the total number of trivalent cations. Four trivalent cations are found in the Y site, and substitutions of R^{3+} in the X site are compensated by substitutions of O^{2-} for OH^-, univalent for divalent cations in the W site, and Al for Si in the Z site.

Table 4.3 illustrates the chemical composition of a pumpellyite from a low-grade blueschist. The analyses of TiO_2 (0.11 wt%), Cr_2O_3 (0.04 wt%) and K_2O (0.03 wt%) originally reported by Bröcker and Day (1995) were omitted from Table 4.3 because further review of the data showed that the X-ray peaks could not reliably be distinguished from background. All other oxides are present in amounts above the detection limit. One of the original 19 analyses was discarded because MgO and CaO were 4.1 and 3.7 standard deviations from their respective mean values. The analysis of total iron, FeO_{tot}, was recalculated as FeO and Fe_2O_3 and H_2O was calculated as 7 OH per formula unit based on the assumptions that there are 16 cations and $x = 1$.

Table 4.3 Chemistry of pumpellyite from low-grade blueschist (Bröcker & Day, 1995, sample 158, Table 4).

$n = 18$	Mean	Standard deviation	RSD	RCE	$\dfrac{RSD}{RCE}$
SiO_2	37.13	0.52	0.014	0.006	2.3
Al_2O_3	23.48	0.47	0.020	0.007	2.9
*Fe_2O_3	4.71	1.29	0.275	—	—
*FeO	4.98	0.87	0.175	—	—
MnO	0.30	0.08	0.261	0.216	1.2
MgO	1.68	0.12	0.073	0.031	2.4
CaO	21.73	0.28	0.013	0.010	1.3
Na_2O	0.19	0.04	0.221	0.183	1.2
Subtotal	94.20	0.56	0.006	—	—
*H_2O	6.41	0.04	0.006	—	—
Total	100.61			—	—
FeO_{tot}	9.22	0.80	0.087	0.027	3.2

Cations normalized to 24.5 O and 16.0 total cations

	Mean	Standard deviation	Standard error
Si	6.00	0.073	0.017
Al	4.48	0.088	0.02
Fe^{3+}	0.57	0.158	0.04
Fe^{2+}	0.67	0.117	0.03
Mn	0.041	0.011	0.003
Mg	0.404	0.029	0.007
Ca	3.77	0.050	0.012
Na	0.059	0.014	0.003
Total	16.000		

The mean includes 18 analyses of several grains in a single polished thin section. Abbreviations are the same as in Table 4.1. The asterisk (*) indicates oxides that were calculated on the assumption of 16 cations and $x = 1$. Note that the standard errors imply that the mean values of most of the calculated cations per formula unit are significant only in the second decimal place.

Table 4.2 Dependence of pumpellyite stoichiometry on the assumed value of 'x'.

O	$20 + x$
OH	$8 - x$
Oxygen basis for normalization	$24 + 0.5x$
Cations in X site	$R^{2+} = 2 - x$
	$R^{3+} = x$
Total R^{3+}	$4_Y + x_X + R + (6 - Si)_X + (6 - Si)_Z$

Subscripts X, Y and Z indicate the sites in which R^{3+} cations are found.

Consequently, the adjusted oxides and total weight per cent are internally consistent with the calculated structural formula.

The relative standard deviation (RSD) of FeO_{tot} is more than three times that expected from counting error alone (RCE). Although this may suggest that some of the differences in composition are real, pumpellyite occurs as very small, fine-grained, polycrystalline aggregates that are difficult to analyze even with extraordinary care. The relatively large standard deviations for several elements may reflect such additional sources of uncertainty rather than analytical significant variations in the composition of the pumpellyite.

Pumpellyite analyses may have totals outside the normally accepted range of $100 \pm 1.5\%$ because of uncertainty in the assumed value of 'x'. The calculated amount of H_2O in this pumpellyite may range from 6.7 to 5.7 wt% for values of x from 0.7 to 1.7, suggesting that the anhydrous total of a perfect analysis should be within the limits 93.8 ± 0.5 wt%. The acceptable limits might vary slightly for a pumpellyite of substantially different composition or unusual value of 'x'. Analyses in the range of $93.8 \pm 2.0\%$ might then be considered acceptable. All 18 analyses summarized in Table 4.3 are within this range.

The stoichiometry of pumpellyite is so variable that tests of good stoichiometry are elusive. Pumpellyite analyses with more than six Si cations are not uncommonly reported and must be considered suspect, but none of the other site occupancies are diagnostic of high quality analyses. The Z position in pumpellyite is usually filled by fewer than $0.2\,Al^{iv}$, but the structural formula would permit the charge deficiency of as many as $2.0\,Al^{iv}$ to be compensated by trivalent cations in the X site. The calculated amount of ferric iron is very sensitive to the assumed value of 'x'. Ferric iron in pumpellyite 158 (Table 4.3) ranges from 0.27 to 1.24 per formula unit as the assumed value of x varies from 0.7 to 1.7. The W site commonly contains slightly less than the ideal $4.0\,Ca$ ions, the remainder being filled with alkali, Mn or Fe^{2+}.

Phyllosilicates. It is also difficult to evaluate the quality of analyses for mafic phyllosilicates. The most common such minerals are trioctahedral chlorite, trioctahedral smectite and corrensite, which is a regular 1:1 interstratification of the former two structures. Analyses of metamorphic chlorites are most commonly reduced to a structural formula unit based on 14 or 28 oxygen (28 or 56 negative charges), e.g. clinochlore $(Mg_5Al)^{vi}(Si_3Al)^{iv}O_{10}(OH)_8$. Aside from the total wt% of the analysis, the only useful stoichiometric criterion for a good analysis is that the total cations, exclusive of Ca, Na, and K, should lie within analytical uncertainty of 10 if the analysis is calculated on a 14 oxygen basis.

In addition, unpublished work in our laboratory suggests that good analyses of mafic phyllosilicates contain no detectable Ti. Many analysts report small amounts of Na, K or Ca in chlorite analyses. In a routine WDS microprobe analysis, typically done with a beam current of 10 nA and a 10-s counting time on peak and background, the detection limit for the alkali oxides is about 0.05–0.1 wt%. Under these conditions, Na_2O and K_2O commonly cannot be reliably distinguished from background and probably should not be reported. Analyses for CaO are commonly above the lower limit of detection. Because there is no obvious site on which to substitute Ca in the chlorite structure, Ca may indicate the presence of a small amount of smectite.

Smectite in very low-grade rocks is very difficult to analyze and there are no useful stoichiometric criteria by which to judge the quality of the results. In the model system MgO–Al_2O_3–SiO_2, the compositions of trioctahedral smectite can be described by reference to the formula of talc, $(Mg^{vi})_3(Si^{iv})_4O_{10}(OH)_2$. Charge deficiencies caused by substitution of trivalent cations (mainly Al) in the tetrahedral site or vacancies in the octahedral site are balanced by the introduction of univalent or divalent cations (Na, K, Ca, Mg) in an interlayer position as well as by substitution of trivalent cations in octahedral sites. Consequently, the structural formula would have the form $X^{IL}(Mg,\ vacancy,\ Al)_3(Si,Al)_4O_{10}(OH)_2$. On an 11 oxygen basis, all Na, K and Ca and all other cations in excess of seven are assigned to the interlayer position (X^{IL}). Because vacancies may be expected, neither an excess nor a deficiency of cations ordinarily found in the octahedral or tetrahedral sites is diagnostic of a poor analysis. Likewise, the stoichiometry of any interlayered chlorite/smectite species, such as corrensite, does not serve as a useful indicator of analytical quality.

4.6 Quantitative application of electron microprobe data

Carefully acquired analytical data may be used to evaluate the degree to which low-grade rocks have approached equilibrium, to investigate the nature of the equilibria that control mineral compositions and the stability of the observed mineral assemblages, and to estimate the physical conditions of metamorphism. The purpose of this section is to illustrate some of the approaches that may be used to meet these goals.

Low-grade mafic rocks commonly contain relict igneous minerals and textures that clearly represent non-equilibrium. Nevertheless, even low-grade metamorphic assemblages rarely violate the phase rule because of the large number of components needed to describe a relatively small number of essential minerals. Achieving a favourable result upon applying the phase rule may not be very informative. Another requirement of equilibrium, which also forms a basis for the metamorphic facies concept, is that there should be a direct correspondence between the bulk composition of a rock, the observed mineral assemblage, and mineral compositions. Graphical illustrations of mineral assemblages in composition space are useful ways to test for this correspondence. Each chemical composition must display only one mineral assemblage, that is, there should be no crossing tielines.

4.6.1 Projections of low-grade mineral assemblages

The principles introduced by Thompson (1957) and Greenwood (1975) can be used to create a variety of useful projections that summarize the relationship between chemical composition space and mineral assemblages of metamorphosed mafic rocks. Projections from chlorite (e.g. Brown, 1977; Liou *et al.*, 1985a; Bröcker & Day, 1995), from epidote (e.g. Harte & Graham, 1975; Laird, 1980; Beiersdorfer & Day, 1995a) and from prehnite (Springer *et al.*, 1992) each have advantages, but no single projection is fully adequate to describe the compositions of the coexisting minerals.

The essence of such projections is rather simple (Thompson, 1957; Greenwood, 1975). The projection of compositions into a smaller, more convenient composition space requires that we choose the components from which to project and the components that define the plane to receive the image of the composition space such that: projection points + image   components = total   independent components.

For example, Fig. 4.12 is a projection from SiO_2 and H_2O to the image plane defined by Al_2O_3, CaO and MgO. The algebra of the projection consists

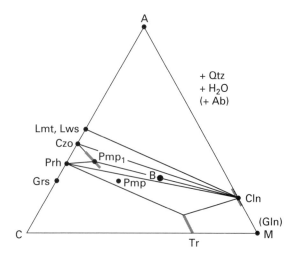

Fig. 4.12 Projection of CMASH minerals to the plane Al_2O_3–CaO–MgO from the ubiquitous phases quartz + H_2O. End-member mineral compositions are summarized in Table 4.4. Glaucophane (Gln) is projected from albite + quartz + H_2O. The ranges of common compositions are shown with the heavy weight lines. The average composition of basalt (LeMaitre, 1976) is shown as the filled circle labelled 'B'. Coordinates of the image plane are: A = Al_2O_3; C = CaO; M = MgO.

simply of removing the projection components from the composition matrix (Table 4.4) and normalizing the remaining image components to 100% for plotting in the ACM plane.

Although it is possible to project from any oxide component, a direct correspondence between the *unprojected bulk chemical composition* of a rock and its *projected mineral assemblage* is only retained if the projection components are present as phases in the system (Thompson, 1957). This is what Greenwood (1975) called a 'thermodynamically legal' projection. Thus, Fig. 4.12 retains the direct correspondence between bulk composition in the model system and mineral assemblages, and its utility as a phase equilibrium diagram, only if quartz and H_2O are present as phases.

The simplest projections, such as Fig. 4.12, are from phases that can be defined as simple oxides. However, as long as the composition of a mineral can be expressed as a linear combination of the projection components and the image components, it can be properly projected. Glaucophane (Gln = $Na_2Mg_3Al_2Si_8O_{22}(OH)_2$) and albite ($NaAlSi_3O_8$) commonly coexist with phases illustrated in Fig. 4.12. It

Table 4.4 Compositions of minerals and exchange operators in the system CMASH.

	Abbreviations	CaO	MgO	Al_2O_3	SiO_2	H_2O	$X_{Al_2O_3}$	X_{CaO}	X_{MgO}
Minerals									
Anorthite	An	1	0	1	2	0	0.50	0.50	0.00
Clinochlore	Cln	0	5	1	3	4	0.17	0.00	0.83
Clinozoisite	Czo	2	0	1.5	3	0.5	0.43	0.57	0.00
Glaucophane	Gln	0	3	1	8	1	0.00	0.00	*1.00
Grossular	Grs	3	0	1	3	0	0.25	0.75	0.00
Laumontite	Lmt	1	0	1	4	4	0.50	0.50	0.00
Lawsonite	Lws	1	0	1	2	2	0.50	0.50	0.00
Prehnite	Prh	2	0	1	3	1	0.33	0.67	0.00
Pumpellyite	Pmp	4	2	2	6	4	0.25	0.50	0.25
	Pmp_1	4	1	2.5	6	3.5	0.33	0.53	0.13
Tremolite	Tr	2	5	0	8	1	0.00	0.29	0.71
Exchange operators	*amh*	0	−1	0.5	0	−0.5	—	—	—
	cm	1	−1	0	0	0	—	—	—
	tk	0	−1	1	−1	0	—	—	—

is possible to display the glaucophane composition on the ACM diagram by projection from albite, in addition to quartz and H_2O. The projection is possible because glaucophane composition can be rewritten as a linear combination of the projection components ($NaAlSi_3O_8$, SiO_2 and H_2O) and the image components (Al_2O_3, CaO, MgO):

$$Gln = 2\ NaAlSi_3O_8 + 5\ SiO_2 + H_2O + 3\ MgO.$$

When projected from Ab + Qtz + H_2O, to the ACM plane, glaucophane will plot at pure MgO (Fig. 4.12). More complex and less obvious projections are best constructed by using techniques of matrix algebra outlined clearly by Greenwood (1975). Such calculations can now be carried out with many spreadsheet programs, as illustrated by Bucher & Frey (1994).

The description of common minerals in low-grade mafic rocks requires at least Na_2O, FeO and Fe_2O_3 in addition to CMASH. Nevertheless, some useful, basic chemographic relationships can be summarized in the simpler system because MgO and Al_2O_3 may be considered as proxies for FeO and Fe_2O_3, and because essential Na_2O occurs only in albite in many low-pressure, low-grade rocks.

Although the Ca–Al silicates show little compositional variation in the CMASH system, Fig. 4.12 illustrates three important solid solutions. The compositions of calcic amphibole and chlorite (clinochlore), in the CMASH system, vary primarily

by the tschermaks exchange ($tk = Al_2Mg_{-1}Si_{-1}$), which displaces the compositions along lines of constant CaO. Minor amounts of Ca–Mg exchange ($cm = CaMg_{-1}$) broaden the range of observed compositions slightly along lines of constant Al_2O_3. The composition of pumpellyite varies by three independent substitutions in CMASH. In addition to minor amounts of tk and cm, the dominant substitution can be written:

$$Ca_4Mg_2Al_4Si_6O_{20}(OH)_8 + x(AlMg_{-1}H_{-1})\ \text{or}$$
$$Pmp + x(amh)$$

and is equivalent to the variation described by the conventional structural formula, $Ca_4(Mg_{2-x}Al_x)Al_4$ $Si_6O_{20+x}(OH)_{8-x}$. In the projection from $SiO_2 + H_2O$ (Fig. 4.12), this substitution corresponds to the variation from pumpellyite to clinozoisite. Although the amount of hydrogen in the structure is seldom measured directly, it is commonly estimated that the value of x is equal to one (Pmp_1, Fig. 4.12). Notice that the composition of pumpellyite can lie above or below the Chl + Prh tieline, depending on the Al content of pumpellyite and chlorite. The different chemographies imply different possible reaction relationships.

4.6.2 Projection from chlorite

The effects of ferric iron substitution on mineral assemblages in the system CMASH + Fe_2O_3 are

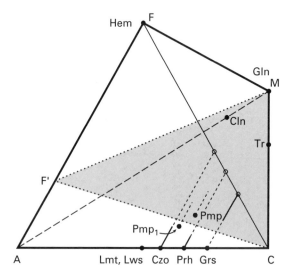

Fig. 4.13 Phases in the system $Al_2O_3 - CaO - MgO - Fe_2O_3$. Dashed lines represent the extent of $AlFe_{-1}$ substitution in Czo, Grs, Prh and Pmp. The heavy weight portion of these lines indicates typical compositions. Garnet solid solutions are typically andradite, rich in ferric iron. The shaded plane McF' is an arbitrary plane of constant $Fe_2O_3/(Fe_2O_3 + Al_2O_3)$.

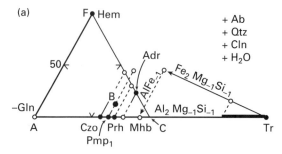

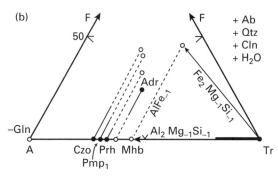

Fig. 4.14 Projections from clinochlore + albite + quartz + H_2O. (a) The ACF image plane. Coordinates of the image plane are: $A = Al_2O_3 - 0.2 MgO - Na_2O$; $C = CaO$; $F = Fe_2O_3$. B is the composition of average basalt (LeMaitre, 1976). (b) The ATrF image plane. Coordinates are: $A = Al_2O_3 - 0.2 MgO + 0.5 CaO - Na_2O$; $Tr = 0.5 CaO$; $F = Fe_2O_3$. In the natural Fe-bearing systems, the MgO is replaced by MgO + FeO + MnO. Gln appears on these diagrams only as the negative projection (−Gln) (Thompson, 1957; Greenwood, 1975). Mhb = magnesio-hornblende; Adr = andradite; others as in Table 4.4.

illustrated in the tetrahedron Al_2O_3–CaO–MgO–Fe_2O_3 (Fig. 4.13), which is a projection from SiO_2 and H_2O. The diagram can be simplified further by projecting mineral compositions from the chlorite composition, clinochlore, to the ACF face of the tetrahedron. Various projections from chlorite have been used to illustrate mineral assemblages (e.g. Brown, 1977; Liou *et al.*, 1985a; Patrick & Day, 1989; Bröcker & Day, 1995). Figure 4.14 illustrates the chlorite projection used by Liou *et al.* (1985a). The advantages of this projection are:

1 chlorite is virtually ubiquitous in low-grade metabasites, so the most common assemblages are easily summarized;

2 many phases of interest lie near or on the plane Al_2O_3–CaO–Fe_2O_3 and change their relative positions very little in the projection;

3 the effects of $AlFe_{-1}$ substitution on the stable mineral assemblages can be illustrated if analyzed compositions are illustrated properly on Fig. 4.14(a) or (b).

The major disadvantages of this projection are:

1 the effects of $MgFe_{-1}$ substitution cannot be illustrated and MgO, FeO and MnO must be combined as a single component;

2 the projected composition of tremolite lies at large negative values of Al_2O_3 and only the negative projection of glaucophane can be shown;

3 the projection does not preserve the $Fe^{3+}/(Fe^{3+} + Al)$ ratios of the projected phases and ferric iron equivalents of aluminous phases do not necessarily project to the CaO–Fe_2O_3 join, making it difficult to use the diagram as a basis for simply sketching possible phase assemblages.

All phases can be plotted in the projection, however, using the plotting coordinates summarized in Table 4.5. Finally, if the Al content of chlorite varies significantly among the samples illustrated, the topology of the projection may not preserve phase equilibrium relations properly.

Table 4.5 Components and plotting coordinates for various projections.

Components		
Projection points	Image plane	Plotting coordinates
Cln + Qtz + Ab + H_2O Liou *et al.* (1985a)* Combine MgO + FeO	Al_2O_3 – CaO – Fe_2O_3	$A = Al_2O_3 - 0.2\,MgO - Na_2O$ $C = CaO$ $F = Fe_2O_3$
Cln + Qtz + Ab + H_2O Liou *et al.* (1985a)* Combine MgO + FeO	Al_2O_3 – Tr – Fe_2O_3	$A = Al_2O_3 - 0.2\,MgO - Na_2O$ $+\ 0.5\,CaO$ $Tr = 0.5\,CaO$ $F = Fe_2O_3$
Tr + Qtz + Ab + H_2O Combine MgO + FeO	Al_2O_3 – CaO – Fe_2O_3	$A = Al_2O_3 - Na_2O$ $C = CaO - 0.4\,MgO$ $F = Fe_2O_3$
Ep + Qtz + Ab + H_2O Harte & Graham (1975) Combine Al_2O_3 + Fe_2O_3	Al_2O_3 – MgO – FeO	$A = Al_2O_3 - 0.75\,CaO - Na_2O$ $F = FeO$ $M = MgO$
Prh + Qtz + Ab + H_2O Springer *et al.* (1992) Combine Al_2O_3 + Fe_2O_3	Al_2O_3 – MgO – FeO	$A = Al_2O_3 - 0.5\,CaO - Na_2O$ $F = FeO$ $M = MgO$
Lws + Qtz + Ab + H_2O Combine Al_2O_3 + Fe_2O_3	Al_2O_3 – MgO – FeO	$A = Al_2O_3 - CaO - Na_2O$ $F = FeO$ $M = MgO$
Lws = Hem + Qtz + Ab + H_2O Bröcker & Day (1995) No combined components	Al_2O_3 – MgO – FeO	$A = Al_2O_3 - CaO - Na_2O$ $F = FeO$ $M = MgO$
Lws + Cl + Hem + Qtz + H_2O Patrick & Day (1989) No combined components	$NaAlO_2$ – MgO – FeO	$N = 2Na_2O$ $F = FeO + 2.5\,CaO + 2.5$ $\quad Na_2O - 2.5\,Al_2O_3$ $M = MgO + 2.5\,CaO + 2.5$ $\quad Na_2O - 2.5\,Al_2O_3$

* Plotting coordinates do not correspond to those used by Liou *et al.* (1985a).

Any inconvenience caused by the large negative plotting coordinates of tremolite can be alleviated by projection from chlorite to the plane Al_2O_3–CaO–Tr (Fig. 4.14b), as suggested by Liou *et al.* (1985a, Fig. 6), but the alternate image plane does nothing to change the fact that ferric iron end-members do not project to the CaO–Fe_2O_3 join. In both versions of the chlorite projection, $AlFe_{-1}$ substitution occurs along lines of constant CaO, but ferric iron end-members do not plot at easily predicted locations. Ferric iron analogues of pumpellyite plot at negative contents of 'A' and ferric iron analogues of tremolite-magnesiohornblende solid solutions plot along the vector describing ferri-tschermaks substitution (Fe_2 $Mg_{-1}Si_{-1}$, Fig. 4.14). Other chlorite projections also suffer from similar deficiencies (e.g. Patrick & Day,

1989, Fig. 6; Bröcker & Day, 1995, Fig. 6a).

A projection from tremolite *would* preserve $Fe^{3+}/(Fe^{3+} + Al)$ ratios because the end-member amphibole contains no aluminium and is common to all possible planes of constant $Fe^{3+}/(Fe^{3+} + Al)$ (CF'M, Fig. 4.13). Amphibole is not ubiquitous in low-grade rocks and it commonly contains a small but significant amount of tschermak substitution ($Al_2Mg_{-1}Si_{-1}$, Figs 4.12 and 4.14). Nevertheless, tremolite projections might be useful for displaying some rocks with low aluminium content.

4.6.3 Projections from calcium–aluminium silicates

The effects of ferrous iron on mineral assemblages

in the system CMASH + FeO can be visualized by expanding Fig. 4.12 to include FeO and projecting from a calcium aluminium silicate, such as lawsonite, clinozoisite or prehnite, to the join Al_2O_3–FeO–MgO. The principal advantages of such projections are:

1 it is possible to illustrate the marked effects of $MgFe_{-1}$ substitution on the mineral assemblages;

2 the MgO/(MgO + FeO) ratios of all phases are preserved because only rarely do the projection phases contain significant amounts of FeO or MgO (cf. Digel & Gordon, 1995);

3 the various possible projection points are common minerals in low-grade rocks.

The main disadvantages are:

1 the effects of $AlFe_{-1}$ substitution are ignored and both Al_2O_3 and Fe_2O_3 must be combined as a single component;

2 pumpellyite and tremolite are a long way from the image plane (Fig. 4.12) so that small variations in composition or analytical error may cause relatively large shifts of the plotted position.

Projections from epidote were introduced by Harte and Graham (1975) and have been used in various forms since then to analyze assemblages in mafic rocks (e.g. Laird, 1980; Beiersdorfer & Day, 1983, 1995a; Bevins & Robinson, 1994, 1995). Projections

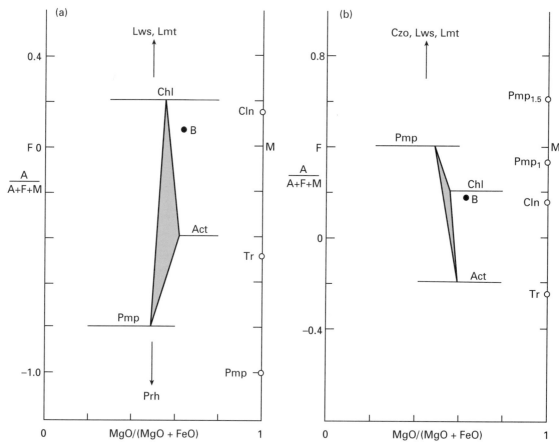

Fig. 4.15 Projections to the image plane Al_2O_3 – FeO – MgO in orthogonal format. The AF and AM sides of the AFM triangle have been rotated into parallelism and the A apex is now a line of constant A/(A + F + M). (a) Projection from clinozoisite + albite + quartz + H_2O. Coordinates of the image plane are: A = Al_2O_3 – 0.2 MgO – 0.75 CaO – Na_2O; F = FeO; M = MgO.

(b) Projection from prehnite + albite + quartz + H_2O. Coordinates are: A = Al_2O_3 – 0.5 CaO – Na_2O; F = FeO; M = MgO. In natural, ferric iron-bearing systems, the Al_2O_3 is replaced by Al_2O_3 + Fe_2O_3. The filled circles labelled 'B' represent the composition of average basalt (LeMaitre, 1976).

from lawsonite and prehnite have been used less commonly (Patrick & Day, 1989; Springer *et al.*, 1992; Bröcker & Day, 1995). Figure 4.15 illustrates projections from epidote and prehnite, for common compositions of coexisting chlorite, actinolite and pumpellyite (Beiersdorfer & Day, 1995a) and for average basalt (LeMaitre, 1976). It is apparent from Fig. 4.12 that lawsonite and laumontite will project from clinozoisite to the alumina apex, that prehnite projects away from the A–M line and cannot be represented in the AFM plane, and that pumpellyite and tremolite project from epidote to large negative values of Al_2O_3. In Fig. 4.15, tielines to lawsonite

and laumontite are lines of constant MgO/(MgO + FeO) extending to A = 1 and the calcium-poor parts of tielines from prehnite to other phases are lines of constant MgO/(MgO + FeO) extending to infinitely negative values of A (Fig. 4.15a).

Most of these projections fail to provide a rigorous view of composition space because it is necessary to combine two or more oxides as a single component (e.g. $Al_2O_3 + Fe_2O_3$ or FeO + MgO + MnO). Perhaps the most successful projection from a calcsilicate was illustrated by Bröcker and Day (1995) (Fig. 4.16). They projected blueschist assemblages from Ab + Qtz + Lws + Hem + H_2O to the image plane $Al_2O_3 -$ FeO – MgO. The projection preserves the ratio of FeO and MgO and does not require combining Al_2O_3 and Fe_2O_3, because excess hematite is present and all phases are saturated with Fe_2O_3. The presence of this additional phase in all the assemblages plotted greatly enhances the utility of the projection.

4.6.4 Algebraic methods

Graphical methods of testing the correspondence among bulk compositions and mineral compositions are appealing because it is possible to visualize the results. Most mineral assemblages, however, are not saturated with respect to enough major components that projections can be applied rigorously. For example, all the projections except those proposed by Patrick and Day (1989) and Bröcker and Day (1995) are undersaturated with respect to either Fe_2O_3 or FeO and require combining components in order to display in two dimensions. Even projections that are faithful reproductions of composition space commonly are able to illustrate only phase relations in a small part of the composition space available to natural rocks. For example, the lawsonite projections used by Patrick and Day (1989) and Bröcker and Day (1995) (Fig. 4.16) are useful only for those assemblages saturated with respect to Fe_2O_3. Phase relations in the hematite-free part of composition space cannot be represented rigorously.

In such circumstances, algebraic methods offer the possibility of rigorously testing whether an assemblage is compatible with the hypothesis of equilibrium or whether it preserves a reaction relationship among the phases. Greenwood (1967, 1968) discussed the so-called *n*-dimensional tieline problem in detail and several authors have applied

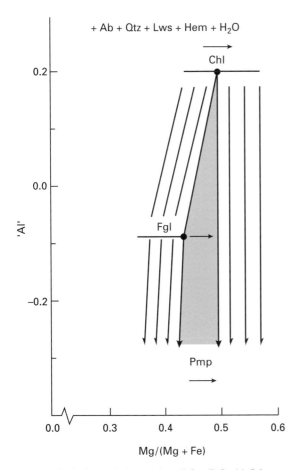

Fig. 4.16 Projection to the image plane $Al_2O_3 -$ FeO – MgO from albite + quartz + lawsonite + hematite + H_2O. Coordinates of the image plane are: A = $Al_2O_3 -$ CaO – Na_2O; F = FeO; M = MgO. Horizontal arrows indicate the direction in which the phases change composition with increasing grade of metamorphism.

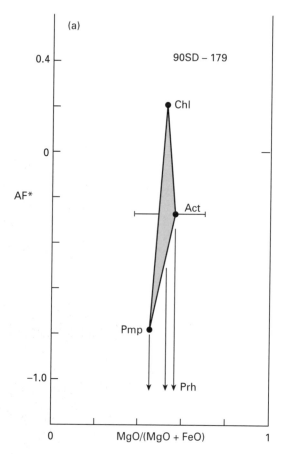

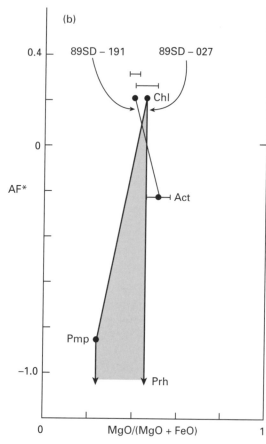

Fig. 4.17 Epidote projection of mineral assemblages from Flin Flon metabasalts (Digel & Gordon, 1995). (a) The assemblage Chl + Act + Pmp + Prh from the transition zone at the actinolite isograd (90SD-179). (b) Assemblages from above (Chl + Act; 89SD-191) and below (Prh + Pmp + Chl; 89SD-027) the actinolite isograd. AF* indicates that Fe_2O_3 has been combined with Al_2O_3 in the A coordinate.

least-squares regression techniques in order to determine whether a reaction or mass balance relationship exists among mineral phases in multidimension composition spaces (Lang & Rice, 1985; Giaramita & Day, 1991).

The most powerful and convenient technique for determining the existence of reaction relationships is known as 'singular value decomposition' (Fisher, 1989). One advantage compared to least-squares regression methods is that it can be applied to underdetermined systems (i.e. more phases than

analyzed elements). Several authors have used this approach to analyze high-grade mineral assemblages (Giaramita & Day, 1991; Gordon *et al.*, 1991). Recently, Digel and Gordon (1995) applied singular value decomposition to pumpellyite-bearing metabasalts.

Digel and Gordon examined three samples from the prehnite–pumpellyite to greenschist facies transition in the Flin Flon belt, Manitoba, Canada. Figure 4.17 illustrates these assemblages in a projection from Ab + Qtz + Ep + H_2O. Sample 90SD-179 comes from the transition between the Prh + Pmp and Act zones of metamorphism. In addition to the projection phases, the sample contains the assemblage Chl + Act + Pmp + Prh. Is this four-phase assemblage a stable assemblage with two or more phase rule degrees of freedom or does it preserve a reaction relationship among the phases? Using the singular value decomposition methods outlined by Fisher (1989), Digel and Gordon (Table 5, 1995) found six independent mass balance relationships among the phases. These mass balances may not correspond to any reaction that took place in the rock, but they imply that, within analytical uncertainty, tielines among the phases present in this sample intersect in the complete composition space. Given the number of components and phases in the rock, this result might not have been expected. One crossing tieline can be seen in Fig. 4.17(a), where the Chl + Prh tieline clearly intersects the phase space defined by Pmp + Act + Chl and crosses Pmp + Act. In fact, the reported analytical uncertainty in MgO/(MgO + FeO) of Act is so large that one could not reject the hypothesis that Chl, Act and Pmp are colinear in this projection. If so, there must be a reaction relationship such that Act = Pmp + Chl.

Singular value decomposition is also useful for comparing assemblages from different rocks. Figure 4.17(b) illustrates the assemblages in samples 89SD-191 and 89SD-027 from above and below the actinolite isograd, respectively. Digel and Gordon (Table 6, 1995) applied singular value decomposition to the phase composition matrix of the combined assemblages of the two rocks. They found three independent mass balance relations that satisfied the requirement that all phases from one sample were on the opposite side of the equation from all phases of the other sample. This result implies that the composition spaces defined by the two samples intersect in composition space. That is, 'tielines' intersect in n-dimensional composition space within analytical uncertainty. This result is expected if the actinolite isograd is, in fact, the expression of a reaction and not simply the result of more magnesian compositions in the actinolite-bearing metabasalts. Because of the large analytical uncertainty attached to MgO/(MgO + FeO) of chlorite in sample 89SD-027 (Fig. 4.17(b)), it is ambiguous from the projection alone whether or not the tielines are required to cross.

The status of ferric iron in these assemblages is not clear. Ordinarily, Al_2O_3 and Fe_2O_3 are combined as a single component in projections from epidote (Fig. 4.17). It would be natural to inquire, therefore, whether the four-phase assemblage illustrated in Fig. 4.17(a) could be stable if Fe_2O_3 were considered an independent component. That is, would the tielines 'uncross' if a dimension were added to the diagram in order to illustrate Fe_2O_3? In principal, an answer to this question could be found by including Fe_2O_3

for each phase in the composition matrix for singular value decomposition. Instead, Digel and Gordon (1995) chose to regard O_2 as a mobile component that could exchange freely with a coexisting fluid phase. All iron was expressed as FeO and is illustrated this way in Fig. 4.17. Although the necessary calculations have not been done, it is unlikely that consideration of ferric iron would change the conclusion that a mass balance exists in the assemblage 90SD-179. The large analytical uncertainties suggest that we could not reject the hypothesis that the three phases have identical MgO/(MgO + FeO) ratios. Consequently, MgO and FeO would function as a single component and there would still be four phases (Chl + Act + Pmp + Prh) in a three-component system (Al_2O_3–Fe_2O_3–FMO).

4.6.5 Petrogenetic grids

Petrogenetic grids are useful tools for illustrating the physical and chemical conditions of metamorphism. The basic principles that govern the disposition of univariant curves at invariant points were outlined by Schreinemakers (1915–1925) and their application to the construction of phase equilibrium diagrams

Table 4.6 Reactions in the system Na_2O – CaO – MgO – Al_2O_3 – SiO_2 – H_2O (NCMASH).

1 Anl + Qtz = Ab
2 Jd + Qtz = Ab
3 Gln + Lws = Pmp + Chl + Ab + Qtz
4 Gln + Czo + W = Pmp + Chl + Ab
5 Pmp + Chl + Qtz = Czo + Tr + W
6 Pmp + Qtz = Prh + Czo + Chl + W
7 Pmp + Qtz + W = Prh + Chl + Lmt
8 Pmp + Chl + Qtz + W = Tr + Lmt
9 Tr + Lmt = Prh + Chl + Qtz + W
10 Pmp + Tr + Qtz = Prh + Chl + W
11 Tr + Czo + W = Prh + Chl + Qtz
12 Prh + Lmt = Czo + Qtz + W
13 Pmp + Lmt = Czo + Chl + Qtz + W
14 Lws + Qtz + W = Lmt
15 Pmp + Lws = Czo + Chl + Qtz + W
16 Gln + Lws = Czo + Chl + Ab + Qtz + W
17 Czo + Gln + Qtz + W = Tr + Chl + Ab
18 Pmp + Gln + Qtz + W = Tr + Chl + Ab

The high-pressure assemblage is on the left side for most of these unbalanced equilibria. The formula unit for analcime (Anl) is $NaAlSi_2O_6.H_2O$. The compositions and abbreviations for other minerals are listed in Table 4.4.

has been explored by Zen (1966), Day (1972, 1976) and Zen and Roseboom (1972), among others. Liou *et al.* (1985a, 1987) used these principles to construct a petrogenetic grid that was calibrated with three kinds of experimental data:

1 experiments on the end-member composition of single mineral phases;
2 the results of experiments on natural basaltic rocks;
3 experiments on equilibria in a model, Fe-free basaltic system (NCMASH). The results of this effort are reproduced in Fig. 4.18 as modified from Beiersdorfer and Day (1995).

The major advantages of this grid are that it portrays the most important mineral assemblages in low-grade mafic rocks. The assemblage pumpellyite + albite + chlorite is stable over a wide *P–T* range, limited by reactions 3, 4, 5, 6, 7 and 8 (Table 4.6). Glaucophane + lawsonite and glaucophane + epidote are stable at high pressures. At slightly lower pressures, pumpellyite + actinolite are stable with chlorite, whereas prehnite + actinolite + chlorite are stable at low pressure in the lightly stippled region (Fig. 4.18). The grid allows a wide region for the stability of epidote + chlorite + quartz ±

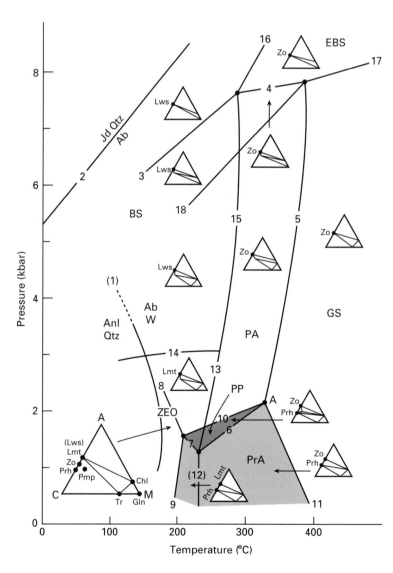

Fig. 4.18 Schematic *P–T* grid for low-temperature mafic rocks (modified after Liou *et al.*, 1985a, 1987). Facies abbreviations: BS, blueschist; EBS, epidote blueschist; GS, greenschist; PA, pumpellyite actinolite; PP, prehnite pumpellyite; PrA, prehnite actinolite; ZEO, zeolite. Mineral abbreviations as in Table 4.4. Only minerals stable with chlorite are illustrated in the compatibility diagrams.

pumpellyite ± prehnite limited at low temperatures by reactions 12, 13 and 14.

The grid as illustrated has several inadequacies. It oversimplifies natural assemblages by ignoring Fe as an additional component. For example, the assemblage implied by invariant point A (Prh + Pmp + Chl + Act + Ep + Ab + Qtz + H_2O) becomes divariant if the components FeO and Fe_2O_3 are included with NCMASH. Liou et al. (1985a, 1987) handled this inadequacy by estimating the displacement of various equilibria caused by Fe^{3+} substitution. The very small region for the stability of prehnite + pumpellyite, outlined by reactions 6, 7 and 10 (Fig. 4.18), is a major concern because this assemblage is so common in nature. The very small stability region for Prh + Pmp, in contrast to the grid presented by Liou et al. (1985a, 1987), is a consequence of illustrating reaction 10 with a positive slope at invariant point A (Fig. 4.18). If chlorite has an Al content similar to clinochlore, as is typical of chlorite coexisting with epidote and pumpellyite (Beiersdorfer & Day, 1995), reaction 10 must be a dehydration reaction as written in Table 4.6 and is likely to have a positive slope. Another, less serious, defect is that the first appearance of Pmp + Chl + Qtz in nature is unlikely to occur by reaction 8, especially because the first appearance of actinolite is usually at higher grades than the first appearance of pumpellyite (e.g. Springer et al., 1992).

The advent of internally consistent sets of thermodynamic data for end-member mineral compositions (Berman, 1988; Holland & Powell, 1990) permits the direct calculation of mineral equilibria and the possibility to resolve some uncertainties in the petrogenetic grid. The enthalpies, entropies, heat capacities and molar volumes included in an internally consistent data set can be used to recover the experimental results from which they were derived or to calculate equilibria for which no experimental calibration is available. Considerable care must be used when adding new data to the set to be sure the internal consistency of the remainder of the data is not affected. Evans (1990) and Frey et al. (1991) used this approach to study equilibria in the NCMASH system. Both studies utilized the internally consistent data of Berman (1988) and GE0CALC software (Berman et al., 1987). Thermodynamic data for glaucophane and pumpellyite are not part of the database and

were added to the database by the authors, but from different sources. Consequently, although both studies are internally consistent, they may not be consistent with one another.

A comparison of Fig. 4.18 with calculations in the Fe-free system NCMASH (Evans, 1990; Frey et al., 1991) gives mixed results. The calculations show that neither clinozoisite nor zoisite have a stability field with glaucophane in the iron-free system (Evans, 1990) and that the overlap between the clinozoisite stability field and the field for Prh + Pmp + Chl is almost non-existent (Frey et al., 1991, Fig. 3). On the other hand, calculations in the Fe-free system confirm a very small field for Prh + Pmp + Chl at low pressure (Frey et al., 1991, Fig. 3). However, even calculations in the iron-bearing system suggest that the width of the Prh + Pmp + Chl field is only about 25°C, extending between c. 3 and c. 4 kbar (cf. Fig. 4.19).

The effects of Fe on equilibria in NCMASH can be included in the calculations, if appropriate activity models are available for the solid solutions. Unfortunately, the thermodynamic properties of most of the solid solutions involved in these equilibria are poorly known. The two studies have used different solution models for several minerals and have made calculations for a considerably different range of end-member activities. Although they have used the same model for pumpellyite activity, the range of pumpellyite activities for which calculations were made do not overlap (Evans, 1990, Table 3; Frey et al., 1991, Fig. 6).

Figure 4.19 is a schematic summary of calculations made with average activities of the end-members in the system NCMASH (Frey et al., 1991, Fig. 5; modified by Beiersdorfer & Day, 1995). Most of the equilibria studied by Evans (1990) lie in the field of glaucophane + zoisite stability above the stability field of pumpellyite. Figure 4.19 retains many of the features shown in Fig. 4.18. In the presence of Ab + Chl + Qtz + H_2O, prehnite stability is restricted to low pressure, pumpellyite is stable at higher pressures, and the region of prehnite + pumpellyite stability is very limited. There are also several notable differences. The stability field for Pmp + Chl + Qtz becomes very narrow with increasing pressure and the lower temperature limit of pumpellyite stability between 4 and 7 kbar is the reaction Lws + Tr = Pmp + Chl + Qtz + H_2O. The equilibrium Pmp +

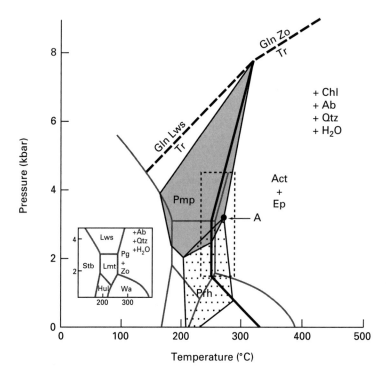

Fig. 4.19 Schematic petrogenetic grid calculated in the system NCMASH with average activities of the end-members (modified after Frey *et al.*, 1991 and Beiersdorfer & Day, 1995). The inset shows the stability fields of lawsonite and various zeolites. The upper pressure stability of tremolite in the presence of excess Chl, Ab, Qtz and H_2O is shown as the heavy dashed line. The steep, heavy weight line shows the lower thermal stability of Czo/Ep. The dashed rectangle shows the possible range of locations for assemblage A for various possible choices of end-member activities.

Chl + Ab + Qtz + H_2O = Gln + Lws (reaction 3, Fig. 4.18) is restricted to an interval too small to illustrate in the uppermost termination of the pumpellyite field. The calculations apparently allow almost no stability field for the assemblage Pmp + Chl + Ab + Qtz + Gln + Lws, which is, nevertheless, widespread in the northern Coast Ranges of California (Bröcker & Day, 1995). Finally, the lower temperature stability of clinozoisite/epidote is indicated by the bold lines passing through the pumpellyite and prehnite stability fields and appears to offer only a limited temperature interval for pumpellyite + epidote, which is very common in nature.

The major advantage of such calculated grids is that the internally consistent thermodynamic database forms an objective basis for estimating the relative stability of mineral assemblages, permitting a good general overview of the stability fields of assemblages that are essential to understanding the metamorphic facies. They are less useful for understanding the details of observed assemblages, however, because we are attempting to portray as univariant lines and invariant points, mineral assemblages that are controlled by high variance,

continuous equilibria in nature. For example, the assemblage Pmp + Prh + Act + Ep + Chl + Ab + H_2O, which is illustrated as 'invariant point' A in Figs 4.18 and 4.19, must be divariant in the system NCMASH with FeO and Fe_2O_3 as additional components. Frey *et al.* (1991) illustrated the high variance nature of these equilibria by repeating calculations for various choices of end-member activities. The heavy dashed box surrounding invariant point A (Fig. 4.19) shows the calculated location of the assemblage for a wide range of possible activities. The variation of activities that leads to the expanded *P–T* field for assemblage 'A' also permits a much wider stability field for the various key assemblages as shown in Fig. 4.20.

It is clear from the results of these repeated calculations that the overlap in *P–T* space among critical assemblages is extensive. The prehnite–pumpellyite metagreywacke facies was defined by Coombs (1960) to include assemblages that might coexist with quartz + prehnite + chlorite or quartz + albite + pumpellyite + chlorite *without* zeolites, lawsonite or jadeite. According to results shown in Figs 4.19 and 4.20, zeolites or lawsonite may be stable throughout

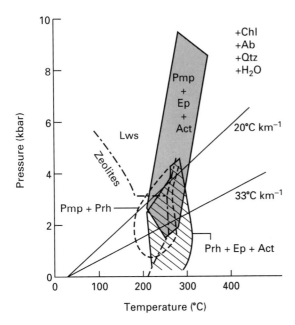

Fig. 4.20 Permissible *P–T* fields of key assemblages for various choices of end-member activities. (Modified after Frey *et al.*, 1991 and Beiersdorfer & Day, 1995.)

most of the stability fields of these two assemblages. Likewise, in rocks of appropriate bulk composition, actinolite may be stable with prehnite and/or pumpellyite over an extensive *P–T* interval. The conflict between these results and the observation of large areas containing the defining assemblages for the prehnite–pumpellyite facies has yet to be satisfactorily resolved.

4.6.6 Thermobarometry

Although the principles and practice of thermobarometry are well known (e.g. Essene, 1982, 1989), there have been relatively few attempts to calculate metamorphic conditions from mineral equilibria in low-grade metamorphic rocks. Notable exceptions, of course, have been attempts to estimate metamorphic pressures using the equilibrium albite = jadeite + quartz (e.g. Essene & Fyfe, 1967).

Thermobarometry in high-grade metamorphic rocks is commonly based on the partitioning of a pair of elements between coexisting phases (e.g.

Fe–Mg between garnet and biotite) (cf. Essene, 1982, 1989). In very low-grade rocks, few systematic data of this kind have been documented and no useful thermobarometer of this kind has been proposed. Maruyama and Liou (1985) used the partitioning of Mn/Mg between pumpellyite and chlorite to infer relative grade in the pumpellyite–actinolite facies (see also Sakakibara *et al.*, 1997). The abundance of Mn is commonly very low or near minimum detection limits, and the uncertainties of the analyses are not reported by most authors. Under these circumstances, it is not possible to demonstrate that observed differences in partitioning are analytically significant. There have also been suggestions that Fe/Mg partitioning between chlorite and actinolite may be indicative of differences in metamorphic pressure (Maruyama *et al.*, 1983; Cho & Liou, 1988; Terabayashi, 1988). Pressure does not notably affect Fe/Mg partitioning in other solid solution series that are used as thermobarometers and, as pointed out by Terabayashi (1988), the possible effects of unknown amounts of Fe^{3+} have not been considered. Consequently, neither of these proposed partitioning schemes have been widely used.

The most successful recent attempts at thermobarometry in very low-grade rocks are based on the internally consistent thermodynamic data sets of Powell and Holland (1988) and Berman (1991) and multiequilibrium calculations. The existence of internally consistent data sets has allowed a considerable expansion of the number of mineral equilibria that can be used for effective thermobarometry and the number of equilibria that can be investigated simultaneously in a single rock.

Figure 4.21 illustrates possible calculated equilibria among end-member components of minerals in a hypothetical mineral assemblage (Berman, 1991). Each of these curves is the locus of solutions to equations having the form: $\Delta G° = RT \ln K$, where $\Delta G°$ is the standard-state free energy of reaction among end-members at *P* and *T*, R is the gas constant and K is the equilibrium constant. Calculations typically are made for all reactions among end-members for which reliable thermodynamic data are available and for which reliable mineral compositions in the rock have been determined. The calculated equilibria will intersect at a single point provided that (i) the thermodynamic parameters of the end-members are perfectly known; (ii) the models relating the

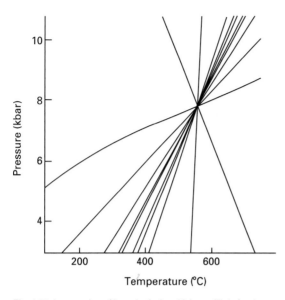

Fig. 4.21 Intersection of hypothetical multiple equilibria for the ideal case in which thermodynamic data are perfect, all analytical data are accurate and all minerals equilibrated at the same *P–T*. (After Berman, 1991.)

thermodynamic properties and compositions of solid solutions are perfect; (iii) the composition of the mineral phases are accurately determined; (iv) equilibrium was attained among all minerals at the same *P–T* (Berman, 1991). Equilibria from different end-member systems may be combined in the same diagram, but should yield the same result if the activity compositions for the solid solutions involved are perfectly known and the other criteria are met. For example, the equilibrium curve for the reaction Pmp + Chl = Ep + Act + H_2O, might be calculated, with appropriate activity terms, in either CMASH or CFASH. If all criteria for a perfect result are satisfied, the calculated equilibrium curves should coincide. The application of such calculations to natural assemblages therefore serves as a basis for testing the reliability of thermodynamic data as well as the extent to which equilibrium was attained in the assemblage (Berman, 1991).

The basic concept of thermobarometry using multiple equilibria has been used in different ways by Powell and Holland (1988) and Berman (1991). If important deviations from the ideal conditions occur, numerous intersections among the various equilibria will occur and it is commonly unclear

which intersections may give more reliable results. Powell and Holland (1988) recognized that only a fraction of the equilibria in such a system are independent. That is, most of the reactions can be calculated by taking linear combinations of a few independent equilibria. Thus, the dependent reactions provide no new or useful information. Berman (1991) noted that several sets of independent reactions could be chosen, each of which would yield different results. Because it is not clear which independent set of equilibria might give the best pressure and temperature, he prefers to find the average pressure and temperature of all calculated intersections.

Powell *et al.* (1993) and Digel and Gordon (1995) used equilibria in the assemblage Prh + Pmp + Ep + Chl + Ab + Qtz + H_2O (invariant point 'A', Fig. 4.19) to estimate metamorphic conditions of low-pressure metabasalts. Powell *et al.* (1993) identified two low-variance assemblages within a single sample of basaltic hyaloclastite from the Abitibi greenstone belt, Ontario, Canada. In domains that are relatively rich in devitrified material, the sample contains Pmp + Prh + Ep + Chl + Qtz, corresponding to reaction (6) at invariant point A (Table 4.6, Fig. 4.18). In domains that contain less devitrified material, less than 2 cm away, Act + Prh + Ep + Chl + Qtz appears to be the stable assemblage and corresponds to reaction 11 in Fig. 4.18. Chlorite compositions are similar in the two domains but prehnite and epidote compositions are not. The authors concluded that the sample did not equilibrate at the scale of the thin section but that local equilibrium was attained within the subdomains.

Calculated pressures and temperatures for these two equilibria are shown in Fig. 4.22 as dashed lines. Because the two assemblages do not contain the same composition minerals, it is not appropriate to treat the two assemblages as a single assemblage corresponding to invariant point A. However, under the assumption that the two subdomains equilibrated at the same *P–T*, it is possible to recover the *P–T* of the sample from the intersection of the two calculated curves. All calculations were carried out with end-member minerals in the CMASH system (Powell *et al.*, 1993).

Digel and Gordon (1995) studied the transition from prehnite–pumpellyite to greenschist facies in metabasalts from the Flin Flon area, Manitoba,

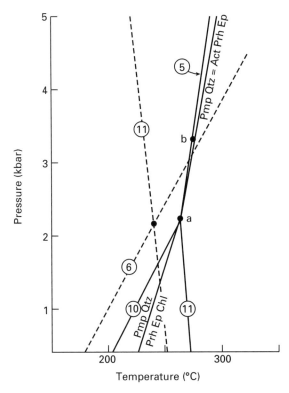

Fig. 4.22 Calculated pressure and temperatures for low-grade metabasalts. Dashed lines show the intersection of two pseudo-univariant equilibria (Powell *et al.*, 1993, Fig. 5(a)). Solid lines and point (a) show equilibria calculated in CMASH for the pseudo-invariant assemblage Pmp + Prh + Chl + Act + Ep + Ab + Qtz (Digel & Gordon, 1995, Fig. 8). Point (b) shows the *P–T* returned by program INTERSX (Berman, 1991) for all equilibria in the same sample calculated in the system CMASH + FeO. Numbered reactions are listed in Table 4.6. Note that the calculated position of reaction 11 is substantially different for the two samples, as expected for minerals of different composition.

Canada. Sample 90SD-179 is a pillowed volcanic rock from the transition zone between the two mapped facies and contains the low-variance assemblage Pmp + Prh + Ep + Chl + Act + Ab + Qtz, corresponding to assemblage A (Fig. 4.19). Figure 4.22 (point a) shows the intersection of the equilibria as calculated in the system CMASH using GE0CALC software. It is important to remember that the assemblage has two or more degrees of freedom and that intersection is invariant only by restriction of the activities to the values specified. Also illustrated (point b) is the *P–T*

returned by Berman's (1991) computer program INTERSX, which calculated the average *P–T* of all intersections of reactions calculated using both Mg end-members, in the system CMASH, and Fe end-members, in the system CFASH. Given the large uncertainties in the thermodynamic properties of several of these phases, especially pumpellyite, it is reassuring that the uncertainty in the calculations seems to be about 1 kbar.

It appears that we can be cautiously optimistic about the future of this approach to understanding low-grade assemblages down to perhaps as low as 200°C. The possibility of calculating pressures and temperatures is best in portions of *P–T* space where several equilibria intersect, such as near invariant point A (Fig. 4.18). In appropriate rocks, we may also have an opportunity to calculate equilibria in the high pressure portion of Fig. 4.18. For example, Bröcker and Day (1995) have used the equilibrium Pmp + Chl + Ab + Qtz + H_2O = Gln + Lws (reaction 3, Fig. 4.18) to estimate the pressures in metagrey-wackes in very low-grade, high *P/T*, metasedimentary rocks in the northern Coast Ranges of California. At intermediate pressures, sufficient equilibria have not been identified to lend much hope that we can pursue this approach.

4.7 Summary

In this chapter, we have shown that petrological studies of very low-grade metabasites can be conducted using many of the same research methods that are routinely employed with higher grade equivalents. However, disequilibrium in low-grade metabasites (e.g. persistence of primary minerals and/or metastability of secondary minerals) and large uncertainties in the thermodynamic proper-ties of some key minerals in these rocks makes thermobarometry a non-routine undertaking.

Nonetheless, careful study of their textures and mineral parageneses, a task greatly aided by modern electron microscopes, BSE detectors and energy dispersive X-ray detectors, can go a long way towards evaluating the most stable assemblages in these rocks. High quality, quantitative analyses of the minerals in low-grade metabasic rocks can be obtained routinely with wavelength dispersive electron micro-probe techniques. Equilibrium among coexisting minerals can generally be demonstrated using appropriate thermodynamically valid projections that graphically display the correlation between bulk rock composition and mineral assemblages. Algebraic methods, such as singular value decomposition, can be employed to test rigorously whether an assemblage of minerals is compatible with the hypothesis of equilibrium, or conversely, exhibits a reaction relationship. Natural low-grade metabasite mineral assemblages can be compared with model petrogenetic grids, constructed with new, internally consistent thermodynamic data sets for low-grade metamorphic minerals. Recent activity–composition models for key low-grade minerals, especially pumpellyite, for the first time are allowing P–T estimates for natural, very low-grade metabasite mineral assemblages.

5   Patterns of regional low-grade metamorphism in metabasites

D. Robinson and R. E. Bevins

5.1 Introduction

Igneous rocks are widespread in the upper parts of the Earth's crust in a variety of tectonic settings including convergent (arc/collision), extensional (ocean ridge) and within plate (flood basalt of both continental and oceanic affinities). Of the great variety of igneous rocks, those of basaltic composition provide a diverse system for the most varied secondary mineralogical development in response to fluid/rock interaction at low temperature ($<c.$ 300°C). In recent years such systems have attracted much attention through the application of experimental analysis (e.g. Liou *et al.*, 1991), thermodynamic techniques utilizing geothermobarometry (Powell *et al.*, 1993), petrogenetic grids (Liou *et al.*, 1985a; Frey *et al.*, 1991) and chemographic projections (Springer *et al.*, 1992) allowing study of low-grade metabasites on a more quantitative basis than has been previously possible (see Chapter 4). In addition the widespread geological investigation through the Deep Sea Drilling Project (DSDP), and more latterly the Ocean Drilling Program (ODP), has highlighted that the upper ocean crust is extensively recrystallized under conditions of low-grade metamorphism (e.g. Alt *et al.*, 1986a; Chapter 6). Thus the low-grade metamorphism of basaltic rocks is the most widespread, but perhaps least understood, of any metamorphic process operating in the Earth's crust. Accordingly, in recent years it is such basaltic rocks that have been amongst the most studied and provided the greatest advances in understanding processes of low-grade metamorphism. Thus, although intermediate igneous and volcaniclastic and related rock types provided the basis for the original low-grade metamorphic studies (e.g. Coombs, 1954, 1960), the focus of attention in this chapter is purposely concentrated on the basaltic system.

Although the low-grade metamorphism of igneous rocks is not regarded as a main field of metamorphic endeavour, there is much of interest and excitement that offers a different perspective and insight into the metamorphic process that cannot be achieved in investigations at higher grades of metamorphism. Some of the principal contrasts between lower and higher grades of metamorphism of basaltic rocks were reviewed by Liou *et al.* (1987). At low grades of metamorphism the principal effect is one of hydration to convert dominantly anhydrous mineral assemblages into ones involving phases with substantial hydroxyl component, contrasting markedly with the dominant dehydration process at higher metamorphic grades. Fluid/rock interaction is thus of critical importance in terms of the fluid origin, as well as access to, and movement through rocks undergoing low-grade metamorphism.

At higher metamorphic grades it is a basic tenet that the primary metamorphic assemblage is one that has equilibrated at peak metamorphic conditions in relation to the chemical system represented by the host whole rock. The partial recrystallization, often with a majority of relict igneous products, at very low to low grades of metamorphism means that such a basic assumption is clearly not immediately applicable. This raises obvious questions such as: is it indeed possible to apply the quantitative techniques of modern petrology to low-grade metamorphic rock systems; are there subdomains within a typical whole rock sample that may be considered as equilibrium domains; and if subdomains are present at what level do they converge into a whole rock dominated system (cf. Chapter 4). The patterns of regional low-grade metamorphism highlighted in the following sections emphasize the recent advances made in understanding the metamorphic process at low temperatures. These studies show that metamorphic patterns at subgreenschist facies can indeed show systematic changes indicative of metamorphic equilibration and that answers to some of the questions posed above can be provided with some confidence.

5.2 Regional low-grade metamorphism

5.2.1 Extensional settings—Welsh Basin

The Welsh Basin is part of the British Caledonides and although pumpellyite has been known from the region for many years (Nicholls, 1959), some authors have regarded it even until relatively recently as part of the so-called 'non-metamorphic' Paratectonic Caledonides (Soper, 1980), contrasting with the 'metamorphic' Orthotectonic Caledonides. Indeed, it is most likely that the 'anomalous chlorite' described from the region by Reed (1895) was in actual fact a record of pumpellyite some 30 years before its formal description by Palache and Vassar (1925). Following extensive research, from the late 1970s onwards, into secondary mineral assemblages in metabasic igneous rocks (see Chapter 4) and clay mineralogy of associated mudrocks (Chapters 2 and 3), the very low to low-grade metamorphic character of the Welsh Basin is well established and the region is known as a classic area for low-grade, low-pressure metamorphism in an extensional setting. The basin has been a testing ground for the diastathermal model of low-grade metamorphism (Robinson, 1987) that invokes a genetic link between the burial style of metamorphism, intense magmatism and deep-seated extensional processes associated with high heat flow. The model applied to the region by Bevins and Robinson (1988) has proved to be controversial and there has been an active and continuing debate of its particular merits.

The Welsh Basin is thought to have been generated as a result of lithospheric extension in a continental margin setting, linked to south-easterly subduction of Iapetus Ocean crust beneath the north-western margin of Gondwanaland (Kokelaar *et al.*, 1984). This intracontinental basin has a Lower Palaeozoic infill that was deformed as a result of Acadian deformation and basin inversion during the early Devonian. The Lower Palaeozoic succession (Fig. 5.1a) is dominated by hemipelagic mudrocks associated with turbiditic sandstones; thick volcanic horizons with related high-level intrusions are abundant particularly in the Ordovician sequences. Volcanic activity was chiefly bimodal basic–acidic; as a consequence basalts, dolerites and gabbros are widely developed across the Welsh Basin area. A composite thickness of 15 km of strata is exposed in the original depocentre region, while only a 4-km thickness is present at the eastern margin of the basin. This marginal area was for the whole period a stable shelf, receiving very little sediment input and lacking the abundant volcanism present in the basin.

Documentation of the regional extent of the low-grade metamorphic character of the succession was undertaken in the late 1970s and early 1980s with assemblages variably containing pumpellyite, prehnite, epidote, actinolite and stilpnomelane (+ ubiquitous phases of Ab + Chl + Ttn + Qtz) being reported across the whole region (Bevins, 1978; Roberts, 1979, 1981; Bevins & Rowbotham, 1983). Initially, however, the mineral distribution pattern was rather confusing, with regional and local variations, such that in certain areas there is a close spatial relationship between samples containing assemblages such as prehnite + pumpellyite; prehnite + pumpellyite + stilpnomelane; prehnite + pumpellyite + actinolite; and actinolite, in most cases along with epidote and the ubiquitous phases. These mineralogical contrasts were attributed to various causes including change in grade, high f_{CO_2} inhibiting characteristic Ca–Al silicate development, whole rock effect, and the scale of equilibrium/disequilibrium mineral associations (Bevins & Rowbotham, 1983; Bevins & Merriman, 1988). The subgreenschist prehnite–pumpellyite and prehnite–actinolite facies were recognized, as well as the pumpellyite–actinolite facies (Roberts, 1979) on the basis of diagnostic assemblages and chlorite/actinolite Mg/Fe distribution coefficients (Bevins & Robinson, 1993). The close juxtaposition of these facies within a localized area and the presence of the pumpellyite–actinolite facies appeared anomalous, the latter facies generally associated with collisional tectonic regimes involving higher pressure such as the European Alps (e.g. Coombs *et al.*, 1976), contrasting with the extensional setting proposed for the Welsh Basin.

An important advance in understanding the distribution of these very low-grade facies in metabasites was provided by a detailed study of the 100-m thick

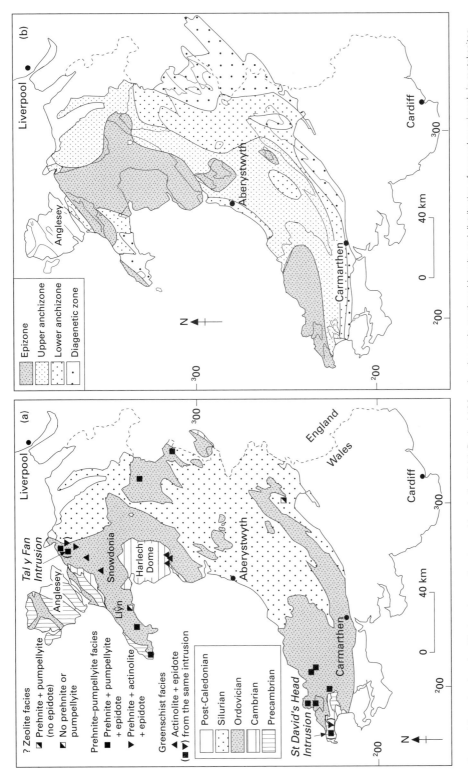

Fig. 5.1 Maps showing simplified geology and metamorphism in the Welsh Basin. (a) Map showing main stratigraphic units and distribution of secondary minerals in metabasites. (b) Map showing illite crystallinity zones. (After Robinson & Bevins, 1986 and Bevins & Robinson, 1993.)

Tal y Fan intrusion in North Wales (Fig. 5.1a; Bevins & Merriman, 1988) that allowed the effect of whole rock control to be identified. The intrusion shows a range of compositions, from highly mafic gabbros through normal gabbros and dolerites to ferrodolerites, that has an influence not only on secondary mineral chemistry, but also on the distribution of secondary minerals, as shown in Fig. 5.2. The pattern of whole rock MgO/(MgO + FeO) [X_{MgO}] ratio through the intrusion is mirrored by that of chlorite indicating their close correlation. A strong link between chlorite and whole rock X_{MgO} ratios was demonstrated even more convincingly by Bevins et al. (1991a). Further discussion of the relationship between X_{MgO} in mafic phyllosilicates and whole rock is presented in section 5.3.2. In addition, Fig. 5.2 shows that when whole rock X_{MgO} ratios are low (<c. 0.55), prehnite occurs in association with pumpellyite while in rocks with higher X_{MgO} ratios (>c. 0.55), prehnite occurs in association with actinolite. Bevins and Merriman (1988) interpreted these assemblages as being diagnostic of the prehnite–pumpellyite facies and the prehnite–actinolite facies respectively.

A more rigorous analysis of the secondary para-geneses of the Welsh Basin was undertaken by Bevins and Robinson (1993) by application of chemographic projection techniques (see Chapter 4). This allowed variations in assemblages occurring in response to whole rock chemistry to be separated from those due to changes in grade. In those areas of the Welsh Basin where pelitic rocks have illite crystallinity values characteristic of diagenetic zone and anchizone conditions (Fig. 5.1b), associated metabasites contain prehnite ± pumpellyite ± epidote in association with the ubiquitous phases (Ab + Chl + Ttn + Qtz), suggestive of the prehnite–pumpellyite facies. With reference to the NCTiFMASH system, and projecting from quartz, water, titanite, albite and epidote, those metabasites with prehnite + pumpellyite + epidote have subparallel pumpellyite–chlorite tielines (Fig. 5.3a), reflecting a range of $K_{DMg–Fe}$ values from 0.27 to 0.58 ($\bar{x} = 0.42$; $1\sigma = 0.112$). The disposition of these tielines and the low variability in chlorite analyses in individual samples suggests that there is an approach towards an equilibrated system.

It should be noted, however, that the tielines in Fig. 5.3(a) are drawn relative to the mean composition

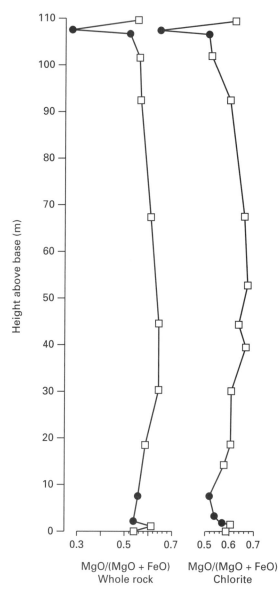

Fig. 5.2 Section through the Tal y Fan intrusion, North Wales, showing the variations through the intrusion between whole rock X_{MgO} and that of chlorite, and variation in the two assemblages of Pmp + Prh (●) and Act + Prh (□) (+ Ab + Chl + Qtz). (Modified after Bevins & Merriman, 1988b.)

of pumpellyite and chlorite in each sample. Although the chlorite shows homogeneous compositions, those of the pumpellyite are much less so, and they have in fact quite variable compositions as represented by the one sigma error bars in Fig. 5.3(a). Thus if

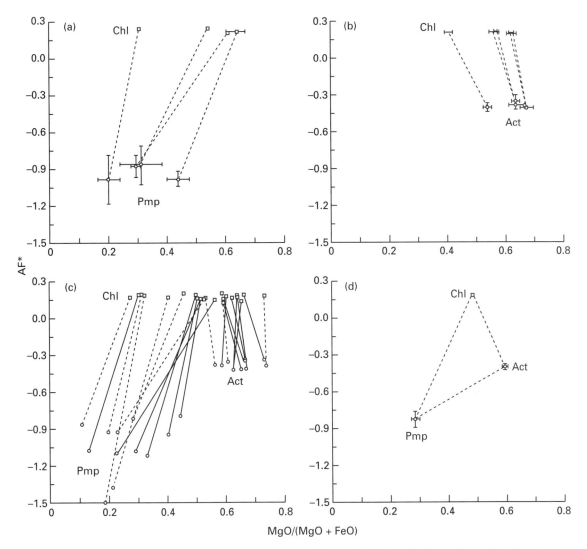

Fig. 5.3 Epidote projections for the Welsh Basin metabasites (after Bevins & Robinson, 1993). $AF^* = 100(Al_2O_3 + Fe_2O_3 - 0.75\,CaO - Na_2O + 0.75\,TiO_2)/(Al_2O_3 + Fe_2O_3 - 0.75\,CaO - Na_2O + 0.75\,TiO_2 + FeO + MgO)$. All samples + Ab + Ttn + Qtz. (a) Pumpellyite–chlorite relationships for metabasites from the diagenetic and anchizone regions shown in Fig. 5.1. (b) Actinolite–chlorite relationships for metabasites from the epizone region shown in Fig. 5.1. (c) Pumpellyite–chlorite and actinolite–chlorite relationships for metabasites from the epizone regions shown in Fig. 5.1 having the assemblages Ep + Prh + Pmp + Chl and Ep + Prh + Act + Chl. Solid lines are for samples from the Tal y Fan intrusion. (d) Pumpellyite–chlorite–actinolite relationships for metabasite sample PCG21 from the epizone region of South Wales (Fig. 5.1).

tielines were to be drawn between individual co-existing chlorite and pumpellyite crystals there would be great variability in their disposition — a general pointer to disequilibrium. This can be demonstrated by using unpublished data from the St David's Head intrusion in south-west Wales (Fig.

5.1a). In this layered intrusion there is a range in composition from olivine-rich mafic gabbros through to granophyric gabbros and late aplites. As with the Tal y Fan intrusion, metamorphic mineral assemblages are again dictated by whole rock compositions. In the most mafic rocks, actinolite, prehnite, chlorite and talc are present, while in the normal gabbroic rocks, pumpellyite, prehnite, chlorite and epidote are the characteristic phases. In the more silicic rocks, pumpellyite is absent and the secondary mineralogy is dominated by prehnite and epidote. Analyses reveal a considerable range in pumpellyite compositions, both within and between samples; in fact the most extreme spread in composition is present within an area of one thin section only some 2 mm across. In this area pumpellyite MgO/(MgO + FeO) ratios range from 0.27 to 0.55, although chlorite shows a restricted range between 0.47 and 0.51 (Fig. 5.4). These variations result in a range of K_D pumpellyite–chlorite values of 0.37–1.31. Critically, there is no phase domain effect in these compositional ranges as all are from small crystals in plagioclase feldspar, with chlorite in immediate association. It appears that at this grade pumpellyite may crystallize with variable compositions, possibly depending in part on local site and fluid compositional variation, but perhaps related also to lower diffusion kinetics in pumpellyite relative to chlorite, so that, unlike chlorite, it is unable to homogenize its composition.

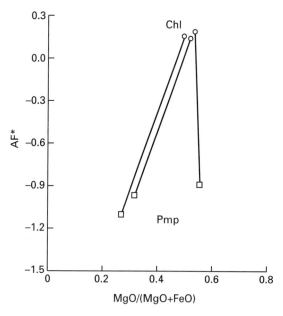

Fig. 5.4 Epidote projection for sample RD 7 from St David's Head intrusion (Fig. 5.1), showing chlorite–pumpellyite relations for different domain areas in one thin section.

With increase in grade, marked in pelitic rocks by epizone illite crystallinities (Fig. 5.1b), metabasites show complex assemblages including some that show no change from pumpellyite + prehnite (+ epidote and ubiquitous phases), while others have prehnite + actinolite assemblages and yet others show actinolite in the absence of either prehnite or pumpellyite. Those metabasites in which prehnite and pumpellyite assemblages are absent show near-parallel actinolite–chlorite tielines in the epidote projection (Fig. 5.3b), reflecting $K_{DMg–Fe}$ of 1.26–1.65 ($\bar{x} = 1.39$; $1\sigma = 0.134$). This feature, along with homogeneous mineral compositions (indicated by small one sigma analytical spreads in individual samples) and the sympathetic relationship between the X_{MgO} of minerals and whole rock, indicates a close approach to an equilibrated system. This actinolite + chlorite assemblage is diagnostic of the greenschist facies.

The other metabasite assemblages from the epizone have either (i) pumpellyite + prehnite or (ii) prehnite + actinolite, and represent transitional assemblages between the prehnite–pumpellyite and greenschist facies assemblages detailed above. Metabasites with X_{MgO} ratios in chlorite of < c. 0.54 contain prehnite, pumpellyite and epidote, while those with ratios of > c. 0.54 contain prehnite, actinolite and epidote (both + Ab + Chl + Ttn + Qtz) as shown in Fig. 5.3(c). In the former, pumpellyite–chlorite tielines are, in the main, broadly parallel, with $K_{DMg–Fe}$ values in the range 0.36–0.78 ($\bar{x} = 0.52$; $1\sigma = 0.140$), similar to values in metabasites in the diagenetic zone and anchizone. In the more mafic metabasites, chlorite–actinolite tielines are steeper (Fig. 5.3c) than those for the greenschist facies metabasites (Fig. 5.3b), reflecting lower $K_{DMg–Fe}$ values, in the range 0.93–1.40 ($\bar{x} = 1.17$; $1\sigma = 0.162$). In Fig. 5.3(c), tielines shown in solid are all samples from the Tal y Fan intrusion (only c. 100 m in thickness) and so these samples must have developed under the same P–T conditions. The contrasting assemblages of the prehnite–pumpellyite and prehnite–actinolite facies at this transitional zone are thus a function of varying whole rock control (X_{MgO}) and are not related to changing P–T conditions. The Welsh Basin thus provides a field demonstration that facies apparently occupying unique areas of P–T space and separated by univariant reaction curves in the end-member NCMASH system (Liou *et al.*,

1985a) do in fact overlap extensively in *P–T* space when the system is expanded to take account of Fe as an additional component (Frey *et al.*, 1991).

Of several hundred metabasite samples examined from Wales there is only one with the assemblage prehnite + pumpellyite + epidote + actinolite (+ Ab + Chl + Ttn + Qtz), defining a three-phase field in the epidote projection (Fig. 5.3d). This has a chlorite X_{MgO} ratio of 0.50 that lies very close to the dividing value of 0.54 identified above as separating the contrasting assemblages. This sample in fact has an assemblage that is equivalent to the pseudo-invariant point (CHEPPAQ) in the petrogenetic grid for the NCMASH system (Liou *et al.*, 1987; Frey *et al.*, 1991; Powell *et al.*, 1993). The uniqueness of this sample suggests it recrystallized at restricted *P–T* conditions. The *P–T* conditions for the pseudo-invariant point have been calculated using the TWQ software of Berman (1991) for the ideal NCMASH system and for the mineral compositions from this sample (Fig. 5.5). For the end-member system the pseudo-invariant point lies at *c.* 3.6 kbar and 315°C, but when the activities based on the mineral compositions are utilized the derived *P–T* conditions are lowered to *c.* 1.1 kbar and 278°C (Robinson *et al.*, 1998a, in review). The above estimate of *P–T* conditions gives a geothermal gradient of *c.* 60°C km⁻¹, that is similar to the estimate of 50°C km⁻¹ of Bevins and Merriman (1988), and equates well with the low-pressure facies series deduced from the clay mineral studies of Robinson and Bevins (1986). The high heat flow, along with a primary depth control on the metamorphic grade (Robinson & Bevins, 1986; Bevins & Robinson, 1988; Roberts *et al.*, 1991), were two of the main features used to link the low-grade metamorphism to a principally burial-type origin (Robinson & Bevins, 1986; Bevins & Robinson, 1988). However, the strong cleavage development in the basin is not a feature that can be equated with a burial style of metamorphism, but accords more with a traditional continental thickening process. These two conflicting scenarios for the metamorphic process can be resolved through the diastathermal model of metamorphism (Robinson, 1987), applied to the Welsh Basin by Bevins and Robinson (1988). Applying the thermal models of Royden *et al.* (1980) and McKenzie (1981), it is envisaged that an early, syndepositional high heat flow occurs in relation to diapiric rise of mantle material in a marginal

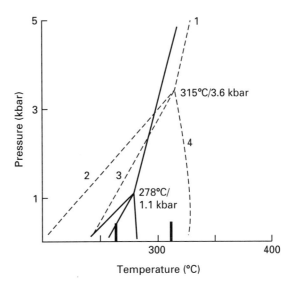

Fig. 5.5 Petrogenetic grid for low-grade metabasites in the CMASH system, projected from chlorite, quartz and H₂O. Dashed lines show grid for unit activity for all phases and solid lines represent grid calculated with activities based on mineral compositions from sample PGC21. Determined using the TWQ software of Berman (1991). Labelled reactions are: (1) Pmp + Chl + Qtz → Act + Ep + F; (2) Pmp + Act + Qtz → Pr + Chl + F; (3) Pmp + Qtz → Pr + Ep + Chl + F; (4) Pr + Chl + Qtz → Act + Ep + F. Two vertical lines on *x*-axis represent mean chlorite temperatures of 263°C and 311°C for subgreenschist and greenschist facies, respectively; data from Bevins & Robinson (1993) and Robinson *et al.* (1998a, in review).

back-arc or continental margin setting linked to lithospheric extension. The primary heat flux is associated with the extensional process and thus the main burial-style metamorphism is developed prior to the onset of the Acadian deformation and cleavage development associated with basin inversion during the early Devonian. This is supported by the initial development of the highest grade mineral present in the pelitic rocks, chloritoid, as a precleavage phase (Bevins & Robinson, 1988; Brearley, 1988; Roberts *et al.*, 1989), thereby suggesting that the main heat source resulting from the 'deformation itself' (Roberts & Merriman, 1985) is not sustainable. The diastathermal model has been controversial and subject to criticism by Roberts *et al.* (1989, 1991, 1996) who supported a model linking the main metamorphism to a climactic event associated with the Acadian deformation. They argued for a syntectonic metamorphic peak as shown by the occurrence of belts having high illite crystallinities and strong cleavage associated with high strain during the Acadian deformation; they proposed that the extensional model is insufficient to give enhanced thermal fluxes for the lifetime of the basin (Roberts *et al.*, 1989, 1991, 1996). However, quantitative modelling of the extensional process giving rise to the stratigraphic succession developed in the Welsh Basin within the time constraints from its inception to its inversion suggest that a progressive temperature increase reaching a maximum of *c.* 250–280°C at the time of inversion was possible as a result of the extensional process alone (Robinson *et al.*, 1998a, in review). Whatever the outcome of the competing models, and it is probable that the best scenario will involve the better parts of both models involving the acceptance of two metamorphic episodes with an earlier diastathermal event overprinted by a later deformation event, the controversy has undoubtedly led to the metamorphism of the Welsh Basin being examined from an ever greater critical viewpoint that has been only to the benefit of advancing understanding of low-grade metamorphic processes, and establishing the region as a type area for low-grade metamorphism.

5.2.2 Extensional settings — Andean Basins

The Andean mountain chain, running from Colombia to Patagonia, is unique amongst orogenic belts for the abundance of volcanic rocks and for the widespread occurrence of low-grade metamorphic sequences (Aguirre *et al.*, 1978; Offler *et al.*, 1980; Aguirre & Offler, 1985; Levi *et al.*, 1989). Indeed, and as detailed by Levi *et al.* (1989), the low-grade metamorphic character of this belt was identified originally by Darwin during his world voyage with HMS Beagle (Darwin, 1846). Several field and petrographic-based Andean studies in the 1970s and 1980s documented a full range of low-grade regional metamorphic facies, all of a non-deformational character, where grade increases with depth, and so have been equated with a burial-type metamorphic process (Aguirre, 1993). Of particular interest in the regional metamorphism of the Andes is the interpretation placed on the metamorphic style in terms of (i) the control played by local variation in rock permeability, (ii) the use of mineral chemistry as a guide to metamorphic style, (iii) the episodic

nature of the metamorphism, and (iv) the link between facies series in different areas and their relation to large-scale geodynamic tectonic controls in relation to marginal basin formation.

Although low-grade metamorphic sequences have been reported from many parts of the Andean range, it is the E–W section in the central Andes to the east of Santiago that is one of the most well documented (Fig. 5.6). In this part of the Andes there are successive volcanic sequences, with minor interbedded sedimentary associations, ranging from Triassic to Quaternary in age. The sequences do not form a continuous stratigraphic succession but developed in successive overlapping basins with unconformities separating the different sequences (Vergara & Drake, 1979). Within these basins the infilling successions have abundant volcanic and volcaniclastic rocks that are predominantly basic-intermediate in the early Cretaceous but intermediate-acidic in the postearly Cretaceous sequences. The rocks show variations in mineralogy and mineral abundance with stratigraphic age, detailed by Levi *et al.* (1989) and shown here in Fig. 5.7. In the youngest Quaternary to late-Miocene sequences, smectite and the low-temperature zeolites mordenite and heulandite are the dominant secondary minerals. The early Miocene to late Cretaceous rocks show a change in mineralogy to higher temperature layer silicate and zeolite minerals with chlorite and laumontite, and the incoming of prehnite, pumpellyite, epidote and rare actinolite. The early Cretaceous-Jurassic units contain abundant occurrences of prehnite, pumpellyite, epidote and actinolite. These mineralogical variations have been interpreted as representing a change from a low-temperature zeolite facies (mordenite subfacies) through high-temperature zeolite facies (laumontite subfacies) into prehnite–pumpellyite, pumpellyite–actinolite and greenschist facies (Figs 5.6 and 5.7).

Related regional scale differences are also developed in a Cretaceous section of the Coastal Ranges *c.* 30 km west of Santiago. The rocks here are a *c.* 9-km thick sequence of flood basalts with localized intercalated volcaniclastic beds and ignimbrites. The numerous basalt flows, initially regarded as spilitic in character, vary from 10 to 40 m in thickness and have non-amygdaloidal bases that grade upward into extensively amygdaloidal flow centres to highly amygdaloidal and brecciated flow tops (Levi *et al.*,

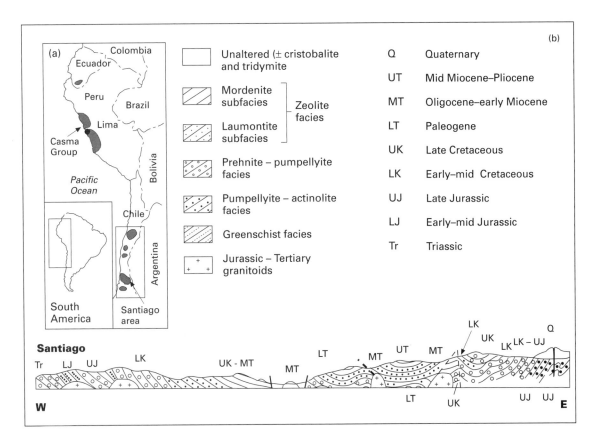

Fig. 5.6 Map of low-grade metamorphic areas in the Andes of South America. (a) Outline map showing main areas of published work. (b) Diagrammatic E–W section of the region around Santiago showing distribution of low-grade metamorphic facies in Triassic–Quaternary sequences. (Modified after Levi *et al.*, 1989.)

1982). In the uppermost 300 m of the succession (A, Fig. 5.8), actinolite with epidote assemblages were interpreted as greenschist facies. This upper part of the sequence unconformably overlies a *c.* 9000 m mid to early Cretaceous sequence in which there is a continuous zeolite through prehnite–pumpellyite to greenschist transition with depth (B and C, Fig. 5.8).

The metamorphism in the areas around Santiago contrasts with that developed in the Cretaceous Casma Group of the Central Andes, exposed in the coastal region of Peru (Figs 5.6 and 5.9). Here a 3000-m succession of basaltic to andesitic rocks is separated into several units by regional unconformities. The rocks show assemblages representative of zeolite through prehnite–pumpellyite to greenschist facies transitions occurring within individual units, but with breaks in metamorphic facies at unconformities separating the units. An example of this feature is shown in Fig. 5.9, where there are four stratigraphic units, of thicknesses from *c.* 250 to 1500 m, with each showing the above facies transitions and a facies break at each bounding unconformity.

Besides occurring on regional and subregional scales as described above, variations in mineral assemblage and facies type also occur on local, within lava flow scales. Within the Coastal Ranges section in central Chile, Levi (1969) and Levi *et al.* (1982) recorded mineralogical changes from flow bases to flow tops (e.g. up to 30-m scale) involving increases in (i) the proportion of secondary minerals, (ii) albitization, and (iii) the proportion of epidote to prehnite + pumpellyite (Fig. 5.10). These changes, recording an apparent upwards (although of local scale) increase in grade, were shown to be correlated

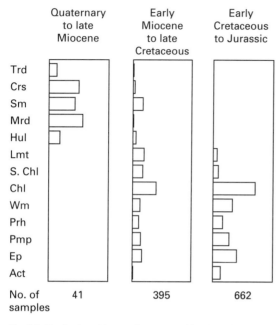

Fig. 5.7 Distribution of low-grade metamorphic minerals in Jurassic to Pleistocene sequences in the Santiago section of the Central Andes. Mineral determinations made using X-ray diffraction and petrography, with bar lengths proportional to percentage of samples in which the mineral is present. (After Levi *et al.*, 1989.)

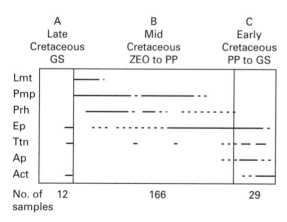

Fig. 5.8 Distribution of low-grade metamorphic minerals in the Cretaceous section of the Coastal Ranges *c.* 30 km west of Santiago. Upper unit A (*c.* 300 m thick) overlies a *c.* 9000 m unit B in which the metamorphic grade at its upper contact is lower than in A. (After Levi *et al.*, 1989.)

with changes in flow morphology, from massive flow bases to increasingly amygdaloidal and brecciated flow tops.

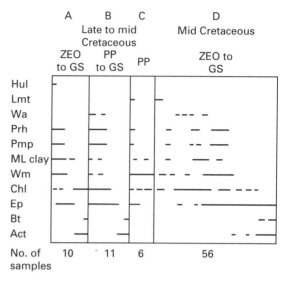

Fig. 5.9 Distribution of low-grade metamorphic minerals in the Cretaceous Casma Group of coastal Peru, *c.* 200 km north of Lima. There are repeated metamorphic breaks between the four units such that strata at higher grade overlie rocks of lower grade at each unconformable contact. Units A–D are 360, 440, 240 and 1470 m thick respectively. (After Levi *et al.*, 1989.)

The metamorphic variations documented in the Andean range are founded largely on petrographic information, with relatively few mineral chemical data. The most mineral chemical data available at present are for pumpellyite, for which variation in its chemistry has been attributed to variations in whole rock composition and the geodynamic setting in which metamorphism took place (Aguirre, 1993). Fe-rich pumpellyite has been linked to low-pressure settings with steep thermal gradients ($> 100°C\,km^{-1}$), and high Al contents equated with higher pressure settings developed under moderate thermal gradients ($20–30°C\,km^{-1}$; Aguirre, 1993).

The metamorphic character of the Andean chain sets this mountain belt apart from most others in the dominance of subgreenschist to greenschist facies assemblages in rocks ranging in age from Jurassic through to late Neogene. These low grades and the absence of a deformational fabric point to a model of metamorphic development alternative to the traditional one of crustal thickening in convergent settings. Particular features of this metamorphism are (i) its clear burial character with an absence of deformation fabrics, (ii) the episodic nature, with

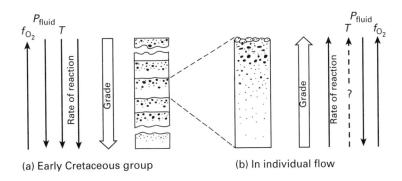

Fig. 5.10 Schematic representation of metamorphic variation within the early Cretaceous Group of the Coastal Ranges near Santiago. This shows metamorphic variation over the whole 9-km thickness and also on the scale of a single flow within the sequence. (After Levi *et al.*, 1982.)

(a) Early Cretaceous group

(b) In individual flow

metamorphic breaks at unconformities, and (iii) the timing of metamorphism closely following formation of the volcanic sequences. The stratigraphic record of the different successions and the geochemical character of the volcanic sequences suggest formation in extensional (marginal or back-arc) basins initially in intracontinental settings. The various facies series developed in the different basins have been used to define geothermal gradients operative during the metamorphism. In the Andean and Coastal Ranges regions of central Chile, the zeolite through prehnite–pumpellyite to greenschist facies series is developed through stratigraphic sequences of several thousand metres and (based on the estimated temperature differences of the facies present at the base and top of the sequences) the

thermal gradient is postulated as being less than *c.* 30°C km^{-1} (Aguirre *et al.*, 1989). By contrast, in the Casma Group of Peru the same facies series is developed over much more limited stratigraphic ranges of *c.* 300–1500 m, and so thermal gradients of up to 300°C km^{-1} have been postulated (Aguirre & Offler, 1985; Aguirre *et al.*, 1989).

An overall geodynamic model for the evolution of these contrasting metamorphic styles was developed involving extension being driven by a back-arc convective cell that resulted in thinning of the overlying crust with extensional basin formation (Fig. 5.11). These basins were either of an 'aborted' type or marginal basin proper, in which the crust does not rupture in the first case but does so in the second case, resulting in generation of abundant

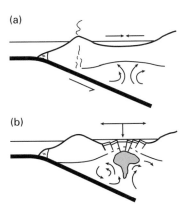

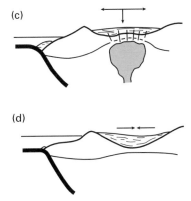

Fig. 5.11 Schematic representations of extensional models of marginal basin generation as applied to the Andes during the Cretaceous. (a) Rapid plate convergence with arc development and initiation of back-arc mantle uprise. (b) Diapiric mantle uprise in back-arc position resulting in melting of lower crust, bimodal volcanism and extensional back-arc basin formation. (c) Steepening and blockage of subducting slab with infilling of extensional basin. (d) Decoupling of oceanic plate and minor period of continental margin compression causing basin inversion. (After Aguirre *et al.*, 1989.)

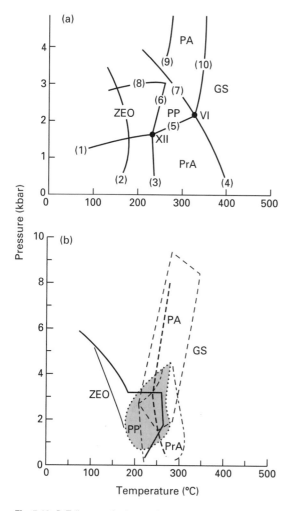

flood basalts. The metamorphism thus owes its origin ultimately to the enhanced heat flow linked to mantle upwelling, with higher heat flows in the case where the crust ruptures. These enhanced heat fluxes are limited to the period of extension, following which the thermal anomaly decays as the upwelled mantle cools and contracts back to its equilibrium state. In this scenario the metamorphism is an integral part of the basin formation and the finite time of the extension means that the metamorphism will develop largely in response to falling temperatures and will be of restricted duration; hence the facies series will be of limited extent. The burial metamorphism is thus envisaged as an integral part of an 'Andean tectonic cycle' lasting *c.* 40 My.

The interpretation placed on the origin of low-grade Andean metamorphism and the link between the various styles of burial metamorphism and major geodynamic processes represents a novel and innovative model. It has been developed on metamorphic facies distributions defined largely on petrographic information, and then using those facies to define the *P–T* ranges of metamorphism. Advances made in low-grade metamorphism since publication of the Andean models show that petrographically defined facies types can be an unreliable means of determining *P–T* conditions. The work of Liou *et al.* (1985a) showed an apparent simple pattern of subgreenschist facies distribution relative to *P–T*, as shown in Fig. 5.12(a), when calculated for the NCMASH system involving end-member phases. Frey *et al.* (1991), however, demonstrated that when the grid is calculated for average, rather than unity, activities for end-member components, utilizing a number of literature sources for *c.* 40–100 individual analyses, there is considerable overlap in *P–T* space between the different facies (Fig. 5.12b). The important feature of these results relative to the Andean model is that a zeolite to greenschist range taken as representative of a temperature difference of *c.* 250°C (Aguirre & Offler, 1985) might realistically represent little or no actual change in temperature. The marked differences between the postulated thermal gradients of the Casma Group (>*c.* 300°C km⁻¹) as opposed to the Coastal Ranges (<*c.* 30°C km⁻¹) may be an artefact of variations in mineral composition in shifting facies boundaries in *P–T* space. Accordingly, in the light of these

Fig. 5.12 *P–T* diagrams for low-grade metabasites showing low-grade metamorphic facies. ZEO = zeolite; PrA = prehnite–actinolite; PP = prehnite–pumpellyite; PA = pumpellyite–actinolite; GS = greenschist facies. (a) Petrogenetic grid determined for NCMASH system. Modified after Liou *et al.* (1987). (b) Petrogenetic grid determined for NCMASH system, using average activities derived from mineral compositions in five metabasites having different whole rock compositions. (After Frey *et al.*, 1991.) Reactions: 1, Prh + Chl + Lmt = Pmp + Qtz + H_2O; 2, Am + Qtz = Ab + H_2O; 3, Lmt + Prh = Cz + Qtz + H_2O; 4, Prh + Chl + Qtz = Cz + Tr + H_2O; 5, Pmp + Qtz = Cz + Prh + Chl + H_2O; 6, Lmt + Pmp = Cz + Chl + Qtz + H_2O; 7, Prh + Chl + H_2O = Pmp + Tr + Qtz; 8, Lmt = Lws + Qtz + H_2O; 9, Lws + Pmp = Cz + Chl + Qtz + H_2O; 10, Pmp + Chl + Qtz = Cz + Tr + H_2O.

variables, the Andean metamorphic facies variations need to be reassessed on the basis of a large database of microprobe mineral data applying modern techniques of petrogenetic grids, thermodynamic evaluation of *P–T* conditions and chemographic projections, all as a means of validating the previous large variations in facies series determined for the Andean range. Such work is at present in progress (Morata *et al.*, 1997; Belmar *et al.*, 1997; Robinson *et al.*, 1997b); whatever the eventual outcome, the exciting potential is that the controls of low-grade metamorphism, whether at a local or global scale, will be further understood.

5.3 Subaerial flood basalt sequences

In some continental flood basalt sequences there is a regional development of subgreenschist assemblages that show increasing grade with depth of burial and as such show similarities with the burial metamorphism of Coombs. The first studies of low-temperature alteration in basaltic sequences were those of Walker in Ireland and Iceland (1960a,b) and later in Scotland (1971), who showed depth-related zonations of zeolite assemblages representative of temperatures less than *c.* 150°C. More recent works on this and other flood basaltic piles, including the Deccan basalts, the Keweenawan rift sequence of north-eastern USA (Schmidt, 1990, 1993) and the Zig-Zag Dal Basalts of eastern North Greenland (Bevins *et al.*, 1991b), have reported higher grade assemblages with laumontite, prehnite, pumpellyite, epidote and actinolite. Subsequent work on Icelandic sequences has utilized material from wells drilled in connection with geothermal exploitation. Kristmannsdóttir and Tómasson (1978) identified from low-temperature geothermal areas, five depth-related zeolite zones (Fig. 5.13), namely (i) chabazite, (ii) mesolite/scolecite, (iii) stilbite, (iv) laumontite, and (v) scattered analcime and/or wairakite. In addition they identified high-temperature geothermal zones in which systematic

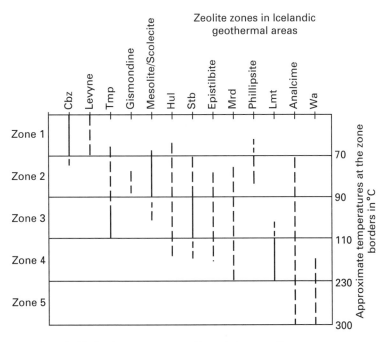

Fig. 5.13 Zeolite zones in Icelandic geothermal areas. (After Kristmannsdóttir & Tómasson, 1978.)

Zone 1: Chabazite
Zone 2: Mesolite/Scolecite
Zone 3: Stilbite

Zone 4: Laumontite
Zone 5: Zone with no zeolites except scattered analcime or wairakite

changes in zeolite type with depth were difficult to detect, with the passage to metabasites containing analcime and wairakite and eventually to epidote and prehnite. This was followed by the detailed analysis of the Nesjavellir geothermal field by Schiffman and Fridleifsson (1991), who in particular studied the smectite to chlorite transition. The overall zonation identified by Kristmannsdóttir and Tómasson (1978) has been shown to be broadly applicable to many subsequent studies including the recent investigation of early Tertiary flood basalts from central East Greenland (Neuhoff *et al.*, 1997). In a section 3000-m thick they reported the presence of seven mineral zones, namely (i) zeolite-free, (ii) chabazite + thomsonite, (iii) analcime, (iv) mesolite + scolecite, (v) heulandite + stilbite, (vi) laumontite, and (vii) prehnite + K-feldspar + quartz. Further documentation of the character of hydrothermal metamorphism of these basaltic rocks is given in Chapter 6.

These sequences have proved important in that they show the development of subgreenschist metamorphism in the absence of deformation and development of metamorphic fabrics. In addition these lava piles often preserve excellent primary flow morphological characteristics with marked contrasts in primary porosity and permeability over scales of only a few metres. Recent studies of these sequences have highlighted the influence of the nature of fluid/rock interaction on the development of subgreenschist metamorphic assemblages and these features are examined in two areas below, namely eastern North Greenland and Minnesota, USA.

5.3.1 Eastern North Greenland

The Late Proterozoic, Zig-Zag Dal Basalt Formation is believed to be a remnant of a larger tholeiitic flood basalt province that covered much of eastern North Greenland. The flat-lying, undeformed sequence consists of some 80 flows, individually reaching up to 100-m thickness, that have been variably affected by low-temperature metamorphism (Bevins *et al.*, 1991b). Not unexpectedly in only a *c.* 1300-m sequence, no assemblage variation with depth was recorded, although there is marked variation on a flow to flow basis through the succession as shown in Fig. 5.14. Here flows with ± pumpellyite ± prehnite + chlorite or regular mixed-layer chlorite/

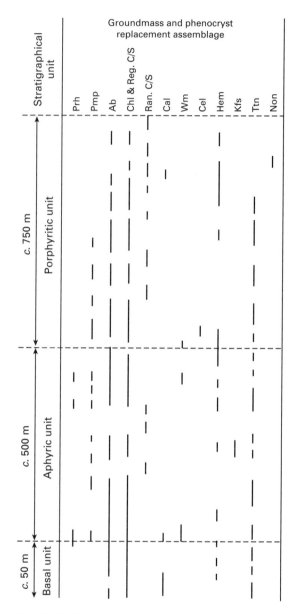

Fig. 5.14 Distribution of low-grade minerals in lavas of the Zig-Zag Dal Basalt Formation, eastern North Greenland. (Modified after Bevins *et al.*, 1991b.)

smectite and with extensive albitization of the primary feldspars are interbedded with flows in which there is very restricted recrystallization, preservation of primary feldspar compositions and a secondary mineralogy consisting in many cases of random mixed-layer chlorite/smectite alone (Bevins *et al.*, 1991b). A detailed analysis of the mafic phyllosilicates in these rocks was undertaken by Robinson *et al.* (1993), applying deconvolution analysis to the XRD patterns. From this survey they highlighted the contrast between the flows with random mixed-layer chlorite/smectite with up to 25% chlorite layers, and those with mixtures of regular mixed-layer chlorite/smectite (with corrensite) and chlorite.

The variation in metamorphic recrystallization and assemblages was linked to differences in the physical characteristics of the flows, with the thinner, more jointed and vesicular flows showing the greater recrystallization and varied assemblages, while the thicker flows have restricted recrystallization and secondary mineral development. This difference in metamorphic style was attributed to contrasts in flow permeability so that there was marked difference in the degree of fluid access into and within the flows (Bevins *et al.*, 1991b).

5.3.2 North Shore Volcanic Group, Minnesota

The Keweenawan Middle Proterozoic sequence is the northern exposed part of the Midcontinent Gravity High running NE–SW across the states of Minnesota, Wisconsin, Michigan, Iowa and Kansas in the USA. This mafic sequence consists of over 400 000 km^3 of intrusive and extrusive igneous rocks that are exposed best along the shore of Lake Superior, where they form the North Shore Volcanic Group (NSVG) and the Duluth Complex (Fig. 5.15). Many studies of the igneous, economic and metamorphic characters of the sequence have been undertaken, prompted by the rich economic deposits of Cu in the Keweenaw Peninsula of Michigan (Amstutz, 1977) and Cu, Ni and platinoids in the plutonic parts. Aspects of the metamorphism of this sequence have received attention in various works over many decades from the early studies of Pumpelly (1871) to Cornwall (1951a,b,c,1956). However, it was not until the work of Jolly and Smith (1972), Jolly (1974) and later Schmidt (1990, 1993) and Schmidt and Robinson (1997) that the uniqueness of the sequence for understanding some of the processes involved in the onset of metamorphism was demonstrated.

The Keweenawan rocks occupy a broad synclinorium underlying Lake Superior with the North Shore providing a continuous section through *c.* 8 km of mafic lava flows, from the oldest in the south around Duluth to the youngest around Tofte to the north-east (Fig. 5.15). The flows range mainly

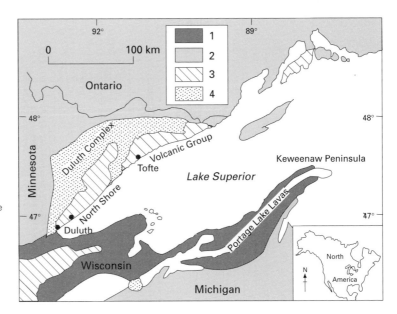

Fig. 5.15 Simplified geological map of the region around the western part of Lake Superior showing the regional position of the North Shore Volcanic Group (NSVG). 1, Upper Keweenawan strata; 2, pre-Keweenawan strata; 3, Keweenawan lavas (NSVG); 4, intrusive complexes. (After Schmidt, 1993.)

between 5 and 25 m in thickness and the great majority were erupted subaerially, as shown by the remarkable preservation in these *c.* 1.2-Ga flows of features such as ropy surfaces (Schmidt, 1990). Many of the flows show well developed morphological features, often with highly amygdaloidal flow tops and bases, while flow centres are massive. These contrasts provided marked differences in porosity and permeability that was to be a crucial control on the succeeding metamorphic development. The metamorphism of the NSVG was studied in detail by Schmidt (1990) who generated a prodigious database of the petrographic character of primary and secondary minerals, their mineral chemistry and whole rock chemistry including REE and isotopic data.

From this study Schmidt (1990, 1993) recognized the mineralogical variations over the 8-km sequence shown in Fig. 5.16, from which five zonal divisions were identified:

Zone 1: thomsonite–scolecite–smectite;

Zone 2: heulandite–stilbite–smectite;

Zone 3: laumontite–chlorite–albite;

Zone 4: laumontite–chlorite–albite ± pumpellyite ± prehnite;

Zone 5: epidote–chlorite–albite ± actinolite.

The upper part of the NSVG zonation compares well with that presented by Kristmannsdóttir and Tómasson (1978) from Iceland (Fig. 5.13). The NSVG section, however, is more extensive, and as such it represents one of the most continuous and best developed field examples of a zeolite to greenschist facies transition. Comparison between mineral occurrences in the NSVG zones with those in modern geothermal systems and experimental mineral stability data, led Schmidt and Robinson (1997) to estimate that temperatures varied from below *c.* 150°C for zones 1 and 2, to over *c.* 275°C for zone 5. As shown on Fig. 5.16, zones 1–3 are developed within the upper 1 km of section, while zone 3 extends to an estimated 6-km depth, and zone 5 is only developed at the maximum observed depth of 8 km.

Superimposed on the regional changes in mineralogy are variations reflective of a more local, within-flow scale alteration. Schmidt (1990, 1993) showed that in highly amygdaloidal flow tops with an originally high porosity/permeability, there is more intensive albitization of primary feldspar,

Fig. 5.16 Distribution of minerals in the North Shore Volcanic Group with respect to position in overall 8-km thick sequence, zones 1–5 and flow morphology. Amygdaloidal flow top zones shown by dark ornamentation; massive flow areas shown with vertical rules; and flow bottom areas by diagonal rules. (After Schmidt, 1993.)

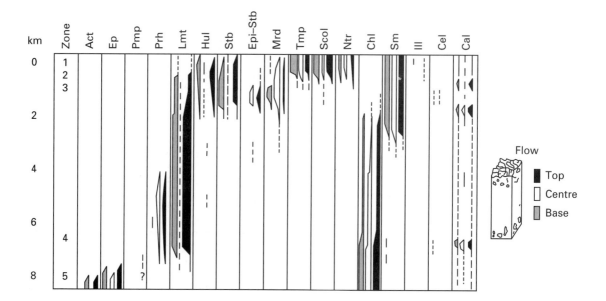

more abundant zeolites, and phyllosilicates have a chloritic composition. In contrast the massive flow centres of low porosity/permeability have restricted albitization, few zeolites and phyllosilicates of a smectic composition. Detailed examination of the phyllosilicates in relation to changes in flow morphology (Schmidt & Robinson, 1997), recognized more definitive mineralogical differences that provide valuable evidence for controls other than P–T on the metamorphic process. In zones 1 and 2, smectite is the dominant phyllosilicate in all morphological zones of an individual flow and it is of a near end-member saponite composition with X_{MgO} of 0.9. In zones 3 and 4, smectite still dominates in the massive flow parts, but it has lower X_{MgO} values (c. 0.5), more in line with whole rock values. By contrast, in the more amygdaloidal flow parts, phyllosilicate assemblages are polymineralic but with corrensite in zone 3 and then chlorite in zone 4 dominating the assemblages. Figure 5.17 is a schematic diagram showing these mineralogical changes in relation to flow morphology for examples from zones 2 and 4.

Combining data from the NSVG, eastern North Greenland and the Welsh Basin it can be determined that there is a definitive evolution in the X_{MgO} of mafic phyllosilicates with increasing grade of metamorphism from zeolite to greenschist facies, demonstrated in Fig. 5.18. At lowest grades in the NSVG the dominant mafic phyllosilicate is smectite in which the X_{MgO} is high (>c. 0.8) and is not correlated with the whole rock value (Group A samples in Fig. 5.18). Within Group A samples, those smectites from zone 1 of the NSVG have highest values of 0.9, but there is a trend to lower values of c. 0.8 in zone 2 as shown by the stippled range in Fig. 5.18. These data are for smectite in various locations in feldspar, interstitial and after mafics and so their restricted composition is not a domain effect. With increase of grade in the NSVG to zones 3 and 4 (Group B samples), corrensite and chlorite become the dominant mafic phyllosilicates and these span a broad range of X_{MgO} from 0.8 to 0.4, with an increasing trend towards a strong correlation with the whole rock values. Samples of Group C (approximately prehnite–pumpellyite facies) and Group D samples (approximately greenschist facies) represent further increases in grade in which chlorite is the dominant mafic phyllosilicate. Here there is a strong correlation with whole rock X_{MgO} approaching

a near 1:1 character (Fig. 5.18). In Group C phyllosilicates, the main discrepancy is for four samples from flow D910 in zone 5 of the NSVG, which have chlorite X_{MgO} of c. 0.6, that are much higher than their whole rock values (c. 0.3). These data indicate the generally high variance situation for low-grade metamorphism of metabasites at prehnite–pumpellyite and low greenschist facies, where whole rock composition plays an important role in controlling mineral distribution and their chemistries (Bevins & Robinson, 1993). The strong correlation with whole rock for samples in Groups C and D develops even though the rocks are only partially recrystallized and many primary minerals remain as relict phases. This suggests that fluid/rock interaction has been sufficiently pervasive at these grades, such that it has 'sampled' all parts of the whole rock giving a chemical fingerprint that in effect is in equilibrium with the whole rock even though these rocks are not totally recrystallized. The Group B samples include data from smectite and corrensite/chlorite (four solid circle symbols, Fig. 5.18) that have closely similar X_{MgO} values even though they developed in different morphological parts of the same flow (see previous paragraph). This shows that the smectite remaining as a 'relict' phase in massive flow parts, has in fact continued to undergo crystal chemical change to a lower X_{MgO} (from the higher values of >0.8 in smectite of Group A samples), but has not recrystallized to a chloritic phase as in the amygdaloidal flow top zone. A possible explanation to account for such differences is that there has been diffusive as opposed to advective fluid flow in different morphological parts of the flows, allowing differential transport of 'nutrients' (e.g. Al) that control the phase change (Schmidt & Robinson, 1997).

These recent studies on thick basaltic piles provide a useful framework in which to discuss varying features relating to the character of metamorphic processes from a mineralogical to regional scale. Features related to those described from the above examples of Greenland and Minnesota have also been described from basaltic sequences of oceanic islands that are reviewed in Chapter 6 (section 6.6). The nature of the fluid/rock interaction is a dominant control on metamorphic processes and while much interest at higher grade is connected with the movement of fluid in response to dehydration/decarbonation, the main interest here is with the

Zone 2: Hul–Stb–Sm flow F177

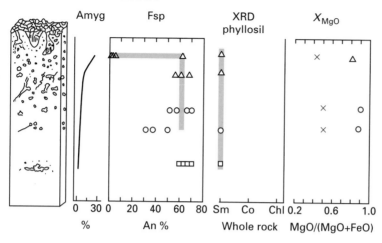

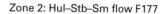

Zone 4: Lmt–Ab–Chl±Prh±Pmp flow LW7

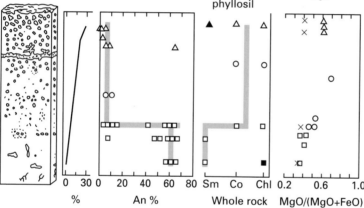

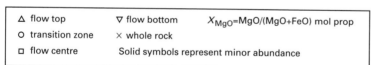

Fig. 5.17 Diagrammatic representation (not to scale) of morphological patterns of flows from zones 2 and 4 showing variation in amygdule (Amyg) percentage, An content in feldspar and X-ray diffraction-determined phyllosilicate type (Sm = tri-smectite; Co = corrensite; Chl = chlorite). (After Schmidt & Robinson, 1997.)

△ flow top	▽ flow bottom
○ transition zone	× whole rock
□ flow centre	

$X_{MgO} = MgO/(MgO + FeO)$ mol prop

Solid symbols represent minor abundance

process of hydration and the means of access of the fluid to the rocks. In addition, and as detailed in Chapter 4, there is active speculation and controversy regarding the mineralogical character and factors. (P–T, kinetics, fluid/rock ratios) controlling the smectite to chlorite transition. One model views the transition as a continuous mixed-layer chlorite/smectite series linking smectite and chlorite (e.g. Reynolds, 1988; Bettison-Varga *et al.*, 1991; Robinson *et al.*, 1993), while an opposing model is of a discontinuous change with discrete phases of smectite, corrensite and chlorite or mixtures with corrensite (Shau *et al.*, 1990; Shau & Peacor, 1992). Linked to this is the interesting speculation (Schiffman, 1995) that mixed-layering of chlorite/smectite with low chlorite contents (<*c*. 40%) does indeed exist, but that it represents a metastable product that crystallizes in response to low integrated water/rock

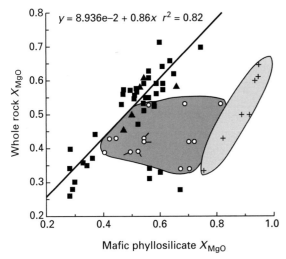

$y = 8.936e{-}2 + 0.86x \quad r^2 = 0.82$

Whole rock X_{MgO}

Mafic phyllosilicate X_{MgO}

Fig. 5.18 Plot of X_{MgO} ratios in mafic phyllosilicates versus whole rock. Group A samples are from metabasites with smectite and abundant zeolites from zones 1 and 2 of the North Shore Volcanic Group (NSVG); group B samples are from metabasites with dominant corrensite and chlorite with abundant zeolites from zones 3 and 4 of the NSVG; group C samples are from metabasites with dominant corrensite and chlorite with prehnite and pumpellyite and very rare zeolites, including samples from zone 5 of the NSVG, the Welsh Basin and eastern North Greenland. Group D samples are from metabasites with dominant chlorite with actinolite, no zeolites, and are from eastern North Greenland and Welsh Basin. Linear regression is for samples of groups C and D, except for four NSVG samples (with tails) offset from the curve with chlorite X_{MgO} values of c. 0.6. NSVG data from Schmidt (1990) and Schmidt and Robinson (1997); Welsh Basin data from Bevins and Robinson (1993), Bevins and Merriman (1988) and R.E. Bevins and D. Robinson (unpublished); Greenland data from Bevins *et al.* (1991) and Robinson *et al.* (1994).

ratios. Examples of such chlorite/smectite mixed layering have been reported from samples in which recrystallization is incomplete and accompanied by patchy albitization of primary plagioclase (Shau & Peacor, 1992). In contrast, where there is extensive recrystallization and albitization, indicative of higher integrated fluid/rock ratios, such mixed-layer chlorite/smectite is rare and discrete phases of corrensite and/or chlorite are dominant (Schiffman & Staudigel, 1994; Schiffman, 1995).

The particularly valuable features of the NSVG are the continuous nature of the metamorphic progression, as well as the availability of local scale morphological variations, that provide a rare

opportunity for these different aspects to be assessed. Although Schmidt and Robinson (1997) identified mafic phyllosilicates whose XRD patterns were representative of mixed-layer chlorite/smectite, there was, despite the continuous regional nature of the metamorphic succession, a compositional gap in chlorite/smectite at low percentages of smectite layers (c. 20–50%). Furthermore, although XRD data suggested a more continuous chlorite/smectite series for smectite compositions in the range 50–90%, detailed microprobe data in a critical sample showed that only chlorite and corrensite were present, whose compositions were analytically distinct at the two sigma level. These results were thus interpreted as more supportive of a discontinuous smectite to chlorite transition involving discrete corrensite, rather than that of the continuous mixed-layer model.

In zones 1 and 2 of the NSVG on a local, within-flow scale, the massive flow centres have extremely limited recrystallization and no albitization compared to extensive recrystallization with much albitization in flow tops (Fig. 5.17). Despite these differences, indicative of minor and major fluid access, the dominant phyllosilicate present is smectite. In zone 3 and above, the phyllosilicate present varied according to whether the material was from a massive flow centre, smectite, or from flow tops, corrensite and chlorite. Thus the interpretation placed on these mineral distributions was that the primary control on the change from smectite to corrensite or chlorite is temperature. Once above the estimated threshold temperature step of c. 150°C for the change from zones 2 to 3 the degree of fluid/rock interaction becomes an additional control on the development of phyllosilicate minerals.

To date the smectite to chlorite transition has been considered as either a continuous mixed-layering chlorite/smectite series or a stepped smectite–corrensite–chlorite transition, taken respectively as representative of disequilibrium and equilibrium progressions. Most recently, however, Robinson and Santana de Zamora (1998, in press) have reported on the smectite to chlorite transition from the Chipilapa geothermal system in El Salvador. This geothermal system provides an example of a different phyllosilicate progression involving a direct step from smectite to chlorite without examples of mixed layering or corrensite. The smectite to chlorite transition thus has been reported to proceed in three modes of progression and now cannot be regarded as a simple contrast between a continuous chlorite/smectite, disequilibrium and discontinuous smectite–corrensite–chlorite equilibrium transition. A recent review of clay mineral thermometry by Essene and Peacor (1995) proposed that clay mineral progressions (e.g. smectite to illite), assemblages and indicators, such as illite crystallinity, do not reflect equilibrium reaction series, but instead record reaction progress controlled more by kinetic than thermodynamic (principally P–T) effects. Robinson *et al.* (1998, in review) have proposed that the three styles of progression for the smectite to chlorite transition offer some of the best evidence in support of an Essene and Peacor (1995) model for this transition. In addition, Robinson *et al.* (1998b, in review) have demonstrated that kinetic effects related to deformation, geothermal gradient and grain size (surface energy) are minor compared to the degree of recrystallization associated with the magnitude of the fluid/rock ratios. This suggests that high fluid/rock ratios linked to advective fluid transport developing in areas of high porosity and permeability would promote dissolution and supply the necessary nutrients to allow the change in chemistry required to promote the direct smectite to chlorite transition. In contrast, in areas of low permeability and porosity, a diffusion process may be more prevalent and thus provide a kinetic constraint on the recrystallization process (Robinson *et al.*, 1998b, in review; Schmidt & Robinson, 1997).

Schmidt (1990, 1993) identified the principal control on the regional zonation of the NSVG as being depth, so that the metamorphism was ascribed to a burial metamorphic event linked to extension in a rift setting. A burial character is supported by the regional scale of the mineral zonation, a lack of metamorphic fabric and the irregular distribution of the zeolites on a lateral scale. However, the sharp change from a zeolite-dominated assemblage from zone 3 to a +/– prehnite/pumpellyite assemblage in zone 4 and to epidote–actinolite in zone 5 within a few hundred metres thickness (cf. fig. 5.2 in Schmidt, 1990 and fig. 5 in Schmidt, 1993) represents an extremely sharp progression from the zeolite to greenschist facies. Such a sharp transition might be a function of whole rock chemistry linked to the almost complete overlap of the subgreenschist

facies in P–T space (Frey *et al.*, 1991). As shown in Fig. 5.15, however, the NSVG is juxtaposed on its northern boundary with the Duluth Complex, which is shown as being intruded at the base of the NSVG (Green, 1982). This leads to speculation that, although there is a definitive regional effect with a depth-controlled increase in metamorphism, the telescoping of the transition from high-temperature zeolite to greenschist assemblages might be influenced by proximity to the Duluth plutonic complex (S. Th. Schmidt, personal communication). This complex perhaps provided a heat engine to drive fluid flow through the regionally extensive NSVG giving rise to the observed metamorphic patterns.

5.4 Convergent settings

Very low-grade metamorphism is most well known for the burial style of metamorphism, although such low grades are also found in the traditional convergent settings involving major crustal thickening. The contrast between the extensional and convergent setting is that in the latter there is a continuum into higher grades of metamorphism (Robinson, 1987).

5.4.1 Canadian Shield, Abitibi terrane

For the past two decades it has been standard procedure at higher grades of metamorphism to apply equilibrium thermodynamics in terms of petrogenetic grids, phase equilibria, geothermobarometry and chemographic projections in order to elucidate P–T–X conditions of metamorphism. Since recognition of the first subgreenschist facies, it has been widely regarded, however, that at low grades of metamorphism equilibrium conditions are not established and so application of such techniques is unjustified. Over the last decade, however, it has become increasingly apparent that there is some uniformity at low grades and that in certain circumstances the application of such techniques is possible. Here, in studies in eastern Canada and the Sierra Nevada in California, are some examples of the first application of such techniques to low-grade metamorphic rocks.

The Abitibi Belt in Canada is well known as an Archean granite-greenstone belt with high-grade metamorphism dominated by granulite facies rocks. In the southern part of the belt, however, tholeiitic basalts and komatiites occur as part of an island arc sequence (Hodgson *et al.*, 1990). In these rocks there is a transition from subgreenschist through greenschist to amphibolite facies.

Petrogenetic grids established for the CMASH system (Liou *et al.*, 1987; Frey *et al.*, 1991; Powell *et al.*, 1993) identify a pseudo-invariant point (CHEPPAQ) that was used by Carmichael (1991) to separate a higher pressure from a lower pressure bathozone. These two bathozones were recognized by the respective assemblages of Pmp–Ep–Act–Qtz and Prh–Chl (Fig. 5.19). However, as the positions of the pseudo-invariant point and reaction curves vary with mineral composition (Frey *et al.*, 1991; Powell *et al.*, 1993), the activities of the end-member

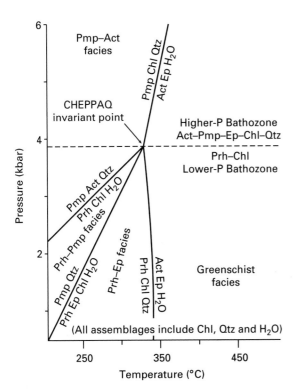

Fig. 5.19 Petrogenetic grid determined for low-grade metabasites in CMASH system, projected from chlorite, quartz and water. The chlorite, H_2O, epidote, prehnite, pumpellyite, actinolite and quartz (CHEPPAQ) pseudo-invariant point separates a higher pressure from a lower pressure bathozone. (After Powell *et al.*, 1993.)

mineral species have to be considered in order to determine the specific topology of the grid and derive quantitative data of the *P–T* conditions of the bathozones. Using this approach, Powell *et al.* (1993) showed that the south to north subgreenschist to greenschist facies transition in the region took place at *c.* 250°C and 2–2.5 kbar under a geothermal gradient of *c.* 35°C km⁻¹. In addition they were able to show that assemblages diagnostic of the pumpellyite–actinolite facies and of the higher pressure bathozone had minerals richer in iron than other areas. This resulted in displacement of the CHEPPAQ invariant point *c.* 1 kbar and so the *P–T* conditions of metamorphism need not have been at a higher pressure than that of the surrounding areas with assemblages diagnostic of the prehnite–pumpellyite facies.

Digel and Gordon (1995) undertook a related study of the metamorphic transition from prehnite–pumpellyite through greenschist to amphibolite facies in the Flin Flon region of Manitoba, Canada that is part of the early Proterozoic Trans Hudson orogenic belt. The succession is dominated by an island arc style volcano-sedimentary succession but with many basaltic units. Digel and Gordon (1995) examined two low-variance samples representative of the prehnite–pumpellyite and greenschist facies and applied numerical analysis of the phase equilibria in these rocks using singular value decomposition (Greenwood, 1968; Fisher, 1989; Chapter 4). Mass balances for the multicomponent and multiphase mixtures of the two samples (Table 5.1) were calculated based on the mineral analyses from the rocks. For the prehnite–pumpellyite facies sample, the mass balances correspond with the equilibria for the prehnite–pumpellyite to greenschist facies transition as modelled in the NCMASH system of Liou *et al.* (1985a). The greenschist sample generated three mass balances, one resulting from differences in epidote and chlorite compositions as a result of an exchange reaction. Thus, the two samples were taken as having equilibrated under different external conditions of *P–T* (Digel & Gordon, 1995). However, the *P–T* conditions at which these two samples equilibrated were calculated using the TWQ program (Berman, 1991) giving values that are within error of each other at 3.4 (±1.7) kbar, 280 (±40)°C and 2.8 (±0.4) kbar, 290 (±10)°C respectively (Digel & Gordon, 1995). Analytical uncertainties also result

Table 5.1 Mass balances at the prehnite- and pumpellyite-out, actinolite-in isograd (a) for the single sample, 90SD-179*, and (b) for 89SD-027$_{(1)}$ and 89SD-191$_{(2)}$*.

(a) 1 1.0 Act + 14.3 Pmp = 29.3 Prh + 4.49 Chl
    2 1.0 Chl + 6.20 Prh = 0.68 Act + 5.54 Ep
    3 2.74 Ep + 6.73 Prh + 1.0 Chl = 4.77 Pmp
    4 1.0 Chl + 50.2 Pmp = 7.69 Act + 91.9 Ep
    5 12.3 Ep + 1.0 Act + 0.89 Prh = Pmp
    6 2.54 Ep + 6.71 Prh + 1.0 Chl = 4.66 Pmp

(b) 1 1.0 Chl$_1$ + 10.4 Ep$_1$ + 0.03 Ab = 1.05 Chl$_2$ + 0.07 Act + 10.2 Ep$_2$
    2 0.10 Chl + 9.1 Ep$_1$ + 1.0 Prh + 0.10 Ab = 0.10 Act + 9.90 Ep$_2$
    3 1.0 Chl$_1$ + 16.4 Pmp + 22.3 Ep$_1$ + 0.13 Ab = 3.49 Act + 51.3 Ep$_2$

---

* Quartz and H$_2$O are present in excess.

in error in the determined P–T conditions and at the two sigma level these were estimated as ±1.5 kbar and ±30°C (Digel & Gordon, 1995).

It has been generally well known for some time that CO$_2$ has an important effect on the stability of the subgreenschist facies Ca–Al hydrous silicates, with high CO$_2$ resulting in suppression of the characteristic index minerals (Seki & Liou, 1981). To date, however, most thermodynamic analysis of CO$_2$ in low-grade metabasites has relied on schematic diagrams and Schreinemakers rules, coupled with general field observations. Digel and Ghent (1994) made a more quantitative analysis of the importance of H$_2$O–CO$_2$ fluid mineral equilibria in the transition from the prehnite–pumpellyite to greenschist facies metabasites for the same Flin Flon region as the work of Digel and Gordon (1995). Using the TWQ program, Digel and Ghent (1994) established T–X_{CO_2} diagrams in the CMASH–CO$_2$ system that showed, for temperatures up to 400°C and over the pressure range 1.5–4.5 kbar (Fig. 5.20), the mole fraction CO$_2$ should be below 0.002 for the stability of prehnite and pumpellyite. At CO$_2$ mole fractions above 0.002, the non-diagnostic assemblage epidote + chlorite + calcite would develop.

These initial results relating to P–T conditions are consistent with general predictions for low-grade metamorphism. It should be borne in mind,

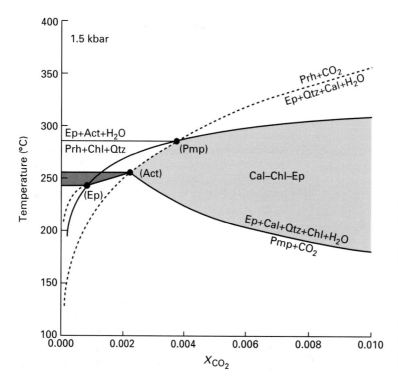

Fig. 5.20 Temperature vs. X_{CO_2} plot for the CMASH–CO$_2$ system at fixed pressure of 1.5 kbar for the prehnite–pumpellyite to greenschist facies transition, using mineral activities for a low-grade metabasite sample from the Flin Flon region, Manitoba. Darker shaded area shows that the stability field for Prh + Pmp + Chl assemblage is limited to X_{CO_2} values of <0.002. Lighter shading represents the large stability field for non-diagnostic Cal + Chl + Ep field developed at higher X_{CO_2}. (After Digel and Ghent, 1994.)

however, that errors associated with such approaches can become much more relevant at such low *P–T* conditions. One major area of concern is that thermodynamic data for most of the phases involved at low-metamorphic grade are of the lowest reliability of all metamorphic minerals (Berman, 1988; Holland & Powell, 1990). In addition activity–composition relationships of these minerals are largely unconstrained by experimental or other approaches and so the simplest ideal site mixing models are typically assigned (Berman, 1988; Holland & Powell, 1990; Digel & Gordon, 1995). Accordingly, the errors expected from such approaches must always be borne in mind as they can be greater than the postulated range of all the subgreenschists facies.

5.4.2 Northern Sierra Nevada

The Northern Sierra Nevada complex, lying *c.* 150 km

north-east of the San Francisco Bay area, consists of four NW–SE trending lithotectonic belts of Palaeozoic to Mesozoic age. This complex has undergone at least three orogenic events in the Devonian, Permo-Triassic and Jurassic, followed by intrusion of the Sierra Nevada batholith. All units are variably metamorphosed as a result of these events and although the metamorphic character of the region is not well understood the main details were reviewed by Day *et al.* (1988). The Nevadan orogeny imparted a regional fabric to many of the rocks and resulted in the amalgamation of the four units that are bounded by major NW-trending faults. The westernmost unit, known as the Smartville Complex, is interpreted as a rifted volcanic arc complex with volcanic, hypabyssal and plutonic components. The complex has not developed the regional Nevadan fabric and is interpreted as a window through the regional event into the non-deformational, burial

Table 5.2 Mineral assemblages in rocks from Smartville Complex, California, showing decrease in number of assemblages in any one sample as metamorphic increases from prehnite–pumpellyite to greenschist facies. (After Springer *et al.*, 1992.)

Sample	Zone	Rock type	Mineral assemblage	
			Relict	Metamorphic (Ab + Ttn + Wm + Mag + Qtz ± Sulph)
712	PrP	volcaniclastic	Cpx, Pl	Chl + Ep + Prh, PmP + Prh, Chl + PmP, PmP + Prh + Ep
715	PrP	mafic intrusive	Cpx, Pl	Chl + Prh + Pmp, Chl + Ep + Pmp, Pmp in Pl
715A	PrP	mafic intrusive	Cpx, Pl, Ox	Chl + Ep + Am, Chl + Prh + Am, Pmp in Pl, Prh in Pl
715B	PrP	volcaniclastic	Cpx, Pl	Chl + Prh + Am 1 + Am 2
795	PrP	mafic intrusive	Cpx, Pl, Ox	Chl + Pmp + Ep, Chl + Prh + Pmp, Chl + Prh + Ep, Pmp in Pl
7136	PrP	mafic intrusive	Cpx, Pl, Ox	Chl + Am, Chl + Ep + Prh + Am, Chl + Ep + Pmp, Chl + Prh + Pmp
7140	PrP	flow	Cpx, Pl	Pmp + Ep + Prh, Chl + Pmp + Ep, Chl + Pmp + Prh, Pmp in Pl
7143	PrP	mafic intrusive	Cpx, Pl, Ox	Chl + Prh + Pmp, Chl + Prh + Am, Chl + Pmp + Ep, Chl + Am + Pmp + Prh
7147	PrP	mafic intrusive	Cpx, Pl	Chl + Ep + Prh, Chl + Pmp + Prh, Chl + Ep + Pmp, Chl + Cal + Ep + Prh
7154	PrP	mafic intrusive	Cpx, Pl, Ox	Chl + Pmp + Prh + Am, Chl + Ep + Am
L19	PrP	flow	Cpx, Pl	Chl + Pmp + Ep, Chl + Pmp + Prh, Cal + Prh + Pmp, Cal + Ep + Prh
L43	PrP	flow	Cpx, Pl	Chl + Ep + Am + Prh, Chl + Ep + Pmp + Cal + Prh, Pmp in Pl, Chl + Pmp + Ep
L56	PrP	flow	Pl, Ox	Chl + Ep + Cal, Chl + Ep + Grt + Cal
L146	PrP	volcaniclastic	Cpx, Pl	Chl + Prh, Chl + Am, Pmp in Pl
L155	PrP	mafic intrusive	Cpx, Pl, Ox	Chl + Am + Prh + Ep, Chl + Prh + Ep, Chl + Pmp + Prh + Ep
710	PrA	tuffaceous chert	Cpx	Chl + Prh + Ep + Am
711	PrA	volcaniclastic	Cpx, Pl, Ox	Chl + Prh + Am + Ep
B19A	PrA	mafic intrusive	Cpx, Pl, Ox	Chl + Prh + Cal, Chl + Am + Cal, Chl + Prh + Am + Cal
754	GS	flow	Cpx, Pl, Ox	Chl + Ep + Am
760	GS	volcaniclastic	Cpx, Pl	Chl + Ep + Am
RM28	GS	flow	Cpx, Pl, Ox	Chl + Ep + Am
B13	GS	volcaniclastic	Cpx, Pl, Ox	Chl + Ep + Am
B23	GS	mafic intrusive	—	Chl + Ep + Am

Am = amphibole; OX = iron oxide mineral; Wm = white mica; Sulph = minerals.

metamorphism developed during arc construction (Springer *et al.*, 1992).

Petrographic analysis of over 200 thin sections by Springer *et al.* (1992) identified assemblages diagnostic of the prehnite–pumpellyite, prehnite–actinolite and greenschist facies; assemblages were defined where minerals were in mutual contact, in the same unzoned amygdule or vein, or if they were in the same irregular patch of matrix interstitial to relict minerals. Even with the above criteria used to define assemblages, uncertainty may still remain about the relationship of certain mineral occurrences. For example in their sample 715, pumpellyite occurs isolated in the interiors of plagioclase grains (Table 5.2) and whether to treat this as a unique assemblage or regard it as part of the main assemblages is not easy to resolve. On this basis, the assemblages listed in Table 5.2 were recognized in samples from the three facies and, as seen, up to four different assemblages were recorded within the same thin section. The interpretation was that local differences in effective bulk compositions within a thin section allowed the development of these different assemblages (Springer *et al.*, 1992), especially in the case of coarser grained rocks. The assemblages are most varied for the prehnite–pumpellyite facies, decreasing in the prehnite–actinolite facies and reduced to one in greenschist facies samples (Table 5.2). This reflects the increasing degree of recrystallization in the rock with grade (Springer *et al.*, 1992) and as a result of an increase in diffusion kinetics the rock system reached a stage at which on the thin section scale it may be regarded as an equilibrated system.

The link between rock/domain composition and mineral assemblage was qualitatively tested by the above petrographic analysis, but a more rigorous assessment of mineral chemistry can be aided by chemographic projections of compositional space from minerals that are ubiquitously present (Beiersdorfer & Day, 1995b). A projection in the eight component NCTiFMASH system, from the low-grade ubiquitous phases albite, quartz, water and titanite plus epidote onto the 'image plane' AF_2O_3–FeO–MgO has been developed, as detailed, in Chapter 4, and is known as the 'epidote projections'. In the epidote projection for the three facies from the Smartville Complex, there is general concordancy between the tielines for coexisting minerals such as

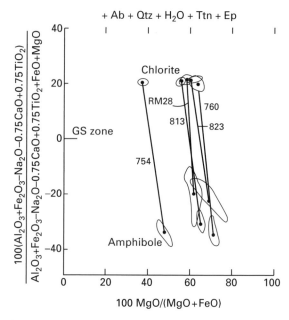

Fig. 5.21 Epidote projection for greenschist facies assemblages recorded from the Smartville Complex, California, showing the regularity of the tielines joining coexisting chlorite and Ca-amphibole. Circles enclose mineral compositions from the same thin section. (After Springer *et al.*, 1992.)

for the greenschist zone as shown in Fig. 5.21. Here the sympathetic variation in the X_{MgO} ratio between the chlorite and amphibole identifies a direct relationship between the X_{MgO} ratio of the minerals and whole rock. Thus, the consistency between the tielines is a general pointer to equilibrated systems, whereas crossing tielines are considered indicative of non-equilibrated systems.

However, the interpretation of crossing tielines is complex and various factors can lead to crossing tielines including: the combination of system components such as treating all Fe as FeO when, if analytical techniques allowed, they should be treated as separate FeO and Fe_2O_3 components; different generations of minerals in the samples; and metamorphism at different $P–T$ conditions. Examples of these features can also be seen in rocks from the Smartville Complex. Springer *et al.* (1992) reported crossing tielines involving minerals whose petrographic/chemical features, such as different coloured and/or chemistries of chlorite and amphibole suggested more than one generation of the mineral.

More specific examples of such features were reported by Bevins and Robinson (1995) in metabasites of the Eastern Belt of the Northern Sierra Nevada Complex. Some metabasites from the belt show a strong Nevadan foliation and were metamorphosed at pumpellyite–actinolite facies. The Nevadan fabric is defined by alignment of actinolite and chlorite as shown in Fig. 5.22(a), but also present are certain pumpellyite crystals which are in the form of elongate crystals aligned oblique to the fabric. As shown in Fig. 5.22(b), pumpellyite in the central field of view is zoned, with a light-coloured core and darker rim, indicative of Fe-rich and Mg/Al-rich parts, respectively. The large (dark) elongate grain of pumpellyite, aligned oblique to the fabric, at centre top of the image is the same colour and thus composition as the rim of the crystal below it (described above). When plotted on the epidote projection this sample shows crossing tielines with respect to the main pumpellyite compositions (Fig. 5.23), although chlorite–actinolite tielines for this and related samples for the area show closely similar trends, as reflected in their restricted range of K_{DMg-Fe} values between 1.73 and 1.82. The tielines for the chlorite–pumpellyite pairs, however, are more complex, showing several broadly similar positive slopes with K_Ds of 0.56–0.61, and two sharply cross-cutting negative slope tielines with K_Ds of c. 1.1. In the case of the zoned pumpellyite in sample 1747, it is the core composition that is comparable

(a)

(b)

Fig. 5.22 Photomicrograph (a) and back-scattered electron microscope (b) images from metabasite sample 1747 from the Eastern Belt of the Sierra Nevada, California. (a) Intricate pumpellyite crystal (centre) with Fe-rich core overgrown by colourless radiating Mg/Al-rich rim. Also discrete Mg/Al-rich crystal at centre top. Field of view 0.42 mm across. (b) Back-scattered image of (a). Central pumpellyite crystal with lighter core (Fe-rich) overgrown by darker Mg/Al-rich rim. Discrete Mg/Al-rich crystal at top centre lies oblique to the main Nevadan fabric that is defined by orientation of actinolite (needles of white crystals) and chlorite (darker laths). Black is predominantly quartz.

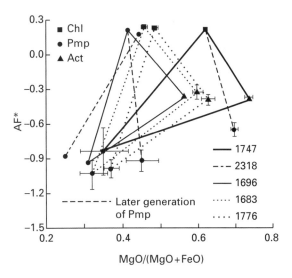

Fig. 5.23 Epidote projection for metabasites from the Eastern Belt of the Sierra Nevada, California. Tielines are for means of analyses from individual samples. Dashed tielines represent a later generation of Mg/Al pumpellyite. (After Bevins & Robinson, 1995.)

with pumpellyite from other samples, while the overgrowing rim and grains oblique to the fabric (Fig. 5.22b) are the compositions that relate to the crossing tielines (Fig. 5.23). The crossing chlorite–pumpellyite tieline points to a disequilibrium relationship and in combination with the textural features, was interpreted as representing overgrowth of the pumpellyite core and Nevadan fabric at a later or separate stage of metamorphism under static conditions, perhaps related to tectonic loading in the region following emplacement of the Eastern Belt as a series of thrust slices (Bevins & Robinson, 1995).

5.5 Conclusions

This review of the regional field character of low-grade metabasites has shown that over the past decade or so there have been most useful additions to the database of petrographic and microprobe information that have applied the research methods outlined in Chapter 4. These works have shown the benefits of applying more quantitative chemographic and thermodynamic methodologies to such rocks. The potential in the near future is for further rapid advance as these techniques are further refined with

particular application to the problems in the low-grade metamorphic field.

It has been shown that it is possible to recognize and distinguish between the effects of domain influence, whole rock chemistry and $P–T$ on the distribution of metamorphic facies. The next step is to understand and link variations in metamorphic facies distributions in any area in response to $P–T$, with the aim of establishing $P–T–t$ paths and correlating these to large-scale geodynamic processes. The study of the smectite to chlorite transition in recent years has resulted in different models of the mineralogical character of this transition and the controlling factors on its evolution. It is especially fascinating and exciting at the present time with the speculation as regards the respective roles of thermodynamic effects as opposed to kinetic effects, in respect of temperature, deformation, surface energy and style of fluid/rock interaction, in terms of diffusive and advective transport. The one certain feature to arise out of these ongoing studies is that there is much potential for increased effort in research in these fields and that our understanding of low-grade metamorphic processes in low-grade metabasites will increase even more rapidly than it has done over the past decade or so.

6 Very low-grade hydrothermal metamorphism of basic igneous rocks

J. C. Alt

6.1 Introduction

Oceanic crust is made up of variably hydrothermally altered and metamorphosed basaltic rocks and comprises three fifths of the Earth's crust, making these the most abundant low-grade metabasites on Earth and therefore, the main focus of this chapter. Heat released during the formation and cooling of oceanic crust and lithosphere drives circulation of seawater through the crust, causing hydrothermal metamorphism from very low-grade up to amphibolite facies conditions. Mass transfer and isotopic exchange between seawater and oceanic crust in these systems result in the formation of massive sulphide deposits and significantly influence the elemental and isotopic compositions of seawater and the crust. This chemical exchange is an important part of global geochemical cycles in the hydrosphere/atmosphere system, but is also a link between the surface and interior of the earth via recycling of altered ocean crust in subduction zones, which affects volcanism in island arcs and contributes to heterogeneities in the mantle.

Very low-grade metamorphism of volcanic rocks was reviewed by Liou *et al.* (1987), but significant advances in understanding of hydrothermal processes have taken place since, particularly in hydrothermal metamorphism of oceanic crust, the transition from smectite to chlorite in various settings and in establishing new petrogenetic grids for basic rocks (e.g. Frey *et al.*, 1991; Alt, 1995; Schiffman & Day, 1995; Schiffman & Staudigel, 1995). This chapter begins with a discussion of the structure of ocean crust and submarine hydrothermal systems, then continues with the secondary mineralogy and bulk rock chemical changes observed in different stratigraphic sections and ages of oceanic crust in order to document the processes of very low-grade hydrothermal metamorphism and how these evolve

with time and space. For these purposes, the crust is divided into the volcanic section, the upper sheeted dyke complex and the combined lower dyke complex plus plutonic section. Variations in very low-grade metamorphism among modern and fossil oceanic crust (ophiolites), oceanic islands and island arcs are discussed, as well as brief comparisons with some subaerial geothermal systems in different rock types.

The term 'hydrothermal metamorphism' as used in this chapter refers to the formation of metamorphic minerals resulting from the interaction of a rock with circulating heated aqueous fluids. The terms alteration and metamorphism are used interchangeably in this chapter, although in other papers alteration has been used to refer to the partial recrystallization of rocks at lower temperatures ($<100°$ down to $0°C$), whereas metamorphism has been restricted to the formation of minerals that characterize various metamorphic facies at higher temperatures and pressures. As will be shown, however, hydrothermal metamorphism commonly leads to partial recrystallization, and kinetic and disequilibrium effects can be important, particularly at lower grades. The conditions of alteration and metamorphism that are discussed here range to low temperatures ($<50°C$), which in sediments might be considered diagenetic. These processes in ocean crust involve large-scale convection of heated fluids, however, and can involve enormous fluxes of fluids through the rocks (up to thousands of times the rock mass), so they are better considered as hydrothermal rather than diagenetic.

6.2 Oceanic crust

6.2.1 Oceanic crustal structure, porosity and permeability

An idealized section of oceanic crust has a layered

structure, with volcanic rocks underlain by a sheeted dyke complex and subjacent gabbroic rocks. The volcanic section is 300–800 m thick and consists of pillow lavas, massive flows and breccias (Donnelly *et al.*, 1979; Auzende *et al.*, 1989; Francheteau *et al.*, 1992; Alt *et al.*, 1996a,c). This grades downward over a 200–500-m thick transition into 100% dykes of the underlying sheeted dyke complex, which are the feeder conduits to the overlying volcanic rocks. The dyke complex is generally 700 m to more than 1 km thick, although estimates range to as little as 350 m thick locally (Auzende *et al.*, 1989; Francheteau *et al.*, 1992; Alt *et al.*, 1996c). Gabbroic rocks comprise the lower ocean crust for an average crustal thickness of 7 km (White *et al.*, 1992).

Seismic evidence indicates the presence of a magma chamber beneath much of the fast-spreading EPR (East Pacific Rise). This magma 'chamber' is really a melt lens, up to a few hundred metres thick, <1–2 km wide and at 1.6–2.4 km depth (Fig. 6.1) (Detrick *et al.*, 1987; Sinton & Detrick, 1992). It is underlain by a zone of crystal mush and hot rock (Fig. 6.1) (Sinton & Detrick, 1992). Similar magma chambers have been imaged at intermediate-rate spreading centres (Morton & Sleep, 1985; Rohr *et al.*, 1988), at the ultrafast spreading southern EPR (Detrick *et al.*, 1993) and the Lau Basin back-arc spreading centre (Collier & Sinha, 1992).

A sill-like melt zone has been geophysically imaged on the Reykjanes Ridge (Sinha *et al.*, 1997), but magma chambers have not been defined in other areas along slow-spreading ridges (Sinton & Detrick, 1992; Solomon & Toomey, 1992). Locally on the MAR (Mid-Atlantic Ridge), however, low-velocity zones at about 3–3.5 km below the seafloor are interpreted to be recently solidified intrusions, where the rock is still hot and is cracked or undergoing cracking (Fig. 6.1; Purdy & Detrick, 1986; Kong *et al.*, 1992). In other places, earthquakes indicate that the entire crustal section may be cooled to within the brittle behaviour field (Huang & Solomon, 1988; Toomey *et al.*, 1988). Large-throw normal faults locally disrupt the idealized layered crustal structure that was outlined above and may act as fluid pathways to cool the crust (Fig. 6.1). In some cases such faulting can lead to local exposure of mantle peridotites along slow-spreading ridges (Karson, 1990).

The porosity and permeability of ocean crust exert strong controls on hydrothermal circulation

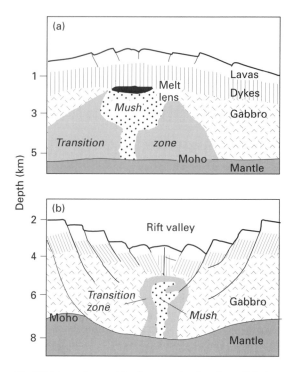

Fig. 6.1 Interpretive models of magma chambers beneath fast-spreading (a) and slow-spreading (b) ridges. In (a), a thin lens of basaltic melt at 1–2 km beneath the ridge axis grades downward into a partially solidified mush, and is surrounded by solidified hot rock. Idealized layered crustal structure is relatively intact. In (b), a steady-state magma chamber is lacking and a dyke-like mush zone forms small, sill-like intrusive bodies. Crustal structure is more disrupted by deep normal faulting than at fast-spreading ridges. (After Sinton & Detrick, 1992.)

and alteration and these properties vary with time and depth in the crust. For ODP Hole 504B *in situ* measurements of porosity and permeability are highest in the uppermost 100 m of the volcanic section where abundant fractures and pore spaces remain open (Fig. 6.2). Values then decrease in the lower volcanic section as the result of filling of open spaces with secondary minerals, mainly smectite. Porosity then decreases drastically downward into the dykes.

Also shown in Fig. 6.2 are permeabilities calculated from vein measurements for ODP Hole 504B, DSDP Hole 418A and for the Troodos and Oman ophiolites. Because it was assumed that all veins were open to their full width, these are maximum estimates that are all significantly greater than the *in situ* Hole

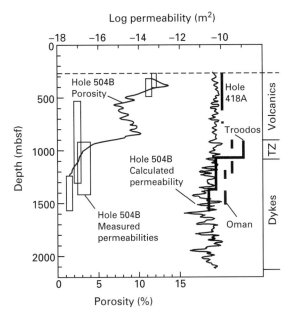

Fig. 6.2 Porosity and permeability of the upper ocean crust with depth (metres below seafloor, mbsf). At left are *in situ* bulk permeabilities (rectangles) and apparent bulk porosities (solid line) for ODP Hole 504B in the eastern Pacific. Vertical extents of rectangles indicate depth intervals over which permeability measurements were made (from Becker *et al.*, 1989). At right are permeabilities for Hole 504B and for other ocean drillholes and ophiolites estimated from measurements of veins in drillcore and outcrops, using the model from Norton and Knapp (1977). This model assumes parallel fractures, where $K = nd^3/12$; K = permeability, n = fracture spacing, and d = fracture aperture. (Data from Alt [unpubl. data]; Johnson, 1979; Nehlig & Juteau, 1988; van Everdingen, 1995.)

504B measurements, which reflect 6 million-year-old crust that is already altered and mostly sealed. The calculated permeabilities all exhibit a sharp decrease downward into the dyke section, similar to, but of lower magnitude than the *in situ* data for Hole 504B (Fig. 6.2). These data illustrate an essential point: that the volcanic section is more porous and permeable than the underlying sheeted dykes by one to four orders of magnitude. Thus, lithology largely controls the physical properties of the crust and consequently hydrothermal circulation and alteration. Much greater volumes of fluid can circulate more freely through the heterogeneous, permeable and porous volcanic section than through the more uniform, massive sheeted dyke section.

6.2.2 Heat sources and convection in ocean crust

Heat flow measured near mid-ocean ridges is less than predicted from conductively cooling plate models, indicating that ocean crust is convectively cooled by circulating seawater (Fig. 6.3). Lister (1982) classified convection in the oceanic crust into 'active' and 'passive' regimes (Fig. 6.4). Active circulation is restricted to spreading axes, and is hence commonly referred to as axial convection, where the heat source is a magma chamber or cooling and cracking plutonic rocks. The high-temperature heat source drives forced convection, with high temperatures (>350°C), rapid fluid circulation, focused discharge zones and diffuse recharge zones (Lister, 1982; Fehn *et al.*, 1983). These are the hydrothermal systems that give rise to black smoker hydrothermal vents and massive sulphide deposits at mid-ocean ridges, and cause higher temperature hydrothermal metamorphism of ocean crust (up to amphibolite grade). Given the depth to the top of the heat source of *c*. 1.5–3.5 km and convection cells with an aspect ratio of about 1 (Lister, 1982; Fehn *et al.*, 1983) this indicates that these active axial cells are restricted to within a few km of the spreading axis. Various estimates suggest that about 8–20%

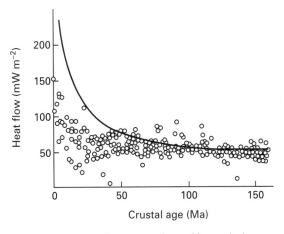

Fig. 6.3 Oceanic heat flow vs. age of crust. Measured values (open circles) for crust <50 Ma in age fall below theoretical conductive cooling curve (heavy line), indicating convective cooling of young ocean crust by circulating seawater. Measured values are averaged over 2 Ma intervals for Atlantic, Pacific and Indian oceans. (After Stein & Stein, 1994.)

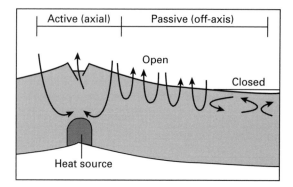

Fig. 6.4 Illustration of convective regimes in ocean crust (after Lister, 1982). Active high-temperature convection at mid-ocean ridges is driven by a magmatic heat source, whereas lower temperature passive circulation off-axis is driven by cooling of the crust and lithosphere. See text for discussion.

of the hydrothermal heat flux occurs through active axial convection, with the remainder occurring in passive systems off-axis (Morton & Sleep, 1985; Mottl & Wheat, 1994; Stein & Stein, 1994). Thus, although the effects of active axial hydrothermal alteration are more dramatic, very low-grade metamorphism of oceanic crust in passive systems off-axis is at least as significant geochemically and physically.

Some early alteration of the uppermost crust occurs in the active axial convection cells, but most axial reactions occur at greater depths and higher grades. Very low-grade alteration and metamorphism of the upper crust occur predominantly in passive circulation systems, also referred to as ridge flank or off-axis systems (Fig. 6.4). Characteristic secondary minerals include smectites, carbonates, and minor Fe-oxyhydroxides, pyrite and zeolites. Passive fluid circulation is driven by the cooling of the ocean crust and lithosphere, and occurs over much larger areas than the higher temperature convection at ridge axes. Passive circulation on ridge flanks begins just outside the axial cell and continues for up to several tens of million years (i.e. out to 100s to 1000s of km from the ridge). Flow rates during off-axis circulation are much slower than in the axial systems and temperatures are much lower, $< 150°C$ and as low as $2°C$ (Fehn *et al.*, 1983; Mottl & Wheat, 1994). The low temperatures, however, require correspondingly huge fluid fluxes: if circulation takes

place at an average of $10°C$, then the fluid flux through ridge flanks is 100 times that through the ridge axes (Mottl & Wheat, 1994). The depth of circulation is highly dependent upon the permeability distribution, but modelling using *in situ* measurements of permeability indicates that convection can occur in the uppermost few hundred metres of basement in cells with very low aspect ratios (Becker *et al.*, 1989; Fisher *et al.*, 1990, 1994). Much of this convection may occur along discrete high-permeability horizons within the basement (Fisher *et al.*, 1994; Davis *et al.*, 1996). Variations in basement topography, sediment thickness and thermal conductivity strongly influence flank circulation, with upflow and high heat flow at basement highs and downwelling and low heat flow at basement lows, where sediments are thicker (Lister, 1972; Davis *et al.*, 1980, 1992a; Langseth *et al.*, 1988; Fisher *et al.*, 1990, 1994). As in the active axial systems, discharge zones are more focused than recharge in passive circulation models, but whereas the axial systems remain fixed by the heat source at the ridge, the off-axis circulation cells move with the crust as it spreads away from the axis (Fehn *et al.*, 1983).

The convective heat loss from ocean crust ceases on average at about 65 Ma (Fig. 6.3; Stein & Stein, 1994), when the mean measured heat flow matches that of the theoretical conductive cooling curve. This has been taken as the age at which the crust is sealed to convection, but 'closed' circulation, where fluids circulate in the basement beneath an impermeable sediment cover but do not remove heat, can continue for millions of years (Fig. 6.4). Thus, very low-grade metamorphism of upper oceanic crust may continue for many tens of million years. Significant variations can occur locally as the result of differences in basement topography and sedimentation rate. For example, where sedimentation is rapid, the average heat flow can reach theoretical conductive values in crust only about 3–6 million years old, but closed convection is still active (Langseth *et al.*, 1988; Davis *et al.*, 1992a). The uppermost few hundred metres of crust can thus be cooled by convection and then reheated as it is buried and insulated by sediments, with hot, young crust that is rapidly buried by sediments reaching higher temperatures (to $> 150°C$) than older, cooler crust that is more slowly buried ($< 50°C$).

6.2.3 Submarine versus subaerial hydrothermal systems

Some fundamental differences exist between submarine and subaerial hydrothermal systems, as illustrated in Fig. 6.5. Subaerial hydrothermal systems are multipass systems, where fluids can recirculate, and the isotherms have a general plume shape where fluids well up above a heat source. At some point in the system boiling can occur, causing precipitation of hydrothermal minerals and partitioning of gases (H_2S, CO_2) into the vapour phase, which can then

(a) Subaerial

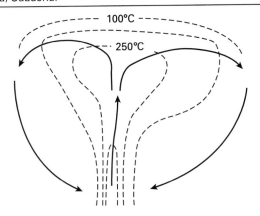

(b) Submarine

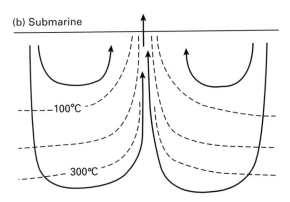

Fig. 6.5 Schematic cross-sections through (a) a subaerial hydrothermal system and (b) an active submarine hydrothermal system at a mid-ocean ridge. Subaerial systems are characterized by multipass, recirculating fluid flow, whereas submarine systems comprise flow-through, single-pass circulation, leading to differences in fluid flow paths (solid lines with arrows) and isotherms (dashed lines). The pressure effect of overlying seawater is also an important difference. See text for discussion.

condense or mix with cooler waters near the surface to produce acid-sulphate or carbonate-rich surface groundwaters. A caprock, produced by precipitation of minerals from hydrothermal solutions, may seal the upper portion of the hydrothermal system.

In contrast, active submarine hydrothermal systems at mid-ocean ridges are open to overlying seawater and are essentially single-pass systems, where seawater flows in, reacts, and then flows out of the system (Fig. 6.5). Because of the lithological and permeability structure of the crust, large volumes of cooler fluids circulate through the shallow crust while only small amounts of fluid penetrate the deeper crust and attain the maximum temperatures. These effects combine to give the general pattern of isotherms and flow paths shown in Fig. 6.5. Because the generation of ocean crust is an integral part of seafloor spreading, crust formed at the spreading axis must move away from the axis. As this occurs, the crust cools and rocks can pass through different circulation regimes shown in Fig. 6.5 and into lower temperature passive circulation off-axis (Fig. 6.4). The high pressures caused by the 2–3.5 km of overlying seawater at mid-ocean ridges make boiling rare in submarine hydrothermal systems, but supercritical phase separation is an important control on the salinity and volatile contents of hydrothermal fluids (Von Damm, 1988, 1995; Edmonds & Edmond, 1995).

6.2.4 Sampling ocean crust

Fault scarps and the surface of basaltic ocean crust are sampled by dredging and from submersibles, but deep ocean drilling provides intact stratigraphic sections of ocean crust. There are only two continuous sections through the volcanic and dyke sections of upper ocean crust. The faulted crustal section at Hess Deep in the eastern Pacific has been sampled by a combination of dredging and submersible work, and by drilling of the lowermost plutonic rocks (Francheteau *et al.*, 1992; Gillis, 1995; Mevel *et al.*, 1996). Many holes have been drilled into the uppermost few hundred metres of the crust and a few into faulted exposures of lower crust, but Hole 504B is the only drilled section through the upper oceanic crust, penetrating through the volcanic section to near the base of the sheeted dyke complex (Fig. 6.6; Alt *et al.*, 1996a). Site 504 is located in 6 Ma

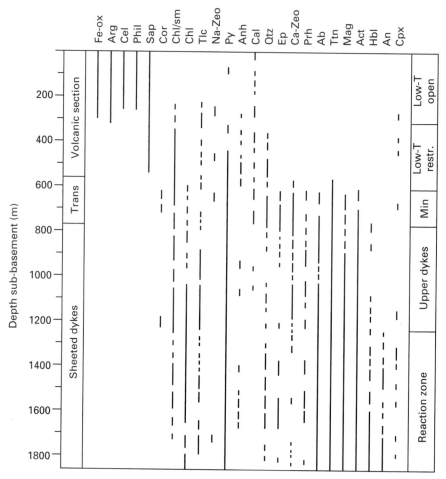

Fig. 6.6 Secondary minerals vs. depth in the upper oceanic crust. Data for ODP Hole 504B (after Alt *et al.*, 1985, 1996a). Depth in metres sub-basement. Min = mineralized with disseminated pyrite; Fe-ox, Fe-oxyhydroxide; Arg, aragonite; Cel, celadonite; Phil, phillipsite; Sap, saponite; Cor, corrensite; Chl/sm, chlorite/smectite; Chl, chlorite; Tlc, talc; Zeo, Zeolite (Na = analcite, natrolite; Ca = laumontite > scolecite >> heulandite); Py, pyrite; Anh, anhydrite; Cal, calcite; Qtz, quartz; Ep, epidote; Prh, prehnite; Ab, albite; Ttn, titanite; Mag, magnetite; Act, actinolite; Hbl, hornblende; An, secondary anorthite; Cpx, secondary clinopyroxene.

crust 200 km south of the Costa Rica Rift in the eastern Pacific (Fig. 6.7). The core has been intensively studied with the most comprehensive mineralogical, chemical and structural analyses of any oceanic section and has become a reference section for upper oceanic crust. The alteration stratigraphy of Hole 504B (Fig. 6.6) indicates that processes change with depth and varying alteration effects in oceanic crust of different ages reveal a sequence of very low-grade metamorphic processes that continue for up to several tens of million years. The origin and evolution of these temporal and spatial variations are discussed in this chapter in terms of mineralogy, chemistry and hydrothermal processes. Results from dredged and drilled lower crustal rocks from Hess Deep in the eastern Pacific and from the south-west Indian Ridge provide many of the constraints on lower crustal hydrothermal metamorphism briefly described in this chapter (Von Herzen & Robinson, 1991; Gillis, 1995; Mevel *et al.*, 1996).

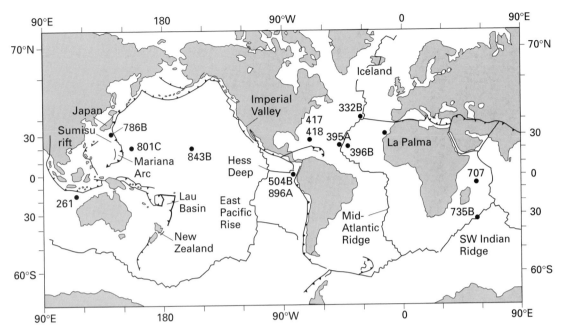

Fig. 6.7 Map showing locations of sites discussed in text. Numbered dots refer to DSDP/ODP drill sites.

6.3 The volcanic section

6.3.1 Initial low-temperature hydrothermal processes in the volcanic section

Alteration effects seen in the youngest rocks (<0.1–2.5 Ma) consist of black bands or haloes, a few mm up to 5 cm wide, subparallel to fractures (Fig. 6.8; Scott & Hajash, 1976; Humphris *et al.*, 1980; Laverne & Vivier, 1983; Bohlke *et al.*, 1984; Adamson & Richards, 1990; Laverne, 1993). The black haloes are characterized mainly by filling of pore spaces with celadonitic phyllosilicates and Fe-oxyhydroxides, although olivine phenocrysts may be partly replaced by these phases. In contrast, the interior portion of the rock is unaltered and pore spaces remain open. These same 'black halo' features are observed in older rocks (up to 170 Ma), but with superimposed alteration effects, indicating that this early alteration is typical of the uppermost ocean crust (Andrews, 1977; Bohlke *et al.*, 1980; Alt & Honnorez, 1984; Alt *et al.*, 1986a, 1992; Alt, 1993).

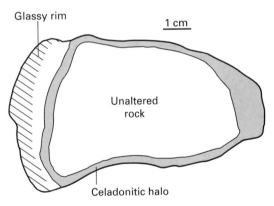

Fig. 6.8 Sketch of young oceanic pillow basalt fragment illustrating initial hydrothermal alteration effects. Black celadonitic haloes occur along exposed surfaces. Impermeable glassy-microcrystalline area immediately beneath glassy pillow rim remains unaltered. See text.

Compositions of the phyllosilicates from the youngest rocks (<0.1 Ma) approach celadonite (discussed p. 176), but X-ray diffraction (XRD) of the <2-μm air-dried fraction reveals weak 14 Å smectite peaks and no mica peaks (Adamson & Richards, 1990). Minerals of similar occurrence and composition in older rocks (110–3 Ma) exhibit dioctahedral mica ± smectite structures (Andrews, 1980; Bohlke *et al.*, 1980; Alt & Honnorez, 1984; Peterson *et al.*, 1986), in some cases with interlayering of mica and smectite (Andrews, 1980; Stakes & Scheidegger, 1981). The term 'protoceladonite' has been widely (mis)used for celadonitic phyllosilicates where XRD data are poor or where K contents are less than for pure celadonite (Donnelly *et al.*, 1979; Berndt & Seyfried, 1986; Peterson *et al.*, 1986; Adamson & Richards, 1990).

The phyllosilicates in the black haloes have lower K contents and higher total Fe and octahedral site occupancies than celadonite (Fig. 6.9). Correlations of decreasing K contents with increasing octahedral site occupancy have been interpreted to reflect physical mixtures of celadonite and saponite (Andrews, 1980; Bohlke *et al.*, 1980; Stakes & Scheidegger, 1981; Teagle *et al.*, 1996). This is reasonable in the older rocks that contain superimposed alteration effects and abundant saponite. Phyllosilicates from the youngest rocks, however, have high total Fe contents and tend toward a mixture of celadonite with Fe-oxyhydroxide (Fig. 6.9b), but nontronite can also be present (Alt & Honnorez, 1984). Thus, the best interpretation is that celadonite ± Fe-oxyhydroxide ± nontronite form initially in very young rocks, which undergo subsequent alteration and abundant saponite forms in the rocks.

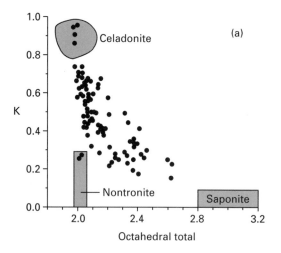

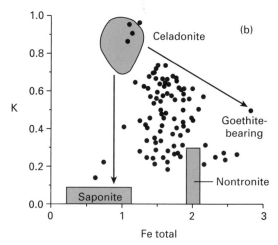

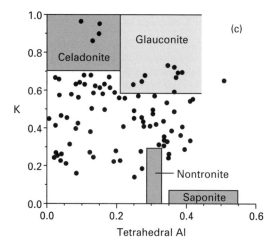

Fig. 6.9 Compositions of celadonitic phyllosilicates from oceanic basalts. The minerals are mostly mixtures of celadonite, nontronite and Fe-oxyhydroxides, but some analyses of older rocks include saponite, and late minerals in the oldest rocks range to true celadonite. High tetrahedral Al contents of some samples fall in the glauconite field. Celadonite and glauconite fields from Buckley *et al.* (1978). Representative analyses taken from various sources (Kempe, 1974; Andrews, 1977; Scheidegger & Stakes, 1977; Seyfried *et al.*, 1978b; Bohlke *et al.*, 1980; Humphris *et al.*, 1980; Laverne & Vivier, 1983; Alt & Honnorez, 1984; Alt *et al.*, 1986a, 1992; Peterson *et al.*, 1986; Adamson & Richards 1990; Alt, 1993; Laverne, 1993; Teagle *et al.*, 1996). All Fe as Fe^{3+}, formulae per $O_{10}(OH)_2$.

The Al contents of celadonitic minerals exhibit a wide range of values, with some samples falling within the field for glauconite (Fig. 6.9c). The rare analyses that do plot in the celadonite field in Fig. 6.9 are from older rocks (50–100 Ma; Kempe, 1974; Scheidegger & Stakes, 1977; Alt, 1993). These are late celadonites at the centre of vesicles, intergrown with late carbonate veins, or in veins that cut late carbonate veins and are optically distinct, having a blue-green colour in contrast to the typical grass-green of the earlier celadonitic phyllosilicates. Five of the 13 celadonite analyses reported by Buckley et al. (1978) are from oceanic basalts and these too are from older rocks (Cretaceous). The distinct composition of the late celadonites indicates influx of basement fluids having compositions distinct from the late carbonate-forming fluids (see section 6.3.4).

The black haloes have grained Fe, K, Rb, H_2O and exhibit elevated Fe^{3+}/Fe^T, $\delta^{18}O$ and $^{87}Sr/^{86}Sr$ compared to unaltered rocks (Humphris et al., 1980; Laverne & Vivier, 1983; Bohlke et al., 1984; Park & Staudigel, 1990). High B and Cs contents of celadonites compared to basalts imply that these minerals contribute to the common enrichment of B and Cs in altered seafloor basalts (Staudigel et al., 1981a,b; Hart & Staudigel, 1986; Berndt & Seyfried, 1986; Teagle et al., 1996).

$\delta^{18}O$ values of 19.5–23.4‰ for celadonite indicate formation temperatures of less than about 40°C (Kastner & Gieskes, 1976; Seyfried et al., 1978b; Stakes & O'Neil, 1982; Bohlke et al., 1984). The alkali enrichment of the black haloes can be accounted for by seawater fluids, but because the host rocks are otherwise unaltered, an external source of Fe is required. A plausible source is distal, mixed hydrothermal fluids at the spreading axis. High-temperature (350°C) hydrothermal fluids originating at depth at mid-ocean ridges (MOR), when mixed with seawater in the subsurface precipitate sulphides and the cooler (10–15°C), mixed fluids remain enriched in Fe and Si and contain elevated alkalis compared to seawater (Edmond et al., 1979a,b). These are the same type of fluids that vent at the seafloor at the Galapagos Spreading Centre and which result in the formation of hydrothermal nontronite and Fe–Mn oxide deposits on the seafloor (Edmond et al., 1979b; Honnorez et al., 1981; Alt, 1988). The formation of black haloes is not necessarily related to sustained venting of hydrothermal fluids, however, but could result from

episodes such as dyking events or sill intrusions that disrupt hydrothermal vents and cause widespread diffuse flow of mixed hydrothermal fluids (Von Damm et al., 1995).

6.3.2 Low-temperature hydrothermal seawater–rock interactions in the volcanic section

Rocks from the uppermost crust greater than 2.7 Ma in age retain black haloes but also exhibit superimposed alteration effects, commonly in cm-scale zonations around veins and fractures, indicating continued hydrothermal reactions during passive off-axis circulation (Bass, 1976; Andrews, 1977, 1980; Bohlke et al., 1980; Honnorez et al., 1983; Alt & Honnorez, 1984; Alt, 1993; Alt et al., 1986a, 1996a; Teagle et al., 1996). A typical such zonation is shown in Fig. 6.10, where a late carbonate vein is surrounded by an Fe-oxyhydroxide zone, a celadonite zone and then the remainder of the rock consists of a saponite zone. In the Fe-oxyhydroxide zone, olivine is partly to totally replaced and pores filled by Fe-oxyhydroxide and Fe-oxide (goethite, hematite and X-ray amorphous Fe-oxyhydroxide), ± saponite ±

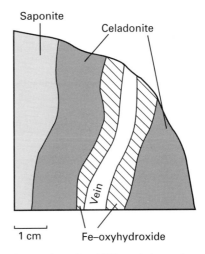

Fig. 6.10 Sketch of sample exhibiting typical secondary mineral zones resulting from low-temperature seawater hydrothermal alteration. Fe-oxyhydroxide, celadonite and saponite (host rock) zones occur along fracture filled with late calcite (after Andrews, 1977, 1980). Variations in relationships among mineral zones, however, indicate that Fe-oxyhydroxide and saponite zones are overprinted upon early celadonite zones (see text).

celadonitic phyllosilicates. Titanomagnetite is intensively maghemitized and primary sulphides are oxidized. The mineralogy of the celadonite zone is similar, but the abundance of Fe-oxyhydroxides is much lower and celadonite greater. In the saponite zone olivine is partly to totally replaced and pores are filled by saponite ± talc ± pyrite ± calcite. Titanomagnetite is less intensively maghemitized and primary and secondary sulphides are disseminated in the groundmass. Secondary minerals typically comprise 5–15% of the rocks, although abundances may be higher locally, especially in breccias or fault zones (up to 40–90%; Andrews, 1977, 1980; Bohlke et al., 1980; Alt et al., 1996a).

Saponites in submarine rocks are trioctahedral smectites, having variable Al_2O_3 and interlayer cation contents and ranging to low-charge Mg-smectites (stevensite; Fig. 6.11). Talc is locally present as indicated by XRD (Alt et al., 1986a). FeO/(FeO + MgO) is variable, ranging from 0.1 to 0.5 for saponites associated with oxidation zonations. Measured values of Fe^{3+}/Fe^T for saponites range up to 0.8 (Alt & Honnorez, 1984), but these high values reflect postsampling oxidation and essentially all iron is ferrous in saponites (Andrews et al., 1983).

Figure 6.10 shows a typical zonation, but relationships among the different zones vary: the Fe-oxyhydroxide zone may be superimposed upon or interior to the celadonite zone; or the celadonite zone may be absent (Alt et al., 1996a). These geometries indicate that the Fe-oxyhydroxide zone is the result of a later, superimposed alteration process reflecting interaction of rock with cold, oxidizing seawater (Andrews, 1977, 1980; Bohlke et al., 1980, 1981, 1984; Staudigel et al., 1981b; Alt & Honnorez, 1984; Alt et al., 1986a, 1992, 1996a,c; Alt, 1993). These effects occur in the upper 300–475 m of the volcanic section (Fig. 6.6), where access of cold seawater to the crust is relatively unrestricted (Andrews, 1977; Honnorez et al., 1983; Alt & Honnorez, 1984; Alt et al., 1986a,b; Gillis & Robinson, 1988). Alteration temperatures estimated from oxygen isotopic data for smectites and carbonates are typically 0–50°C (Lawrence, 1979, 1991; Muehlenbachs, 1979; Lawrence & Drever, 1981; Staudigel et al., 1981b; Bohlke et al., 1984; Alt et al., 1986b, 1992). This type of very low-grade metamorphism has variably been referred to as low-temperature alteration, the seawater zone, seafloor weathering, oxidative diagenesis or alteration, oxic alteration and oxidizing seawater alteration. The term seafloor weathering, however, is better reserved for rocks that remain exposed to seawater at the seafloor for millions of years (Thompson, 1973, 1983; see p. 180).

Glass at pillow rims is altered to 'palagonite' (hydrated glass), smectite (mainly saponite and Al-saponite) and zeolites (mainly phillipsite, but also analcite and rare chabazite, gismondine and faujasite) and cemented by these same minerals ± celadonite and Fe-oxyhydroxides (Melson & Thompson, 1973; Andrews, 1977; Juteau et al., 1979; Honnorez, 1981; Staudigel & Hart, 1983; Teagle et al., 1996). Chemical changes during palagonitization involve hydration, gains of K, Rb and Cs and losses of other cations (e.g. Si, Al, Mn, Mg, Ca, Na, P, Zn, Cu, Ni, REE), although the latter are retained locally in replacement products and cements (Honnorez, 1981; Staudigel & Hart, 1983).

Andrews (1980) proposed a mechanism whereby the mineralogical zonation is controlled by decreasing f_{O_2} during progressive reaction as fluids penetrate into the rock from fractures containing circulating seawater. The presence of uniform mineral chemistry within each zone led to the inference of an infiltration metasomatic process, but this treatment requires equilibrium whereas low-temperature alteration is characterized by incomplete mineral

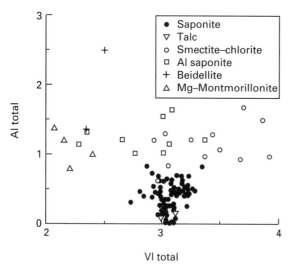

Fig. 6.11 Compositions of saponitic phyllosilicates from oceanic basalts (see text for discussion). All Fe as Fe^{2+}, formulae per $O_{10}(OH)_2$.

reactions and disequilibrium. Moreover, the concentric alteration zonation within individual rock fragments requires that fluids would have to move radially inward and it is impossible to transport sufficient fluid into the rock to account for the observed chemical changes in the rocks (Fig. 6.8 and pp. 179–180). Therefore, these alteration zones must form by diffusion and reaction within the rock, with circulation of seawater fluids along fractures.

Basement fluids sampled from ridge flank systems fall into two regimes: (1) open flow of large volumes of cold (<30°C) oxidizing seawater; and (2) more restricted flow of evolved fluids (low Mg) at higher temperatures (40–150°C; Mottl & Wheat, 1994). The alteration described here is consistent with the lower temperature, high water/rock ratio regime. Bulk rock chemical changes include variable enrichments of Fe^T, H_2O, K, Li, Rb, Cs, B and U contents, losses of S and Ca and increased Fe^{3+}/Fe^T, $^{87}Sr/^{86}Sr$, $\delta^{11}B$ and $\delta^{18}O$ (e.g. Fig. 6.12; Andrews, 1977, 1980; Muehlenbachs, 1979; Humphris et al., 1980; Thompson, 1983;

Laverne & Vivier, 1983; Bohlke et al., 1984; Alt & Honnorez, 1984; Hart & Staudigel, 1982; Gillis & Robinson, 1988; Adamson & Richards, 1990; Ishikawa & Nakamura, 1992; Alt, 1993, 1995; Staudigel et al., 1995; Alt et al., 1996c; Teagle et al., 1996). The Fe-oxyhydroxide and celadonite zones typically exhibit the greatest changes for Fe^T, Fe^{3+}/Fe^T, H_2O, the alkalis, S and Ca, reflecting alteration of primary phases and abundances of secondary minerals. In some cases losses of Mg from the alteration haloes may account in part for slight gains of Mg in the saponite zones (Andrews, 1977; Bohlke et al., 1980, 1981). Although Mg may be lost from individual rock samples, it is retained and even gained by the crust as smectite in veins (Alt et al., 1996c). Assuming that the measured oxidation and uptake of alkalis by the rocks results from quantitative extraction of dissolved oxygen and alkalis from seawater, minimum integrated seawater/rock mass ratios of the order 10–100 for the upper volcanic section are required (Alt et al., 1986b).

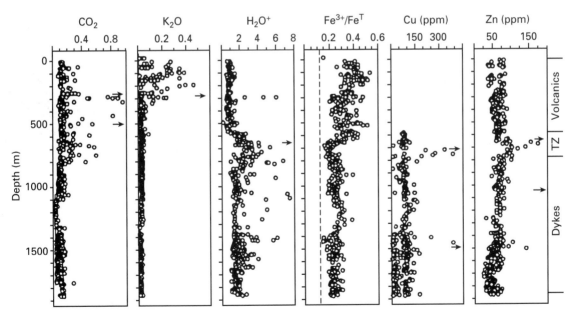

Fig. 6.12 Bulk rock geochemical data for ODP Hole 504B. Lithostratigraphy at right and depth in metres sub-basement (beneath 274.5 m sediment) at left. Arrows indicate samples that plot off-scale. The upper volcanic section exhibits the effects of open seawater circulation, the lower volcanic section is affected by more restricted circulation and smaller chemical changes, the transition zone and upper dykes by subgreenschist to greenschist grade hydrothermal metamorphism, and the lower dykes by greenschist to amphibolite-grade hydrothermal metamorphism. The transition zone (TZ) is mineralized, and lower dykes have lost sulphur and metals in a high temperature (c. 400°C) deep 'reaction zone'. See text for discussion. After Alt et al. (1996a). Dashed line indicates unaltered MORB Fe^{3+}/Fe^T.

Variations in the temperature and intensity of seawater circulation lead to differences in alteration effects in the upper crust. In 13 Ma Atlantic crust (DSDP Hole 396B), intensive circulation of cold seawater resulted in the formation of wide (2 cm) Fe-oxyhydroxide zones, where olivine is leached and pseudomorphed by goethite, resulting in the loss of olivine constituents from these zones (Si, Mg, Co and Ni) and significant gain of P (Bohlke et al., 1980, 1981). Minimum cumulative seawater/rock mass ratios of 3000 are calculated for these zones assuming quantitative uptake of P from seawater (Bohlke et al., 1981).

DSDP Hole 417A, in 110 Ma Atlantic crust, penetrated a basement hill that remained uncovered by sediment for 10 Ma and served as an outflow site for fluids circulating in the surrounding basement (Donnelly et al., 1979). Focusing of fluid flow at this site and entrainment of seawater led to extreme alteration of these rocks at temperatures of up to 50°C (Donnelly et al., 1979; Muehlenbachs, 1979; Alt & Honnorez, 1984; Bohlke et al., 1984). The rocks record early alteration stages in the presence of dark celadonitic haloes and Fe-oxyhydroxide zones, but rather than saponite, the rocks contain abundant K-feldspar, beidellite, Fe-beidellite and Al-saponite (Alt & Honnorez, 1984). The breakdown of plagioclase provided a source of Al for formation of the aluminous smectites. Glass at pillow rims is also replaced by aluminous smectites (Juteau et al., 1979; Alt & Honnorez, 1984). Beidellite and Fe-beidellite have high Al_2O_3 contents (up to 32 wt%; Fig. 6.11). XRD indicates a dioctahedral smectite structure, but these minerals have high octahedral site occupancies (c. 2.5 per $O_{10}(OH)_2$; Fig. 6.11), which may offset the high tetrahedral layer charge (0.65–0.95; Alt & Honnorez, 1984).

At Site 417A the breakdown of primary phases and formation of secondary minerals led to whole rock gains of K, Rb, Cs, Li, B, P and H_2O, increased Fe^{3+}/Fe^T, $^{87}Sr/^{86}Sr$ and $\delta^{18}O$ and loss of Ca, Mg, Si and locally Al and Na (Donnelly et al., 1979; Staudigel et al., 1981a,b; Alt & Honnorez, 1984). Chemical changes are locally extreme, e.g. gains of > 5 wt% K_2O, losses of 4 wt% MgO and total oxidation of iron for some samples. Minimum cumulative seawater/rock mass ratios of 2000–4500 are calculated assuming quantitative uptake of P from seawater and oxidation of Fe.

In the deeper, less altered rocks from Hole 417A, Al-saponite takes the paragenetic place of saponite as observed in more typical basement sections. Similarly, alkalic basalts from Hole 801C in 170 Ma Pacific crust locally exhibit more intensive alteration of plagioclase and pyroxene, with consequent development of Al-saponite (Alt et al., 1992). Al-rich saponites also occur in association with altered glass at pillow rims (Melson & Thompson, 1973; Teagle et al., 1996). High fluid fluxes cause the breakdown of plagioclase and unstable glass reacts rapidly, providing sources of Al. The Al-saponites have broad 060 peaks at 1.505–1.538 Å, indicating an intermediate or mixed dioctahedral–trioctahedral smectite structure (Alt & Honnorez, 1984). Al contents are high (8–16 wt% Al_2O_3, 0.7–1.2 Al per $O_{10}(OH)_2$) and octahedral site occupancies are typically around 2.5 atoms per $O_{10}(OH)_2$ (Fig. 6.11; Melson & Thompson, 1973; Andrews, 1980; Alt & Honnorez, 1984; Alt et al., 1992; Alt, 1993). It is possible that such occurrences are mixtures between saponite and montmorillonite–beidellite.

The most extreme examples of this type of low-temperature seawater alteration are in rocks exposed to seawater at the seafloor for millions of years and which has been termed 'seafloor weathering' (Thompson, 1973, 1983). Secondary mineralogy is generally poorly documented in these early studies, which focused more on bulk geochemical changes, including increases in alkalis, P, LREE, U, $^{87}Sr/^{86}Sr$ and $\delta^{18}O$ and losses of Mg, Ca, Si, Co and Ni (Muehlenbachs & Clayton, 1972; Hart, 1973; Thompson, 1973, 1983; Hart et al., 1974; Ludden & Thompson, 1979).

Based on observation of alteration effects in rocks of different ages, it can be concluded that the above very low-grade metamorphic processes occur within a few million years (13–2.5 Ma) of crustal formation. Subsequent formation of later carbonates and zeolites in veins also takes place, however (see following sections) and seawater fluids can circulate within basement for several tens of million years, suggesting that interactions can continue for longer periods. Taking into account uncertainties in crustal age and in Rb–Sr isochrons for smectite and celadonite veins from altered ocean crust, celadonite and smectite can form any time up to about 23 Ma after crustal formation, although the best estimate is generally within 10–15 Ma (Staudigel & Hart, 1985;

Hart & Staudigel, 1986; Waggoner, 1993). Similar ages of smectite formation within 18–10 Ma of crustal formation are obtained by K–Ar dating (Peterson *et al.*, 1986).

6.3.3 Restricted low-temperature hydrothermal interactions in the volcanic section

Seawater fluids evolve as they circulate downward and react with basalts so the deeper portions of the volcanic section lack the oxidation effects observed in the upper 300–475 m of the crust (Andrews, 1977; Honnorez *et al.*, 1983; Alt & Honnorez, 1984; Alt *et al.*, 1986a). Seawater circulation can also evolve with time: circulation slows and fluids evolve to more reacted compositions as the crust is buried by sediments and fractures are sealed with secondary minerals. Thus, the effects of restricted seawater circulation can be superimposed on previous oxidation effects in the upper crust, or may occur in rocks where early seawater oxidation was lacking. Depending upon the temperatures of reaction and fluxes of fluids, this process can lead to the formation of random chlorite/smectite and what might be more traditionally called the beginning of hydrothermal metamorphism.

In the upper volcanic section of Hole 504B and elsewhere, early celadonite and Fe-oxyhydroxide in veins and lining pores are followed by saponite veins or pore fillings, representing later restricted seawater alteration and more reducing conditions (Seyfried *et al.*, 1978b; Andrews, 1980; Bohlke *et al.*, 1980; Honnorez *et al.*, 1983; Alt & Honnorez, 1984; Alt *et al.*, 1986a, 1992; Alt, 1993). The lower portions of the volcanic section affected by restricted seawater circulation are generally 10–20% recrystallized, but locally rocks are more intensely altered (up to 40–100% secondary minerals in breccias and fault zones). Olivine is replaced and pores and fractures are filled by saponite ± calcite ± pyrite. Trace disseminated pyrite is common, plagioclase may be partly altered to saponite or Al-saponite and talc is present locally. Glass at pillow rims is altered to palagonite, saponite, Al-saponite and zeolites (mainly phillipsite) and cemented by saponite, zeolites and calcite.

Saponites are similar to those in the upper volcanic section, but have generally higher FeO/(FeO + MgO)

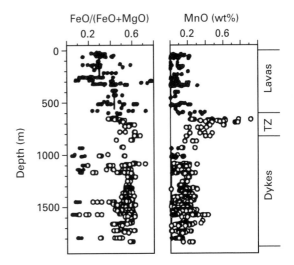

Fig. 6.13 Compositions of saponite, chlorite and chlorite/smectite with depth in ODP Hole 504B. The mean FeO/(FeO + MgO) (wt %) content (vertical lines) of saponite in the upper volcanics, affected by open, oxidizing alteration is lower than that of saponite in the lower volcanics, which underwent more restricted low-temperature alteration. Open circles indicate smectite, chlorite and mixed-layer chlorite/smectite, solid circles indicate talc-bearing samples. Depth in metres below sub-basement. TZ, transition zone. (After Alt *et al.*, 1996a.)

values (0.2–0.7, Fig. 6.13), reflecting incorporation of greater amounts of ferrous iron under the more reducing conditions (Andrews, 1980; Alt *et al.*, 1986a). Minor chlorite layers are present locally in smectite, as indicated by XRD and higher Al and Fe contents of phyllosilicates (Fig. 6.11; Alt *et al.*, 1985; Stakes & Scheidegger, 1981). The presence of minor chlorite/smectite in the lower volcanic section of Hole 504B is consistent with oxygen isotopic data for phyllosilicates that indicate higher temperatures there (>100°C, Fig. 6.14). Saponite-rich pillow breccias from the East Pacific Rise that contain chlorite/smectite also formed at somewhat elevated temperatures (130–170°C; Stakes & Scheidegger, 1981; Stakes & O'Neil, 1982). In contrast, the lower volcanic section of the 550 m deep Hole 332B in 3.5 Ma Atlantic crust lacks any chlorite/smectite (Andrews, 1977). This site is located in a low-heat flow area where basement outcrops allow access of seawater to cool the crust, so this site has had a lower temperature history than the more heavily sedimented Site 504. Similarly, the only chlorite

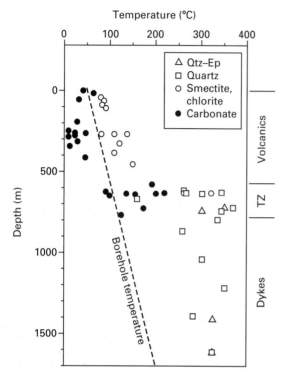

Fig. 6.14 Measured and estimated temperatures in Hole 504B. Metamorphic temperatures estimated from oxygen isotopic analyses of secondary minerals, assuming equilibrium and using various experimentally and empirically calibrated mineral–water (quartz, smectite, chlorite and calcite) and mineral–mineral (quartz–epidote) oxygen isotopic fractionations. (Data from Alt *et al.*, 1985, 1986b, 1989, 1995, 1996b; J. C. Alt, unpublished data.) The mineral data define the steeply stepped fossil thermal gradient coinciding with a metamorphic boundary and the lithological transition from lavas to dykes (see Fig. 6.6). Present measured conductive thermal gradient is also shown. See text for discussion.

reported from the lower volcanic sections of Holes 417D and 418A in old Atlantic crust is associated with coarser grained portions of massive units and represents chlorite formed during initial penetration of seawater into the cooling rocks (e.g. Bohlke *et al.*, 1980; Alt & Honnorez, 1984). The lower volcanic sections at the latter sites underwent restricted seawater alteration at temperatures less than about 70–80°C (Muehlenbachs, 1979; Bohlke *et al.*, 1984; Staudigel *et al.*, 1981b). These data suggest that chlorite layers start to appear in smectite in altered ocean crust at temperatures of about 100°C.

The temperatures of restricted seawater alteration can thus vary from near 0 up to 150–200°C, depending upon the age at which the crust is blanketed with sediment, which is a function of basement relief and sedimentation rate. Lawrence and Drever (1981) also showed that temperatures of smectite formation estimated from oxygen isotopes can vary locally: in 7 Ma Hole 395A in the Atlantic, saponites typically formed at 30–60°C, but one low–$\delta^{18}O$ smectite is interpreted to reflect higher temperature (160°C) immediately beneath an intrusive sill.

Trace amounts of secondary pyrite associated with saponite in veins and rocks are characteristic of restricted seawater alteration (Andrews, 1979; Hubberten, 1983; Alt *et al.*, 1989). The rocks lose some sulphur via degassing during crystallization, but subsequent alteration results in dissolution of igneous sulphides and precipitation of secondary pyrite under limited oxidation conditions (Andrews, 1979; Hubberten, 1983; Alt *et al.*, 1989). Sulphur may even be gained by the rocks locally (Teagle *et al.*, 1996).

Carbonates are commonly associated with saponite, but also form later (see pp. 183–194). Calcite is the most common carbonate, although aragonite is present in some cases (Bass, 1976; Andrews, 1977; Lawrence, 1979, 1991; Bohlke *et al.*, 1980; Lawrence & Drever, 1981; Alt & Honnorez, 1984; Teagle *et al.*, 1996) and siderite is rare (Alt *et al.*, 1992; Burns *et al.*, 1992). Calcite has Mg contents that range up to eight mole per cent $MgCO_3$, but most contain less Mg, indicating formation from seawater fluids having lowered Mg/Ca ratios (Bass, 1976; Bohlke *et al.*, 1980; Alt & Honnorez, 1984; Teagle *et al.*, 1996).

Fibrous carbonate ± saponite veins are common in upper oceanic basement. In some cases these may reflect crystal growth simultaneous with crack opening (Bohlke *et al.*, 1980), but in others these have been interpreted as crack-seal veins indicative of fluid overpressures (Harper & Tartarotti, 1996). This implies that while circulation occurs through some portions of basement, other parts may be sealed and have high pore pressures. Such veins are common in older crust (Alt & Honnorez, 1984), but have not yet been described from crust younger than 6 Ma.

Chemical changes for bulk rocks affected by restricted seawater circulation are generally small

(Fig. 6.12) and include slight increases of Fe^{3+}/Fe^T, H_2O, CO_2, $^{87}Sr/^{86}Sr$ and $\delta^{18}O$ (Andrews, 1977, 1980; Staudigel et al., 1981b; Stakes & Scheidegger, 1981; Bohlke et al., 1984; Alt & Honnorez, 1984; Alt et al., 1986a,b, 1992; Alt, 1993). Some of the oxidation of iron, however, may be the result of postsampling oxidation of iron in saponite, which oxidizes readily upon exposure to air (Andrews et al., 1983). Li substitutes for Mg in saponite leading to high Li contents of saponites and concomitant increases in bulk rocks (Berndt & Seyfried, 1986). Although the K content of saponite is typically low, it is greater than tholeiitic basalts, leading to slightly elevated whole rock K contents. In the more intensely recrystallized rocks, Mg may be gained and Ca lost. Even though individual rocks may not gain Mg, the abundant saponite in veins and cementing breccias results in a net uptake of Mg throughout the upper crust during restricted seawater circulation (Andrews, 1980; Stakes & Scheidegger, 1981; Alt et al., 1996c).

6.3.4 Ageing of the volcanic section: late carbonates and zeolites

Carbonates and zeolites are typically the last phases to form in veins and vugs of oceanic volcanic rocks, although they also form earlier in the sequence (Bass, 1976; Andrews, 1977; Bohlke et al., 1980; Staudigel et al., 1981a; Stakes & Scheidegger, 1981; Alt & Honnorez, 1984; Alt et al., 1986a, 1992; Alt, 1993; Teagle et al., 1996). The change from formation of Fe-oxyhydroxides and phyllosilicates to carbonates and zeolites reflects a change to more alkaline conditions. Extraction of OH^- by precipitation with Mg to form clays or Fe to form oxyhydroxides maintains slightly acidic conditions, but as Mg is depleted in solution hydrolysis reactions become more important, pH increases and carbonates and zeolites precipitate (Bass, 1976; Bohlke et al., 1980; Stakes & Scheidegger, 1981; Alt & Honnorez, 1984). This interpretation is consistent with the generally low-Mg calcites that are observed filling fractures (Bass, 1976; Bohlke et al., 1980, 1984; Alt & Honnorez, 1984). As reaction of seawater proceeds with basalt, Ca may also be released to solution, further decreasing Mg/Ca ratio of circulating seawater fluids. Late carbonates may also contain elevated Fe and Mn, consistent with formation from more reacted, reducing fluids (Alt & Honnorez, 1984;

Burns et al., 1992; Alt et al., 1992; Teagle et al., 1996), and in some cases pyrite is associated with late carbonate (Alt et al., 1992; Alt, 1993).

Oxygen and carbon isotopic analyses of vein carbonates generally indicate formation from fluids having normal seawater carbon isotopic compositions at temperatures of 0–70°C (Lawrence, 1979, 1991; Muehlenbachs, 1979; Lawrence & Drever, 1981; Bohlke et al., 1984; Alt et al., 1986b, 1992; Alt, 1993). Less common low-$\delta^{13}C$ vein carbonates reflect incorporation of sedimentary organic carbon (Alt et al., 1992), or fractionation of carbon isotopes resulting from continued closed system precipitation of carbonates (Lawrence, 1991). Teagle et al. (1996) documented two generations of carbonates in veins of 6 Ma Hole 896A: an early, lower temperature (25–35°C) generation having low Mg, Fe, Mn and high Sr contents formed during open circulation of seawater, and later carbonates that formed at higher temperatures (50–70°C) from more reacted seawater fluids that had low Mg/Ca and Sr/Ca ratios and contained elevated Fe and Mn. The latter are consistent with formation under the present 'reheated' ridge flank conditions at that site. Staudigel et al. (1981b) identified two different regimes of carbonate formation in 110 Ma Site 417/418: most carbonates there formed from seawater-like solutions, but carbonates intergrown with smectites and in breccias formed from more highly reacted solutions.

The consistency of initial $^{87}Sr^{86}Sr$ ratios of vein carbonates from 110 Ma Atlantic Sites 417 and 418 and the coincidence of this ratio with that of 110 Ma seawater was interpreted to indicate that carbonates precipitated within a few million years of formation of the crust (Hart & Staudigel, 1978; Richardson et al., 1980; Staudigel et al., 1981a,b). Subsequent revisions of the magnetic anomaly time scale and of the curve for the Sr isotopic composition of seawater through time have extended the uncertainty to within 40 Ma of crustal formation (Fig. 6.15; Staudigel & Hart, 1985). Calcite from these sites has Sr/Ca ratios lower than predicted for equilibrium with seawater; this requires input of basaltic Ca into circulating seawater solutions, but without significant input of basaltic Sr, in order to lower the fluid Sr/Ca ratio without changing its Sr isotopic ratio (Hart & Staudigel, 1978; Staudigel et al., 1981a,b).

Using the same arguments as above, Sr isotopic data for one other MOR site and one on the Ninetyeast

Ridge (a hotspot trace) in the Indian Ocean indicate formation of carbonates within about 10–15 Ma of crustal formation (Hart & Staudigel, 1986; Hart *et al.*, 1994). At two other sites in the Indian Ocean carbonates formed for up to about 27–40 Ma after formation of the crust (Fig. 6.15; Burns *et al.*, 1992). These durations are minima, however, because a large basaltic component in solution (lowering the Sr isotopic ratio of the fluid) is indicated by the widely varying Sr contents and high Fe and Mn of the carbonates. At several other MOR and hotspot sites the carbonates have initial $^{87}Sr^{86}Sr$ ratios lower than seawater for any time between the formation of the crust at those sites and the present, clearly indicating a significant component of basaltic Sr in solution so the Sr dating technique is not applicable (Fig. 6.15; Hart & Staudigel, 1986; Waggoner, 1993; Hart *et al.*, 1994; Teagle *et al.*, 1996).

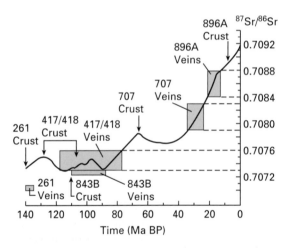

Fig. 6.15 Illustration of Sr isotopic dating of vein carbonates from oceanic basalts. Arrows indicate ages of DSDP/ODP oceanic crustal sites. Boxes indicate the intersection of $^{87}Sr/^{86}Sr$ ratios of vein carbonates from each site projected from the vertical axis onto the Sr isotope curve for seawater (heavy line). Projection of these intersections onto the horizontal (time) axis is interpreted to give the age of formation of the carbonates, assuming the minerals precipitated from seawater. This approach works for most data from some sites (e.g. 417/418, 707 and 3.2 Ma 332B not plotted), but data from other sites (261, 896A and others), as well as additional data from Site 417/418 (not plotted) indicate significant basaltic Sr (0.7026) in solution, rendering this technique inapplicable in these cases: e.g. $^{87}Sr/^{86}Sr$ of carbonates from Hole 896A intersect the seawater curve at ages older than the 6 Ma crust. (Data from Staudigel *et al.*, 1981b; Hart & Staudigel, 1986; Burns *et al.*, 1992; Waggoner, 1993; Teagle *et al.*, 1996.)

Carbonate veins are generally late and continue to form for several tens of million years, consequently bulk rock CO_2 contents and the amount of carbonate in veins are greater in older crust (Staudigel *et al.*, 1981a; Alt, 1993; Alt *et al.*, 1996a). Staudigel *et al.* (1989) calculated that the amount of CO_2 taken up annually during very low-grade metamorphism of oceanic basalts may be equal to, or be even greater than the amount of CO_2 outgassed at mid-ocean ridges each year. If this is true, then this should have implications for global cycling of CO_2 (Staudigel *et al.*, 1989).

Zeolites are also typically late-forming phases, occurring in and along veins and cementing and replacing basaltic glass at pillow rims and in hyaloclastite breccias of the volcanic section. Phillipsite is the most common zeolite, but analcite and natrolite are not uncommon and mesolite, thompsonite, chabazite, gismondine, faujasite and stilbite also occur (Bass, 1976; Andrews, 1977; Juteau *et al.*, 1979; Bohlke *et al.*, 1980; Honnorez *et al.*, 1983; Alt & Honnorez, 1984; Alt *et al.*, 1986a; Teagle *et al.*, 1996). Phillipsite in low-grade seafloor metabasites mostly contains subequal amounts of Na and K, with very little Ca, but range to K-rich (Pritchard, 1979; Bohlke *et al.*, 1980).

Sodic zeolites, analcite and natrolite, occur locally in veins and vugs of Holes 896A and 332B (Alt *et al.*, 1996a; Teagle *et al.*, 1996) and are especially common in the highly altered rocks and hyaloclastites of 110 Ma Hole 417A (Alt & Honnorez, 1984). Analcite typically has end-member Na formulae, but a Ca-bearing variety also occurs rarely (Alt & Honnorez, 1984). Natrolite, phillipsite, chabazite and thompsonite are also locally present in these rocks. Analcites are enriched in K, Rb and Cs relative to basalt, contributing locally to the elevated alkali contents of the rocks (Staudigel *et al.*, 1981a,b; Alt & Honnorez, 1984). Some low-grade metabasites dredged from the seafloor also contain analcite and/or natrolite filling voids and replacing plagioclase and exhibit increases in whole rock Na, H_2O and $\delta^{18}O$ (Miyashiro *et al.*, 1971; Muehlenbachs & Clayton, 1972; Melson & Thompson, 1973; Shido *et al.*, 1974).

In Hole 504B a 20-m thick zone of abundant cm-sized veins at 300 m contains variable combinations of analcite, natrolite–mesolite, thompsonite, gyrolite, apophyllite, calcite and aragonite, plus earlier chlorite/smectite, celadonite and Fe-oxyhydroxides

(Honnorez et al., 1983; Alt et al., 1985). These same phases replace plagioclase, pyroxene and olivine in cm-size alteration haloes along the veins. Formation of these minerals has been attributed to focused flow of late-stage, ridge flank fluids at temperatures of c. 50°C, as indicated by fluid inclusions in analcite (Alt et al., 1986a).

Zeolites generally form late, from evolved, alkaline fluids that are depleted in Mg (Bass, 1976; Bohlke et al., 1980; Honnorez et al., 1983; Alt & Honnorez, 1984). These minerals are associated with smectites and carbonates that formed at temperatures of about 0–60°C, implying similar temperatures for the zeolites (Andrews, 1977; Juteau et al., 1979; Bohlke et al., 1980, 1984; Honnorez et al., 1983; Alt & Honnorez, 1984; Alt et al., 1986a; Teagle et al., 1996). In Hole 504B zeolites are zoned with depth: phillipsite occurs in the upper volcanic section, minor natrolite occurs in the lower lavas, then laumontite, scolecite and trace analcite and heulandite are late minerals in rocks of higher metamorphic grade in the transition zone and upper dyke section (Fig. 6.6). The distribution of zeolites in low-grade rocks from ocean crust thus reflects a combination of the effects of varying temperature and evolution of fluid compositions (K/Na). Zeolites are minor to trace phases, however, so use of the term zeolite facies is not appropriate.

The $^{87}Sr/^{86}Sr$ ratios of analcites and natrolite range from near seawater to more basaltic values (Richardson et al., 1980; Teagle et al., 1996). Analcites from Hole 417A fall on the same Rb/Sr isochron as smectites and celadonites (Richardson et al., 1980), but petrographically are clearly later (Alt & Honnorez, 1984). Because of errors on the isochron, however, the data require only that analcite formed within 3 Ma of the smectite. In Hole 417A the abundant zeolites reflect the extensive breakdown of plagioclase and glass to provide Al and Si, whereas the intensive uptake of K by earlier fluid–rock interactions led to elevated Na/K of fluids and formation of abundant sodic zeolites (Alt & Honnorez, 1984).

6.4 The transition zone and upper sheeted dyke complex: the transition to low-grade hydrothermal metamorphism

The transition from the volcanic section to the sheeted dyke complex coincides with radical changes in the grade of hydrothermal metamorphism of ocean crust, but this boundary has only been described from two places in the oceans: Hess Deep and Hole 504B in the eastern Pacific. In a c. 70-m interval within the basal lavas of Hole 504B titanite appears, the proportion of chlorite layers in chlorite/smectite increases, minor laumontite appears and quartz, pyrite and anhydrite become more abundant in veins (Fig. 6.6; Alt et al., 1985, 1986a). Narrow (1 mm) zones of wallrock along phyllosilicate veins are more intensely altered, but recrystallization of the remainder of the rock is much like the lower volcanic section (c. 20% secondary phases).

Beneath this zone, intensely hydrothermally metamorphosed and mineralized pillow basalts appear abruptly at 620 m in the lithological transition zone (Fig. 6.6). These rocks are recrystallized to chlorite, albite, titanite and quartz, and contain abundant veins of chlorite, quartz and sulphides (Alt et al., 1985, 1986a; Honnorez et al., 1985). These rocks were hydrothermally metamorphosed in a subsurface mixing zone, where hydrothermal fluids in the dykes mixed with much larger volumes of cooler seawater fluids circulating in the overlying more porous and permeable volcanic section. Oxygen isotope ratios and fluid inclusions indicate temperatures of 250–350°C within the mineralized zone (Fig. 6.14; Honnorez et al., 1985; Alt et al., 1986b). The transition from very low-grade to low-grade hydrothermal metamorphism in these rocks thus spans a temperature range from <150° to 250°C or greater.

The remainder of the transition zone and upper dyke section at Site 504 range from slightly (10–30%) to intensively recrystallized (30–100%) in patches and haloes around chlorite ± actinolite veins and zoned amygdules. Primary mineralogy controls the secondary mineralogy: glass at pillow rims is replaced by chlorite and titanite; plagioclase is partly to totally replaced by albite–oligoclase, minor laumontite, heulandite, prehnite and chlorite. Olivine is replaced by chlorite, smectite/chlorite, talc ± quartz; pyroxene is only slightly replaced by actinolite and local actinolitic- and magnesio-hornblende; and titanomagnetite is replaced by titanite. Disseminated pyrite is common. The increased intensity of recrystallization along veins and surrounding amygdules illustrates the importance of porosity and permeability controls on fluid access to the rock and hydrothermal metamorphism.

Veins record the evolution of fluid compositions and hydrothermal processes and indicate that mineral assemblages in recrystallized rocks represent the superimposed effects of varying processes and conditions (Honnorez *et al.*, 1985; Alt *et al.*, 1986a,b). Chlorite and actinolite veins and (sub)greenschist alteration assemblages formed during interaction with seawater at *c.* 250°C to >300°C. Next, disseminated sulphide mineralization in the transition zone and quartz, epidote and sulphides in veins formed at 250–350°C from more highly reacted, evolved hydrothermal fluids that were similar to black smoker vent fluids (depleted in Mg and enriched in Si, Ca and metals). Subsequently, penetration of seawater resulted in precipitation of anhydrite in cracks and local replacement of plagioclase by anhydrite. Finally, retrograde zeolites (laumontite and scolecite), prehnite and calcite formed in fractures at lower temperatures (<250°C), from more evolved hydrothermal fluids (increased pH and Ca/Mg). Vein calcites provide evidence for the lowest temperatures (100–200°C) in this zone (Fig. 6.14; Alt *et al.*, 1986b). In contrast to the reheating of the volcanic section, there is no evidence that the transition zone and upper dyke section were ever cooled below their present temperatures of 100–150°C (Fig. 6.14).

Epidote from the upper dykes has Fe/(Fe + Al) ratios of 0.16–0.20, whereas prehnite exhibits a range of 0.01–0.20. Ishizuka (1989) showed that prehnite and epidote coexisting with laumontite have higher Fe/(Fe + Al) than prehnite and epidote associated with actinolite. The Fe/(Fe + Al) ratios of prehnite and epidote coexisting with laumontite are close to the predicted equilibrium exchange for these minerals (Bird *et al.*, 1984), suggesting that these associations may represent equilibrium assemblages. The actinolite assemblage reflects higher temperatures, but the Fe contents of prehnite and epidote can also decrease as the oxygen fugacity decreases (Liou *et al.*, 1983). Epidote occurrences and compositions also vary widely in Hole 504B. For example, epidote crystals commonly exhibit patchy zoning, with Fe/(Fe + Al) ratios varying by up to 0.09 within a single crystal and small amounts of clinozoisite are also present (Alt *et al.*, 1985a). Therefore, caution must be used when applying mineral equilibria to secondary mineral assemblages in these rocks, which generally reflect disequilibrium

and reaction relationships between primary and secondary minerals and among various secondary minerals (Alt *et al.*, 1985).

Although there is a general trend from smectite in the volcanic section to chlorite in the transition zone and upper dykes, phyllosilicates in the latter sections range from chlorite to chlorite/smectite, corrensite, chlorite/corrensite, talc, talc/chlorite and saponite, reflecting the variable intensity of hydrothermal alteration and recrystallization (Fig. 6.6; Alt *et al.*, 1985, 1986a; Shau & Peacor, 1992). As the proportion of smectite layers in chlorite increases, the minerals exhibit increasing Si and Mg and decreasing Al and Fe contents (Fig. 6.16). In several intervals the rocks are only slightly altered, with olivine replaced by chlorite/smectite and talc and with very few veins or fractures in the rocks (Alt *et al.*, 1985, 1989a). The observed hydration of olivine to talc requires only addition of water to the rocks, whereas replacement by chlorite in more highly altered rocks requires further interaction and breakdown of plagioclase to provide the necessary alumina. Thus, superimposed upon the general increase of temperature downward recorded by the transition from smectite to chlorite, are variations in fracturing which controlled fluid access and hence alteration and phyllosilicate mineralogy.

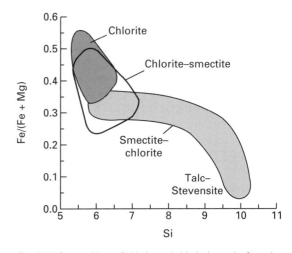

Fig. 6.16 Compositions of chlorite and chlorite/smectite from the upper ocean crust, illustrating general trends of decreasing Fe/(Fe + Mg) and increasing Si contents from chlorite to smectite (after Alt *et al.*, 1985, 1996a). Formulae contents calculated based on total layer charge of 56, all Fe as Fe^{2+}.

The rocks exhibit striking chemical changes: Cu, Zn, Mn and S exhibit strong peaks in the mineralized rocks (Fig. 6.12). H_2O contents, δD and $^{87}Sr/^{86}Sr$ ratios increase significantly and $\delta^{18}O$ decreases as alteration temperatures and the extent of recrystallization increase downward from the top of the transition zone (Alt *et al.*, 1996a). Likewise, K_2O and Fe^{3+}/Fe^T exhibit stepwise decreases downward, as K is leached from the rocks at higher temperatures. TiO_2 underwent limited mobilization to form titanite in the veins. CaO and Na_2O vary locally in response to albitization or zeolitization of calcic plagioclase. FeO^T exhibits local enrichments that coincide with the occurrence of Fe-rich amphibole and chlorite and which may be the result of rocks interacting with Fe-enriched upwelling hydrothermal fluids.

Notably, the rocks do not exhibit increased MgO contents as might be predicted from experiments and vent fluid data which indicate quantitative loss of Mg from seawater to the crust (Seyfried, 1987). Glassy pillow rims and millimetre-sized zones along veins are chloritized, indicating locally high-integrated water/rock mass ratios of *c.* 65 (Alt *et al.*, 1986a,b). In contrast, most of the rock reacted with small volumes of seawater (integrated water/rock mass ratios *c.* 1 indicated by Mg uptake and calculated from oxygen isotope ratios; Alt *et al.*, 1986a,b). There still may be a net uptake of Mg by the upper dyke section, however, via chlorite filling fractures and cementing breccias.

By analogy with Hole 504B, dyke samples from the Atlantic that are partly recrystallized and contain albite, chlorite/smectite, chlorite, ± actinolite ± quartz ± titanite are interpreted to be from the upper dyke complex (Gillis & Thompson, 1993). At Hess Deep, the volcanic-dyke transition is marked by mineralogical changes similar to those in Hole 504B (Gillis, 1995). In all sections of this crustal interval, clinopyroxene is the mineral least affected by recrystallization, most probably reflecting kinetic effects. At Hess Deep the position of the transition from lower-temperature (<150°C) smectite-dominant alteration to subgreenschist and greenschist hydrothermal alteration effects (>250°C) varies with respect to the lithological transition from lavas to dykes depending upon local permeability and hydrothermal circulation (Gillis, 1995).

6.5 The lower sheeted dyke complex and the plutonic section: the roots of submarine hydrothermal systems

As seawater hydrothermal fluids in axial convection cells penetrate deeper into the crust, they react at higher temperatures and under more rock-dominated conditions. It is in these deep (*c.* 2 km) roots of hydrothermal systems, commonly referred to as subsurface 'reaction zones', where hydrothermal fluids are thought to attain their maximum temperatures and acquire their final chemical characteristics before upwelling to vent at the seafloor (Alt, 1995, 1997; Seyfried & Ding, 1995; Von Damm, 1995).

The lower dyke section of Hole 504B was initially metamorphosed in such a zone (summary in Alt *et al.*, 1996a). These rocks have lost sulphur and metals (Figs 6.6 and 6.12) due to the breakdown of sulphides and titanomagnetite at high temperatures (>350°C; Alt, 1995; Alt *et al.*, 1996b,c). The rocks are generally more extensively recrystallized than those from the upper sheeted dykes, and the abundance of amphibole increases downward (Laverne *et al.*, 1995; Alt *et al.*, 1996a,b); Vanko *et al.*, 1996). Clinopyroxene is intensely altered to amphibole (actinolite, actinolitic hornblende, magnesio-hornblende), plagioclase is partly replaced by albite–oligoclase, titanomagnetite is replaced by titanite and olivine is replaced by amphibole or chlorite, and locally by chlorite/smectite, talc, magnetite and sulphides. As in the upper dyke complex, the rocks are more intensively recrystallized (up to 100%) in cm-sized patches surrounding vugs and haloes along amphibole veins. The mineralogy of these more altered rocks differs from the upper dykes; however, plagioclase is partly replaced by secondary anorthite; clinopyroxene is recrystallized to magnesio-hornblende; and retrograde actinolitic hornblende, actinolite and albite are also present. Locally, small amounts of secondary diopside partly replace magmatic augite along amphibole veins.

The secondary mineralogy of the lower dyke complex in Hole 504B indicates greenschist to amphibolite facies conditions and high temperatures (>400°C), but uncertainties in absolute temperature are large (Laverne *et al.*, 1995; Vanko *et al.*, 1996). Lindsley and Anderson's (1983) two-pyroxene thermometer applied to the secondary diopsidic clinopyroxenes gives temperature estimates of 500–700°C (Laverne *et al.*, 1995; Vanko *et al.*, 1996). Geothermometry based on [NaSi]–[AlCa] exchange between magnesio-hornblende and secondary anorthite (Spear, 1980; Holland & Blundy, 1994) indicates similar temperatures of >500°C to >700°C (Laverne *et al.*, 1995; Vanko *et al.*, 1996). Uncertainties arise because of the presence of minor components in the natural minerals not taken into account in the calibration, extrapolation of the geothermometers calibrated at high pressures (2 kbar) to lower pressures (*c.* 0.5 kbar), and the question of whether equilibrium was attained (Vanko *et al.*, 1996). Amphiboles vary in composition with occurrence, with actinolitic amphibole in veins and magnesio-hornblende restricted to the alteration haloes. This may reflect retrograde vein minerals or dependence of mineralogy on the site of nucleation (vein vs. replacement).

The fluxes of hydrothermal fluids through the lower dyke complex were low, as indicated by Sr isotope ratios and only slight chemical changes (Alt *et al.*, 1996a,b). Subsequent retrograde alteration in the lower dykes was similar to metamorphism in the upper dykes (Fig. 6.6) and resulted in the formation of actinolitic hornblende, actinolite, albite, chlorite, chlorite/smectite, talc and later minor epidote and prehnite. Locally low porosity and permeability restricted fluid fluxes through certain intervals of the lower dykes, which inhibited water–rock reactions and resulted in the formation of talc and chlorite/smectite rather than chlorite in these rocks (Fig. 6.6; Alt *et al.*, 1996a).

Broadly similar hydrothermal effects are observed in the lower portions of sheeted dyke complexes from elsewhere in the oceans. By analogy with Hole 504B, intensively recrystallized dykes from the mid-Atlantic Ridge that contain magnesio-hornblende and have lost metals are interpreted to be from the lower portion of the dyke complex (Gillis & Thompson, 1993). The lower dyke complex from Hess Deep is similarly highly recrystallized, contains assemblages of the greenschist facies and transitional to the amphibolite facies and the rocks have lost Cu and Zn (Gillis, 1995). The rocks were initially altered at high temperatures (>450°C; Gillis, 1995), but superimposed retrograde effects and disequilibrium are common.

Hydrothermal metamorphism in deep 'reaction zones' at the base of hydrothermal cells extends downward into the uppermost plutonic section of the crust. In the fault-exposed section at Hess Deep, *c.* 100 m of gabbros beneath the sheeted dyke complex are hydrothermally metamorphosed to amphibolite and transitional greenschist to amphibolite facies conditions, and have lost Cu and Zn, as the result of intense hydrothermal alteration overlying an axial magma lens (Gillis, 1995). Gabbronorites beneath these rocks are less intensely recrystallized, with alteration mainly restricted to the formation of amphibole in microfracture networks. This occurred as hydrothermal fluids penetrated downward along cracks through the solidified melt lens when the lower crust cooled and moved off-axis. Maximum temperature estimates are up to 650–800°C for initial penetration of hydrothermal fluids and formation of secondary clinopyroxene, and >450°C for formation of amphibolite facies mineral assemblages (Gillis, 1995). For the deeper Hess Deep section, beneath the melt lens, Manning *et al.* (1996) used statistical techniques and the Holland and Blundy (1994) plagioclase–amphibole geothermometer to obtain estimates of 716 ± 8°C for the temperature of initial penetration of fluids into the lower crust. These authors also combined thermal models of spreading centres to suggest that initial metamorphism of the lower crust at Hess Deep was rapid (<6000 years) and occurred 1–4 km off-axis.

Because of fundamental differences in crustal accretion structures, large-scale faulting at slow spreading centres results in high-temperature plastic deformation and hydrothermal metamorphism that is not observed at fast spreading ridges (Mevel & Cannat, 1991; Gillis & Thompson, 1993; Gillis, 1995). The best example of lower crustal alteration processes at a slow-spreading ridge is the 500-m-long gabbroic section from ODP Hole 735B in the south-west Indian Ocean. Here, brittle faulting at the ridge crest penetrated below the brittle-ductile transition and caused early plastic deformation and the formation of foliated amphibolites at 700–600°C (Mevel & Cannat, 1991; Vanko & Stakes, 1991). Further cooling of the rocks allowed brittle deformation and crack formation, penetration of hydrothermal fluids, and formation of hornblende and sodic plagioclase in near-vertical veins and surrounding wallrock at temperatures of 500–600°C (Mevel & Cannat, 1991; Vanko & Stakes, 1991). Retrograde hydrothermal metamorphism occurred along veins and local breccia zones at amphibolite grade (550–400°C) to greenschist (<400°C) and lower temperatures (Vanko & Stakes , 1991). As at Hess Deep, the upper 200 m of the Hole 735B section is more intensely metamorphosed and has low Cu and S contents as the result of leaching of metal sulphides from the rocks during hydrothermal metamorphism (Alt & Anderson, 1991).

6.6 Hydrothermal upflow zones and formation of seafloor sulphide deposits

Hydrothermal fluids in deep reaction zones are near the critical point of seawater and the large increase in buoyancy forces fluids to flow rapidly upward to vent at the seafloor. Because of the water depths (2.5–3.5 km) and the high hydrostatic pressures (*c.* 0.5 kbar) at the base of submarine hydrothermal systems, subcritical boiling is rare (Butterfield *et al.*, 1990; Von Damm *et al.*, 1995). Phase separation at supercritical conditions is common, however, and differential migration and mixing of the higher- and lower-salinity fluid components leads to variable salinities of hydrothermal fluids, which range from 6% to 200% that of seawater (Von Damm, 1988, 1995; Butterfield & Massoth, 1994; Edmonds & Edmond, 1995; Von Damm *et al.*, 1995). The low-salinity phase is enriched in volatile components (CO_2, H_2S), whereas the high-salinity phase is depleted in gases, which has significant affects on metal/sulphur ratios of fluids and the formation of metal sulphide and oxide deposits at the seafloor (Edmond *et al.*, 1979b).

Deep, focused upflow zones have not been sampled from the seafloor, but in ophiolites are characterized by epidosites in the lower sheeted dyke complex, where the rocks are recrystallized to granular epidote + quartz (Richardson *et al.*, 1987; Schiffman *et al.*, 1987; Harper *et al.*, 1988; Nehlig *et al.*, 1994). Epidosites are depleted in metals and sulphur and were altered by Mg-depleted, Ca-enriched hydrothermal fluids at temperatures of 350–440°C and at high integrated fluid/rock ratios (*c.* 500–1000) in the basal portion of focused upflow zones (Richardson *et al.*, 1987; Harper *et al.*, 1988; Schiffman & Smith, 1988; Bickle & Teagle, 1992; Alt, 1994). Permeability discontinuities may be important for focusing fluid flow

and metasomatism in order to generate epidosites (Rose & Bird, 1994; Rose, 1995).

In ophiolites, deep, focused upflow zones are inferred to grade upward from epidosites into quartz + epidote veins and then to quartz + sulphide (± epidote) veins (Haymon *et al.*, 1989; Nehlig *et al.*, 1994). The shallowest portions of upflow zones, the stockwork feeder zones for massive sulphide deposits, have been sampled from the seafloor as well as from ophiolites. Basalts in the stockworks are totally recrystallized to chloritic, illitic or paragonitic assemblages (+ quartz and sulphides) via interaction with upwelling hydrothermal fluids and variable proportions of seawater, at temperatures of 250–360°C (Embley *et al.*, 1988; Zierenberg *et al.*, 1988; Richards *et al.*, 1989; Alt & Teagle, 1997; Honnorez *et al.*, 1997; Teagle *et al.*, 1997). The central cores of the stockworks are more siliceous and chlorite is less abundant than on the margins, reflecting the presence of upwelling hydrothermal fluids and more intense hydrothermal alteration in the centre. Illite or paragonite is present reflecting control by hydrothermal fluids, which are enriched in alkalis and depleted in Mg, and the rocks lose Mg to hydrothermal fluids. Chlorite commonly varies from more Fe-rich in the inner portions of the stockwork to more Mg-rich on the margins. This typically results from the greater involvement of Mg-bearing seawater on the periphery of the stockwork, although the Fe/Mg of chlorite is a complicated function of temperature, the Fe/H_2S of hydrothermal fluids, and pathways of cooling and mixing with seawater (Saccocia & Seyfried, 1994).

Where hydrothermal fluids reach the seafloor, cooling and mixing with seawater result in the formation of metal sulphide deposits at mid-ocean ridges. These range from tiny deposits to much larger, up to several million tons (Hannington *et al.*, 1995). The mineralogy and processes of formation of these sulphide deposits and reactions within sulphide chimneys venting hydrothermal fluid have recently been reviewed by Hannington *et al.* (1995) and Tivey (1995). The TAG hydrothermal mound on the mid-Atlantic ridge is probably the best documented deposit, having been drilled by the Ocean Drilling Program as well as sampled by submersible. Here, 360°C hydrothermal fluids vent from sulphide chimneys atop the 200-m diameter mound, which is composed of various sulphide + quartz + anhydrite breccias (Edmond *et al.*, 1995; Humphris *et al.*, 1995). Entrained seawater cools the mound and mixes with hydrothermal fluids, resulting in a zone-refining process within the deposit whereby anhydrite, quartz, pyrite and chalcopyrite precipitate and sphalerite dissolves (Edmond *et al.*, 1995; Tivey *et al.*, 1995; Alt & Teagle, 1997; Teagle *et al.*, 1997). The lower temperature mixed fluids vent at silica-sulphide chimneys at temperatures of 260°C, or exit the mound as ubiquitous diffuse low-temperature (<50°C) flow (James & Elderfield, 1996). Similar processes occur in the more typical, much smaller sulphide deposits on the seafloor, but these deposits are more dominated by sulphide chimneys (Hannington *et al.*, 1995; Tivey, 1995).

Where upwelling fluids are not sufficiently focused to vent at the seafloor, they mix with seawater and deposit sulphides in the subsurface, as in the transition zone of Hole 504B and elsewhere (Gillis, 1995; Honnorez *et al.*, 1985). This type of subsurface disseminated mineralization is common in ophiolites (Harper *et al.*, 1988; Gillis & Robinson, 1990; Alt, 1994; Nehlig *et al.*, 1994; Gillis, 1995) and may be typical of ocean crust. Depending on the Fe/H_2S ratio of the hydrothermal fluids, subsequent cooler mixed fluids may contain significant amounts of Fe, Mn and Si, and lead to the formation of low-temperature hydrothermal Fe–Mn oxide or Fe-silicate (nontronite) deposits at the seafloor.

6.7 Seawater–crustal chemical fluxes at mid-ocean ridges and cycling at subduction zones

Early studies of seafloor alteration of oceanic basalts revealed that this process could have a significant effect on the composition of seawater (Thompson, 1973) and it was subsequently realized that subduction of altered crust could contribute to metasomatism of subarc mantle and to the formation of heterogeneities in the mantle. Comprehensive review of these subjects is beyond the scope of this chapter, but several recent papers discuss seawater–crustal chemical fluxes at mid-ocean ridges and altered crustal compositions relative to subduction recycling (e.g. Alt, 1994; Mottl & Wheat, 1994; Spivack & Staudigel, 1994; Kadko *et al.*, 1995; Staudigel *et al.*, 1995; Alt *et al.*, 1996c) and some of the main points are summarized here.

Water–rock interactions in active axial hydrothermal systems occur mainly in the sheeted dykes and upper gabbros at high temperatures, whereas lower temperature alteration in passive, off-axis systems takes place mostly in the volcanic section. For some elements the effects of low-grade alteration on ridge flanks are opposite to those of high-grade axial processes and, while the chemical fluxes for both are large, these effects may offset each other. For example, the alkalis, B and $\delta^{18}O$ decrease in the lower crust via axial hydrothermal metamorphism, but increase in the upper crust in passive systems on ridge flanks and the effects essentially cancel each other. It is the balance between these processes that buffers the oxygen isotopic composition of the oceans (Muehlenbachs & Clayton, 1976). For S the trends are reversed but the net result is similar (Alt, 1994). For other elements, the chemical changes are essentially unidirectional during both axial and flank processes, with the only difference being a matter of the degree of change between axial and flank systems, so the effects are cumulative (e.g. H_2O, $^{87}Sr/^{86}Sr$, Ca, Si, Mg and U). The combined effects of active axial hydrothermal metamorphism and lower temperature alteration in passive ridge flank systems exert significant controls on the Mg content of seawater and on the Sr isotopic compositions of the oceans (Bickle & Teagle, 1992; Mottl & Wheat, 1994). Despite the very small chemical changes in passive ridge flank systems, the enormous fluid fluxes make these processes important (Mottl & Wheat, 1994).

The volcanic section, affected by very low-grade hydrothermal metamorphism, is the main carrier of the seawater signature (e.g. elevated alkalis, B, $\delta^{18}O$, $^{87}Sr/^{86}Sr$, CO_2), so it may be this portion of the crust that contributes most to subduction zone metasomatic fluids (Staudigel *et al.*, 1995; Alt *et al.*, 1996a). Mineral dehydration reactions at progressively greater temperature and pressure lead to dewatering and loss of various elements during subduction metamorphism (Peacock, 1990). The overlying mantle is serpentinized and dragged downward by the subducting slab, to dewater and melt the overlying peridotite where temperatures are great enough. Serpentinite diapirs also rise up through the forearc mantle and protrude onto the seafloor, as in the Mariana forearc (Fryer *et al.*, 1990). Here, fluids derived from dehydration reactions in the subducting slab, 30 km below, vent at the seafloor (Mottl, 1992) and minerals such as lawsonite and sodic pyroxene in the serpentinite muds at the seafloor attest to high-pressure metamorphic reactions at depth (Maekawa *et al.*, 1992). Ultimately, however, some of the seawater signature added to ocean crust through very low-grade hydrothermal metamorphism is carried into the mantle to contribute to heterogeneities in the contents of alkalis and U and the isotopic compositions of O, Sr and S.

6.8 Ophiolites

Ophiolites are fragments of ocean crust exposed on land, but in contrast to mid-ocean ridges, most ophiolites were generated by seafloor spreading above subduction zones. The Troodos ophiolite, Cyprus, is one of the best studied sections and very low-grade hydrothermal metamorphism of the volcanic section is quite similar to that of oceanic crust (Gillis & Robinson, 1985, 1990; Bednarz & Schmincke, 1989; Alt, 1994). The upper 20–300 m, or 'seawater zone,' of the volcanic is characterized by pervasive low-temperature (<100°C) seawater hydrothermal alteration and oxidation, with Fe-oxides, saponite, K-feldspar, calcite, phillipsite and analcite filling fractures and vesicles, replacing glass and olivine and some plagioclase phenocrysts.

The heterogeneous alteration of the underlying 'low-temperature zone' extends to near the base of the volcanic section. Pillow lavas are variably altered, with mesostasis replaced by smectite and glass replaced by smectite, phillipsite and analcite. The less permeable massive flows are generally less altered, with celadonite, saponite, clinoptilolite, mordenite and calcite filling pores and replacing the rock along flow margins and fractures, which are lined with celadonite, Fe-oxyhydroxides, saponite, sepiolite, calcite and opal. This alteration is analogous to the restricted seawater circulation in oceanic crust (Bednarz & Schmincke, 1989; Gillis, & Robinson, 1990). K/Ar and Rb/Sr dating of secondary minerals indicates that mineral precipitation can continue for ≥40 Ma (Staudigel *et al.*, 1986; Gallahan & Duncan, 1994).

An approximately 100-m-thick 'transition zone' marks the change from very low-grade metamorphism in the volcanic section to low-grade metamorphic rocks that are totally recrystallized to chlorite,

epidote, albite and quartz. The transition occurs near the top of the dykes and is characterized by increased recrystallization, the presence of chlorite/smectite, the appearance of laumontite, albite and quartz and the disappearance of the other zeolites. Temperatures estimated for this transition are about 100–200°C (Gillis & Robinson, 1990). The transition is quite similar to that within oceanic Hole 504B, but both the lavas and the underlying low-grade rocks in the ophiolite are much more pervasively altered than in oceanic crust, reflecting greater fluid fluxes through the ophiolite (Bickle & Teagle, 1992; Alt, 1994).

In contrast to Hole 504B and the Troodos ophiolite, the volcanic sections in other ophiolites are more intensely altered at higher grades (chlorite or chlorite/smectite, albite, epidote, prehnite, pumpellyite) and alteration assemblages exhibit more gradual changes downward (Spooner & Fyfe, 1973; Coish, 1977; Spooner *et al.*, 1977; Liou, 1979; Liou & Ernst, 1979; Stern & Elthon, 1979; Cocker *et al.*, 1982; Evarts & Schiffman, 1983; Ishizuka, 1985; Bettison & Schiffman, 1988; Harper *et al.*, 1988; Bettison-Varga *et al.*, 1991; Pflumio, 1991; Schiffman *et al.*, 1991). Such differences reflect generally shallower or more continuous thermal gradients in these settings compared to Troodos and mid-ocean ridges (Beiersdorfer, 1993; Schiffman & Staudigel, 1994). These can result from burial of ophiolite lavas by thick sequences of volcaniclastic sediments, which seal the basement from free access of seawater and convective heat removal (Schiffman *et al.*, 1991). Similar processes occur at modern sediment-covered seafloor spreading centres, where sediment–sill complexes and basaltic basement are more intensely hydrothermally altered at higher grades than normal mid-ocean ridge crust (Curry & Moore, 1982; Davis *et al.*, 1992b; Stakes & Schiffman, 1996). Faulting may also play a role in the ophiolites: a lack of deep normal faults to focus fluid flow has been suggested to result in more pervasive alteration of volcanic sections in some ophiolites (Schiffman *et al.*, 1991). Multiple episodes of intrusion and volcanic activity produce superimposed and more intense alteration and metamorphic effects in some ophiolite volcanic sequences (Pflumio, 1991) and in some cases the effects of superimposed regional metamorphism may be important.

6.9 Very low-grade metamorphism at oceanic islands

6.9.1 Iceland

Iceland lies on the mid-Atlantic Ridge so it is part of the mid-ocean ridge system, but its hotspot origin has led to its subaerial position. Exposed crust in Iceland ranges from 16–0 Ma and is comprised of flood basalts, central volcanoes and active rifts (Palmason *et al.*, 1979). Superimposed on regional burial metamorphic zones are hydrothermal systems associated with intrusions and central volcanoes in the active rift zone (<0.7 Ma; Palmason *et al.*, 1979). Low-temperature areas (<100°C) contain relict higher temperature minerals, but active high-temperature (>300°C) hydrothermal systems display consistent mineral zonations that correlate with temperature (Tómasson & Kristmannsdóttir, 1972; Kristmannsdóttir, 1975, 1978; Palmason *et al.*, 1979). Hydrothermal fluids range from seawater in the Reykjanes peninsula to meteoric waters at higher elevations inland (Arnorsson *et al.*, 1978). Rock types include hyaloclastites and lavas of tholeiitic or olivine tholeiitic compositions, with rare acid rocks and sediments locally. Alteration of the rocks depends on mineralogy, texture and permeability: glass, olivine and matrix materials are replaced by phyllosilicates, but pyroxene and plagioclase are more resistant.

Mineralogical zones in active high-temperature Icelandic hydrothermal fields are mainly a function of temperature, and the four zones are shown in Fig. 6.17. Zeolites occur in the upper portions and extend to varying depths in different holes. At less than 100°C, zeolite mineralogy corresponds to rock type: tholeiitic basalts contain mordenite, heulandite, stilbite and epistilbite, whereas olivine tholeiites contain chabazite, thompsonite and mesolite–scolecite (Kristmannsdóttir, 1978). Laumontite replaces the other zeolites at 100–120°C, wairakite appears at 180°C and all other zeolites disappear by *c.* 200°C. Opal is present below 100°C, whereas quartz is present throughout the higher temperature zones. Calcite is present throughout all zones, but is most abundant in tuffs of zones I and II. Prehnite first appears between the zeolite and epidote zones and is present in varying amounts at greater depths.

Fluid compositions, rock texture and permeability play critical roles in secondary mineralogy and

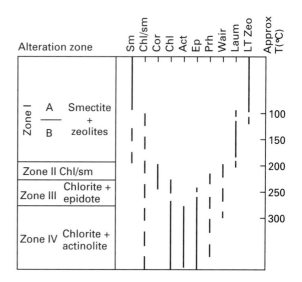

Fig. 6.17 Secondary mineral depth zonation in active Icelandic hydrothermal fields. (After Kristmannsdóttir, 1975, 1977, 1978, and Schiffman & Fridleifsson, 1991.) LT Zeo, low-temperature zeolites; Wair, wairakite; other abbreviations as in Fig. 6.6 (see text).

hydrology. As seawater enters the system at Reykjanes it is heated on the margins of the hot hydrothermal system, resulting in the precipitation of anhydrite that seals the system from further ingress of seawater (Tómasson & Kristmannsdóttir, 1972). Epidote first appears in tuffaceous sediments and breccias, but disappears in the less permeable layers, its occurrence only becoming continuous at temperatures >260–270°C. It was suggested that the initial appearance of epidote in tuffaceous material is permeability related (Tómasson & Kristmannsdóttir, 1972), but the extent of water/rock interaction and fluid evolution may play a role as well (Alt *et al.*, 1986a, 1996b). Ragnarsdóttir *et al.* (1984) showed that fluids in the chlorite–epidote zone at Reykjanes are close to equilibrium with the phases present and that seawater-derived fluids are highly reacted and have low Mg contents, consistent with this suggestion. Elderfield and Greaves (1981) showed that palagonitized hyaloclastites are in Sr isotopic equilibrium with hydrothermal fluids, whereas crystalline rocks from deeper in the system are only about 50% equilibrated, indicating that mineralogy, rock texture and permeability exert important controls on fluid/rock interactions.

Other minerals present include albite, which has no clear relation to depth and sporadic K-feldspar (Tómasson & Kristmannsdóttir, 1972; Kristmannsdóttir, 1975, 1978). The presence of hematite in upper layers and of secondary magnetite at greater depths suggests decreasing f_{O_2} with depth. (Ragnarsdottir et al., 1984). The presence of actinolite is controlled by temperature, and it only occurs in a few areas at temperatures greater than 280–300°C.

Phyllosilicates provide perhaps the most continuous and sensitive record of very low-grade to low-grade hydrothermal metamorphism. Their distribution is mainly a function of temperature, but fluid fluxes, rock texture and permeability are important controls. Smectite is present up to 200°C, with montmorillonite in the upper 50–100 m and Fe-saponite at greater depths (Kristmannsdóttir, 1977, 1978). Thermal and infrared analyses, however, indicate the presence of minor chlorite layers in smectite at low temperatures (Kristmannsdóttir, 1977). Electron microprobe analyses also reveal that trioctahedral smectite-rich chlorite/smectite is present rather than smectite at temperatures > 170°C, with a progressive increase in the amount of chlorite layers at higher temperatures (Fig. 6.18; Schiffman & Fridleifsson, 1991). These authors also show that both di- and trioctahedral smectites are present at shallower depths and lower temperatures (less than 500 m and 170°C).

Kristmannsdóttir (1977, 1978) showed that chlorite appears in the range 230–270°C (250–500 m depth) and is present to the maximum depths. The zone between the disappearance of smectite and appearance of chlorite (i.e. between 200 and 230–270°C) comprises a transition zone, where random chlorite/smectite occurs. In their more detailed work on the Nesjavellir field, Schiffman and Fridleifsson (1991) showed that the transition zone (here at 240–270°C) contains smectite, random chlorite/smectite and corrensite, with the proportion of chlorite increasing downward through this zone. At temperatures greater than 270°C chlorite and minor random chlorite/smectite are present. The Mg/(Mg + Fe) of the phyllosilicates varies widely and is independent of mineralogy and temperature, suggesting a bulk rock compositional control (Schiffman & Fridleifsson, 1991).

Jakobsson and Moore (1986) showed temperature and kinetic effects on hydrothermal alteration of

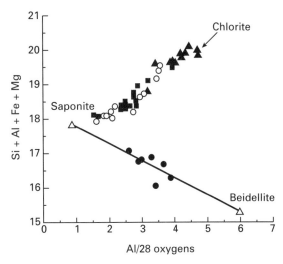

Fig. 6.18 Compositions of phyllosilicates from the active Nesjavellir hydrothermal area in Iceland illustrating the smectite to chlorite transition (after Schiffman & Fridleifsson, 1991). Solid circles are mixtures of di- and trioctahedral smectites from zone IA (Fig. 6.16), solid squares are chlorite/smectite from zone IB, open circles are chlorite/smectite from zones II and III, and solid triangles are chlorite from zones III and IV. Open triangles are end-member saponite and beidellite. Smectite transforms to chlorite via a gradual increase in chlorite layers, then discrete corrensite and finally chlorite appear.

basaltic glass by studying rocks from the volcanic island of Surtsey several years after its initial formation. Samples from a 181-m drill hole reaching temperatures of 150°C revealed that palagonitization of glass in basaltic tuff is strongly temperature dependent, the rate doubling for every 12°C increase. Above 120°C, the edges of olivine crystals are replaced by nontronite, with the clay rim doubling in thickness for every 8°C increase. Alteration of the coarser unconsolidated tephra was much less advanced, however, consistent with a textural control of fluid/rock interaction. Phillipsite and analcite are also present, with phillipsite abundance decreasing at temperatures above 100°C, whereas analcite abundance increases progressively above 60–80°C up to the maximum temperature, 150°C.

The general fossil hydrothermal zonation in the 2-km IRDP drillcore from eastern Iceland is similar to that of the active systems outlined above, but the older rocks (9 Ma) and continuous core provide a complex history that is missed in the usual drill cuttings from active systems (Kristmannsdóttir,

1982). In the smectite zone, open space fillings are important and the sequence from early smectite to later calcite, quartz and zeolites reflecting evolution of fluid compositions is similar to that observed in submarine basalts (Mehegan *et al.*, 1982). Similarly, the sequence to progressively lower temperature phases in veins of the epidote zone (1, epidote; 2, prehnite; 3, laumontite) indicates cooling of the higher temperature system as is recorded in submarine dykes (Exley, 1982; Mehegan *et al.*, 1982; Alt *et al.*, 1986a). Host rock composition exerts a strong mineralogical control on secondary phases, with adularia, wairakite and pumpellyite restricted to silicic pyroclastic rocks (Viereck *et al.*, 1982).

6.9.2 La Palma

La Palma is a Pliocene seamount, with an extrusive section of alkaline pillow lavas and volcaniclastites (Schiffman & Staudigel, 1994, 1995). The volcanic rocks exhibit a complete low-grade series from zeolite to prehnite–pumpellyite to greenschist facies (Fig. 6.19). Phyllosilicates grade from smectite to chlorite through a transitional corrensite zone approximately 350 m thick, although the steepness of metamorphic gradients within the island varies. The first appearance of prehnite, epidote and corrensite at 700–1050 m implies temperatures of 225–250°C for this depth. The layer silicates fill veins and vesicles, replace glass, mesostasis and phenocrysts and, as in oceanic basalts, they precede other minerals in sequential vesicle fillings. The smectites have high noninterlayer cation totals, suggesting the presence of minor chlorite layers or of Mg in interlayer positions. Phyllosilicates display a wide variation in $Mg/(Mg + Fe)$, but a general decrease in this ratio from smectite to corrensite to chlorite. This is in part due to bulk rock compositional control, but also reflects fluid evolution from seawater (Mg-rich) at low temperatures to hydrothermal fluids (Fe-bearing, low Mg) at higher temperatures. In contrast to the gradual transition via increase of chlorite layers in random chlorite/smectite in Iceland, the stepwise transition from smectite to corrensite to chlorite at La Palma is explained as the result of greater fluid fluxes at La Palma (Schiffman & Staudigel, 1994, 1995). Shau and Peacor (1992) similarly suggest that corrensite is the stable phase in samples from oceanic Hole 504B that have experienced large fluid fluxes,

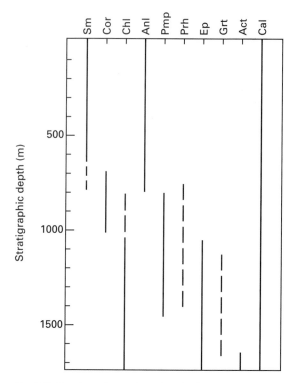

Fig. 6.19 Zonations of secondary minerals with depth in hydrothermally metamorphosed basalts from the oceanic island of La Palma (Schiffman & Staudigel, 1994, 1995). The section contains a complete low-grade sequence from zeolite to prehnite–pumpellyite to greenschist facies, in contrast to the sharp metamorphic boundary resulting from the stepped thermal gradient in ocean crust from ODP Hole 504B (contrast with Fig. 6.6).

whereas random chlorite/smectite occurs in samples that reacted with small amounts of fluids.

6.10 Island arcs and back-arc basins

6.10.1 Submarine rocks

A complete review of island arcs and back-arc basins is beyond the scope of this chapter, but observations on modern submarine arcs and back-arc basins are discussed. Hole 786B in the Izu-Bonin forearc penetrates >700 m into basement and provides a complete volcanic section bottoming out in mineralized and intensely hydrothermally metamorphosed and mineralized dykes (Alt *et al.*, 1998). Volcanic rocks (boninites, bronzite andesites, dacites and rhyolites) are 5–60% altered to celadonitic minerals, saponite, Mg-montmorillonite, Fe-oxyhydroxides, calcite and phillipsite (Fig. 6.20). The abundance of unstable glassy material and mafic phases led to more intensive recrystallization than in typical MOR sections. The basal dykes (below 650 m) are intensely altered to corrensite, smectite, quartz, albite, K-feldspar, calcite and traces of talc and chlorite. The rocks in a 5-m-thick mineralized zone are completely recrystallized to chlorite (± trace chlorite/smectite), metastable intermediate Na–K mica, quartz, albite, K-feldspar, sulphides and rare magnetite. This section is thus similar to oceanic Hole 504B and the Troodos ophiolite, having a stepped thermal gradient co-

inciding with the transition from volcanics to dykes. Oxygen isotopic data for phyllosilicates, quartz and carbonates indicate low-temperature alteration of the volcanic section at <80°C, temperatures of about 150°C for the corrensite-bearing dykes and mineralization of the chlorite-bearing rocks at approximately 250°C (Alt *et al.*, 1998). Because of the high volatile contents of arc magmas, input of magmatic fluids into hydrothermal systems in arc environments can be important (Gamo *et al.*, 1997), and sulphur isotopic data suggest the presence of magmatic sulphur gases in the Izu-Bonin forearc system (Alt *et al.*, 1998).

Yuasa *et al.* (1992) describe pumpellyite-bearing rocks from the Sumisu back-arc Rift (just west of Hole 786B) and from a nearby seamount. One sample of crystal-lithic tuff from Hole 791 in the Sumisu Rift contains pumpellyite with epidote, albite and chlorite. Two dredged andesite breccias from Ohmachi seamount contain pumpellyite, chlorite, albite, epidote, titanite and quartz. The pumpellyite is Al-rich and similar in composition to that from prehnite–pumpellyite facies rocks, although the pumpellyite in the arc rocks does not coexist with prehnite. Both sites are interpreted to represent rifted arc crust, but the presence of pumpellyite suggests a similarity to sites having continuous geothermal and metamorphic gradients such as in La Palma (Fig. 6.19), in contrast to the stepped gradient documented in the rifted early arc of Hole 786B (Fig. 6.20). Pumpellyite is extremely rare from oceanic crust, with only one other documented occurrence (from the Vema fracture zone in the Atlantic; Mevel, 1981). Factors influencing formation of pumpellyite include fluid composition (Mevel, 1981; Harper, 1995) and pressure, which may be generally too low for stability of pumpellyite in mid-ocean ridge systems, resulting in the formation of common minor prehnite instead (Frey *et al.*, 1991).

Other shallower (70–278 m) penetrations of the Mariana and Izu-Bonin forearcs indicate typical seawater oxidation and more restricted circulation effects (Natland & Mahoney, 1981; Fryer *et al.*, 1990; Taylor & Fujioka, 1990). The rocks are slightly altered to smectite, celadonite, Fe-oxyhydroxides, phillipsite and local palygorskite. Basic rocks recovered from shallow (35–323 m) sections in the Lau and Mariana back-arc basins also exhibit alteration effects similar to those observed in MOR crust (Parsons *et al.*, 1992).

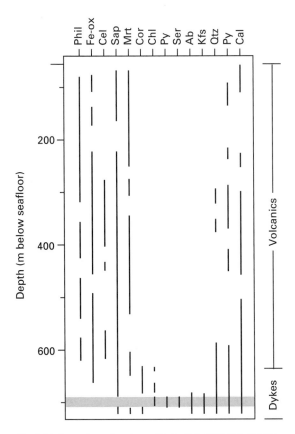

Fig. 6.20 Secondary mineralogy with depth in ODP Hole 786B in the Izu-Bonin forearc. Note the stepped metamorphic and inferred thermal gradients (over 50 m) that coincide with the lithological change from lavas to basal dykes at 625 m, as in ocean crust from Hole 504B (Fig. 6.6), and in contrast to oceanic islands (Iceland and La Palma, Figs 6.17 and 6.19). Shaded zone is mineralized interval. (After Alt *et al.*, 1998.)

These rocks are generally only slightly recrystallized (filled vesicles and fractures ± altered glass, meso-stasis and olivine) to smectite, Fe-oxyhydroxides, aragonite, phillipsite and natrolite, representing typical oxidizing seawater alteration. Hole 458 in the Mariana Trough back-arc penetrates 35 m into basalts where the uppermost 20 m is intensely altered to chlorite, quartz and pyrite at temperatures up to 200°C as the result of flow of hydrothermal fluids beneath the sediment cap (Natland & Hekinian, 1981). Beneath these rocks there is a zone where rocks exhibit seawater oxidation effects, then unaltered basalts.

6.10.2 Japan

Japan is an active island arc with common active and fossil hydrothermal systems. Rocks range from andesites to acid volcanics, tuffs and sediments, with minor basalts, so these systems differ somewhat from others discussed in this chapter. Boiling can occur in these subaerial systems and oxidation of H_2S may lead to acid surface waters, further differentiating these systems from the submarine rocks discussed previously. As discussed in examples from ocean islands and mid-ocean ridges, however, mineral zonations with depth in Japanese hydrothermal systems depend primarily on temperature, but factors such as rock and fluid compositions, permeability and kinetics are important.

Liou *et al.* (1985b) reviewed results for the Onikobe geothermal system. The rocks include andesitic to dacitic flows, pumice, tuffs and sediments. Smectite ± mordenite ± clinoptilolite characterize the upper 200 m. Chlorite/smectite (corrensite + chlorite + smectite) is present at depths > 150 m to the bottom of the drillholes. Chlorite/smectite overlaps with smectite, but smectite disappears at 200 m and the proportion of chlorite in chlorite/smectite then increases continuously with depth. Laumontite is present from 50 m to 300 m and minor yugawaralite may be present at 150–700 m, but wairakite is the main zeolite at depths greater than 100–150 m. Fe-rich prehnite and epidote occur sporadically in the wairakite-mixed-layer zone, with their Al contents increasing downward as temperatures increase. Liou *et al.* (1985b) emphasized the controlling influence of rock composition on phyllosilicate sequences, at Onikobe and elsewhere: illite–smectite and illite are restricted to more felsic rocks, with the proportion of illite increasing downward at Onikobe. Within the basalts at constant bulk rock composition, increasing Al contents from smectite to corrensite to chlorite require corresponding changes in the Al contents of the coexisting Ca-aluminosilicates. Shikazono (1984) also showed that epidote compositions in geothermal areas commonly reflect the Fe contents of the host rocks.

Inoue and Utada (1991) summarized their work at the Kamikita area. Here, basaltic and andesitic flows and volcaniclastics overlie a dacitic intrusion heat source. The rocks are only partly recrystallized: glass, groundmass and orthopyroxene are replaced by secondary phases, but clinopyroxene and plagioclase are more resistant. As in other areas, temperature is a major mineralogical control, but rock composition, texture and permeability are important. Zone I is characterized by smectite: saponite is present in mafic and intermediate rocks and montmorillonite in felsic rocks. Zone II comprises smectite ± zeolites at temperatures less than 100°C: stilbite and heulandite occur in mafic rocks and mordenite in felsic rocks. Zone III is characterized by the presence of corrensite and laumontite, albitized plagioclase and, in the deeper portion, epidote. Zone IV contains chlorite and epidote, which is typically compositionally zoned; albite is present, but corrensite and laumontite are absent. Rock composition strongly influences mineralogy: in contrast to basalts and andesites, mudstones in zone IV are recrystallized to illite, chlorite and quartz; and phengitic mica coexists with chlorite in felsic rocks. Zone V contains biotite and actinolite and is restricted to rocks of the dacitic intrusion. Temperature is the main control on the mineral depth zonation, with the smectite to chlorite transition in these rocks taking place stepwise, via an intermediate corrensite zone at 100–200°C. Porosity and permeability are also important controls, however, with less expandable chlorite/smectite in porous breccias and hyaloclastites (greater fluid flux and extent of fluid–rock interaction) and more expandable minerals in massive lavas (lower fluid flux, less fluid–rock reaction). Smectite, corrensite and chlorite are interpreted to be pure phases, although corrensite coexists with either or both.

Other active and fossil hydrothermal areas in Japan exhibit generally comparable mineral zonations and controls on very low-grade hydrothermal metamorphism. In the Yugawara geothermal area, zonations from smectite, to chlorite/smectite, to chlorite and from mordenite to laumontite + yugawaralite to wairakite (+ prehnite) occur (Oki *et al.*, 1974). Minerals are slightly different in the felsic systems of the Kuroko ore deposits, but they nevertheless exhibit zonations from montmorillonite, to illite/smectite, to chlorite + sericite + quartz associated with the massive sulphide deposits (Sudo & Shimoda, 1978; Date *et al.*, 1983; Inoue, 1987). Illite/smectite undergoes a continuous transformation of increasing proportion of illite with depth and temperature (Inoue, 1987). Similar minerals and zonations are developed in felsic rocks (rhyolitic pyroclastics) that are not associated with ore deposits: from sericite + chlorite at 300°C to smectitic assemblages at lower temperatures (Masuda *et al.*, 1986). Local basaltic rocks in these felsic systems display the typical trioctahedral smectite to corrensite to chlorite and mordenite ± clinoptilolite to laumontite zonations (Inoue, 1987; Meunier *et al.*, 1991).

6.11 Examples of sediment-hosted geothermal systems

Earlier work on geothermal systems in sedimentary and other rock types has been reviewed by Browne (1978). The mineralogy and compositions of rocks in the sedimentary systems differ from those in basic volcanic rocks, leading to significant differences in metamorphic minerals and mineral reactions. Many of the hydrothermal processes and controls on metamorphism are similar throughout these different systems, however, and it is these similarities and controls on mineralogy that are emphasized in two examples here. Metamorphic mineral zonations in the sedimentary systems reflect the temperature distributions, but the compositions of rocks and fluids, fluid fluxes, rock texture and permeability are also important.

The Imperial Valley in southern California and Mexico is an extensional basin filled with deltaic, lacustrine and evaporitic sediments, with a basaltic intrusion at depth (Bird *et al.*, 1984; Elders *et al.*, 1984; Schiffman *et al.*, 1984). Several mineralogical zones have been identified with increasing depth and temperature in these geothermal systems (McDowell & Elders, 1980; Schiffman *et al.*, 1984; Cho *et al.*,

1988). A shallow zone characterized by the presence of montmorillonite or illite/smectite ± carbonates or kaolinite occurs at temperatures <150–190°C, and the proportion of illite layers in illite/smectite increases with depth and temperature. This interval is underlain by a higher temperature zone characterized by chlorite ± calcite or illite, which gives way downward to biotite and clinopyroxene zones in the Salton Sea geothermal system (McDowell & Elders, 1980; Cho *et al.*, 1988), or to biotite and calcsilicate zones at Cerro Prieto (Schiffman *et al.*, 1984).

As in basic volcanic rocks, permeability is an important control on metamorphic reactions in the sedimentary systems. Mineral isograds in the shales are at systematically greater depths and temperatures than in the sandstones, as the result of kinetic inhibition of metamorphic mineral reactions in the less permeable shales (McDowell & Elders, 1980; Bird *et al.*, 1984; Elders *et al.*, 1984; Cho *et al.*, 1988).

The compositions of bulk rocks are important influences on mineralogy and mineral compositions. Illite/smectite is present in the sedimentary systems rather than the chlorite/smectite in basic volcanic rocks, but in both systems the proportion of smectite decreases with increasing temperature and depth. In the Salton Sea, sandstones have different bulk compositions than shales, leading to different mineral assemblages in the two rock types. Systematic variations in the composition of epidote at Cerro Prieto are associated with the appearance or disappearance of other phases (e.g. increasing Fe content downward through the prehnite zone, decreasing Fe contents downward through the biotite zone; Schiffman *et al.*, 1984).

Variations in f_{O_2}, f_{S_2}, pH and CO_2 can also lead to significant differences in mineralogy and mineral compositions within geothermal systems. In contrast to the Salton Sea and the basaltic systems discussed previously, calcsilicates are dominant and phyllosilicates are minor at Cerro Prieto. Figure 6.21 shows that the pH and P_{CO_2} of fluids are important controls on the metamorphic mineralogy in hydrothermal systems. As the CO_2 content of the fluid increases, calcsilicates become unstable and calcite forms rather than the calcsilicates. Boiling causes partitioning of CO_2 and H_2S into the vapour phase, resulting in an increase in pH and the precipitation of minerals such as K-feldspar, quartz and calcite (Fig. 6.21).

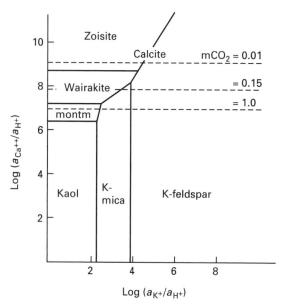

Fig. 6.21 Activity diagram illustrating stabilities of potassium and calcium minerals at 260°C in the presence of quartz. Increasing mCO_2 results in an increase in the stability field of calcite at the expense of calcsilicates (after Browne, 1978). Boiling causes an increase in pH, resulting in the precipitation of K-feldspar and calcite.

The Salton Sea geothermal system has a high f_{O_2}, 4–5 orders of magnitude greater than that at Cerro Prieto (Bird *et al.*, 1984). This leads to high Fe^{3+}/Al and the formation of hematite + Fe-rich epidote in the Salton Sea system, whereas lower Fe^{3+}/Al in Cerro Prieto results in the stability of wairakite + prehnite + Fe-poor epidote (Cho *et al.*, 1988). Although minerals and fluids are close to equilibrium in these systems at 300°C, slight variations in f_{O_2}, f_{S_2}, pH and CO_2 can account for the variations in epidote compositions and multiple vein mineral assemblages that are not in equilibrium (Bird *et al.*, 1984; Caruso *et al.*, 1988). The Salton Sea system is characterized by a deep hot brine overlain by less saline waters, and vertical migration of this brine interface may result in varying mineralogy and mineral compositions.

6.12 Rhyolitic volcanic-hosted geothermal systems

Hydrothermal alteration in geothermal systems associated with rhyolitic volcanic rocks is well documented (e.g. reviews in Browne, 1978; Henley & Ellis, 1983). As in the basic volcanic and sedimentary hydrothermal systems, temperature is the main control on secondary mineralogy, but rock composition, texture and permeability, as well as fluid composition are important influences. These systems consist of recent volcanics and sediments, with an intrusion at several kilometres depth driving circulation of meteoric fluids and with temperatures reaching >300°C at depths of a few kilometres. Boiling of geothermal fluids occurs at some depth, with CO_2 and H_2S partitioning into the vapour causing an increase in pH and deposition of K-feldspar, quartz and calcite. At the surface, the steam condenses or heats cooler ground waters and H_2S is oxidized, leading to acid-sulphate waters and corresponding alteration to low-pH assemblages (alunite, kaolinite, gypsum, opal, Fe-oxides). Similar alteration effects occur in subaerial hydrothermal systems in basic volcanic rocks (e.g. Japan). The subsurface of geothermal systems in silicic volcanic rocks is characterized by more or less horizontally stratified permeable and non-permeable units, cut by subvertical fractures and faults that act as fluid channels (Lonker *et al.*, 1990).

Steiner (1968) documented the depth and temperature zonation of secondary minerals in the rhyolitic system at Wairaki, New Zealand, where montmorillonite and mordenite occur at temperatures of 100–120°C. At increasing temperature illite/smectite occurs, with the proportion of smectite decreasing until only illite is present at 230°C. Mordenite disappears and laumontite occurs from 150° to 200°C and wairakite is present at higher temperatures. In contrast to Wairaki, at the nearby Broadlands–Ohaaki geothermal area high P_{CO_2} inhibited the formation of calcsilicates and calcite is abundant instead (Fig. 6.21; Henley & Ellis, 1983; Lonker *et al.*, 1990). Less permeable units more closely reflect interactions of the rocks with deep higher temperature fluids, whereas the more permeable flow horizons are affected by boiling and changes in fluid compositions (Lonker *et al.*, 1990). The former contain illite and rare K-feldspar, whereas the latter contain abundant K-feldspar and calcite, with only minor illite. Variations in the compositions of the phyllosilicates also occur between the permeable flow zones and more impermeable surrounding strata (e.g. lower $Fe/(Fe + Mg)$ and Al in chlorite and decreased $Fe/(Fe + Mg)$ and K in illite in the flow zones).

6.13 Summary and conclusions

Very low-grade to low-grade hydrothermal metamorphism of ocean crust occurs in the upper portions of active axial hydrothermal systems and throughout passive, off-axis systems, which occur mainly in the uppermost (volcanic) part of the crust. Multiple alteration stages are attributed to specific conditions and circulation regimes, which can continue off-axis for up to several tens of million years.

Early alteration of the upper volcanic section is characterized by celadonitic phyllosilicates + Fe-oxyhydroxides from distal or mixed, low-temperature hydrothermal fluids. Subsequent circulation of seawater results in limited recrystallization and the formation of saponite and Fe-oxyhydroxides at temperatures $< 50–100°C$. Locally greater seawater fluxes can lead to the breakdown of plagioclase and formation of more aluminous smectites. Seawater fluids evolve with time and depth in the crust, leading to more restricted circulation and more reducing fluids and the formation of saponite and pyrite throughout the volcanic section. Depending on the age of crust and the temperatures (up to 150°C in young crust) chlorite/smectite may also be present. Continued fluid evolution (depletion of Mg, increased pH) leads to formation of late zeolites and carbonates in veins and rocks, at temperatures of near zero up to about 70°C. Carbonates may continue to form for up to tens of million years, leading to significant uptake of CO_2 by the upper crust.

At the base of the volcanic section there is a steep transition from very low- to low-grade hydrothermal metamorphism over approximately 100 m, where the proportion of chlorite in chlorite/smectite increases, titanite and laumontite appear and the abundance of quartz increases. Rocks intensively altered to chlorite, albite, epidote, quartz and titanite $\pm$ actinolite appear abruptly and were hydrothermally altered in a mixing zone between axial hydrothermal fluids in the dykes and seawater fluids in the overlying volcanic section at temperatures of 250–350°C. The depth of this boundary varies, however, depending upon the local permeability structure. Hydrothermal metamorphism in the upper dykes is heterogeneous, with more intense recrystallization in areas of fracturing and greater fluid fluxes. Cooling of the system and evolution of fluid compositions are recorded in vein sequences, from early Mg metasomatism and formation of chlorite and actinolite, to later quartz + epidote $\pm$ sulphides from Mg-depleted fluids, to local later zeolites (laumontite and scolecite) and minor prehnite from highly evolved, lower temperature fluids. Local anhydrite reflects recharge of seawater into the system.

Fluids become more rock dominated as they penetrate deeper in the axial system and leach metals and sulphur from what is termed the deep 'reaction zone' in the lower dykes and uppermost gabbros at temperatures $> 350°C$. Initial hydrothermal metamorphism is localized along fractures, at amphibolite-grade conditions (hornblende + plagioclase), with superimposed pervasive retrograde alteration. Hydrothermal fluids rapidly ascend to vent at the seafloor and form massive sulphide deposits, or to mix with seawater in the subsurface. Stockwork feeder zones beneath sulphide deposits are variably silicified and hydrothermally altered to chloritic and illitic or paragonitic assemblages (+ quartz and pyrite).

Comparable hydrothermal processes occur in back-arc basin rifts, in some ophiolites (e.g. Troodos) and in some oceanic forearc areas. Oceanic islands and island arcs generally exhibit similar mineral zonations, but display more continuous thermal and metamorphic gradients than the stepwise gradients seen in ocean crust. There is some evidence in ocean islands and island arcs for temporal sequences of mineral formation and for fluid evolution as seen in ocean crust, but these processes are not as well documented as for ocean crust. Boiling and steam condensation can lead to local differences in these systems relative to the submarine systems.

The steepness and continuity of metamorphic gradients has been linked to the extent of rifting, where mid-ocean ridges and back-arc basins are at the extreme of rifting with steep, discontinuous or step-like gradients, rifted arcs are intermediate and nonrifted island arcs and seamounts are at the other extreme of shallow, continuous thermal gradients (contrast Figs 6.6 and 6.20 with Figs 6.17 and 6.19; Beiersdorfer, 1993; Schiffman & Staudigel, 1994). This is reasonable, as it is 100% crustal extension that produces a sheeted dyke complex and the volcanic layer of oceanic crust: this lithological structure in turn controls permeability and results in a steep thermal gradient and the resultant alteration boundary that coincides with the lithological boundary. In contrast, central volcanoes are fixed in space allowing much longer durations of high-temperature hydrothermal systems than at mid-ocean ridges, where crust is continually moved out of the high-temperature axial hydrothermal cell. This situation may be complicated, however, by effects such as rapid burial and insulation by sediments leading to high temperatures but shallower thermal gradients at sediment-covered spreading centres.

Differences in metamorphic gradients and the structures of hydrothermal systems may account for the lack of pumpellyite in oceanic systems (and Iceland), in contrast to its common occurrence in many ophiolites and other ocean islands (Liou *et al.*, 1987; Schiffman *et al.*, 1991). Other factors include fluid compositions (Mevel, 1981; Harper, 1995) and pressure, which may be generally too low for stability of pumpellyite in oceanic systems, resulting in the formation of common minor prehnite instead (Frey *et al.*, 1991).

The depth zonation of zeolites in hydrothermal systems in basaltic rocks is generally consistent with experimental data and observations on other systems (Liou *et al.*, 1987, 1991), but in submarine systems zeolites are minor secondary phases and form from late-stage, evolved fluids, so the application of the term 'zeolite facies' to these rocks is problematic. The observed depth zonations of phyllosilicates and calcsilicates in various systems are primarily a function of temperature, but fluid and rock compositions, permeability and rock mineralogy and texture are also important. Variations in fluid fluxes and the intensity of fluid–rock interaction are particularly important for phyllosilicate mineralogy. With increasing temperatures and depths smectite transforms to chlorite via increasing chlorite layers in random smectite/chlorite, then corrensite forms and finally chlorite appears, in some cases with significant overlap between mineral zones (e.g. Figs 6.17, 6.19 and 6.20; Liou *et al.*, 1985b; Inoue, 1987; Inoue & Utada, 1991; Meunier *et al.*, 1991; Schiffman & Fridleifsson, 1991; Shau & Peacor, 1992). Where fluid fluxes are low, however, kinetic effects are important and smectite or random smectite/chlorite may form rather than corrensite or chlorite. This

sort of heterogeneity of metamorphism and kinetic effects is particularly common in oceanic crust (e.g. Fig. 6.6).

Rocks in the various types of hydrothermal systems discussed in this chapter are typically incompletely recrystallized and superimposed hydrothermal stages or events and evolving or fluctuating fluid compositions lead to multiple generations and varying compositions of secondary minerals. Thus, equilibrium is attained only locally and the application of petrogenetic grids and metamorphic facies terminology to hydrothermally metamorphosed rocks must be made with suitable caution (Liou *et al.*, 1987; Frey *et al.*, 1991). Mineral–fluid equilibrium is a more important consideration than bulk rock equilibrium.

In subaerial geothermal systems in andesitic volcanoes, acid volcanic terranes and sedimentary basins, P_{CO_2} clearly can be an important control on mineralogy, with high P_{CO_2} resulting in formation of calcite rather than calcsilicates. This process does not generally appear to be important in hydrothermal systems in oceanic crust, oceanic islands or back-arc basins, although off-axis fluids percolating through sediments dissolve carbonates, contributing to the formation of late carbonates in uppermost ocean crust. Because of the high pressures caused by overlying seawater, boiling is generally not important in seafloor hydrothermal systems, but plays a role in some areas of Iceland and Japan and is significant in subaerial geothermal areas in various settings. Phase separation at supercritical conditions, however, is an important control on the salinity and volatile contents of submarine hydrothermal fluids.

One area of future research is to further understand and quantify the effects of porosity and permeability on metamorphic grade. Linked to this is the quantification of fluid fluxes during hydrothermal metamorphism and how this affects metamorphic reactions and grade. Estimating the temperatures of very low-grade metamorphism is also commonly problematic, particularly in the interesting range from 100° to 300°C where these kinetic controls are important. Dating the formation of hydrothermal minerals would be a great advance, enabling quantification of the temporal evolution of temperatures and the chemistry of seafloor and subaerial hydro-thermal systems.

The temporal variations in submarine hydrothermal systems are beginning to be documented (e.g. Von

Damm *et al.*, 1995), but this is far from being well understood: how exactly do intrusion, dyking and eruption events disrupt the permeability, temperature and flow structure of hydrothermal circulation, and how does this affect fluid compositions and fluid/rock interactions at depth? Predictive models, based on vent fluids and experimental and theoretical considerations, for minerals in deep 'reaction zones' in the roots of submarine hydrothermal systems have not been fully resolved with observations of rocks from this portion of the crust. Moreover, it is still not clear how high-temperature hydrothermal reactions (up to 700°C) in the plutonic section relate to hydrothermal fluids that vent at the seafloor.

The global magnitudes of the heat, fluid and chemical fluxes at different temperatures in seafloor systems remain controversial. In particular, understanding the partitioning of these fluxes between active, high-temperature systems at ridge axes and the lower temperature, passive systems off-axis is critical because of the many offsetting chemical effects. Moreover, we still do not know the proportions of fluid and heat that are transported in the lower and higher temperature portions of active, axial hydrothermal systems (e.g. Fig. 6.5).

7 Isotopic dating of very low-grade metasedimentary and metavolcanic rocks: techniques and methods

N. Clauer and S. Chaudhuri

7.1 Introduction

The radiogenic isotope systematics of very low-grade metamorphic rocks are often complex because of the presence of minerals of different origins that may be intimately mixed: those from original protoliths as well as those newly formed. The systematics may even be more complex in metasedimentary rocks than in metavolcanic rocks, because of the potential multifarious characteristics of the detrital minerals, especially when they have similar grain sizes. As routine preparation and separation techniques seldom give pure mineral fractions in the micron size, the individual isotopic signatures of fine-grained components are often difficult to identify. While the isotope systematics of the authigenic minerals may also be very sensitive to discrete thermodynamic modification, which can modify the isotopic signature of the rocks, the extent of the related isotopic modifications will depend on the fluid access to, and movement in the host rocks, both being controlled by the rock porosity and permeability.

The widely varied mineralogical and textural characteristics of metasedimentary rocks are especially critical in transitional domains such as in that of very low-grade metamorphism. Although in many situations of low-grade metamorphism, the tendency is towards a simplification of the mineral assemblages and parageneses (e.g. Dunoyer de Segonzac, 1969; Velde, 1985), and an increase in the size of the clay to mica particles (e.g. Weaver & Broekstra, 1984), the formation of authigenic mica-type sheetsilicates is attendant with complex geochemical and isotopic signatures. In fact, the true mineral indicators of very low-grade conditions are those precipitating directly from fluids, or those deriving from complete reconstitution of detrital and diagenetic minerals of the rocks being metamorphosed. The isotope

systematics of these authigenic minerals, however, depend also on the conditions at which they formed. A dissolution–precipitation process in a porous environment, such as in sandstones, has open-system potentials which allow supply of the chemical elements that are needed for the mineralogical adjustment to the metamorphic conditions, and release of the inherited radiogenic isotopes. By contrast, a transformation process in a restricted environment, such as in shales, may reuse all or most of the available chemical elements, including the inherited radiogenic isotopes. With such a process, only limited input or release of elements and release of inherited isotopes may take place.

In the last two decades or so, several isotope-oriented studies have provided detailed information about progressive thermally induced transition from mixed-layer illite/smectite to illite–mica; some of these crystallo-chemical changes having been induced by overburden (Perry & Turekian, 1974; Aronson & Hower, 1976; Hunziker et al., 1986; Reuter, 1987; Awwiller, 1993; Ohr et al., 1994; Schaltegger et al., 1994; Clauer et al., 1995). Among these studies, that on the illite-to-mica transition in a shale-to-slate rock sequence from the Glarus Alps in Switzerland, reported

1 a change from $1M_d$ to 2M polytype with increase in K_2O content;

2 a change in the morphology from irregular to euhedral particles;

3 a decrease in the K–Ar and the Rb–Sr dates towards a concordance at epizone grade; and

4 a crystallo-chemical reorganization of the particles without oxygen exchange, suggesting complete recrystallization at $260 \pm 30°C$ (Hunziker et al., 1986).

This reorganization of illite into mica apparently caused continued repartition of Ar and K in the mineral lattice, as can be seen in the progressively changing form of the $^{40}Ar/^{39}Ar$ data patterns (Fig. 7.1).

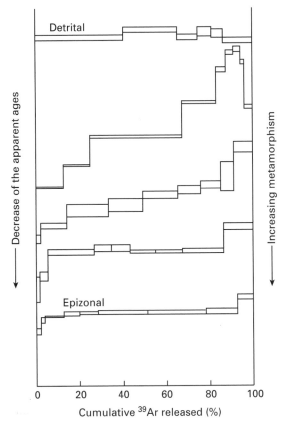

Fig. 7.1 Evolution of the $^{40}Ar/^{39}Ar$ patterns from a detrital illite (top) to an authigenic mica (bottom) along the illite–mica transition in the Glarus region of the Alps. (After Hunziker *et al.*, 1986.)

Initially, the detrital illite displays a plateau with a detrital apparent age. This plateau, that is indicative of a fairly homogeneous repartition of the initial Ar isotopes throughout the mineral particles, is followed, when the reorganization progresses, by 'staircase' patterns which are typical of mixing of inhomogeneous mineral grains or inhomogeneous parts within the particles. The staircase patterns progressively flatten before epizonal grade, which is again indicative of a progressive homogeneous repartition of the Ar isotopes in the particles, but not necessarily complete removal of the inherited Ar isotopic signature. The complete removal of the inherited isotopic signature seems to be only effective at epizonal grade, where K–Ar, Rb–Sr and $^{40}Ar/^{39}Ar$ ages are identical.

Volcaniclastic rocks may also be potentially valuable for dating low-grade metamorphic processes, because they are basically free of detrital components. Differences in the isotopic systematics between mica-type minerals of metapelites and metavolcanics were documented in a study of such rocks in the Rheinische Schiefergebirge of Germany (Reuter, 1987; Reuter & Dallmeyer, 1989). Mica separates of the metatuffs with sizes from 0.6 to 20 µm, yielded fairly constant K–Ar and $^{40}Ar/^{39}Ar$ ages. In contrast, isotopic dates for the same size fractions of the associated metapelites displayed positive correlations with grain size, illustrating the influence of detrital precursors. Volcaniclastic rocks, however, are much less common than sedimentary rocks, and they even may not be as well suited for dating paleothermal events. In thick tuff beds (> 1 m), illitization can be retarded by the need for potassium diffusion into the bed and the obtained age may then be younger relative to any thermal event (Altaner *et al.*, 1984; Clauer *et al.*, 1997). Also in bentonite units, whole rock dating is not suitable, as any primary biotite may yield an age characteristic of deposition, while the illite component may yield an age that corresponds to a late thermal overprint (Kunk & Sutter, 1984; Elliott & Aronson, 1987; Toulkeridis *et al.*, in press).

The above studies offer opportunities for further understanding of the behaviour of the isotope systematics in illite- and mica-type minerals of meta-sedimentary and metavolcanic rocks. The purpose of the present contribution is to examine in detail three major constraints in isotopic dating of very low-grade mineral indicators, namely (i) the importance of the mineral separation and characterization for consistent interpretation of isotopic results; (ii) the specific aspects of isotopic dating of mica-type fractions; and (iii) the potential mechanisms of radiogenic isotope homogenization induced by very low-grade metamorphic conditions in different rock lithologies. The aim is not only to review critically available studies in the literature, but to present a coherent model of mineral isotope systematics at very low-grade metamorphic conditions, with a guide to evaluate reliable data. Fission-track dating, which is not a basic isotopic dating technique, will also be examined because it represents a powerful methodology, especially when combined with K–Ar dating of authigenic illite, for deciphering the thermal evolution of slightly metamorphosed rocks (Roden *et al.*, 1993).

7.2 Importance of mineral separation and characterization

The small size of most very low-grade mineral indicators, such as the sheetsilicates, is a major hindrance to the separation and characterization of single mineral generations. Although progress has been made in respect of related analytical techniques in recent years, concerns still remain because inappropriate separation has the potential to cause significant biases in isotopic data of mica-type materials. Other minerals, such as salts (Brookins, 1980; Baadsgaard, 1987) and sulphates (Shanin *et al.*, 1968; Blanco *et al.*, 1982), may also be isotopically dated, but they raise related problems for separation or for characterization as do the sheetsilicates. This technical aspect of separation and characterization may be easier to solve with sanidine-type feldspars of volcanic origin or with chlorite particles formed in the cleavage planes of deformed rocks (Fitch *et al.*, 1969), because these minerals can be found as larger grains or flakes.

7.2.1 Separation

Separation of monomineralic fine-grained sheet-silicates is almost impossible with routine techniques, because of the very small size of the particles. Routine methods are also lacking for high-quality size fractionation of fine-grained material. Granulometric separations are still the most widely used, including gravity separation in distilled water followed by ultracentrifugation. These grain-size separations can now be improved by high-grade magnetic separations which allow mineral separations in subfractions: nonmagnetic minerals can be well separated from magnetic ones (Tellier *et al.*, 1988; Hillier, 1993a). Such magnetic separations have the potential to improve the quality of the grain-size separation, as authigenic non-magnetic illite may be separated from detrital more-magnetic phengite, biotite or chlorite that have the same particle size. It is inherent to the routine separation techniques that the greater the size of the separated clay fractions, the higher the chances for the presence of 'contaminant' minerals which will bias the isotopic results by increasing or decreasing the ages of the authigenic clay or mica particles.

The potential contamination of any size fraction, which is especially crucial in the case of metasedi-mentary rocks, is increased by the use of conventional crushing techniques that lead to uncontrolled size reduction of the framework grains of the rocks, mixing artificially produced clay-sized materials with the natural clay- to mica-type components. Thus, clay-type material, or other minerals from crushed samples, may be easily contaminated by fine fragments of naturally coarser mineral grains, such as framework feldspar, muscovite and biotite. The degree of contamination through crushing was examined by Liewig *et al.* (1987) who compared K–Ar data of clay fractions from identical sandstones after crushing and freezing–thawing disaggregation.

The age bias introduced by framework grains in sheetsilicate fractions has also been documented in hydrothermally affected mica-type fractions of metasedimentary rocks (N. Clauer & J. Mendez Santizo, unpublished data). Subfractions taken from several samples along a 20-m-long transect across a major fault system in the Lodève uranium district in France, yielded a large range of K–Ar dates depending on the distance to the fault and on the grain size of the fractions (Table 7.1). X-ray diffraction (XRD) of size fractions obtained by ultracentrifugation after crushing, showed no contaminant but the Na_2O contents were abnormally high for pure illite fractions. Such anomalous Na_2O contents suggested the occurrence of minute amounts of feldspar which, obviously, were below the XRD detection limit. This possibility was checked by dating pure feldspar splits of the rocks and by calculating the feldspar amounts in the size fractions on the basis of the Na_2O contents. The feldspar fraction being older than the illite fraction, the age calculations induced systematic decreases in the measured K–Ar dates of the fractions. The same samples were then disaggregated by freezing and thawing, several subfractions were extracted and the finest were K–Ar dated. No Na_2O was detected in the subfractions after the freezing–thawing disaggregation and the new K–Ar results were found to be identical to those previously calculated after mass–balance determination of the contaminant feldspar contents in the illite fractions (Table 7.1). The fact that K–Ar ages of mineral fractions may clearly depend on the preparation technique, emphasizes the importance of using careful separations, as well as the importance of sample characterizations having an accuracy that goes beyond the potential of XRD.

Table 7.1 Comparison of K–Ar data on sheetsilicate and feldspar fractions after crushing and freezing–thawing disaggregations. (N. Clauer & J. Mendez Santizo, unpublished data).

Samples	Size fractions (µm)	Na$_2$O (%)	K$_2$O (%)	^{40}Ar (%)	Rad. ^{40}Ar (10^{-6}cm^3 g^{-1})	Measured t (Ma $\pm$ 2σ)	Calculated t (Ma)
Crushed samples							
LDET	<0.4	0.29	6.43	78.07	30.24	140 ± 4	128
LD11	<0.4	0.22	6.10	76.97	22.31	110 ± 3	100
LD22	<0.4	0.24	7.09	90.15	60.53	247 ± 6	238
Feldspars							
2948	80–100	n.d.	4.82	45.22	47.71	284 ± 13	
	100–125	n.d.	5.45	52.43	53.38	281 ± 12	
2949	125–200	n.d.	9.56	85.32	93.67	281 ± 8	
Disaggregated samples							
LDET	<0.2	b.d.l.	6.36	71.65	28.07	132 ± 5	
LD11	<0.2	b.d.l.	6.04	58.84	20.67	103 ± 4	
LD22	<0.2	b.d.l.	7.05	85.99	55.95	231 ± 8	

n.d. stands for not determined and b.d.l. for below detection limit. For the calculated t, see text.

The presence of other mineral components in sheetsilicate size fractions does not, of course, automatically mean that the isotopic ages will be biased: feldspar, for instance, can also form in very low-grade metamorphic conditions. Age biases occur only if the fine contaminants are representative of coarser detrital minerals, and the potential for incorporating such ground detrital grains is greater when analyzing whole rocks or coarse-size fractions, such as bulk <2-µm fractions, instead of 10 times smaller subfractions for instance. From the analytical point of view, systematic separation of very fine clay fractions (in the tenth of a micron range) has the greatest potential for providing ages that are the closest to the periods of crystal precipitation and growth.

7.2.2 Characterization

Probably much more critical than for other minerals is the characterization of sheetsilicate minerals. It represents, together with the separation, a 'fundamental' step in the proper selection of size fractions for isotopic analysis. The purpose is not to detail here techniques that have already been reported elsewhere and will be reviewed in the present volume. Clay-to-mica fractions can be appropriately identified (Kisch, 1983) and even quantified in separated fractions by XRD (e.g. Brindley & Brown, 1984; Moore & Reynolds, 1989). Additional par-

ameters such as the illite crystallinity index (Kübler, 1966) and polymorphic typology (Bailey, 1988), which are also obtained by X-ray analysis, represent basic information in isotopic dating of mica-type materials. High crystallinity indices and mixtures of several types of polymorphic forms suggest incomplete isotopic homogenization of illite and the isotopic values may reflect this inhomogeneity. As most of the conversion from smectite- to illite-type minerals occurs in the diagenetic domain and during the transition into low-grade metamorphism, techniques such as deconvolution (e.g. Lanson & Besson, 1992) or simulation (Reynolds, 1985) of XRD patterns and precise identification of XRD peak location (Środoń, 1984), provide means that are well suited for the best identification of the mineral phases present in the separated fractions, and evaluation of the extent to which the smectite to illite conversion proceeded. XRD is also a helpful means to identify potential contaminants in the size fractions and it may allow their quantification. The study discussed above, however, with sheetsilicates yielding anomalous Na$_2$O contents, indicates that mineral phases amounting to less than *c.* 3% of the analyzed fractions, may not be detected by XRD: undetected detrital constituents by XRD may simply be below the detection limit of the technique.

For isotopic dating of very low-grade metamorphic rocks, it is also worth examining the morphology and typology of the mica-type particles by scanning electron microscopy (SEM), and to compare them to the XRD data. Many studies on metasedimentary rocks have shown, for instance, that the cleavage dissects the rocks and separates microlithons of different sizes. The original detrital micas seem to be preserved in the microlithons areas, whereas those in the cleavage fabric are mechanically rotated into the cleavage planes where new crystals of mica-type are also growing (e.g. McPowell, 1979). As cleavage development is often contemporaneous with thermal pulses, it can be a valuable qualitative indicator of the amount of strain developed in very low-grade metamorphic rocks. Dating the newly formed sheetsilicates can, thus, give an insight into recrystallization dynamics of these minerals subjected to combined effects of heat and deformation. This, however, is only possible if the small sheetsilicates of the cleavage can be separated without being contaminated by either the micas of the microlithons or those of the cleavage planes, which will need careful SEM observations of rock chips and size-fraction separates. Alternative approaches to date cleavage formation in metasediments were made by applying the $^{40}Ar/^{39}Ar$ methodology on whole rock slates and phyllites (Wright & Dallmeyer, 1991; Dallmeyer & Wright, 1992) or the Rb–Sr methodology on distinct petrographic domains of folds (Turner et al., 1994). In the former case, the whole rock incremental $^{40}Ar/^{39}Ar$ analyses are believed to document a geographic diachronism in cleavage development during orogeny, that allowed calculation of convergence rates. In the latter case, phyllosilicate-rich and quartz-rich domains of folds were dated by the Rb–Sr method, and gave ages older than that of pervasive deformation that were related to thermal pulses linked to intrusive plutonic rocks. This age difference was thought to relate either to propagation of the deformation across the fold belt, or to its episodic occurrence. It seems, however, that such promising applications may need more case studies as they are controversial (Findlay, 1992; Wright & Dallmeyer, 1992; Preiss, 1995; Turner et al., 1995). It may also be mentioned here that timing of uplift and cooling of metasediments can be evaluated by combined K–Ar and fission-track datings of schists (Adams & Robinson, 1993).

Transmission electron microscopy (TEM) is useful for the observation of particle shapes and verification of the presence or absence of even small amounts of contaminants. It is known that straight particle edges are typical for authigenic sheetsilicates, whereas irregular edges are typical for detrital particles variously affected by dissolution processes (Hunziker et al., 1987). TEM observations may, therefore, provide definitive information on paragenetic relationships in heterogeneous mineral assemblages.

7.2.3 Summary

The quality of the separation and characterization of clay- to mica-type fractions determines the quality and reliability of their isotopic dating. Separation of < 0.2 μm-sized mica-type materials is recommended, because such fine-grained aggregates may be devoid of, or contain negligible amounts of contaminants. What will be shown later is that isotopic analyses of several size fractions for any sample, with an upper grain size limit set at 6 μm, provide an instructive control of the reliability of the entire set of isotopic dates. This procedure constrains the crystallization time if identical ages are obtained on several size fractions of the same sample; it also gives information about the extent of isotopic homogenization by indicating the size fraction at which occurrence of detrital grains begins to bias the data.

7.3 Principles of isotope dating

Radiogenic isotope geochemistry is well known for high-temperature minerals, and has been reviewed in several publications (e.g. Faure, 1986; DePaolo, 1988; McDougall & Harrison, 1988). In contrast, radiogenic isotope geochemistry of clay-type minerals has only progressed substantially in recent years, as a result of advances in understanding the specific mineralogical characters, analytical techniques and isotopic methodologies (Clauer & Chaudhuri, 1995). The radiogenic isotope geochemistry of clay-sized material occurring at the surface and subsurface of the Earth is complex, mainly because these materials have multifarious origins and stabilities when they are of sedimentary origin. Thus, they will retain their isotopic signatures to varied degrees under given circumstances. A brief review of the principles of isotopic dating is provided here to give an understanding of how various radiogenic isotope systems (Rb–Sr, K–Ar, $^{40}Ar/^{39}Ar$, Sm–Nd and U–Th–

Pb) behave in minerals and whole rocks, and to emphasize some specific aspects related to their applications to fine-grained sheetsilicates at very low-grade metamorphic conditions.

7.3.1 Methodology of isotopic dating

The spontaneous decay of an unstable nucleus into a stable configuration is called radioactivity. The decay of radioactive (= parent) nuclei is accompanied by particle and positron emissions, and by heat release. In 'Isotope Geology', the interest is focused on naturally occurring radioactive isotopes whose half-lives are comparable to the age of the Earth. The rate of decay of a radioactive nucleus is proportional to the number of its atoms (N) present at a given moment. It is expressed by:

$$-dN/dt = \lambda N,$$

where λ is the decay constant. Integration of this equation gives:

$$t = 1/\lambda \ln (1 + D/N),$$

where D is the number of radiogenic (= daughter) isotopes formed at a time t. This equation may give meaningful ages for geological materials provided they behave, since formation, as a closed system to both the radioactive and the radiogenic isotopes.

7.3.1.1 The rubidium–strontium method

The Rb–Sr method is based on the natural decay of ^{87}Rb to ^{87}Sr along the following basic equation for age determination:

$$(^{87}Sr/^{86}Sr)_{total} = (^{87}Sr/^{86}Sr)_{initial} + (^{87}Rb/^{86}Sr) (e^{\lambda t} - 1).$$

The initial ^{87}Sr/^{86}Sr ratio needs to be known precisely to determine Rb–Sr ages of terrestrial materials. Because this ratio is not easily obtained, the Rb–Sr 'isochron' dating method plotting ^{87}Sr/^{86}Sr and ^{87}Rb/^{86}Sr ratios in a linear regression, is generally used. Cogenetic samples with the same initial ^{87}Sr/^{86}Sr ratio but with different ^{87}Rb/^{86}Sr ratios, and which evolved in a closed system with time, plot on a linear regression. Such a regression is an isochron, but any line in a diagram is not automatically an isochron, as it can arise from mixing of varied amounts of mineral components that are not isotopically homogeneous. In such cases, the slope of the line has no geological meaning and it should not be identified as an isochron.

7.3.1.2 The potassium–argon method

The K–Ar method is based on the natural decay of ^{40}K to ^{40}Ar and is widely applied to almost any type of K-bearing mineral and rock. As required for all isotopic methods of dating, a closed system is essential. This requirement is especially crucial for this method, because Ar is a noble gas that is not bonded to other atoms or ions in a mineral, which makes it particularly susceptible to loss when temperature increases. The equation of the K–Ar age is:

$$t = 1/\lambda\ (^{40}Ar^*\lambda_{total}/^{40}K\ \lambda_{capt.\ elec.}) + 1.$$

The value of the initial ^{40}Ar/^{36}Ar ratio during crystallization of a mineral is often assumed to be identical to that of the present-day atmosphere at 295.5 (Nier, 1950). In general, the amount of initial Ar is very low in K-rich minerals and, hence, the uncertainty in the value of the initial Ar isotopic ratio has a minor-to-negligible influence on the age calculation of old to very old rocks or minerals (>100 Ma). The initial ^{40}Ar/^{36}Ar ratio can, sometimes, be higher than 295.5, but the consequences of such values can be corrected using the isochron technique. Part of the non-radiogenic Ar in any mineral can be due to contamination from present-day atmospheric Ar that is adsorbed by the mineral particles during preparation and purification of the samples. Adsorption of such atmospheric Ar decreases the accuracy of the data, but its effect can be reduced by preheating the minerals at 80–100°C for several hours under vacuum before Ar extraction.

The isochron approach is not needed for most K–Ar isotope applications. Some studies, however, have reported isochrons with abnormally low or even negative intercept values (Langley, 1978). Clearly, minerals or rocks cannot yield negative initial ^{40}Ar/^{36}Ar ratios and lines with such intercept values in isochron diagrams can only be mixing lines with meaningless slopes. Harper (1970) presented theoretical examples with negative intercepts of ^{40}Ar, suggesting that such values may result from ^{40}Ar loss from cogenetic mineral phases irrespective of the K contents. A more likely cause for a line to yield a negative intercept value in a ^{40}Ar vs. ^{40}K diagram, is mixing of inhomogeneous materials, or differential loss of ^{40}Ar by minerals during recrystallization (Clauer & Chaudhuri, 1995).

7.3.1.3 The ^{40}argon/^{39}argon method

The ^{40}Ar/^{39}Ar technique is a variation of the ^{40}K/^{39}Ar (K–Ar) technique (Merrihue & Turner, 1966), but it has so far been of limited application in studies of very low-grade metamorphic rocks. The principle is that ^{39}K decays to ^{39}Ar by irradiation under fast neutrons in a reactor. The technique has the theoretical advantage that measurement of both the radioactive and radiogenic isotopes is made from the same sample aliquot. This avoids the uncertainty from sample heterogeneity, the radioactive and radiogenic isotopes being measured simultaneously by a single isotopic determination of the Ar with an excellent precision by using the most abundant ^{39}K isotope (^{39}K = 93.26%, while ^{40}K represents only 0.01%). The simultaneous measurement of the parent and daughter isotopes using the same powder split also allows analysis by stepwise progressive heating, leading to a theoretical separation of the Ar released at low temperatures from that released at higher temperatures. The analyses, however, are not without inconvenience and uncertainties that include the effect of recoil on the retention of Ar, the knowledge of the flux of neutrons and the corrections for the production of ^{39}Ar from the neutron irradiation of ^{39}Ca. Recoil effects can be especially significant when the particles are small, because the isotopes which are impacted by the neutrons, are moved in the mineral lattices and can even be expelled from the structure. Also, the amount of Ca in the analyzed minerals has to be known precisely.

Reuter and Dallmeyer (1989) documented the recoil effect on the redistribution of ^{39}Ar from high-K-bearing illite-to-mica mixed with low-K-bearing chlorite and albite. This recoil seems to happen preferentially with particles having large surface areas and irregular grain edges, that had not been altered and/or had not undergone significant recrystallization. The ^{40}Ar/^{39}Ar method may, therefore, be preferentially applied to illite and mica particles of sizes in the 10-μm range, and which crystallized in temperature conditions that are above those of

the irradiation in the reactor. To determine precisely the recoil effect, encapsulation of the samples in small quartz vials under vacuum is recommended, as it ensures retention of the recoiled ^{39}Ar which can be determined by breaking the vial in the vacuum system, before heating the aliquot (McConville *et al.*, 1988; Foland *et al.*, 1992; Dong *et al.*, 1995).

An advantage of the ^{40}Ar/^{39}Ar method lies in the potential of *in situ* measurements of grains by coupling the mass spectrometer to a laser microprobe (York *et al.*, 1981; Sutter & Hartung, 1984; Kelley *et al.*, 1994). Application to fine-grained minerals of low-grade metamorphic rocks, however, is not straightforward since neutron activation potentially causes recoil by mean distances that are of the same order as the thickness of the individual particles (Turner & Cadogan, 1974). Some initial results of laser ^{40}Ar/^{39}Ar dating of illite from a uranium deposit in northern Saskatchewan (Canada) indicated that the *in situ* ^{40}Ar/^{39}Ar data are more scattered than the corresponding K–Ar data from extracted clay particles (Bray *et al.*, 1987), which may be due, at least partly, to uncontrolled recoil. On the basis of laser ^{40}Ar/^{39}Ar dating of clay particles from bentonites and shales, Dong *et al.* (1995) argued that ^{39}Ar loss during irradiation was controlled by release from low-retentivity sites of the illite-type particles. This claim does not take into account two analytical problems which have not yet been addressed:

1 the fact that irradiation in a reactor may have a combined effect on clay particles: the neutron activation removing preferentially ^{39}Ar and the ambient temperature in the reactor removing preferentially radiogenic ^{40}Ar ; and
2 the potential effect of the size and power of the laser beam on the mineral grains surrounding the impacted illite particles.

Clearly, a consolidated methodological basis is needed before routine application of the ^{40}Ar/^{39}Ar method on fine-grained very low-grade metamorphosed sheetsilicates, can be expected. Preliminary data of an illite-to-mica transition have shown, for instance, that sheetsilicates of very low-grade metamorphic conditions yield complex age patterns (Fig. 7.1), and a strict control of the laser ^{40}Ar/^{39}Ar dating has still to be investigated.

7.3.1.4 The samarium–neodymium method

Sm and Nd are rare-earth elements (REE) with similar chemical properties due to a similar electronic configuration. Despite close similarities in the behaviour due to 3+ oxidation states, these elements can fractionate depending on their complexing properties with organic and inorganic ligands and on structural properties of the host minerals. The processes controlling mobilization and fractionation of the REE in sedimentary to very low-grade metamorphic environments have been discussed in a number of articles (e.g. review by Chaudhuri et al., 1992).

Sm has two natural radioactive isotopes, ^{147}Sm and ^{148}Sm. Because the half-life of ^{148}Sm is too long for isotopic dating of terrestrial materials, only the decay of ^{147}Sm to ^{143}Nd is used. Because the half-life of ^{147}Sm is also very long, the Sm–Nd method is best suited for dating old materials (> 500 Ma) using the following equation:

$$(^{143}Nd/^{144}Nd)_{total} = (^{143}Nd/^{144}Nd)_{initial} + (^{147}Sm/^{144}Nd)\,(e^{\lambda t} - 1).$$

Linear regressions may also be constructed for cogenetic samples by using the ^{143}Nd/^{144}Nd ratio as the ordinate and the ^{147}Sm/^{144}Nd ratio as the abscissa in an isochron diagram.

Sm–Nd model ages are often used to determine the genetic history of sedimentary materials, in assuming that the Nd isotopic composition of a crustal material derived from a reference source, had a linear growth in the ^{143}Nd/^{144}Nd ratio during the last 4.5 Ga. Calculations of such model ages, however, have not yet provided useful data on very low-grade metamorphic rocks.

7.3.1.5 The uranium–thorium–lead method

Lead is widely distributed in terrestrial materials as the radiogenic product of U and Th, but it also forms minerals without U and Th. Its isotopic compositions can thus vary widely. In fact, both U and Pb often occur as trace elements in common rocks and minerals which may, then, contain three radioactive isotopes of U (^{238}U, ^{235}U and ^{234}U) and one of Th (^{232}Th). The decays of ^{238}U, ^{235}U and ^{232}Th into the stable ^{206}Pb, ^{207}Pb and ^{208}Pb isotopes occur through a long chain of disintegrations, ^{234}U being an intermediate product in the ^{238}U decay chain. The time-dependent growths of the Pb isotopes are given in the following equations:

$$(^{206}Pb/^{204}Pb)_{total} = (^{206}Pb/^{204}Pb)_{initial} + (^{238}U/^{204}Pb)(e^{xt} - 1),$$

$$(^{207}Pb/^{204}Pb)_{total} = (^{207}Pb/^{204}Pb)_{initial} + (^{235}U/^{204}Pb)(e^{yt} - 1),$$

$$(^{208}Pb/^{204}Pb)_{total} = (^{208}Pb/^{204}Pb)_{initial} + (^{232}Th/^{204}Pb)(e^{zt} - 1),$$

where x, y and z are the decay constants of ^{238}U, ^{235}U and ^{232}Th, respectively. The equations for the growths of ^{207}Pb and ^{206}Pb may be combined to express the ^{207}Pb–^{206}Pb equation:

$$(^{207}Pb/^{204}Pb)_{total}/(^{206}Pb/^{204}Pb)_{total} = (^{235}U/^{238}U)(e^{yt} - 1)/(e^{xt} - 1),$$

where the ^{235}U/^{238}U ratio is a constant value of 1/137.88, assuming a normal U isotopic composition.

Dating of rocks and minerals by the U–Pb isotope method often gives highly discordant ages, as a result of differential losses of radiogenic Pb during further evolution. A reliable age may be obtained by the use of the combined ^{207}Pb–^{206}Pb data. The age of a suite of cogenetic minerals with the same initial Pb isotopic composition can then, theoretically, be determined from slope of a Pb–Pb isochron defined by the ^{207}Pb/^{204}Pb and ^{206}Pb/^{204}Pb ratios. Rocks and minerals may also be dated by applying a model integrating a single or a multiple stage evolution of Pb. The single-stage model assumes that the isotopic compositions of common Pb, such as that in Pb-sulphide minerals, evolved in different environments with different U/Pb and Th/Pb ratios from the same primordial isotopic value at 4.5 Ga until the Pb was separated from the different sources to form Pb minerals. The Pb isotopic compositions of different minerals or rocks would then define an isochron in the ^{207}Pb/^{204}Pb and ^{206}Pb/^{204}Pb coordinates, and an age can be calculated from the slope of this line. The two-stage evolution model of Stacey and Kramers (1975) is also frequently used. It assumes that the Pb evolved first from primordial isotopic value between 4.5 and 3.7 Ga, and then remained in a reservoir until the different types of Pb were separated from the different sources. An isochron defined by the ^{207}Pb/^{204}Pb and ^{206}Pb/^{204}Pb ratios relates to the time elapsed since the Pb trapped in a sample was isolated from second reservoir.

Although very low-grade metamorphic carbonate-type rocks can be dated currently by the U–Th–Pb method, real uncertainties still exist as to whether or not clay minerals have the ability to meet the requirement that their U–Pb systems remained closed to both U and Pb since crystallization. Clay minerals are known for high ion-exchange capacities and this inherent property casts doubts on the usefulness of the Pb-isotope method to studies of clay genesis. Pb-isotopic data on sedimentary to metasedimentary mineral phases are still scarce and the available results do not yet allow the full potential to be assessed for an extended application, as the data may be either higher than the Sm–Nd results on the same clay-to-mica fractions or even scattered (e.g. Gauthier-Lafaye *et al.*, 1996).

7.3.2 Some specific aspects and applications

During the pioneering period of isotope dating of clay- to mica-type minerals, the data often turned out to be equivocal, mainly because of inadequate choices of the suitable clay materials and isotopic methods. The clay-type sheetsilicates were often considered to be too small to be separated into pure fractions or paragenetic assemblages. Also, not much consideration was given, at that time, to the crystallo-chemical characteristics of the clays and to their formation conditions. The most common conclusion was that isotopic dating of these minerals was not possible, because of a combination of (i) preferential loss of radiogenic isotopes (mainly radiogenic ^{40}Ar) due to small particle size, which induces reduction of the ages, and (ii) preferential addition of radiogenic isotopes due to the occurrence of detrital contaminants, which induces increases in the ages. In recent years, there has been much improvement in the selection of materials suitable for isotopic dating and also in the applicability of the analytical techniques. This improvement does not, however, avoid completely the inherent problems of dating very small minerals, especially in studies of very low-grade metamorphic sheetsilicates. Mineral separation and characterization will always depend on parameters such as the lithology of the rocks and the typology of the minerals.

To define the isotopic behaviour of authigenic minerals, it is critical to understand both the physical and chemical conditions of the reactions that allow crystallization of the minerals to be dated, whether thermally or kinetically driven. The progress of any mineral reaction depends on the isotopic equilibrium–disequilibrium relations of the sheetsilicate fractions, and the thermokinetic parameters depend on the relative importance of temperature and time for completion of the reaction. This means in turn that there will never be immutable fixed rules for isotopic dating of very low-grade metamorphic rocks. It is also important to know precisely how laboratory treatments may or may not affect the isotopic systematics of the material studied and what improvements they may provide to the analytical data.

7.3.2.1 Behaviour of radiogenic isotopes under natural conditions

Various studies on the effects of natural weathering on isotopic ages of clay- to mica-type minerals were recently reviewed by Clauer and Chaudhuri (1995). Several investigations have addressed the problem of radiogenic isotope retention and they generally concluded that these radiogenic isotopes are not preferentially lost by the clay-to-mica crystals (e.g. Thompson & Hower, 1973; Bath, 1977; Aronson & Douthitt, 1986). However, the fact that radiogenic ^{40}Ar is most likely capable of diffusing from clay particles in given thermal settings, should be kept in mind for any study of very low-grade metamorphic materials. Evaluation of the effects of long lasting high-temperature conditions on clay-to-mica particles can be made by using an Ar-diffusion code (Huon *et al.*, 1993). Theoretical calculations for sheetsilicates of different sizes subjected to 150°C, suggest that a 0.2–2-μm fraction should lose radiogenic ^{40}Ar only if this temperature is maintained for at least 35 Ma, whereas a <0.2-μm fraction should have lost at least 10% of its radiogenic ^{40}Ar after 5 Ma (Fig. 7.2).

The potential loss of radiogenic isotopes from particles due to their small size has been a persistent question since the beginning of isotopic dating of clay- or mica-type minerals. Laboratory studies were conducted to examine potential diffusion of Ar from clay minerals at varied temperatures and pressures. Most concluded that the Ar retention behaviour of illite is similar to that of glauconite and that these minerals begin to lose Ar only at about 200°C, the rate of Ar diffusion is negatively related to the K$_2$O

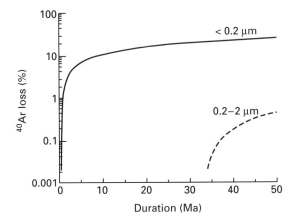

Fig. 7.2 Simulated Ar diffusion from two sheetsilicate size fractions at 150°C as a function of time (N. Clauer & H. Zwingmann, unpublished data).

content and it is much slower under confining water pressure than under vacuum (Odin & Bonhomme, 1982; Zimmermann & Odin, 1982). Continuous significant diffusion of Ar from sheetsilicate minerals under sedimentary to low-grade metamorphic conditions is most unlikely, even for very fine fractions, unless low-grade metamorphic heating proceeds over long periods (>10 Ma). The extent of Ar diffusion, therefore, depends mainly on the thermal history of the studied mineral phase.

The study by Reuter (1987) on contaminant-free epizonal metatuff samples has emphasized the fact that most if not all size fractions (from <0.6 to 6 μm) yield identical ages, providing a 'flat' age distribution relative to size (Fig. 7.3a). Such a distribution of identical ages that are lower than the stratigraphic age (380 Ma) can be assumed to represent the true age of crystallization of the mica-type mineral fractions of this volcaniclastic material. Reuter (1987) also showed that the age distribution pattern remained flat for the progressively coarser-size fractions of the metatuffs studied up to a size of 20 μm, suggesting complete isotopic homogenization for the K–Ar system. Recently, Clauer et al. (1997) found a similar distribution for illite fundamental particles (<0.03 and 0.1–0.3 μm) and for particle aggregates (>1 μm) from a hydrothermally metamorphosed bentonite sample (Fig. 7.3b), suggesting that the formation conditions for the mica-type material are the same for the smallest particles. This latter study also

underscores the fact that fundamental clay-to-mica particles, which represent the smallest individual clay-to-mica particles that can be technically separated, do not preferentially lose radiogenic ^{40}Ar due to their small size.

It may also be mentioned that extremely well crystallized smectite may occur, together with epi-metamorphic micas, in almost pure carbonate rocks such as in the micritic clay massifs of the Doldenhorn nappe, Switzerland (Burkhard, 1988). This smectite is believed to have been subjected to temperatures of 300–350°C for at least 10 Ma. Temperature therefore, seems, not to be a criterion for smectite stability, as it is possible to heat it to 700°C without destroying the mineral structure, nor altering its physical properties (Burkhard, 1988). However, it has not yet been proved if this smectite, when occurring in very low-permeability carbonate rocks, potentially loses or not its radiogenic isotopes during a long lasting thermal setting, such as that described by Burkhard (1988). Ar diffusion calculations have shown that such losses are possible, especially from fine-grained clay particles, if thermal resetting lasts long enough and the rock system is permeable (e.g. Fig. 7.2).

7.3.2.2 Effects of leaching treatments on isotopic signatures

Depending on the host-rock lithology, it is sometimes necessary to apply chemical or mechanical treatments during extraction of the clay- to mica-type minerals. It is, of course, essential that the treatments do not induce preferential removal of the accumulated radiogenic isotopes from the mineral structures. Leaching by acids or other reagents has been commonly used for the removal of interfering soluble minerals or adsorbed elements from clay particles to be isotopically analyzed. But concerns have been raised at various times about the use of acids on minerals to be used for isotopic dating, as leaching may cause simultaneous removal of both parent and daughter isotopes from easily exchangeable sites (surface, edges, lattice defects, expandable layers, etc.), potentially producing unknown effects on the isotopic systematics of the minerals.

Acid leaching experiments of a mixed-layer illite/smectite (Aronson & Douthitt, 1986) and of a diagenetic illite (Clauer et al., 1993) have shown that

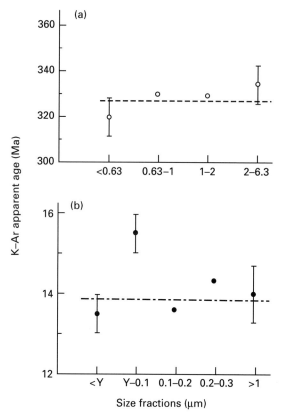

Fig. 7.3 Age distribution of size fractions from volcaniclastic rocks. (a) Epimetamorphic tuff of the Rheinische Schiefergebirge (after Reuter, 1987); (b) a hydrothermal bentonite from the Zempleni Mountains. (After Clauer *et al.*, 1997.)

affecting the isotopic dates. However, only moderate periods (a few minutes) of ultrasonic treatment are recommended to avoid overeffects (Odin, 1982).

7.3.2.3 Relevance of leaching experiments

The Sr trapped by clay- to mica-type minerals during crystallization and the radiogenic [87]Sr formed by decay of the radioactive [87]Rb are fixed tightly in the mineral structure. This can also be assumed for the Sm–Nd method, even if the structural sites for the REE in sheetsilicates are still not identified. In addition, variable amounts of elements can be adsorbed on particle surfaces or located in easily exchangeable sites at the edges of the interlayer positions, or in interparticulate sites. Knowledge of the isotopic signature of such exchangeable elements, Sr for instance, is useful to identify the isotopic compositions of the fluids present during genesis or low-temperature modification, and hence to characterize the chemical environment during crystal growth or further recrystallization. Leaching of clay- to mica-type minerals facilitates, to some degree, separate analysis of elements that are located in different mineral phases, carbonates being often the mineral phase that is mostly present in the leachates. Ohr *et al.* (1994) and Schaltegger *et al.* (1994) have recently identified phosphate-rich soluble phases in leachates that carry most of the REE, silicates included, which means in turn that they are most representative of the Sm–Nd ages when present in the leachates.

A Sm–Nd study of sheetsilicates from carbonates in the Barberton Greenstone Belt (South Africa) has demonstrated that leaching improves methodological aspects, giving rise to a significant improvement in the precision of the age determinations (Toulkeridis *et al.*, 1994). The Sm–Nd isochron defined from untreated <2-μm fractions, yielded a date of 3115 ± 144 Ma with an initial [143]Nd/[144]Nd ratio of about 0.50854, whereas the data points of the corresponding acid leachates, untreated aliquots and acid-treated residues provided a better constrained date of 3104 ± 63 Ma with a similar initial [143]Nd/[144]Nd ratio of about 0.50855. The clay fractions consisted mainly of 2M illite having a morphology typical of authigenic material and a crystallinity index characteristic of greenschist facies conditions. Since the sedimentation time of these carbonates is considered to be between

dilute acids such as 1 mol l[-1] HAc and HCl do not preferentially remove radiogenic isotopes. It has also been shown that Rb–Sr data of illite leached with 1 mol l[-1] HCl acid are concordant with those treated with 1 mol l[-1] NH$_4$OAc (Clauer & Chaudhuri, 1995). Consequently, high [87]Sr/[86]Sr ratios for leachates derived from 1 mol l[-1] HCl treatments are not automatically indicative of selective leaching of radiogenic [87]Sr from a distinctive crystallographic position, such as the interlayer site.

Ultrasonic treatment is the most commonly used method for mechanical purging of impurities associated with sheetsilicate minerals. Ultrasonic treatment essentially fractures mineral grains without

3259 ± 4 and 3227 ± 4 Ma (Kröner *et al.*, 1991), the Sm–Nd date of *c.* 3.1 Ga was interpreted as recording a thermal event.

Leaching clay-to-mica material is technically of interest, because the spread of the data points is increased in isochron diagrams. Any single data point for an untreated sample on an isochron line (U, Fig. 7.4) provides two additional data points in the form of leachate and residue points on the line, on either side of the untreated sample (L and R, Fig. 7.4). Depending on the dating methods, the patterns change with the leachate data point being either close to the *y* intercept for the Rb–Sr and Pb–Pb methods, or distant for the Sm–Nd method. This distribution also provides information about the mineral sites containing most of the elements studied: most of the Rb for a carbonate–silicate mixture is in the latter and most of the Sr in the former, explaining the pattern. For the Sm–Nd method, phosphate-rich leachates will contain most of the REE and the highest Sm/Nd ratio, relative to the silicates, which again explains the pattern. Relative to ^{204}Pb, the ^{207}Pb and ^{206}Pb isotopes are less abundant in the carbonate leachates than in the silicate residues. The Sr or Nd ratios of the leached soluble mineral phases may also provide useful geochemical information. Leaching of the mica-type material from the carbonate rocks of the Glarus Alps that served for the illite-to-mica transition (Hunziker *et al.*, 1986), for instance, indicates an increase in the $^{87}Sr/^{86}Sr$ ratio from about 0.70870 ± 0.00005 (2σ external) in the diagenetic domain to as high as 0.72125 ± 0.00020 in the epizone, that is accompanied by a significant decrease in the Sr content and a significant increase in the Rb content (N. Clauer, unpublished data). This trend in the $^{87}Sr/^{86}Sr$ ratios supports the hypothesis of a progressive Rb and Sr exchange along with increasing degree of sheetsilicate recrystallization, between these silicates and their carbonate host rocks. This exchange probably occurred within limited rock volumes, because an open system behaviour would probably have allowed the excess of ^{87}Sr to be released to the migrating fluids.

7.4 Fission-track dating

Fission-track analysis has become a very powerful tool to delineate time–temperature postformational histories of rocks from sedimentary basins, because

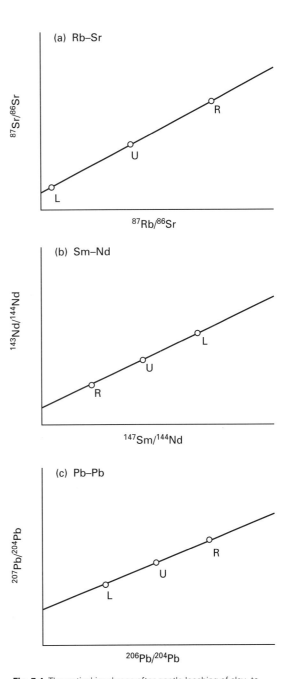

Fig. 7.4 Theoretical isochrons after gentle leaching of clay- to mica-type size fractions for (a) Rb–Sr, (b) Sm–Nd and (c) Pb–Pb methods. L, U and R stand for leachate, untreated and residue, respectively. (After Toulkeridis & Clauer, 1997.)

of its ability to provide a chronology of events that might have occurred in a basin (e.g. Green *et al.*, 1989; Naeser *et al.*, 1989). Initially, fission tracks were used in studies on meteorites, lunar samples and basement rocks; the first suggestion to apply the method to sedimentary rocks was made much later (Naeser, 1979). Decay of atoms by spontaneous fission is associated with several natural isotopes, such as ^{238}U, ^{235}U and ^{232}Th. The occurrence of these spontaneous fissions in solids results in formation of tracks in the medium, as energy from the charged particles is transferred to atoms of the solid medium. In natural materials, fission tracks may be alluded solely to the decay of ^{238}U, as tracks produced by fissions of other isotopes have been considered negligible (Price & Walker, 1963). The tracks created by fissions are distinguished from other etch marks related to crystal imperfections by their tubular shapes.

The fission-track method for dating minerals or establishing thermal histories of rocks, such as in sedimentary basins, requires thorough knowledge of the track density, that is the number of tracks per square centimetre of the investigated material, created by the spontaneous fission of ^{238}U. The production of fission tracks is a function of both U concentration and time. Hence, the determination of fission-track data requires that (i) the U concentration be sufficiently high to produce tracks that might be counted with a high degree of reliability, usually greater than 10 tracks per square centimetre, (ii) the tracks have been stable since the time of production and have not been altered or destroyed by heating or weathering, (iii) the materials are free of inclusions or lattice imperfections to allow clear identification of fission-related tracks, and (iv) the U distribution in the material is uniform, as the ^{238}U concentration has to be determined from a different portion of the same sample. A major problem of dating minerals by the fission-track method lies with the fading of tracks due to annealing of the damages at elevated temperatures. The fission-track dates are thus to be interpreted as 'minimum' dates and not necessarily as crystallization ages. The uncertainty about the dates may be further compounded by several other factors such as a lack of full knowledge about the neutron dose used for determining U concentration of the sample by induced fission method and errors in counting the naturally produced

tracks. Derivation of the age equation formulated by Faure (1986), gives:

$$t = 1/\lambda_a \ln (1 + (\rho_s/\rho_i) (\lambda_a/\lambda_f) \phi \sigma I)$$

where λ_a is the decay constant of ^{238}U by alpha emission $(= 1.55125 \times 10^{-10}y^{-1})$, λ_f is the decay constant of ^{238}U by spontaneous fission $(= 8.46 \times 10^{-17}y^{-1})$, ρ_s is the area density of spontaneous-fission tracks, ρ_i is the area density of induced-fission tracks, ϕ is the thermal neutron dose (neutrons per square centimetre), σ is the cross-section for induced fission of ^{235}U by thermal neutrons $(= 580.2 \times 10^{-24}cm^2)$ and I is the atomic ratio of $^{235}U/^{238}U$ $(= 1/137.88)$.

Technically, the inner surfaces of the separated grains are polished and etched with appropriate reagents, which means that they are prepared under controlled conditions making the tracks more visible under an optical microscope. The measurement of the U concentration requires counting of fission tracks produced by induced fission of ^{235}U by irradiation of the sample with thermal neutrons. A common procedure is to first erase the spontaneous fission tracks by heating and then expose the sample to thermal neutrons to produce induced fission tracks from decay of ^{235}U. The U concentration is known from the density of the induced fission tracks, after taking into account the thermal neutron flux and the irradiation time. The induced fission tracks may be recorded by using an external detector held against the sample surface of which the tracks are counted. This methodology requires correction among the tracks and the U content, and the way to correct has been a matter of debate. Hurford and Green (1982) also proposed that rigorous calibration of the method was needed against minerals of known ages obtained by isotopic methods.

Although annealing could be a major problem for dating of the crystallization of minerals by the fission-track method, the different characteristics of annealing various minerals may be used advantageously to reconstruct the thermal history of rocks. Annealing rates are dependent on the type of mineral, to some degree on its composition, and also on the temperature. Hence, the track records of two minerals from the same rock could record two discordant fission-track dates, because they differ in their retention of tracks at some elevated temperatures, and consequently give detailed information about the thermal history of the host rock. On the other hand, concordant fission-track dates from track records of two minerals are suggestive of crystallization of the host rock.

A variety of minerals subject to cooling at different rates have been investigated with respect to their track-annealing properties. Apatite, zircon and titanite are among some common sufficiently high U-bearing minerals, that have received considerable attention in this respect, with potential for application to the reconstruction of the thermal evolution of sedimentary basins. The tracks of these minerals anneal at very different rates that are a function of temperature and duration of heating. One mineral may lose all tracks and another none at all at a given temperature of heating and a given duration. For example, apatite fission tracks having mean lengths of 16 μm, are annealed when temperature increases from 20°C to about 150°C, when heating duration lasts from a few hundred thousand to a few million years. Titanite, on the other hand, experiences no annealing effect by the same thermal episode. Reported mineral closing temperatures have been occasionally found to be discordant, which have been attributed to a variety of factors in etching conditions, such as the etching solution used and the duration of the exposure.

The distribution of measured lengths of etchable, confined and spontaneous tracks in minerals allows assessments of the magnitude and type of cooling of host rocks that were subjected to paleotemperatures higher than those of the present day (e.g. Crowley, 1985; Gleadow *et al.*, 1986). Quantitative analysis of track annealing has been developed, mainly for apatite in sedimentary environments (e.g. Gleadow *et al.*, 1986, Hurford *et al.*, 1991), but also for zircon (e.g. Tagami *et al.*, 1990; Yamada *et al.*, 1995). Interpretations have been greatly facilitated by the development of predictive modelling (e.g. Willett, 1992). For instance, time–temperature relationships and distributions of fission-track lengths for samples having theoretically experienced heating events that followed a cooling period, were modelled (Fig. 7.5). Depending on the climax temperature of the secondary heating period, the model can predict the distribution of the track lengths and the average ages, which certainly is of help for the interpretation of natural track-length patterns and fission-track ages. The distribution patterns combining single-grain ages

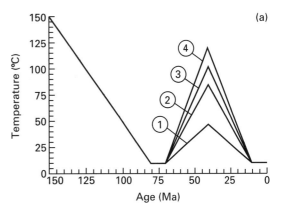

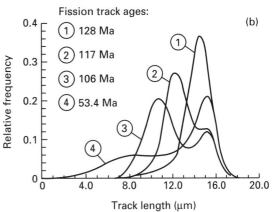

Fig. 7.5 Simulated time–temperature evolution of a U-bearing mineral having undergone a thermal event after cooling (a), with the predicted distribution of the track lengths and fission-track ages (b). The late thermal event is considered to have been of the same duration for the different examples, but its climax temperature changed from 45°C for (1), to 85°C for (2), to 103°C for (3), to 121°C for (4). The fission-track closure temperature was evaluated to be approximately 140 Ma. (After Willett, 1992.)

and fission-track ages in crystals, also yield sensitive indications of the thermal evolution of the host rocks.

Reconstruction of thermal evolutions of sedimentary basins is through annealing models that describe the relationship between temperature, time and length of fission tracks in one or several of the minerals mentioned (e.g. reviews in Naeser & McCulloh, 1989). Such models are derived mainly from laboratory experiments that analyze the length of the tracks in samples that have been heated at given temperatures over given periods of time. These suggest that annealing of tracks in apatite can take place at 105°C for a 100-Ma heating period, to 150°C for a 0.1-Ma heating period. Zircon yields a less constrained cooling temperature of either 240 ± 40°C (Harrison *et al.*, 1979) or 290 ± 40°C (Tagami & Shimada, 1996), whereas titanite tracks start to fade at 290 ± 40°C (Harrison *et al.*, 1979). It may, however, be mentioned here that these latter 'closure temperatures' are still poorly constrained. As extrapolations for the critical temperatures are needed, some disagreement exists about the modelling to be used, as the exact temperature at which minerals lose or retain their tracks depends strictly on the cooling rate to which the minerals are subjected. An example for the fading of fission tracks in apatite and titanite which is based on measured and extrapolated data, can be constrained by two lines for each mineral (Fig. 7.6). The lines labelled 0% give the time–temperature relation at which no tracks are annealed while those labelled 100% give that at which all tracks are lost. For instance, if a temperature of 175°C is applied to a detrital apatite for 1 million years, it will lose all tracks, while losing none at 50°C. On the other hand, annealing of tracks in titanite will start if a temperature of 250°C is applied during the same period of 1 million years. If cooling occurs, tracks will start to accumulate in apatite at 175°C and this will continue until 50°C. In this case, the effective temperature recorded by the fission-track data is the value at which half of the tracks are preserved. Recent reviews by Wagner and Van den Haute (1992) and by Ravenhurst and Donelick (1992) may be useful for additional information about the methodology and its application potentials.

7.5 Rock lithologies and isotopic homogenization

Mica-type materials of different rock lithologies having different physical properties, are expected to respond differently when subjected to any tectono-thermal disturbance. Comparative studies of isotopic systems of micas from associated rocks of different types were made on metapelites and metaarkoses from the Schwarzwald in Germany (Brockamp *et al.*, 1987), metapelites and metatuffs from the Rheinische Schiefergebirge in Germany (Reuter, 1987), and metapelites and metabasites from the Erquy region in France (Clauer *et al.*, 1985). Such comparisons are of importance in order to establish if there is coherent behaviour of isotopic systems for the same minerals.

7.5.1 In sheetsilicates of metapelites and metaarkoses

A hydrothermal uranium deposit with extensive sericitization occurs in Carboniferous metaarkoses and metapelites near Baden-Baden in Germany. It was dated at 150 Ma on the basis of U-chemical age determinations of pitchblende, and formed at temperatures of 240–290°C (Brockamp *et al.*, 1987). The deposit itself, designated as the ore zone, is surrounded by three less intensely altered zones, namely the sericite, albite and weakly altered zones. The <2-µm mica fractions of the metaarkoses from these three zones yield a mean K–Ar age of 155 Ma, agreeing with the U-chemical age of the pitchblende. By contrast, K–Ar dates of the <2-µm mica fractions from metapelites increase from 170 Ma in the ore zone, to 180 Ma in the sericite zone, to 185 Ma in

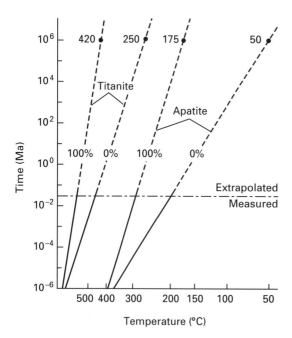

Fig. 7.6 Fading of fission tracks in apatite and titanite from combined effect of temperature and time. See explanation in the text. (After Faure, 1986.)

the albite zone and to 190 Ma in the less altered zone. The systematically higher K–Ar dates for the micas in the metapelites, compared to those in the metaarkoses, imply that the <2-μm fraction of the metapelites carries a K–Ar isotopic memory whose value increases from the sericite zone to the weakly altered zone, clearly indicating an increasing occurrence of relict detrital components. This detrital isotopic signature, however, can be eliminated by using a finer particle size for the micas of the metapelites. The <0.6-μm mica fraction of a metapelite from the sericite zone yields a K–Ar age of 150 Ma, which is identical to the pitchblende age and to the K–Ar ages of the <2-μm fraction of the metaarkoses from the sericite, albite and weakly altered zones, while the <2-μm fraction of the same metapelite yields the higher K–Ar date of 180 Ma.

7.5.2 In sheetsilicates of metapelites and metatuffs

A Devonian sequence of interbedded metapelites and metatuffs from eastern Rheinische Schiefergebirge in Germany, was variously metamorphosed at anchi- to epimetamorphic grades. A slaty cleavage of variable intensity developed discontinuously in the metapelites and continuously in the metatuffs. The <2-μm fractions of both rock types consist of illite with small amounts of chlorite and quartz (Reuter, 1987). The 2–6.3-μm fractions of both rock types contain also up to 40% quartz and subordinate amounts of albite, with some calcite in the metapelites. The illite in the metapelites is of the 2M type, whereas that in the metatuffs is of both the 2M and 1M$_d$ types.

The K–Ar dates of different size fractions from an anchimetamorphic metapelite sample distinctly decrease as a function of decrease in size. The dates decrease from about 405 to 345 Ma for sizes ranging from 6.3 to <0.4 μm, giving an inclined age spectrum relative to particle size (curve labelled AMP in Fig. 7.7). By contrast, the K–Ar dates for the fractions ranging from 6.3 to <0.6 μm of a corresponding anchimetamorphic metatuff are all within analytical uncertainty between 335 and 315 Ma, giving a flat age spectrum for the size fractions (line labelled A + EMT in Fig. 7.7). The invariant K–Ar data of the metatuffs with respect to changes in the particle sizes, suggests that a distinct metamorphic recrystallization

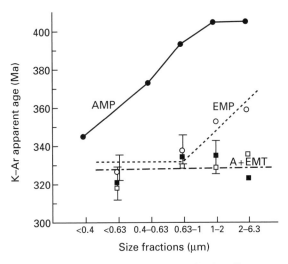

Fig. 7.7 Inclined and bench-type age spectra for sheetsilicate fractions of anchi- (AMP) and epimetamorphic (EMP) metapelites, and flat age spectra for the same fractions of associated anchi- (AMT) and epimetamorphic (EMT) metatuffs from Rheinische Schiefergebirge (after Reuter, 1987). The anchimetamorphic samples are shown by the open symbols and the epimetamorphic ones by the closed symbols.

of the metatuffs occurred at about 325 ± 5 Ma and that isotopic homogenization is evident up to, at least, the 6.3-μm size fraction.

The inclined spectrum of the K–Ar data for the fine fractions (<0.6–2 μm) of the metapelite is suggestive of incomplete homogeneity, even for the finest analyzed size fraction. The difference between this age spectrum and the flat one of the associated metatuff reflects differences in the lithologically controlled metamorphic recrystallization of the sheet-silicates. Based on the metatuff K–Ar age of 325 ± 5 Ma, it is clear that even the very fine <0.4-μm fraction of the anchimetamorphic metapelite inherited an isotopic memory, thus giving a higher isotopic date. The anchimetamorphic conditions were obviously insufficient to obliterate completely this inherited isotopic signature in the clay-type material of the metapelite.

Increase of the metamorphic grade from anchizone to epizone does not result in a change of the K–Ar data of the metatuff size fractions within analytical uncertainty (line labelled A + EMT in Fig. 7.7), which confirms a Hercynian metamorphic recrystallization. On the other hand, there is a marked change in the metapelite size fractions with increase in meta-morphic grade (dashed line EMP in Fig. 7.7). The two finest fractions (<0.63 and 0.63–1 μm) yield identical ages as the metatuff fractions of the same size (325 ± 5 Ma), while the coarser size fractions (1–2 and 2–6.3 μm) give higher dates (350 and 355 Ma) which are close to that of the finest size fraction (<0.4 μm) of the corresponding anchizonal metapelite. The change from inclined to bench-type pattern shows that isotopic homogeneity in the mica-type particles is being progressively achieved in the fine size fractions of the metapelite samples that underwent epizonal recrystallization.

7.5.3 In sheetsilicates of metapelites and metabasites

The upper part of the Cambrian Spilitic Group of Erquy in France consists of metabasite pillow lavas with intercalated sedimentary units (Clauer et al., 1985). The <2-μm sheetsilicate fraction of the sedimentary unit consists mainly of illite and chlorite with subordinate amounts of smectite or mixed-layer illite/smectite; the illite crystallinity index is indicative of epizone conditions. The Rb–Sr isochrons of the metasedimentary whole rocks and of their <2-μm fractions gave a date of 494 ± 11 Ma with an initial ^{87}Sr/^{86}Sr ratio of 0.7052 ± 0.0005 (2σ). The K–Ar analyses of the <2-μm fractions consisting of 100% mica gave identical dates of about 485 Ma. These ages are identical to the Rb–Sr isochron age of 482 ± 10 Ma with an initial ^{87}Sr/^{86}Sr ratio of 0.7055 ± 0.0002 for the surrounding metabasites (Vidal, 1980).

The Rb–Sr and K–Ar data of the size fractions in the metasedimentary unit and the Rb–Sr data of the enclosing metabasites, suggest that the formation of the micas was chemically dependent on the metamorphic process, as the formation ages and initial Sr isotopic compositions are identical for both types of material, even at whole rock scale. The low initial ^{87}Sr/^{86}Sr ratios of both lithologies indicate that the metavolcanic rocks acted as the monitoring reservoir for the Sr of the associated metasediments.

7.5.4 Summary

The K–Ar systematics of illite- to mica-type minerals

in metapelites and in slates do not relate to identifiable events in the history of the minerals, when they were subjected to metamorphic conditions of epizonal grade or less. On the other hand, those of corresponding materials in porous sandstones–greywackes and in metavolcanics devoid of detrital components do relate to identified events. The K–Ar data of clay-type materials from shales suggest that recrystallization during diagenesis occurs in a restricted chemical system, inducing variable excesses of radiogenic ^{40}Ar during the process. A similar behaviour was also reported for metacarbonates (Huon et al., 1994). Lack of information about these excesses invalidates the K–Ar dates of clay-type minerals in such rocks for dating purposes; inheritance of detritus from the protolith being the main cause.

K–Ar data are also useful to determine the volu-metric extent of isotopic homogenization in minerals by analyzing several size fractions of a sample. If the analyses are made on small particles so that contamination by detrital framework components can be avoided, the data often correspond to true crystallization or recrystallization ages.

7.6 Metamorphic degree and extent of isotopic homogenization

The studies discussed above have shown that metamorphism of epizonal grade is required to homogenize the K–Ar isotopic system of detrital illite in metapelites, but that it occurs at lower grade when the host rocks are more porous or devoid of detrital micas. It is thus essential to determine (i) if this observation can be applied to other isotopic systematics, (ii) to what extent progressive hom-ogenization can affect the isotope systematics of sheetsilicates depending on the type of the host rocks, and (iii) if whole rocks can be completely homogenized and at what grade.

7.6.1 In sheetsilicates of shales

Ohr et al. (1994) published a very detailed Sm–Nd and Rb–Sr isotopic study on <0.2-μm fractions of an Ordovician argillaceous-rock transect from Wales in the UK that ranged from diagenetic to epizonal conditions. The fractions consisted of illite and chlorite, considered both to be of detrital and diagenetic origin, with subordinate amounts of quartz, albite

and muscovite. The size fractions were leached with dilute (1 mol l^{-1}) HCl or HAc, which allowed separate analysis of the soluble and sheetsilicate components of the fractions. Leaching resulted in Sm/Nd fractionation between leachates and residues, that was a function of grain size and metamorphic grade. The leachate–residue assemblages gave Sm–Nd ages of 453–484 Ma for uncleaved Welsh mudrocks of diagenetic degree, and as low as 413 Ma for anchizonal to epizonal slates. The higher ages agree well with the biostratigraphic age of the rocks and the younger age was assigned by the authors to the Acadian deformation which was determined at 390 Ma by other authors. These Sm–Nd ages were obtained from individual leachate–residue assemblages of the size fractions and not from an isochron, which indicates a lack of isotope homogeneity at whole rock scale, in both the diagenetic and epizonal rock samples. These data are also not straightforward as can be seen by comparison with the Rb–Sr dates obtained from the same leachate and residue splits:
1 the Rb–Sr dates are significantly younger than the Sm–Nd ones, most being below even the reference age of the Acadian deformation;
2 the Sm–Nd and Rb–Sr data are widely scattered from 627 to 334 Ma and from 430 to 200 Ma, respectively, considering that the age range between deposition and deformation is from 475 to 390 Ma. The very wide scatter in ages for both the Sm–Nd and Rb–Sr methods restricts the usefulness of the results. The dataset would probably have benefited from a more detailed examination of separated material involving complementary studies by XRD, SEM and TEM to identify the morphological characteristics of the mineral components and the importance of the structural deformation on the rocks.

Isotope dating of such fine-grained low-grade metamorphic metasedimentary rocks will become more reliable if the results can be explained and even expected from information concerning the mineralogical–morphological characteristics. This basic requirement is only partly fulfilled here because (i) no detailed electron microscopic study was completed to identify the habit, type(s) and shape(s) of the sheetsilicates in the rocks, and (ii) no clear explanation was offered for the very young Rb–Sr data of some epizonal size fractions. It is also to be expected that the leachate–residue data of epizonal mica-type fractions, with sizes as small as

<0.2 μm, are isotopically homogeneous, at least for the Rb–Sr systematics (e.g. Clauer & Chaudhuri, 1995). Thus, they should plot on a single isochron representative of the recrystallization process instead of giving parallel to subparallel lines, which reflects isotopic inhomogeneity at rock scale. This seems not to be the case for the Welsh data, as the Sm–Nd ages were not obtained from a single isochron and as those ages identified by the authors to represent the Acadian deformation are significantly higher, at 413 Ma, than the age of the Acadian deformation set at about 390 Ma. A similar picture seems to apply, in this publication, to the Rb–Sr dates which were also obtained from single leachate–residue lines and not from an isochron, the ages being in this case at 200 Ma, which is significantly lower than the 390–400 Ma generally assigned to the Acadian deformation.

7.6.2 In sheetsilicates of slates

Slates from Hercynian Belt of north-eastern Morocco were metamorphosed with contemporary development of a flat-lying cleavage during the Middle Devonian and Upper Visean (380–330 Ma). Epizonal conditions and an intense cleavage were recorded in the western part of the belt, while lower grade conditions and less intense cleavage development prevailed to the east (Huon et al., 1993). The size fractions of a slate from the epizonal region consisted of 70–90% 2M illite and 30–10% chlorite. The three smallest fractions (<1, 1–2 and 2–6 μm) yielded nearly identical dates at 291, 295 and 300 Ma, with an average of 295 ± 5 Ma (open circles in Fig. 7.8). These data were interpreted as recording the time of mica crystallization. By contrast, the coarser 6–10-μm fraction yielded a higher K–Ar date of 332 Ma (Fig. 7.8). Equivalent size fractions from a slate of the eastern part of the belt consisted of 60–80% 2M illite and 40–20% chlorite. The K–Ar data of the different size fractions fit an inclined age spectrum with dates of 330 Ma for the <0.5-μm fraction and 410 Ma for the 6–10-μm fraction (closed circles in Fig. 7.8). This distribution differs from that of the epizonal sample, reflecting substantial detrital K–Ar isotopic memory of the precursor minerals in the lower grade region.

The different size fractions of the diagenetic sample display an inclined age distribution relative to the

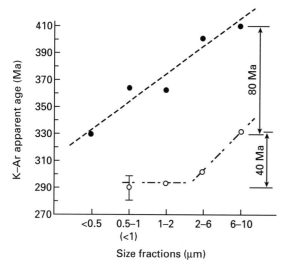

Fig. 7.8 Inclined and bench-type age spectra for sheetsilicate fractions of an anchimetamorphic slate (closed circles) from eastern Hercynian Belt and of an epimetamorphic slate (open circles) from western Hercynian Belt in Morocco. (After Huon *et al.*, 1993.)

size of the fractions, whereas that of the epizonal sample is of the bench type. Also, the difference in the K–Ar dates between the <0.5-μm and the 6–10-μm micaceous fractions from the diagenetic sample is significantly higher (80 Ma) than that from epizone sample (40 Ma), suggesting a more effective influence of the higher metamorphic conditions on the coarser size fractions (Fig. 7.8). These data show that fine fractions of shale-type rocks can yield K–Ar data that approximate the age of metamorphism when of epimetamorphic grade.

7.6.3 In sheetsilicates of metagreywackes

In north-western Morocco, a sequence of Middle Cambrian greywackes underwent progressive diagenetic to epizonal metamorphic recrystallization. Clay fractions from a N–S suite of samples through the prograde sequence consist of illite with subordinate amounts of chlorite in the anchizone, while the epimetamorphic samples contained also smectite (Schaltegger *et al.*, 1994; Clauer *et al.*, 1995). This smectite concentrated mainly in the smallest fractions, was considered to have originated from a late hydrothermal event. The K–Ar data of the different

fractions of the diagenetic samples decrease from about 450–430 Ma to 387–362 Ma with decrease in particle size, giving an inclined age spectrum (closed and open squares in Fig. 7.9). On the other hand, most of the size fractions of an anchimetamorphic sample yield an age of 300 ± 10 Ma giving a bench-type age spectrum (closed circles in Fig. 7.9), as did those of the coarse (1–2 and 2–6 μm) fractions of an epimetamorphic sample (open triangles in Fig. 7.9). For three additional epizonal samples (open circles, closed triangles and closed squares in Fig. 7.9), the K–Ar data display either inclined or a bench-type age spectrum with the finest fractions yielding K–Ar ages of about 210 Ma.

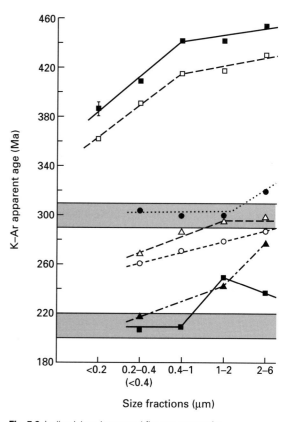

Fig. 7.9 Inclined, bench-type and flat age spectra for sheetsilicate fractions of diagenetic (closed and open squares), anchimetamorphic (closed circles) and epimetamorphic (open triangles) metagreywackes of a transect near Casablanca in Morocco (after Clauer *et al.*, 1995). The two hatched areas outline the average ages of a Hercynian deformation at 300 ± 10 Ma and of a hydrothermal episode at 210 ± 10 Ma.

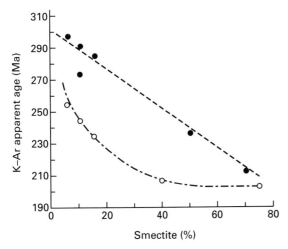

Fig. 7.10 Relationship between apparent K–Ar ages of sheetsilicate fractions of epimetamorphic metagreywackes from the southern region of the transect near Casablanca, Morocco and the content of smectite (after Clauer *et al.*, 1995). The open symbols represent samples containing a mixture of sheetsilicates from the epimetamorphic and hydrothermal parageneses with the older epimetamorphic material having undergone diffusive loss of radiogenic ^{40}Ar. The closed symbols, in contrast, represent similar mixtures with an unaffected Ar budget for the older paragenesis.

The K–Ar dates of the epimetamorphic size fractions containing smectite show a negative correlation with the amounts of smectite, which suggests that smectite is the major mineral of the 210-Ma-old mineral paragenesis. The relationship between the K–Ar data and the smectite content of the epimetamorphic fractions was found to be either linear or asymptotic (Fig. 7.10). Such patterns suggest that the hydrothermal event either induced diffusive loss of radiogenic ^{40}Ar from the Hercynian illite, which resulted in an asymptotic relationship between the K–Ar dates and the smectite contents, or it did not affect the Ar budget of the previously formed illite material, which resulted in a linear relationship between the isotopic dates and the amounts of smectite in the size fractions.

7.6.4 In whole rocks

The debate about the usefulness of whole rock isotopic studies in sedimentary to very low-grade metamorphic environments has been ongoing for some time and it is useful to assess whole rock studies in comparison with those referred to above and involving separated size fractions. These latter studies indicate that isotopic homogenization between detrital precursors and authigenic minerals in variously affected rocks of different lithologies, develops only under specific conditions among minerals from the very fine size fractions. It is thus appropriate to review whole rock isotopic dating in the light of the results obtained from studies of size fractions involving a progressive change from detrital–authigenic mixtures to pure authigenic components.

Pioneer whole rock studies have shown that agreement between the isotopic dates of sedimentary units and their stratigraphic ages are often fortuitous (Compston & Pidgeon, 1962; Chaudhuri & Brookins, 1969; Field & Raheim, 1973), and that there is very little merit in analyzing whole rock samples with the aim of determining sedimentation ages. The concept was that colinear regressions of whole rock data points in Rb–Sr diagrams can arise by a thorough mixing of suspended material (e.g. Cordani *et al.*, 1978, 1985). In fact, thorough mixing might induce isotopic uniformization which is a mechanical process, but not isotopic homogenization which involves a chemical process. Although low-grade metamorphism results in a general simplification of the mineral assemblages of whole rocks, isotopic homogenization is not implicitly realized, even knowing that low-grade metamorphism of sedimentary or volcanic rocks can modify the isotopic systematics of the mineral components towards homogenization (e.g. O'Nions *et al.*, 1973; Gebauer & Grünenfelder, 1974). Linear trends in isochron diagrams have even been reported for low-grade metasedimentary rocks, suggesting that metamorphism can induce isotopic homogenization among the constitutive minerals of whole rocks (e.g. Moorbath *et al.*, 1967; Obradovich & Peterman, 1968). However, such homogenization not only depends on the intensity of the metamorphic imprint, but also on rock lithology, as framework minerals such as muscovite and K-feldspar need a higher metamorphic grade than anchi- to epimetamorphic intensity to have their 'isotopic clock reset' (e.g. Clauer, 1974).

Carbonate rocks were not thought suitable for isotopic dating until marine organisms were found to contain U in their skeletons, which prompted U–Th–Pb dating of ancient carbonates. The first evaluation of dating carbonate rocks by the U–Pb and Pb–Pb methods was not favourable (Doe, 1970). Some two decades later, Moorbath *et al.* (1987) further explored Pb–Pb dating of metacarbonate rocks and since then, several studies reported Pb-isotope analyses of marbles (Jahn, 1988; Jahn *et al.*, 1992; Taylor & Kalsbeek, 1990) and non-metamorphic carbonates (Smith & Farquhar, 1989; Jahn *et al.*, 1990; Smith *et al.*, 1991; DeWolf & Halliday, 1991). Moorbath *et al.* (1987) made various batch dissolutions of metamorphosed carbonate samples and analyzed both leachates and residues. As the isotopic data points defined a line, the results were considered by the authors to date a metamorphic event which might have caused isotopic equilibrium between the carbonate and silicate phases. By contrast, Smith and Farquhar (1989) and Jahn *et al.* (1990) obtained analytical scatters when using the same procedure. These differences in the behaviour of the U–Pb systematics of metacarbonate rocks and other metarocks emphasize the need for more detailed mineralogical and geochemical studies of leached and residual materials to understand why the same procedure on similar rocks gives either a linear array of U–Pb data or scatter plots, which may be indicative of differential origins for the major components of such rocks, as well as differential behaviour during further tectonothermal evolution.

Despite these difficulties associated with whole rock dating of low-grade metamorphic volcaniclastic and metasedimentary rocks, renewed efforts were made to achieve better results from whole rock dating. Ordovician volcanic and high-level intrusive rocks from part of a dominantly clastic Lower Palaeozoic succession of the Welsh Basin metamorphosed at zeolite to low-grade greenschist facies, were dated by the Rb–Sr method (Evans, 1991). Ordovician volcanic rocks yielded a whole rock Rb–Sr isochron age of 399 ± 9 Ma that was interpreted as the age of the deformation-related metamorphism. Among the studied intrusive rocks, three preserved Ordovician ages at 356 ± 75 Ma, and two gave imprecise Devonian ages at 325 ± 127 Ma (Acadian deformation?). The Devonian ages were related to rocks having a well-developed secondary mineralogy, whereas those rocks retaining their Ordovician ages were less extensively recrystallized. In the same region, two more intensively cleaved Ordovician slates gave a Rb–Sr whole rock isochron age of

409 ± 9 Ma (Evans, 1989), which is consistent with the average age of most of the nearby volcanic rocks and also with the recrystallization ages of other cleaved slates from the region (Bath, 1974). In contrast, nearby slates of the same stratigraphic level, but altered to a lower anchizone grade, yielded Rb–Sr ages of 430 ± 9 Ma. A possible reason for these age differences is that the anchizonal slates contain minerals that are still incompletely homogenized relative to the volcanic rocks and the more intensely cleaved slates. Consequently, only the 400-Ma average age for the rocks deformed by the Acadian deformation seems to be reliable by analogy among the different types of rocks.

More recently, Evans (1996) determined whole rock Rb–Sr ages on Ordovician and Silurian metasedimentary rocks from the same Welsh area, along with K–Ar ages of their <2-μm illite fractions. Interestingly, the K–Ar data of the separated illite provided a narrow distribution of 399 ± 3 Ma, which was attributed as the age of the Acadian deformation in the region. In contrast, the Rb–Sr whole rock ages obtained by the isochron method are higher and scattered between 431 ± 10 and 403 ± 15 Ma. The highest of these values is close to those obtained for nearby anchizonal slates (Evans, 1989; Ohr *et al.*, 1994), which favours the occurrence of non-equilibrated detrital components. The progressive Rb–Sr age reduction from 430 to 403 Ma in the whole rocks was thought to measure the duration of the smectite–illite transition. This interpretation was favoured on the basis of geological arguments, not considering possible mixings that could have involved neoformed and detrital framework constituents. Also, it is still not clear why a smectite–illite transition lasting over 25–30 Ma (which is the difference among the scattered Rb–Sr whole rock ages), results in K–Ar ages for the <2-μm illite that are scattered narrowly at 399 ± 3 Ma. Furthermore, the linear correlation between the $^{87}Sr/^{86}Sr$ and the $1/Sr$ ratios of the whole rocks (Evans, 1996, Fig. 3) represents a strong argument in favour of a mixing of components generated in two distinct Sr reservoirs, as shown long ago by Boger and Faure (1974). One reservoir, probably the detrital one, could be characteristic of high $^{87}Sr/^{86}Sr$ and $1/Sr$ ratios, whereas the second, probably the authigenic one, could have yielded lower $^{87}Sr/^{86}Sr$ and $1/Sr$ ratios.

This review on the existing information about whole rock isotopic dating still emphasizes the fact that it is clearly more difficult for isotopic homogenization to occur in a material that is mineralogically highly heterogeneous, such as a whole rock, than in a material that is far less heterogeneous, such as a granulometric fraction, whatever the considered isotopic system. This can be illustrated by a Sm–Nd study of two black-shale rocks from the Lower Proterozoic Francevillian Basin (south-eastern Gabon). This study showed that the Sm–Nd data of different size fractions (up to 0.4 μm), untreated and leached with dilute (1 mol l⁻¹) HCl, defined respectively two Sm–Nd isochrons, one for each sample, with ages of 2099 ± 115 Ma and 2036 ± 79 Ma (Bros *et al.*, 1992). By contrast, the corresponding whole rock data points plot off the isochron lines, clearly indicating that their Sm–Nd systematics are only partly homogenized (Fig. 7.11). Interestingly, the data point of a third organic-rich black-shale whole rock plots between the two previously described Sm–Nd isochron lines, and it defines with the data point of a pure, fracture-filling bitumen sample, a line parallel to the two previous isochrons. The agreement in the Sm–Nd isotopic evolution between the organic-rich material, the

sheetsilicates of the two organic-poor samples and the HCl leachates representing the associated soluble mineral phases, suggests equilibration of the Nd isotopes between these different components during formation of the clay-type minerals. The geochemical arguments in favour of a crystallization age at about 2070 Ma for the sheetsilicates are supported by the isochron trend of the two organic-rich and organic-poor samples, while Sm/Nd isotopic homogeneity at the whole rock scale has obviously not been achieved.

7.7 A model for isotopic homogenization in low-grade metamorphic conditions

The above discussion highlights three types of isotope-data distributions relative to the sizes of the analyzed fractions, suggesting a general picture for isotopic behaviour of sheetsilicates from different rock lithologies subjected to low-grade metamorphic conditions. Subfractions extracted from the <2-μm size fractions of these rocks have the potential to display one of the three different age-distribution patterns (Fig. 7.12) that are described with terms 'borrowed' from Pevear and Vrolijk (1996).

1 An *inclined spectrum* that signifies incomplete

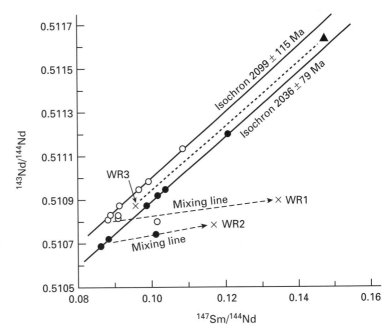

Fig. 7.11 Sm–Nd isochron diagram with data of clay fractions (circles), whole rocks (crosses) and a bitumen sample (triangle) of Francevillian sedimentary rocks, Gabon. The <0.4-μm size fractions of each sample (closed and open circles) fit two parallel isochrons yielding similar ages at about 2050 Ma. The coarser fractions tend towards the data of the corresponding whole rocks. The whole rock enriched in organic matter fits a third line, together with the bitumen sample, that parallels the two isochrons. (After Bros *et al.*, 1992.)

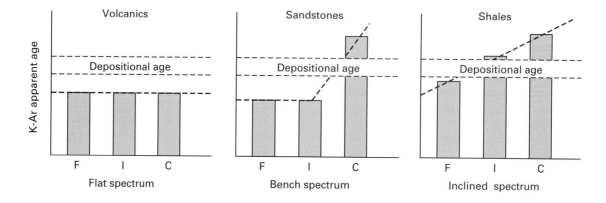

Fig. 7.12 Theoretical flat, bench-type and inclined age spectra relative to the size of analyzed sheetsilicate fractions. The letters F, I and C stand for fine, intermediate and coarse size fractions, normally in the <2-μm range.

isotopic homogeneity in fine-grained clay- to mica-type material. Such behaviour often characterizes the micaceous material of relatively impermeable shale- to slate-type rocks. No meaningful geological age can yet be extracted from such data.

2 A *bench-type spectrum* that outlines partial isotopic homogeneity in the finest subfractions (at least two fractions should yield the same age) of the micaceous material. Such a pattern is also expected for sheetsilicates from relatively impermeable slate-type rocks, but it may also be found for sheetsilicates of permeable low-grade metaarkoses and metasandstones. The flat part of the age pattern can be considered to provide a meaningful geological age for a postdepositional metamorphic/deformational event.

3 A *flat spectrum* that signifies complete isotopic homogeneity at a larger scale in the whole rock volume, for all analyzed subfractions. Such a pattern can be expected in volcaniclastic rocks because of the initial lack of detrital components in the small fractions. It can be found also for the sheetsilicates of porous rocks, such as those already mentioned.

In summary, these patterns depend on two independent parameters: the lithology of the host rocks and the degree of recrystallization. It is therefore essential to analyze several size fractions of any type of rock subjected to a low-grade metamorphism, to ensure that the data generated correspond to a geologically meaningful age.

7.8 Basin evolution based on combined isotopic and fission-track data

The good coincidence between the temperature range of fission-track annealing in apatite grains, over time periods of 1 to 100 Ma, and that in which hydrocarbons are generated, has prompted this method to emerge as a powerful tool in the study of thermal histories in sedimentary basins (e.g. Green *et al.*, 1989). Good dependency between fission-track data and elevation of sample locations were also reported, allows reconstruction of a map with respect to given altitudes (Rahn *et al.*, 1997). Many fundamental studies using this tool have been published in the last decade, but the purpose here is, beyond this fission-track dating aspect, to evaluate combined approaches of this method and isotopic dating of associated authigenic sheetsilicates. The discussion above has emphasized, provided the sample preparation and fraction separation were done carefully, that crystallization of the extracted sheetsilicates is strictly related to the climax of thermally induced events. As the information provided by the fission-track data is often in a lower temperature range, combination of both methods theoretically provides a more complete picture of the thermal evolution of a basin in setting the age(s) of the highest temperature(s) and that (those) at which the fission tracks are annealed in apatite (or zircon) grains. The annealing temperatures for apatite, which range between 105°C for a long heating period and 150°C for a short heating period, clearly depend on the knowledge of the duration of the postthermal history of the sedimentary sequence.

Studies integrating these two dating approaches are still scarce; one being, for instance, that of Pagel *et al.* (1997) on sedimentary rocks of the Ardèche paleomargin in the south-eastern Massif Central in France. The K–Ar dating of sheetsilicates within the lower part of the sequence gave an age of 110–120 Ma corresponding to maximum temperature, which was equated with maximum burial. The sheetsilicates from the fault system observed at the bottom of the sequence formed earlier, at about 190 Ma, while the youngest ages were obtained at

40 Ma when fission tracks were annealed in apatite grains away from a faulting system. Linking this information with vitrinite reflectance and fluid-inclusion microthermometry in anhydrite, quartz, baryte and dolomite, the authors were able to recognize burial lasting for *c.* 100 million years from initial deposition, followed by 75 million years of relatively little uplift and then more rapid uplift for the last 50 million years (Fig. 7.13).

In summary, combined isotopic dating of authigenic sheetsilicates and fission-track dating of detrital apatite in associated rocks, represent a powerful tool that has the potential for integrated reconstruction of the evolutionary history of sedimentary to slightly metamorphosed rock sequences, when used in conjunction with vitrinite reflectance and fluid-inclusion microthermometric data.

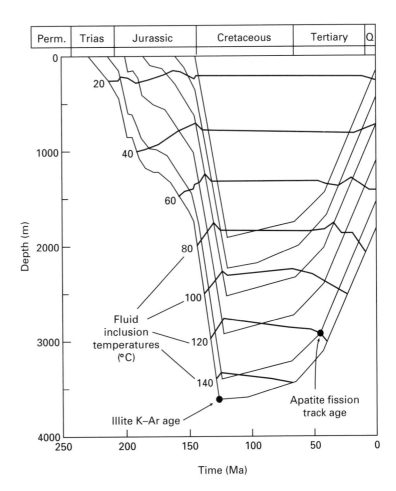

Fig. 7.13 Burial history of the sedimentary rock sequence from the paleomargin of the south-eastern French Massif Central constrained in temperature by fluid-inclusion microthermometry and in time by K–Ar dating of illite and fission-track dating of apatite. Further information is available in the text. (After Pagel *et al.*, 1997.)

7.9 Conclusions

Isotopic dating of clay to mica material from metapelites does not systematically provide realistic ages, in contrast to clay- and mica-type material from metamorphosed sandstones, arkoses and volcaniclastic rocks. Besides temperature, permeability of the host rocks is a major parameter, as it is the determining factor in controlling the fluid fluxes. Examination of available isotopic results from clay- to mica-type fractions suggests that the overall behaviour of the isotopic systematics of micaceous material (at least of the K–Ar systematics) can be organized into three distinct patterns that include a flat, a bench-type and an inclined age spectrum relative to fraction sizes, pointing to complete, partial or an absence of isotope homogenization, respectively. The isotopic data combined with information about rock lithology and recrystallization degree, give important insights into the illite-to-mica transitional process.

Some of the confusing and erroneous statements reported in past literature about clay-to-mica isotope geochemistry should be definitely avoided. Preferential loss of radiogenic isotopes can no longer be used to explain unexpected low ages. Also, as the presence of detrital contaminants in size fractions can be reduced and even avoided by gentle disaggregation of the rocks and separation of the very fine fractions (in the 1/10 of μm), the effect of such contaminants on the isotopic ages of the authigenic components can be controlled by characterizing and analyzing several size fractions. The occurrence of detrital material in authigenic fractions should no longer be the cause for rejection of a basinal evolution model based on isotope dating of fine-grained sheetsilicates in very low-grade metamorphic environments. Should high ages appear to be related to faulty sample separation, these should be carefully characterized and discarded on the basis of mineralogical (and/or geochemical) arguments. It should also be remembered that isotopic data alone are not sufficient to understand metamorphic-related processes inducing reorganizations of rock constituents, which result in the formation of new mineral parageneses.

Reconstruction of the history of sedimentary basins or tectonized rock complexes shall be based on studies combining isotopic dating of authigenic sheetsilicates which provides the period(s) of temperature maxima, and fission-track determinations of detrital apatite (and/or zircon) which complete the reconstruction by constraining the postthermal history of the host rocks.

7.10 Acknowledgements

We are sincerely indebted to M. Frey and D. Robinson for having invited us to contribute to this book and also for their very constructive comments and remarks on previous drafts of this contribution.

8 Application of stable isotope geochemistry to low-grade metamorphic rocks

Z. D. Sharp

8.1 Introduction

The application of stable isotope geochemistry can be divided into two general fields. The first is for thermometry, where the fractionation between any two phases is related to the temperature of their formation. The second is as a tracer of fluid–rock interaction; the isotopic composition of a rock or fluid is used to evaluate the sources of fluids or rock protoliths.

Stable isotope thermometry requires that three conditions are met:

1 The two phases must be able to be separated, purified and analyzed;

2 An independent calibration of the thermometer must be available. The calibration is the relationship between the equilibrium fractionation of the two phases and the corresponding temperature. Empirical, theoretical and experimental methods have been used to ascertain this relationship. Obviously, the use of the thermometer is only as accurate as is the calibration;

3 The two phases must attain isotopic equilibrium at the conditions of 'geological interest'—either peak metamorphic temperature or the time of a hydrothermal interaction, and preserve the peak-temperature isotopic compositions during later cooling.

In low-grade metamorphic rocks, all three of these factors pose problems. Grain sizes are often extremely fine, making mineral separation a formidable and often impossible task, especially where the minerals are often a mixture of detrital and authigenic components. Unless a mineral undergoes complete isotopic re-equilibration, its isotopic composition will have no geological meaning. Some detrital or authigenic materials are often unchanged during low-grade metamorphism, while other phases, such as calcite, may fully re-equilibrate with pore fluids during retrogression. Therefore, application of stable isotope thermometry in low-grade rocks is difficult, and requires an understanding of numerous subtleties in order to obtain meaningful results.

The use of stable isotope geochemistry as a tracer is predicated on different rocks or fluids having distinct and often unique isotopic compositions. Mineral reactions, diagenetic processes and fluid sources can often be constrained by the isotopic compositions preserved in minerals or fluid inclusions. An example comes from an oxygen isotope study of altered volcaniclastic rocks from Honshu, Japan (Kita & Honda, 1987). The rocks were altered over an extended time period, some by hydrothermal fluids and others by diagenetic ones. Both rock types have similar mineralogies, so that conventional methods of distinction are ineffective. Their oxygen isotope compositions differ by 7‰, with the hydrothermally altered samples recording a seawater-derived source, and the diagenetic samples having significantly lower $\delta^{18}O$ values. In another example, Nesbitt and Muehlenbachs (1995) analyzed $\delta^{18}O$ values of vein quartz and calcite, and δD values of associated fluid inclusions from a subgreenschist to amphibolite facies terrane in the Omineca Belt of the Canadian Cordillera. The δD values were mostly below −100‰, requiring a meteoric water source. The $\delta^{18}O$ values were 6–10‰, far higher than expected from meteoric water, and imply a regional oxygen isotope equilibration of the fluids with the host rocks.

Fluid sources in a wide range of rock types can be constrained with stable isotopes, including those associated with low-temperature carbonate diagenesis of sandstones (e.g. Lee, 1987; Macaulay *et al.*, 1993b), hydrothermal systems near oceanic spreading centres (Gieskes *et al.*, 1982; Schiffman & Smith, 1988; Kusakabe *et al.*, 1989), altered ultramafic rocks (Wenner & Taylor, 1973, 1974; Burkhard & O'Neil,

1988) and sandstones and shales (Savin, 1980). In this chapter, these problems are addressed and examples are given to illustrate how stable isotope geochemistry can be used in the study of low-grade metamorphic rocks.

8.2 Terminology

The isotopic ratios of the elements H, C, N, O and S vary in nature as a result of equilibrium and kinetic processes. Kinetic effects are generally larger than equilibrium ones; nevertheless, the total isotopic variation in nature is very small. In order to use the natural isotopic variations that exist to solve geological problems, extremely high analytical precision is required for routine analysis. For example, paleoclimate, oceanographic or high-temperature thermometric studies demand a precision in $^{18}O/^{16}O$ ratios as high as 2.0×10^{-8}, thus making stable isotope geochemistry one of the most precise geochemical measurements in the geological sciences. In order to achieve such a high level of precision, special mass spectrometry and sample preparation techniques have been developed.

The isotopic composition of a phase is reported relative to a standard using the delta notation (McKinney *et al.*, 1950), defined as follows:

$$\delta = \left(\frac{R_x - R_{std}}{R_{std}} \right) \times 1000. \tag{1}$$

In eqn 1, R_x is the ratio of the heavy to light isotope of the sample, and R_{std} is the corresponding ratio of the standard. For H, C, N, O and S, the isotopic ratios are D/H, $^{13}C/^{12}C$, $^{15}N/^{14}N$, $^{18}O/^{16}O$ and $^{34}S/^{32}S$. The isotopic composition is reported in parts per thousand (or per mil ‰), with positive values indicating that the sample has a higher isotopic ratio than the standard and a negative value indicating a lower one.

Precision of isotopic measurements made with modern equipment can be as high as 0.01‰. It is not possible to determine the absolute isotope ratio at this level of precision. By reporting data relative to internationally accepted standards, data can be presented with the necessary precision to address a wide range of geological problems. Oxygen and hydrogen are generally reported relative to standard mean ocean water (SMOW), where the $\delta^{18}O$ and δD values of SMOW are defined as zero. Carbon

is reported relative to PDB (PeeDee Belemnite), where the $\delta^{13}C$ value of PDB = 0. Nitrogen is reported relative to air = 0‰ and sulphur is relative to CDT (Cañon Diablo Troilite) = 0‰. Oxygen is also reported relative to PDB, where $\delta^{18}O_{PDB} = 1.03091(\delta^{18}O_{SMOW}) + 30.91$[1].

The isotope fractionation between any two phases is defined by the following equation:

$$\alpha = \frac{R_a}{R_b} \tag{2}$$

which corresponds to

$$\alpha = \frac{1000 + \delta_a}{1000 + \delta_b} \tag{3}$$

in delta notation. In stable isotope thermometry, the expression

$$1000 \ln \alpha = \frac{a \times 10^6}{T^2} + b \quad (T \text{ in Kelvin}) \tag{4}$$

or a variation of eqn 4 (e.g. Clayton & Kieffer, 1991) accurately describes the isotopic fractionation between any two phases as a function of temperature (Bottinga & Javoy, 1973). Many authors have substituted Δ_{a-b} for $1000\ln\alpha$ in eqn 4 (where $\Delta^{18}O_{a-b} = \delta^{18}O_a - \delta^{18}O_b \approx 1000\ln\alpha$). Although the difference between $\Delta^{18}O_{a-b}$ and $1000\ln\alpha_{a-b}$ is small, there is still no excuse for making this approximation given the wide availability of hand-held calculators.

8.3 Mass spectrometry

With very few exceptions, all stable isotope ratios are measured as gases using a gas source, isotope ratio mass spectrometers (IRMS). Samples are measured relative to a reference gas, whose composition has been determined independently. In the standard inlet system, the sample and reference gas are introduced into separate variable volume reservoirs (bellows).

---

[1] Other standards are SLAP (Standard Light Antarctic Precipitation; $\delta^{18}O \equiv -55.5‰$, and $\delta D \equiv -428‰$ vs. SMOW), GISP (Greenland Ice Sheet Precipitation; $\delta^{18}O \equiv -24.78‰$, and $\delta D \equiv -189.7‰$ vs. SMOW), NBS-19 (a carbonate with a $\delta^{18}O \equiv -2.2‰$, and $\delta^{13}C \equiv 1.95$ vs. PDB). Standards can be obtained from the IAEA in Vienna or NIST in the United States. For a more complete description of isotopic standards for stable isotope geochemistry, see O'Neil (1986a) and IAEA-TECDOC-825 (1995).

The volumes (and relative pressures) are controlled by compressing or expanding the bellows assembly. The gases bleed through a capillary system and into a switching block, a set of four valves that allow for either the reference or sample gas to enter the ionization source of the mass spectrometer while the other enters a waste trap. By repeated switching of the reference and sample gas, the isotopic composition of the sample can be determined relative to the reference gas.

Recently, a new carrier gas method of sample introduction has been developed for isotope ratio mass spectrometers that allows extremely small samples to be analyzed (Matthews & Hayes, 1978). Samples are carried into the mass spectrometer in a He carrier gas stream. This technique, the Isotope Ratio Monitoring Mass Spectrometer system (or IRMMS), is illustrated in Fig. 8.1. Unlike the conventional inlet system, the entire sample is swept into the mass spectrometer as a single pulse. The intensities of the different masses of a gas (e.g. masses 44, 45 and 46 for CO_2) vary with time in a gaussian-like form, and the isotopic ratios are determined by integrating the total area under each peak. Because the pressure of the gas entering the mass spectrometer is controlled by the He carrier gas, and not the sample, sample size is limited only by signal strength. As long as the signal-to-noise ratio is sufficiently high, a precise isotope ratio can be obtained. Many of the new microsampling systems that are being developed take advantage of the IRMMS system. Although applications to low grade

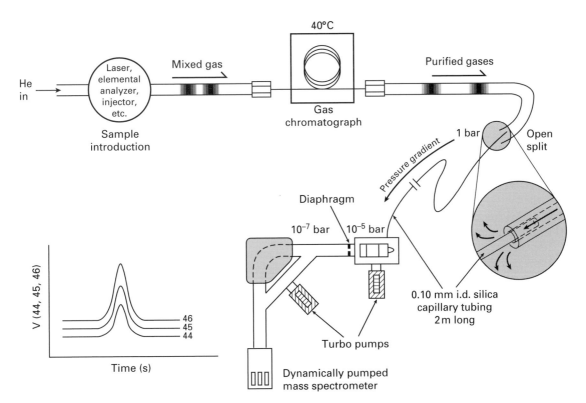

Fig. 8.1 Schematic diagram of the Gas Chromatography, Isotope Ratio Monitoring Mass Spectrometry (GC-IRMMS) system. The sample is introduced into a helium flow at 'Sample introduction,' is entrained in the helium stream and passes through the 'Gas chromatograph'. The different gases are separated in the GC so that each arrives at the 'Open split' at different times. A portion of the gas enters the 2-m-long capillary connected to the mass spectrometer. As the gas enters the mass spectrometer, a gaussian-like peak for all masses (e.g. 44, 45 and 46 for CO_2) is measured as a function of time. The total area under each peak can be compared to similar ratios obtained on a standard gas in order to obtain the isotopic ratio of the unknown.

metamorphic rocks are just beginning, their numbers are sure to increase in the near future.

8.4 Sample extraction techniques

Most phases of geological interest cannot be analyzed directly, but first must be converted to an appropriate gas. The method of conversion must be reproducible at a very high level of precision in order to measure isotopic differences at the natural level. Two extraction 'philosophies' are used; either the sample is converted quantitatively (e.g. 100% conversion to the gas), or it is partially converted with a reproducible fractionation. While the first of these philosophies may sound better, some of the most precise extraction techniques (oxygen from water and carbon and oxygen from carbonates) both employ the second.

8.4.1 Carbonates

The simplest, most elegant and longstanding extraction technique is that for carbonates (McCrea, 1950). Carbonates are reacted with 100% phosphoric acid in an evacuated vessel to produce CO_2 gas that is purified cryogenically and then analyzed in the mass spectrometer. The $\delta^{13}C$ and $\delta^{18}O$ values of the CO_2 gas are related to those of the original carbonate. All of the carbon from the carbonate reacts to form CO_2; however, only two-thirds of the oxygen from the carbonate is incorporated into the CO_2. The $\delta^{18}O$ value of the carbonate is related to that of the CO_2 gas by the equation

$$\alpha_{CO_2-calcite} = \frac{1000 + \delta^{18}O_{CO_2}}{1000 + \delta^{18}O_{calcite}}. \qquad (5)$$

If the alpha value (α) is known, then the $\delta^{18}O$ value of the calcite can be obtained from eqn 5 using the measured $\delta^{18}O$ value of the CO_2 gas. The alpha value for a given carbonate varies as a function of temperature, and for that reason, the reaction must be made at a constant temperature, in either a water bath or constant temperature oven. The actual temperature of reaction is not important, but the alpha value at the reaction temperature must be known. Originally, all reactions were made at 25°C or less over a period of 12h to several days. Reaction temperatures for modern, on-line automated systems

are as high as 80°C in order that the reaction is complete within several minutes.

The carbonate extraction method is one of the most precise and foolproof. The sample sizes for on-line, commercially available extraction systems are limited only by the gas requirements of the mass spectrometer (e.g. Wada, 1988; Dettman & Lohmann, 1995). Numerous descriptions of the carbonate method with minor modifications of the original McCrea have been published (Bowen, 1966; Cornides & Kusakabe, 1977; Videtich, 1981; Coplen et al., 1983; Fairchild et al., 1988; Coleman et al., 1989; Swart et al., 1991). There are many subtle effects which can lead to unwanted fractionations, including those of grain size, mixed carbonates, and the presence of organic matter or sulphur compounds (Charef & Sheppard, 1984; Spangenberg et al., 1995). Other than the problem of mixed carbonates in the same sample, these problems can generally be ignored in low-grade metamorphic rocks. Calcite and dolomite are often present in the same rock sample. If there is an interest in determining the fractionation between calcite and dolomite, such as for temperatures of equilibration (Sheppard & Schwarcz, 1970), stepwise fraction techniques are used (Al-Aasm et al., 1990).

8.4.2 Silicates

The silicate extraction method was first described in the early 1950s (Baertschi & Silverman, 1951; Silverman, 1951) and has been modified by Taylor and Epstein (1962) and Clayton and Mayeda (1963). The technique involves oxidation of a silicate by reaction with fluorine or a halogen fluoride at high temperature which quantitatively liberates O_2 gas from the silicate; e.g.

$$5\,SiO_2 + 4\,BrF_5 \xrightarrow{\ 550°C\ -12\,hrs\ } 5\,SiF_4 + 4\,Br + 5\,O_2. \qquad (6)$$

Generally F_2, ClF_3 or, most commonly, BrF_5 are used as the fluorination agent. The latter two halogen fluorides have the advantage in that they can be transferred within a vacuum line cryogenically, whereas F_2 gas is non-condensible at LN_2 temperatures. Molecular fluorine does have the advantage in that it does not fractionate oxygen isotopes during partial fluorination as the halogen fluorides appear to do (Garlick & Epstein, 1967).

Samples are reacted with the (halogen) fluoride in heated nickel tubes. Total fluorination can take anywhere from several hours for reactive glasses or feldspars to days at temperatures as high as 700–750°C for 'refractory' phases such as kyanite and olivine) (Vennemann & Smith, 1990). Sample sizes of 5–15 mg are generally used in order to avoid a blank contribution, although the system has been miniaturized for analysis of smaller samples (Lee *et al.*, 1980).

Following reaction, the evolved O_2 gas is purified cryogenically and converted to CO_2 by reaction with a hot graphite rod catalyzed with Pt wire. Alternatively, the O_2 gas can be transferred quantitatively by adsorption on a 5-Å molecular sieve cooled to liquid nitrogen temperatures and analyzed as molecular oxygen.

Zeolites and some clays contain exchangeable channel or interlayer water (e.g. Savin & Epstein, 1970a) which does not provide geologically meaningful information. This water must be removed to obtain $\delta^{18}O$ values of the framework oxygen. A number of methods have been described, such as desiccation, partial fluorination and vacuum heating. In the case of zeolites, Stallard and Boles (1989) determined $\delta^{18}O$ values and water contents of zeolites preheated under vacuum in the nickel reaction vessels over a range of temperatures. Although they were able to get reproducible results for samples heated above 150–300°C, they realized that water released during vacuum heating might interact with the NiF coating on the inside of the reaction vessels, thus contaminating the silicate residue. Karlsson and Clayton (1990) found similar contamination effects, and proposed that the samples must be heated in vacuum off-line prior to loading in the nickel vessels.

A recent modification of the fluorination technique for silicates involves using a laser as a heat source (Sharp, 1990, 1992; Elsenheimer & Valley, 1992). The basic fluorination reaction is the same, but in place of nickel bombs heated by external furnaces, the sample is heated internally by the laser. The major advantages of the laser system are that very small amounts of material can be analyzed and fluorine-resistant phases can easily be reacted (e.g. Sharp, 1995). Ideally, *in situ* analyses can be made with the laser system, with the reaction occurring only at the point where the laser beam impinges on the sample surface. Some successful applications of the *in situ* technique have been made with a CO_2 laser (Chamberlain & Conrad, 1991, 1993; Elsenheimer & Valley, 1992; Kirschner *et al.*, 1993), but the most promising results are being made with ultraviolet laser systems (Wiechert & Hoefs, 1995; Rumble & Sharp, 1998). The ultraviolet laser minimizes the thermal halo that is generated during laser interaction with the sample surface. The major disadvantage of the laser technique is that all samples are exposed to fluorine during reaction. If minerals that partially fluorinate at room temperature are included in the sample block, there is a risk of contamination. This problem is especially severe for very low-grade materials such as clays, which are particularly susceptible to room-temperature fluorination.

8.4.3 Sulphides and sulphates

Sulphides are analyzed by conversion to either SO_2 or SF_6 (Rees, 1978; Rees & Holt, 1991). The SO_2 method is the easier of the two, and can be used with most mass spectrometers (Thode, 1949; Thode *et al.*, 1953). The method involves oxidation of the sulphide using either a solid oxidant (CuO, V_2O_5) or O_2 gas via a reaction such as

$$Ag_2S + 2\,CuO \xrightarrow{\text{High T}} Ag + Cu + SO_2. \qquad (7)$$

This reaction must be run at high temperatures to prevent the production of SO_3, which is favoured over SO_2 at lower temperatures. The SO_2 technique has two disadvantages:

1 The $\delta^{34}S$ value cannot be determined unambiguously; there is a contribution to the heavy isotope of SO_2 by oxygen. The two masses measured are 64 and 66; mass 64 is due to $^{32}S^{16}O_2$, while mass 66 can be either $^{34}S^{16}O_2$ or $^{32}S^{16}O^{18}O$. An oxygen correction needs to be made, but this is a relatively straightforward procedure, obtained by analyzing reference standards for which the $\delta^{34}S$ value is known.

2 SO_2 is not a 'clean' gas; it can dirty the source of the mass spectrometer and many laboratory researchers are reluctant to allow it to 'soil' their inlet system.

The SF_6 method (Thode & Rees, 1971; Rees, 1978) involves the fluorination of a sulphide by a reaction of the form

$$Ag_2S + 2\,BrF_5 \xrightarrow{\text{High T}} 2\,AgF_2 + SF_6 + Br. \qquad (8)$$

The SF_6 method requires no correction for isobaric interferences and SF_6 is a clean gas. It is very accurate and can be used to analyze for the rare isotopes ^{33}S and ^{36}S (relevant only to extraterrestrial materials). The problem with the SF_6 method is that it is a more difficult procedure, and requires a specially configured, large radius mass spectrometer.

8.4.4 Graphite and organic matter

The measurement of $\delta^{13}C$ values from graphite and organic matter is one of the oldest and most straightforward techniques developed for stable isotope geochemistry (Murphey & Nier, 1941). The method entails oxidation of organic matter by reaction with a metal oxide or oxygen gas. Samples are placed in silica tubes with the metal oxide (CuO or V_2O_5), evacuated and sealed. The samples are heated at 900–1000°C for minutes to 1 h causing them to oxidize. The products from organic matter include H_2O, which requires an additional off-line cryogenic purification step to separate the CO_2 and H_2O. In order to reduce costs, some laboratories seal the samples in Pyrex© tubes, followed by heating to 550°C for 12 h. Very small amounts of material can be analyzed and precision is very high.

8.4.5 Hydrogen

Hydrogen from water or hydrous minerals or glasses is analyzed in the mass spectrometer as H_2 gas. The extraction methods are therefore ones of quantitatively converting the hydrogen in the mineral to H_2. Two methods have been used for this — uranium reduction and zinc reduction (Bigeleisen *et al.*, 1952; Friedman, 1953; Vennemann & O'Neil, 1993). The highest precision data are obtained with the Bigeleisen method, which involves reduction of H_2O with hot uranium metal. The mineral sample is heated in an induction furnace to temperatures of 1100–1300°C to liberate water and H_2 (the latter is produced by the reaction of water with ferrous iron). The water is passed over a uranium furnace at 800°C, causing reduction. The H_2 gas generated by this reduction is transferred within the vacuum line either by adsorption on activated charcoal cooled to liquid nitrogen temperatures, or with a toepler pump. The precision of the technique can be better than ± 1‰. The obvious disadvantage of this method is that it uses uranium, which is of some environmental concern.

The zinc reduction method is the off-line reduction of water to H_2 + zinc oxide. The sample can be heated in an induction furnace as before; alternatively, the mineral is loaded in a quartz tube and heated with a gas-oxygen flame (Vennemann & O'Neil, 1993). Water and H_2 are liberated during heating. The H_2 is converted to water by reaction with a copper oxide furnace. The water is then transferred to an evacuated Pyrex® tube preloaded with zinc metal. The tube is sealed off and the water–zinc mixture is heated to 500°C to reduce the water to H_2. The method is simple and safe, but the reproducibility, generally of the order of ± 3–5‰, is less than that of the uranium method. In addition, special zinc is required in order to ensure that the reduction of water is quantitative (e.g. Kendall & Coplen, 1985). For both of the hydrogen extraction methods, an equivalent of 1 mg of water is generally needed for analysis.

Hydrogen isotope analysis of smectite is complicated by the easily exchanged interlayer water, which can exchange with the atmosphere in days (Savin & Epstein, 1970a). Heating the samples *in vacuo* at 200°C effectively removes the interlayer water (Marumo *et al.*, 1995) so that the δD value of the structural water can be measured.

8.4.6 Gas chromatography isotope ratio mass spectrometry methods

The GC-IRMS method of stable isotope analysis (Matthews & Hayes, 1978) is an alternative to the conventional extraction line–dual inlet dynamic mass spectrometry technique. The GC-IRMS technique has a number of advantages over the conventional extraction methods. Analysis times are extremely rapid, sample preparation is minimized, sample sizes are greatly reduced, and gas purification is made automatically by the GC column. The disadvantages are that precision is only about ± 0.2‰, and older mass spectrometers often cannot be retrofitted for this technique. The use of the GC-IRMS system works in tandem with a gas extraction technique ('Sample introduction' in Fig. 8.1). These include the elemental analyzer, GC system for analysis of individual organic compounds and laser extraction techniques.

The elemental analyzer is a method for bulk analysis of $\delta^{13}C$, $\delta^{15}N$ and $\delta^{34}S$ values from solids or liquids. The system consists of a pair of in-line furnaces; the first oxidizes the sample and the second reduces the gases generated during the oxidation. After high-temperature oxidation/reduction, the gases pass through the gas chromatograph, separating the CO_2, N_2 and SO_2, and finally into the mass spectrometer for analysis. It is possible to measure $\delta^{13}C$, $\delta^{15}N$ and $\delta^{34}S$ values on the same sample if the mass spectrometer is equipped with the appropriate software to jump rapidly from focusing conditions for N_2, to CO_2, to SO_2.

Organic materials (including graphite), sulphides, sulphates and liquids can all be analyzed with the elemental analyzer. Samples as small as $1\,\mu g$ C equivalent can be analzyed, although there is an unavoidable carbon blank inherent in the system.

The system has also been 'inverted', whereby volatile organic materials are first separated by a gas chromatograph and then oxidized to produce CO_2 and N_2 (Hayes et al., 1990; Freeman, 1991; Merritt et al., 1995). Samples are dissolved in an appropriate solute and injected into the GC system. Individual compounds are separated and oxidized prior to entry into the mass spectrometer. Most studies have concentrated on unmetamorphosed or extraterrestrial materials (Engel et al., 1990), although the metamorphic community is beginning to appreciate this method (e.g. Spangenberg & Macko, 1996).

A laser can be incorporated into the GC-IRMS system for in situ analysis of carbonates, biogenic phosphates and sulphides (Cerling & Sharp, 1996; Sharp & Cerling, 1996; Rumble & Sharp, 1998, and unpublished observations). The laser heats the sample to temperatures in excess of 2000°C. CO_2 (or SO_2) evolved during laser heating is entrained in the helium stream and passes through the GC column and into the mass spectrometer. CO_2 is generated by decarbonation. Sulphide analysis with the laser system is accomplished by heating the sample in an O_2 atmosphere, similar to the static procedure originally described by Crowe et al. (1990) and Kelley and Fallick (1990). The advantage of the GC method is twofold: (i) it is straightforward to purify the SO_2 gas and (ii) the amount of SO_2 that

enters the mass spectrometer is minuscule, so that there is no danger of contaminating the inlet system and source of the mass spectrometer.

The GC techniques have a number of advantages over their conventional mass spectrometric counterparts. Sample sizes are greatly reduced, gas purification is efficient and rapid, and with the laser system, analyses can be made in situ with a spatial resolution of c. $200\,\mu m$ diameter or better. Samples of carbonate less than $1\,\mu g$ can be analyzed. The GC methods are still in their infancy, but will no doubt play a far more important role in the coming years and will be modified for other mineral phases such as silicates and sulphates.

Common analytical methods are presented in Table 8.1. The purpose of Table 8.1 is to convey a sense of the difficulty and precision of each method. The details are enormously variable from laboratory to laboratory, and the table is meant as a rough guideline to anyone considering an isotopic study.

8.5 Isotopic reservoirs

Rocks and fluids have distinct isotopic compositions. The types of interactions that occur during metamorphism can often be identified on the basis of these unique isotopic compositions of rocks or fluids. With a general understanding of the isotopic values of different fluids and rock types, the reader is better able to evaluate the feasibility of an isotope-based research project. Characteristic compositions are outlined in the following section.

8.5.1 Water

8.5.1.1 Ocean water

The major terrestrial water reservoir, by far, is the oceans, comprising 98% of the water in the hydrosphere. The oceans are the source of all meteoric water and the oxygen isotope composition of the ocean is buffered by interaction with mantle-derived material on the ocean floor. Ocean water is, therefore, the reference material for oxygen and hydrogen, with an isotopic composition for both phases defined as zero on the SMOW scale. There are small variations in the $\delta^{18}O$ and δD values of the oceans due to mixing with meteoric water and evaporation (Craig & Gordon, 1965; Redfield &

Table 8.1 Generalized analytical procedures for different materials.

Mineral	Method	No. of analyses per day	Minimum sample size	Difficulty	Precision ‰
Carbonates	Acid digestion	c. 18[1]	1–10 mg[2]	Easy	0.05
Silicates, oxides	Fluorination	5–15	2–10 mg	Difficult	0.15
Silicates, oxides	Laser fluorination	10–20	0.2 mg	Difficult	0.15
Graphite	Oxidation	10–20	c. 1 mg	Straightforward	0.1
Hydrogen in hydrous phases	Conversion to H_2	5–10	10–30 mg	Difficult	3–5
Hydrogen, carbon, oxygen from fluid inclusions	Decrepitation and oxidation (methane) or reduction (H_2O to H_2)	5 (degassing and gas transfer can be time consuming)	Depends on density, c. 1 mg of 'fluid'	Ranges from straightforward to difficult	Generally lower precision than other methods
Phosphates	Wet chemistry; reduction	5–10	10 mg	Difficult	0.2–0.4
Phosphates	Laser	>20	10 µg	Moderate	Lower precision than conversion method

[1] Many more samples can be measured if the laboratory is equipped with an automated extraction line.
[2] Sample sizes of less than 100 µg can be measured with on-line extraction lines (e.g. Dettman & Lohmann, 1995).

Friedman, 1965), but outside ocean studies, these very small variations can be ignored.

The oxygen isotope composition of the ocean is buffered to its present value of zero by high-temperature hydrothermal activity occurring at spreading centres and simultaneous lower temperature alteration farther from the hot spreading zones (Muehlenbachs & Clayton, 1976). Small temporal variations in $\delta^{18}O$ values of the oceans occur in response to addition or removal of large quantities of water to the polar ice caps. The hydrogen and oxygen isotope composition of polar ice is far lower than those of the oceans. In times of large accumulations of ice, the $\delta^{18}O$ values of the oceans rise by 1.3–1.5‰ (Savin & Yeh, 1981), while during warmer times, such as the pre-Oligocene, the lack of ice caps caused a lowering of the $\delta^{18}O$ values of the oceans by c. 1‰. There is some controversy as to whether the isotopic composition has always been buffered to its present value (e.g. Walker & Lohmann, 1989; Gregory, 1991; Holmden & Muehlenbachs, 1993).

8.5.1.2 Meteoric water

Meteoric waters are 'waters that fall from the sky'. The source of all meteoric waters is evaporation from the ocean; the light isotopes preferentially escape into the vapour phase. The factors that control the isotopic compositions of meteoric waters are complex, and not completely understood. The equilibrium fractionation between water and vapour at 25°C is 9.1‰. Water evaporates from the ocean surface with a kinetic fractionation of c. −13‰, larger than the equilibrium fractionation. Condensation of water from vapour, on the other hand, is an equilibrium process. The fractionation is controlled by the temperature of condensation. As water with a higher $\delta^{18}O$ value than the vapour is removed from the system, the remaining vapour becomes successively lighter. Precipitation obeys a Rayleigh fractionation law, with the $\delta^{18}O$ and δD values of precipitation becoming lighter with increasing distance from the oceans, increasing altitude and decreasing temperature (Dansgaard, 1964; Siegenthaler

& Oeschger, 1980). The hydrogen and oxygen isotope compositions of meteoric water obey a simple linear relationship where $\delta D = 8*(\delta^{18}O) + 10$ (Craig, 1961). Waters at high latitude and elevation have $\delta^{18}O$ and δD values that are depleted by over 50 and 400‰, respectively, lower than SMOW.

8.5.1.3 Magmatic and metamorphic water

There have been a number of attempts to define the isotopic composition of waters associated with magmatic and metamorphic rocks (Sheppard, 1986). Most determinations of the isotopic compositions of magmatic/metamorphic waters are indirect. The isotopic compositions of minerals are measured, and using estimated temperatures of formation and available mineral–water fractionation factors, the δD and $\delta^{18}O$ values of the coexisting waters are calculated. More direct evidence is obtained from fluid inclusions or from water vapours collected from volcanic exhalatives. Sheppard's estimates of the magmatic and metamorphic water fields are given in Fig. 8.2.

8.5.2 Rocks

Mantle-derived basalts have a very narrow range of $\delta^{18}O$ values of 6 ± 0.3‰ (Harmon & Hoefs, 1995). The δD values are $c.$ −80‰, although this is less well known (Boettcher & O'Neil, 1980). Terrestrial, low-temperature processes broaden this range, generally causing an enrichment in ^{18}O. Sedimentary rocks are composed of mixed detrital and authigenic components. The detrital components are weathering products such as quartz and clays. Non-carbonate sediments have an average $\delta^{18}O$ value of 18‰ (Savin & Epstein, 1970b,c). Archaean sediments tend to have lower $\delta^{18}O$ values because the weathering products are largely derived from igneous rock fragments (Longstaffe & Schwarcz, 1977). Authigenic sediments can have very high $\delta^{18}O$ values, because they form at low temperatures, where the fractionation between water and minerals is very large. At 25°C, the fractionation between a mineral and water is of the order of 30‰. Generally, the $\delta^{18}O$ values of sedimentary rocks ranges from 8 to 25‰ (Longstaffe, 1983), although extreme values up to 44‰ have been measured (O'Neil & Hay, 1973).

High-temperature hydrothermal interaction with meteoric waters causes a decrease in the $\delta^{18}O$ value of a rock (e.g. Taylor, 1971; Taylor & Forester, 1971). The final $\delta^{18}O$ value of a rock undergoing hydrothermal alteration is a function of the isotopic composition of the fluid, the temperature of interaction and the degree of interaction (fluid–rock ratio). Information on the metamorphic history of a rock can often be obtained by comparing its measured values with the typical 'generic' ones. The following list gives typical oxygen isotope ratios for various rock types:

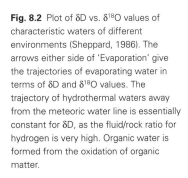

Fig. 8.2 Plot of δD vs. $\delta^{18}O$ values of characteristic waters of different environments (Sheppard, 1986). The arrows either side of 'Evaporation' give the trajectories of evaporating water in terms of δD and $\delta^{18}O$ values. The trajectory of hydrothermal waters away from the meteoric water line is essentially constant for δD, as the fluid/rock ratio for hydrogen is very high. Organic water is formed from the oxidation of organic matter.

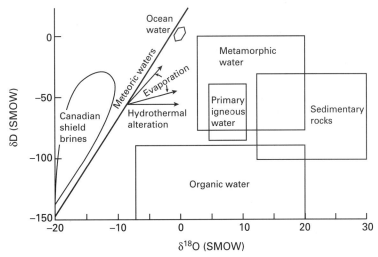

• *Sedimentary carbonate*. The $\delta^{18}O$ value of primary carbonates is 25–30‰ (SMOW). The $\delta^{13}C$ values are −5 to 0‰ (PDB). There are exceptions, such as lacustrine carbonates or those derived from methanogenesis, where $\delta^{13}C$ values can be very high. Low-temperature diagenetic alteration always decreases the $\delta^{18}O$ value of carbonate.

• *Marbles*. The $\delta^{18}O$ values of marbles span a wide range, from less than 10‰ to near sedimentary values. The shift from primary sedimentary values is related to the degree of interaction with metamorphic fluids. Large, massive marbles often preserve their sedimentary $\delta^{18}O$ values, while siliceous marbles, or those interbedded with schists commonly have much lower values.

• *Clastic sediments*. Clastic sediments are often a mixture of detrital and authigenic material. The $\delta^{18}O$ values generally range from 13 to 18‰. Quartz in modern day sediments ranges from 15 to 19‰.

• *Mafic rocks*. Basalts, gabbros, syenites, anorthosites, andesites, etc. have very uniform $\delta^{18}O$ values of 5.5–7.0‰. This reflects the rather constant composition of the mantle reservoir. There is a general correlation between $\delta^{18}O$ and wt% SiO_2, as a result of fractionational crystallization and incorporation of crustal sialic material.

• *Granitoids*. Primary I-type granites have $\delta^{18}O$ values of 8–10‰; S-type granites have $\delta^{18}O$ values of 10–12‰ and higher δD values due to incorporation of sedimentary material (O'Neil & Chappell, 1977). The combination of crustal assimilation and hydrothermal interaction spreads the $\delta^{18}O$ values of granitic rocks to a range of −5 to 15‰ (Taylor, 1978, 1980; Nabelek *et al.*, 1983).

• *Sea-floor basalts*. The $\delta^{18}O$ values of MORB ($\delta^{18}O$ value of 5.7‰) are altered by hydrothermal processes. High-temperature alteration near the spreading ridge causes a depletion in the $\delta^{18}O$ values, while low-temperature alteration leads to an increase of the $\delta^{18}O$ values (Muehlenbachs & Clayton, 1976; Gregory & Taylor, 1981). The overall range is *c.* 2 to >10‰.

• *Eclogites*. The $\delta^{18}O$ values of mantle eclogites have a range similar to sea-floor basalts. The large range is attributed to inheritance of the $\delta^{18}O$ values of their precursors, namely the sea-floor basalts themselves (Garlick *et al.*, 1971; MacGregor & Manton, 1986; Ongley *et al.*, 1987).

• *Veins*. The isotopic composition of quartz veins has been measured in a wide range of rock types and spans a considerable range of close to 20‰ (e.g. Sharp *et al.*, 1993). Veins can be in equilibrium with the host rock, indicating low fluid rock ratios and rock-buffered conditions (Kirschner *et al.*, 1995) or out of equilibrium with the host rock, indicating a non-equilibrium condition and higher fluid rock ratios (Crespo-Blanc *et al.*, 1995). In many cases they provide the most direct measure of the isotopic composition of fluids attending metamorphism.

8.6 Thermometry

8.6.1 Basic fractionation equations

Oxygen isotope fractionation between gases (phases *x* and *y*) obeys the equation

$$1000 \ln \alpha_{x-y} = \frac{a \times 10^6}{T^2} \text{ (T in Kelvin)} \qquad (9)$$

at temperatures above 25°C. Bottinga and Javoy (1973) proposed a general equation for solids,

$$1000 \ln \alpha_{x-y} = \frac{a \times 10^6}{T^2} + b \text{ (T in Kelvin)} \qquad (10)$$

where the *b* term is nonzero for hydrous phases and water. The validity of the constant *b* has recently been questioned (e.g. Chacko *et al.*, 1996), nevertheless, the Bottinga and Javoy formula with its *b* constant continues to receive wide acceptance. Clayton and Kieffer (1991) have expanded eqn 9 to a polynomial form, where

$$1000 \ln \alpha = ax + bx^2 + cx^3; \, [x = 10^6/T^2]. \qquad (11)$$

Other forms of the fractionation equation are simple polynomials of $1/T$ of second order (e.g. Zheng, 1993) to fourth order (Savin & Lee, 1988).

8.6.2 Mineral fractionation factors

Fundamental to the application of stable isotope thermometry is the availability of accurate calibrations of the thermometer. In spite of the enormous effort that has been expended on calibrating stable isotope thermometers, there are still a number of problems and uncertainties (e.g. O'Neil, 1986b; Kyser, 1987). Experimental calibrations for most minerals are lacking below *c.* 500°C (e.g. Bottinga

& Javoy, 1975; Chiba *et al.*, 1989). There are only a few minerals (carbonates, titanium oxide) that have been precipitated at low temperatures in isotopic equilibrium with their host fluid (McCrea, 1950; O'Neil *et al.*, 1969; Bird *et al.*, 1993). Feldspar and quartz have been experimentally studied at somewhat higher temperatures (O'Neil & Taylor, 1967; Clayton *et al.*, 1972), although there are even some discrepancies about a system as 'simple' as quartz and calcite in the range defining low-grade metamorphic rocks (Sharp & Kirschner, 1994).

The problem is that exchange rates for most minerals at low temperature are too slow to be investigated by laboratory experiments. O'Neil and Kharaka (1976) attempted to equilibrate clay minerals and water over a temperature range of 100 to 300°C. Exchange rates were rapid for hydrogen, but negligible for oxygen, even for reaction times approaching 2 years! For most phases, exchange must be made at high temperatures, and the results extrapolated to temperatures appropriate for very low-grade metamorphism. This is a source of considerable uncertainty. In addition, a number of minerals commonly found in low-grade metamorphic rocks, such as clay minerals, are stable only at low temperatures, so that experimental investigations of these phases cannot be made at higher temperatures.

Statistical mechanical calculations of fractionation factors have been made for solids (e.g. Bottinga, 1969; Kieffer, 1982), although for most complex minerals the uncertainties are too high for application to thermometry. More often, the statistical mechanical calculations are combined with experimental data permitting extrapolation of the experimental results to lower temperatures (Clayton & Kieffer, 1991; Chacko *et al.*, 1996). Another 'semitheoretical' approach is to use empirical bond-type calculations (Taylor & Epstein, 1962; Savin & Lee, 1988; Zheng, 1993). The concept is based on the assumption that the fractionation between two minerals can be approximated by the sum of the different bond types that each mineral contains. For example Si–O–Si, Si–O–Al, Al–O–Al, and Si–O–M bonds all have different bond strengths. By assigning values to each of these bonds, fractionation factors for a mineral can be calculated by summing the contributions for each of the different bond types that the mineral contains. The method is not strictly quantitative, but by adjusting the results to match experimental and

natural data, these techniques often provide the only fractionation estimates for a mineral system.

Fractionation factors for low-temperature metamorphic minerals are often best determined from natural samples. This is truly an empirical approach, but Nature provides us with an excellent laboratory. By measuring the isotopic composition of minerals that formed or equilibrated in a 'simple' system, we are in effect measuring the run products of a reaction that may have taken place over many thousands or millions of years, and would be impossible to duplicate in the laboratory. Often, a natural fractionation gives data only at a single temperature, but these data provide a calibration point for the empirical bond-type calculations discussed above. The combination of the bond-type calculations and the natural data allow for an extrapolation over a wider temperature range and to different mineral phases.

Savin and Epstein (1970a) measured the oxygen and hydrogen isotope composition of kaolinite of weathering origin. The delta values have a near constant shift from those of the local meteoric waters. They obtained fractionation factors $\alpha^{ox}_{kaolinite-water} = 1.0265$ and $\alpha^{hy}_{kaolinite-water} = 0.970$ for kaolinite–water at an assumed weathering temperature of 17°C. Eslinger (1971) obtained kaolinite–water fractionation data from hydrothermal kaolinite (150°C) from Broadlands, New Zealand. A compilation of these data with additional unpublished experimental data and the bond-type calculations were presented by Savin and Lee (1988) and are shown in Fig. 8.3. The combination of natural data and empirical calculations provide the best estimates for the kaolinite–water fractionation factors. Savin and Lee (1988) have calculated fractionation factors for pyrophyllite, smectite, mixed-layer illite/smectite, illite, serpentine, talc, chlorite and brucite.

There have been several attempts to estimate temperatures of formation using single minerals. If a mineral has two or more distinct crystallographic sites for oxygen (or hydrogen or sulphur), then the isotope fractionation between the two sites should be related to the temperature of equilibration. Clays are the most studied single mineral thermometer. They have oxygen in hydroxyl and non-hydroxyl sites, and if the $\Delta^{18}O$ value between the two crystallographic sites within the clay mineral is temperature dependent *and* can be measured, then there is the potential for a single mineral thermometer

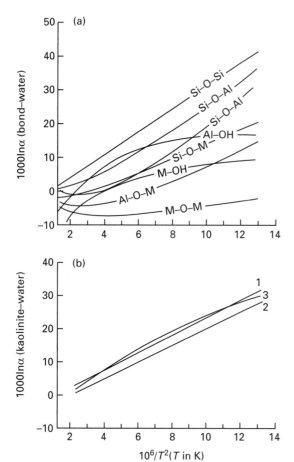

Fig. 8.3 Results of bond-type calculations of Savin and Lee (1988). (a) The oxygen isotope fractionations for different cation–oxygen vs. water bonds. (b) The combination of these bond calculations for the kaolinite–water fractionations (curve 3). Also shown are the empirical kaolinite–water fractionations of Land and Dutton (1978) (curve 1) and Kulla and Anderson (1978) (curve 2). (From Savin & Lee, 1988.)

could be used for thermometry. The $\Delta^{18}O_{clay\ mineral-water}$ and $\Delta D_{clay\ mineral-water}$ values are a function of temperature, and the $\delta^{18}O$ and δD values of meteoric water vary linearly with respect to one another (Craig, 1961). Delgado and Reyes argued that if the $\delta^{18}O$ and δD values of the clay mineral are known, the three remaining unknowns, δD_{water}, $\delta^{18}O_{water}$ and temperature, can be determined from the three independent equations (Fig. 8.4). The approach depends on both hydrogen and oxygen isotopes equilibrating with meteoric water at the same temperature. These conditions are probably rarely all met, so that its applicability may not be widespread in the field of very low-grade metamorphism.

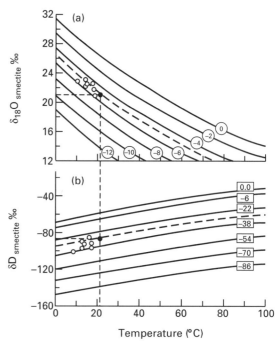

Fig. 8.4 Plot of the $\delta^{18}O$ and δD values of smectite in equilibrium with water as a function of temperature. The curved lines represent constant delta values for the water ($\delta^{18}O$ in circles, δD in boxes). For a measured $\delta^{18}O$ (smectite) value of 21‰ and a measured δD (smectite) value of −85‰, there is only one temperature where the corresponding $\delta^{18}O$ and δD values of the water fall on the meteoric water line (21°C, $\delta^{18}O$ (water) = −5‰, δD (water) = −30‰). (Modified from Delgado & Reyes, 1996.)

(Hamza & Epstein, 1980; Ustinov & Grinenko, 1985; Savin & Lee, 1988; Zheng, 1993; Girard & Savin, 1996). Other minerals, such as alunite (sulphate and silicate sites), can also be used for thermometry. Bechtel and Hoernes (1990) have demonstrated the use of oxygen isotope fractionation between hydroxyl and non-hydroxyl sites in illite for low-temperature thermometry. In a somewhat novel approach, Delgado and Reyes (1996) suggested that the combined δD and $\delta^{18}O$ values of clay minerals that formed in equilibrium with meteoric water

8.7 Volatile loss: dehydration, decarbonation, desulphidation and dehydrogenation

Rocks rarely behave as closed systems during metamorphism. Frequently, fluids infiltrate them, changing their bulk chemistry and isotopic composition (e.g. Valley, 1986; Kerrich, 1987; Burkhard & Kerrich, 1988; Chamberlain & Conrad, 1993; Fein *et al.*, 1994; Chamberlain *et al.*, 1995; Crespo-Blanc *et al.*, 1995). Fluids evolved during moderate- to high-grade metamorphism often invade lower grade overlying units. Variations in the $\delta^{18}O$, $\delta^{13}C$ and δD values of a metamorphic sequence can be used to trace fluid pathways. Stable isotopes are extremely well suited for this purpose, as passage of an H_2O-CO_2 fluid often occurs without any modifications of either mineralogy or bulk rock chemistry. The fluids do, however, affect the isotopic composition of the rock.

Rocks can still behave as open systems in the absence of fluid infiltration, if volatiles are locally generated and then removed from the system. Dehydration or decarbonation reactions are certainly the most volumetrically significant, although the common presence of sulphides along vein walls is testimony to the transfer of sulphur in a fluid phase.

At temperatures of very low-grade metamorphism, the $\delta^{18}O$ value of the aqueous fluid phase almost always concentrates ^{16}O relative to the solid. At higher temperatures, there is a crossover, so that the fluid phase concentrates ^{18}O relative to the solid (Fig. 8.5). CO_2 concentrates ^{18}O relative to solids to a much greater degree than water does (Fig. 8.6). The change in the $\delta^{18}O$ value of the solid is therefore much more pronounced for decarbonation reactions than for dehydration reactions. Deuterium is always concentrated in an aqueous phase relative to hydrous solids at low temperatures. (Note, however, that dehydrogenation—the loss of hydrogen gas from a mineral—leads to a strong deuterium enrichment in the solid phase (Feeley & Sharp, 1996).) The carbon fractionation between CO_2 and calcite ($\Delta^{13}C_{CO_2-calcite}$) is positive above 200°C and negative at lower temperatures. (See Kyser (1987) for a more detailed discussion of stable isotope fractionations between solids and fluid.)

The change in the isotopic composition of a rock during devolatilization (loss of fluid or vapour) depends on a number of factors:

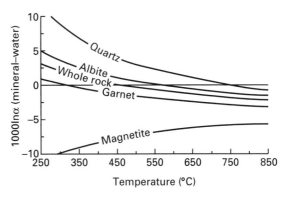

Fig. 8.5 Oxygen isotope fractionation between minerals and water as a function of temperature. Data from O'Neil and Taylor (1967), Bottinga and Javoy (1973), Matthews *et al.* (1979), Lichtenstein and Hoernes (1992). Whole rock data are assumed to be equal to an intermediate feldspar.

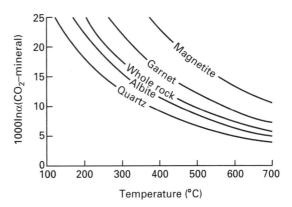

Fig. 8.6 Oxygen isotope fractionation between minerals and CO_2 as a function of temperature. Unlike mineral–water data, the CO_2 is always significantly enriched in ^{18}O relative to the minerals. Fractionation data as in Fig. 8.5 with additional data from Zheng (1994).

1 the fractionation between the solid and fluid (related to temperature);
2 the proportion of solid to fluid;
3 the manner of fluid (vapour) loss—the two end-member processes being 'batch volatilization' and 'Rayleigh fractionation' (Valley, 1986).

Each of these factors can be important. For example, if the mineral–fluid fractionation is zero, the isotopic composition of the rock will not change, no matter how much fluid leaves the system. If only a small portion of an element is lost as a fluid, the isotopic composition of the residual solid will not change, regardless of the magnitude of the mineral–water isotope fractionation.

The end-member 'batch volatilization' is a simple mass balance phenomenon, defined by the relative proportions of fluid and rock, and the isotopic fractionation between them. Geologically, this process can be considered as one where the fluid phase remains in the system until the devolatilization is complete; and only then is the fluid phase removed (Nabelek *et al.*, 1984). The change in the isotopic composition of the rock during batch volatilization is related to simple mass balance considerations by the following equation:

$$\delta_f = \delta_i - (1 - F) \, 1000 \ln \alpha \qquad (12)$$

where δ_f is the final delta value of the rock, δ_i is the initial delta value of the rock, F is the atomic fraction of the element that remains in the system, and $1000 \ln \alpha$ is the equilibrium fractionation between the rock and fluid. The maximum change in the delta value of the solid occurs when the amount of the element remaining in the rock approaches zero, and is identical to the equilibrium fractionation between the solid and fluid phase. Valley (1986) gives an example of batch volatilization of a siliceous dolomite. The devolatilization reaction considered in his example is

$$\text{CaMg(CO}_3)_2 + 2 \text{ SiO}_2 = \text{CaMgSi}_2\text{O}_6 + 2 \text{ CO}_2 \qquad (13)$$
$$\text{dolomite} \qquad \text{quartz} \quad \text{diopside}$$

where $\alpha = 1.0060$ and $\delta^{18}\text{O}_i$ is 22‰. The maximum change in the $\delta^{18}\text{O}$ value of the solid following eqn 12 is 6‰ for $F = 0$ (e.g. no oxygen remaining in solid). The F value, however, for eqn 13 for oxygen cannot go beyond 0.6. That is 60% of the oxygen remains in the rock even if reaction 13 goes to completion. Therefore the maximum change in the $\delta^{18}\text{O}$ value can only be 2.4‰. Note however that for carbon, F can go to zero, as all of the carbon in the products is in the vapour phase.

The other end-member, Rayleigh fractionation, is modelled as the continuous removal of the vapour (or fluid) as it is produced. The result of Rayleigh fractionation is that the isotopic composition of the system is continuously changing. The derivation of the equation that describes Rayleigh fractionation is given by Broecker and Oversby (1971):

$$\delta_f - \delta_I = 1000 \, (F^{(\alpha-1)} - 1). \qquad (14)$$

At high degrees of devolatilization, 'open system' devolatilization has a very large effect on the δ value of the remaining solid (Fig. 8.7a). However, as pointed out by Valley (1986), reaction 13 can only proceed to 40% removal of oxygen (e.g. $F = 0.6$). The difference between batch and Rayleigh fractionation mechanisms will have only a very minor effect on the final $\delta^{18}\text{O}$ value of the rock. For other reactions where the F value can approach zero, the difference between Rayleigh and batch reaction mechanisms can be pronounced. Examples are carbon removal by decarbonation reactions or hydrogen removal accompanying dehydration reactions (Fig. 8.7b). In practice, most geological processes are intermediate between these two end-member cases. Nevertheless, the delta values of rocks can often be approximated by one or the other of the two fractionation models.

8.8 Mineral purification

The philosophy of mineral separation for low-grade metamorphic rocks is very different from higher grade equivalents. Mineral separation of extremely fine-grained rocks, such as slates, certain blueschists and metamorphosed basalts can be next to impossible. There are, of course, some low-grade rocks that are coarse grained, such as metamorphosed granites. However, low-grade metamorphism of coarse-grained rocks with an inherited granulometry poses its own set of problems. In coarse-grained protoliths that have been only slightly metamorphosed, there is the serious question of 'inheritance' of the high-temperature isotopic signatures. Detrital sediments may also retain their inherited isotopic composition. Rocks can be too fine-grained for mineral separation, or too coarse-grained (they have not re-equilibrated at the conditions of the low-grade metamorphism). Generally, neoformed minerals must be analyzed.

8.8.1 Chemical purification

Many phases, such as carbonates, organic matter or graphite, can be analzyed using simple chemical purification techniques, or can be analyzed without any prior separation. Mixed silicate–carbonate samples can be reacted directly with phosphoric acid,

Fig. 8.7 (a) The $\delta^{18}O$ values of coexisting rock and fluid for Rayleigh and batch volatilization as a function of the atomic fraction of the element remaining in the rock (F) (assuming an α (alpha)-value of 1.006 between the two; see Valley, 1986). As devolatilization progresses, the $\delta^{18}O$ value of the rock becomes progressively lighter. In decarbonation reactions, reaction can only proceed to a value of c. 0.6, the calcsilicate limit. The difference between batch and Rayleigh fractionation at this value is less than 0.5‰. For hydrogen and carbon isotope devolatilization, the F values can approach zero, and the difference between Rayleigh and batch fractionation can be extreme. (b) Plot of $\delta^{13}C$ values vs. wt % carbonate in the prograde sequence from the Triassic red-bed formations of the Central Swiss Alps (Sharp *et al.*, 1995). The carbonates have nearly constant $\delta^{13}C$ values of −5‰ (PDB) below the epimetamorphic grade. In the epimetamorphic samples, the $\delta^{13}C$ values decreases to −10‰ (filled circles) following a Rayleigh fractionation pattern. The $\delta^{13}C$ values increase to −5‰ in the mesometamorphic zone (open squares) as a result of fluid infiltration from adjacent black shales during high-grade metamorphism. The systematic change in the $\delta^{13}C$ values as a function of wt % carbonate is consistent with Rayleigh, but not batch devolatilization, indicating that the CO_2 fluids escaped as they were formed.

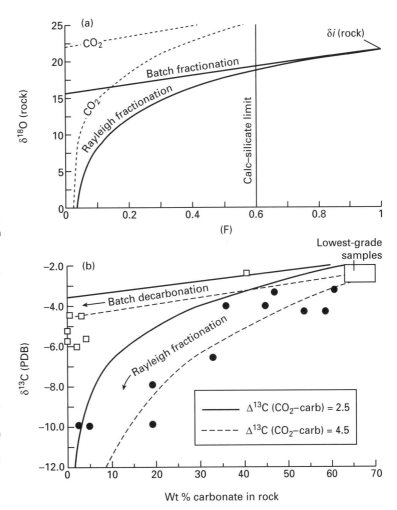

liberating only the carbon and oxygen associated with the carbonate. Care needs to be taken if the samples have mixed carbonates (Al-Aasm *et al.*, 1990), but *generally*, no mineral purification is required. (In some cases, mixtures of carbonates with either organic matter or certain sulphides can cause problems. See Charef and Sheppard (1984) for further details.) Analysis of $\delta^{13}C$ values of organic matter, like carbonates, requires a minimum of sample purification. Carbonate must first be removed by treatment with a weak acid (e.g. acetic acid); the residue, consisting of silicates and the organic matter, is oxidized in the conventional manner. Sulphides also require a minimal amount of treatment, assuming no sulphate is present. Pyrite can be concentrated

by dissolution of the bulk of the silicates with hydrofluoric acid. Hydrogen isotope analyses can be made on bulk rock samples as long as all hydrous phases were formed during metamorphism, and are not inherited from the protolith. Organic matter can also contribute hydrogen to the bulk analysis (Sheppard & Gilg, 1996)—there are a number of methods to remove organic matter, including treatment with H_2O_2, NaOCl, Br_2 water, plasma ashing and step heating (Taieb, 1990; Sheppard & Gilg, 1996). In some cases, however, organic matter will persist in spite of all efforts.

Purification of silicate minerals can sometimes be made using purely chemical techniques that are generally derived from the soil science literature.

Quartz can be purified by treatment with sodium pyrosulphate and hydrofluosilicic acid (Syers *et al.*, 1968). Fusion of powdered rock with sodium pyrosulphate removes most of the minerals except for quartz and feldspar. The residue is then treated with hydrofluosilicic acid to remove the feldspar. The $\delta^{18}O$ value of the feldspar can be calculated from the measured $\delta^{18}O$ value of the quartz–feldspar mixture, the weight loss associated with feldspar removal and the $\delta^{18}O$ value of the quartz residue. The feldspar wt % estimates are probably accurate to within 10%, and the calculated $\delta^{18}O$ values to ± 1‰ (Eslinger & Savin, 1973b). In diagenetic environments, quartz overgrowths can be separated from the detrital quartz grains by chemical techniques (Lee & Savin, 1985).

8.8.2 Fine-grained phyllosilicates: clays

Mineral concentration of different size fractions of fine-grained material is accomplished by settling techniques (e.g. Hower *et al.*, 1976). Grains are disaggregated using ultrasonic methods, and sorted according to size by settling the rock powder through a column of distilled water. The purification of fine-grained phyllosilicates poses a number of perplexing problems. Settling techniques alone are often ineffective in producing monomineralic clay samples (Sheppard & Gilg, 1996), and hydroxides, oxides and organic matter are frequent contaminants. There are a number of chemical treatments that are effective at removing these unwanted materials (Jackson, 1979). Reviews of mineral separation techniques can be found in Sheppard and Gilg (1996) and Ayalon and Longstaffe (1990).

A main advantage of settling technique for mineral purification is that different size fractions of the same material can be obtained. The variation in $\delta^{18}O$ or δD values of a phase as a function of grain size is one of the most direct measures of the degree of attainment of equilibrium. Yeh and Savin (1977) measured the $\delta^{18}O$ values of coexisting quartz and clay from borehole samples from the Gulf of Mexico (Figs 8.8 and 8.9). Sample sizes of >2 µm, 1–2 µm, 0.5–1.0 µm, 0.1–0.5 µm and <0.1 µm were analyzed as a function of depth. In the shallowest samples (1500 m), different size fractions of both quartz and mixed illite–smectite were clearly out of equilibrium, suggesting inherited detrital values from multiple

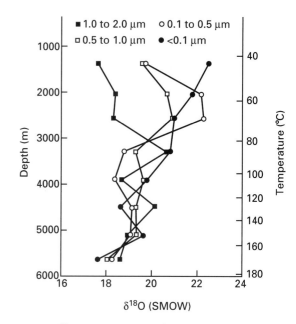

Fig. 8.8 $\delta^{18}O$ values of different clay-sized fractions from a deep well in the Gulf Coast shown as a function of depth and temperature. The initial scatter is due to different sources. At 85°C, the samples are partially equilibrated with each other (and pore fluid); by 150°C, all size fractions are in equilibrium. Temperatures are measured downhole values from CWRU Gulf Coast well 6. (From Yeh & Savin, 1977.)

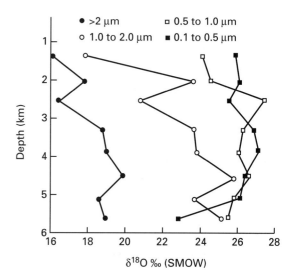

Fig. 8.9 Variations in the $\delta^{18}O$ values of separated size fractions of quartz as a function of depth. From CWRU Gulf Coast well 6 (Yeh & Savin, 1977).

sources. The clays equilibrated with each other by 150°C, while only the finest size fractions of quartz came into equilibrium with the clays. The fine quartz fraction is interpreted as authigenic silica overgrowths on preexisting grains formed during conversion of smectite to illite.

8.8.3 Porphyroblasts, veins and vug fillings

As an alternative to separating fine-grained ground mass minerals, more coarsely crystalline porphyroblasts or veins can be targeted for stable isotope analysis. Veins and porphyroblasts are easily purified by hand picking, and by their very nature, are metamorphic phases — they cannot have any isotope inheritance from their precursors because they form from a fluid phase during the metamorphic event. Coarse-grained materials — either single porphyroblasts or veins — may have fluid inclusions that can be used for independent temperature and pressure estimates. Magaritz and Taylor (1976) used quartz–garnet fractionations from blueschists in the Franciscan Formation, California to estimate temperatures of metamorphism to be 300°C. Blattner (1975) and Kirschner et al. (1995) used coexisting quartz, calcite and feldspar from fissure fillings and veins for temperature estimates of vein formation. Schmidt and Sharp (1993) made careful in situ analyses of quartz and calcite in vug fillings from Upper Proterozoic low-grade metabasites as a monitor of fluid–rock interaction.

8.9 Applications to rocks of sedimentary origin

8.9.1 Diagenesis: general considerations

The early burial diagenesis and incipient metamorphism of sedimentary rocks has been studied extensively by isotopic methods. Sedimentary minerals are frequently out of chemical and isotopic equilibrium, as they consist of mixed detrital and authigenic phases. As sediments are buried and heated, they undergo a series of chemical and mineralogical modifications. Stable isotopes are an important method of monitoring temperatures of reaction, sources of porewaters, and the degree to which equilibrium has been established.

For a complete discussion of diagenesis, the reader is referred to the following review articles (Yeh & Savin, 1976; Eslinger et al., 1979; Savin, 1980; Yeh, 1980; Land, 1984; Longstaffe, 1987, 1989; O'Neil, 1987; Savin & Lee, 1988). A number of simple generalizations can be made regarding isotopic effects during clastic diagenesis, although as is the case with most geological processes, many exceptions exist:

1 Most isotopic modifications that occur during diagenesis are a result of recrystallization, and not simple diffusion.

2 Carbonates form over the full temperature range of diagenesis, from essentially room temperature to low-grade metamorphism and beyond. Original authigenic $\delta^{18}O$ and $\delta^{13}C$ values of carbonates can survive to the highest temperature diagenetic conditions, as shown by the large degree of isotopic variation over small distances. On the other hand, their isotopic composition can easily be disturbed.

3 Quartz is extremely resistant to recrystallization. Diagenetic quartz grains (>2-μm size fraction) retain their $\delta^{18}O$ values to temperatures of at least 170°C (Yeh & Savin, 1977), although the $\delta^{18}O$ values of very fine-grained quartz may be reset (by recrystallization?) as low as 100°C (Eslinger et al., 1979). Significant amounts of quartz overgrowths can form at very low diagenetic conditions. The source of this quartz may be from SiO_2-rich fluids from deeper and hotter lithologies (e.g. shales), with silica liberated during smectite–illite transition or from 'pressure solution' effects (E. Oelkers, personal communication, 1997).

4 Feldspars are generally not 'active' in the lowest diagenetic zones ('passive' zone of Milliken et al., 1981). The formation of albite occurs during smectite/illite conversion at temperatures between 120 and 150°C (Milliken et al., 1981). Oxygen isotope re-equilibration of feldspar can occur as low as 100°C (Shieh, 1983), but generally, only if there is accompanying simultaneous chemical modifications. The $\delta^{18}O$ values will not change by diffusion alone until significantly higher temperatures are reached.

5 Even very fine-grained clay minerals can retain their original $\delta^{18}O$ values for extended periods of time at low temperature (Yeh & Eslinger, 1986). Authigenic recrystallization of fine-grained smectite sometimes occurs at low temperatures with only very minor chemical and isotopic modifications towards equilibrium with their environment, perhaps reflecting low water–rock ratios (Clauer *et al.*, 1990). At temperatures of *c.* 90–100°C, significant oxygen isotope exchange takes place between different size fractions of detrital clays, reflecting re-equilibration during the illite–smectite transition (Yeh & Savin, 1977). Hydrogen isotope exchange occurs at far lower temperatures — coincident with the beginning of the illite–smectite transition (Yeh, 1980). Kaolinite can form at very low temperatures by influx of meteoric water (Lawrence & Taylor, 1972); thus the $\delta^{18}O$-δD values of kaolinite have been used to infer early fluid histories and/or mixing between different fluid reservoirs in buried sediments (Fallick *et al.*, 1993; Macaulay *et al.*, 1993a).

8.9.1.1 Silica

Quartz is extremely resistant to low-temperature exchange. The $\delta^{18}O$ value of fine-grained quartz does not change in the sedimentary environment (Savin & Epstein, 1970b; Lawrence & Taylor, 1972; Clayton *et al.*, 1978), and even moderate heating may not affect the $\delta^{18}O$ value of detrital quartz. In present day geothermal systems, quartz grains may resist exchange at temperatures of 250°C (Clayton *et al.*, 1968; Eslinger & Savin, 1973a; Clayton & Steiner, 1975), although the heating of these present-day samples may be short lived.

In contrast to quartz, biogenic silica, manifest as opal, cristobalite, chert, etc. undergoes rapid isotopic exchange at very low temperatures. Murata *et al.* (1977) studied various forms of biogenic silica from the Monterey Shale Formation (California), identifying a number of isotopic effects that accompanied chemical and structural modifications (Fig. 8.10). With increasing depth, the silica is transformed from biogenic opal to disordered cristobalite, ordered cristobalite and finally microcrystalline quartz. The $\delta^{18}O$ values of samples within each group are relatively constant, but decrease in discreet intervals at the polymorphic transitions[1].

O'Neil (1987) presented a number of 'rules of thumb' that can be applied to cherts during diagenesis

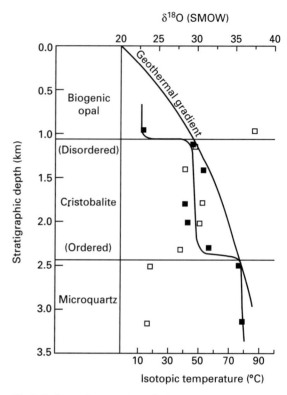

Fig. 8.10 Oxygen isotope values of silica (open squares) as a function of depth in the Monterey Shale, California, USA. For each polymorph of silica, and the $\delta^{18}O$ values are constant. The changes in the $\delta^{18}O$ values occur only with polymorphic transition. Temperature estimates (solid squares) were obtained assuming a $\delta^{18}O$ value of pore fluid equal to 0.0‰. (After Murata *et al.*, 1977.)

(where chert is defined as microcrystalline quartz with or without structural water). Samples that retain their microcrystalline character are least likely to have undergone diagenesis — grain coarsening indicates diagenetic reaction with fluids. Authigenic quartz has a very high $\delta^{18}O$ value, and *all* diagenetic processes lower it. Hydrogen isotope analysis of structural water can (potentially) be used to indicate the source of a diagenetic fluid.

Far more important than diffusional exchange of quartz is chemical precipitation of quartz overgrowths on detrital grains. Yeh and Savin (1977) proposed that most changes in the $\delta^{18}O$ values of quartz under diagenetic conditions are related to quartz overgrowths. On the basis of measured $\delta^{18}O$

---

[1] The analysis of biogenic silica for stable isotopes is complicated by the rather complex extraction methods necessitated by dissolved water and hydroxyl (Labeyrie, 1972; Kolodny & Epstein, 1973; Labeyrie & Jullet, 1982; Matheney & Knauth, 1989).

values of quartz and *assumed* $\delta^{18}O$ values of the pore water, Milliken *et al.* (1981) estimated quartz overgrowth formation temperatures in a stable isotope study of Tertiary sandstones of the Frio Formation, Texas, to be 70–90°C. Microscopic textural evidence supports their conclusion.

Longstaffe (1987) found evidence for low-temperature quartz precipitation in a study of the Cretaceous Belly River Formation, Western Canada, where quartz and chlorite were the first minerals to form. Similar results were obtained on Proterozoic quartzites from West Africa, where up to 27% quartz cement has been precipitated (Girard & Deynoux, 1991). Their temperature estimates are less than 70°C, and as low as room temperature. The source of quartz is local, on the basis of hydrological considerations.

Recently, ion microprobe techniques have been developed that allow for *in situ* determinations of $\delta^{18}O$ values of quartz on a 20–30-µm diameter spot (Hervig *et al.*, 1995; Graham *et al.*, 1996). A number of important conclusions can be drawn from these studies. First, the range in the $\delta^{18}O$ values of both the detrital and authigenic quartz is far larger than previously assumed. Variations of > 20‰ are seen in the detrital quartz grains, indicating a mixed source of sedimentary (chert), igneous and metamorphic grains. Temperature estimates are difficult to make, due to the uncertainty in the $\delta^{18}O$ values of the porewaters; however, the spread in the data requires that either the temperatures of formation were quite variable, or the $\delta^{18}O$ values of the fluids fluctuated due to mixing of meteoric and basinal waters.

8.9.1.2 Clays

Authigenic clays have distinct $\delta^{18}O$ and δD values related to whether meteoric or oceanic water is involved in their formation (Savin & Epstein, 1970a). The temperature interval over which oxygen and hydrogen isotope re-equilibration occurs has been well established. In young marine sediments, no oxygen isotope exchange and no hydrogen isotope exchange occurs over 2–3 million years, excepting the < 0.1-µm size fraction (Yeh & Epstein, 1978; Yeh & Eslinger, 1986). The degree of isotopic equilibration increases with increasing depth and temperature. In a hydrogen isotope study of a thick sequence of shales from the Gulf of Mexico, Yeh (1980) found

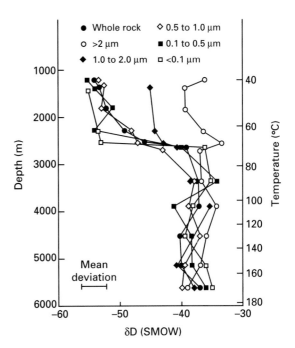

Fig. 8.11 Variations in the δD values of different size fractions of clay minerals as a function of depth from a deep well (CWRU 6) in the Gulf Coast. Original isotopic heterogeneities due to different sources are eliminated by 80°C for all size fractions due to exchange with porewaters. (After Yeh, 1980.)

that hydrogen isotope equilibrium between all size fractions of clay minerals occurred with the onset of the conversion from smectite to illite at about 80°C (Fig. 8.11). In a correlation between downhole temperature estimates from quartz–illite oxygen isotope temperature estimates and measured borehole temperatures, the two agree only above 100°C (Fig. 8.12). At a depth corresponding to 100°C, a significant fraction of smectite begins to convert to illite (Hower *et al.*, 1976). Concurrent with illite formation is an increase in the production of authigenic quartz—hence isotopic equilibrium between the two phases is established.

8.9.2 Diagenesis: examples

In a detailed stable isotope study of Tertiary sandstones from the Frio Formation, Brazoria County, Texas, Milliken *et al.* (1981) determined the oxygen isotope composition of detrital quartz, quartz overgrowths, kaolinite, albite, calcite and dolomite

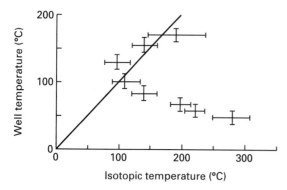

Fig. 8.12 Temperature estimates for quartz–illite thermometry from well CRWU 6 in the Gulf Coast plotted against measured downhole temperatures. Isotopic temperatures were calculated with the finest quartz size fraction (0.1–0.5 μm) and the <0.1-μm clay fraction. Below 100°C, the temperature estimates from the isotope thermometer are much higher than real temperatures, reflecting an inheritance of detrital quartz. Above 100°C, there is good agreement between the two temperatures indicating that oxygen isotopes equilibrated between the two phases. (After Hower *et al.*, 1976.)

in order to better understand the temperature intervals of their formation. Using $\delta^{18}O$ values for modern subsurface waters, and the measured $\delta^{18}O$ value of the overgrowths, they estimated quartz overgrowth formation temperatures to be 70–90°C. Microscopic textural evidence supports the conclusion that quartz precipitated at low temperatures and shallow levels. Quartz overgrowths thus formed in the zone of passive diagenesis, below temperatures at which significant mineral reaction occurs. The source of silica at these low temperatures was not related to mineral breakdown. Instead, Milliken *et al.* (1981) proposed that it was probably derived from fluids expelled from underlying, hotter units that were undergoing active diagenesis. An alternative explanation is that quartz is locally dissolved as a result of pressure solution (Robin, 1978), or related to small potential differences in silica at quartz–clay mineral interfaces. Although temperatures continued to increase to 150°C or more in the Frio Sandstone, no additional quartz precipitation occurred with increased burial and heating. Kaolinite formed between 90 and 120°C, albite at >150°C, while carbonates formed over the entire diagenetic range, from 70 to 150°C (Fig. 8.13). The scatter in the $\delta^{18}O$ and $\delta^{13}C$ carbonate values illustrates that early formed authigenic carbonates did not undergo

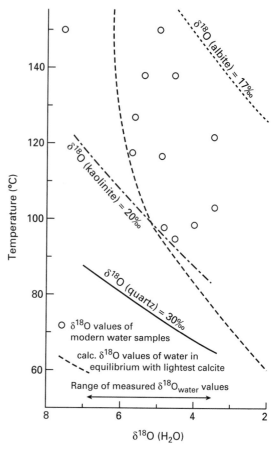

Fig. 8.13 Temperature estimates for the formation of authigenic minerals for the Frio formation, Texas (Milliken *et al.*, 1981). Temperature estimates are plotted vs. the assumed $\delta^{18}O$ values of formation waters. The temperature estimates are determined from the measured $\delta^{18}O$ values for each group of minerals and the modern subsurface waters, assuming that the modern waters have the same $\delta^{18}O$ values as those responsible for the formation of the authigenic minerals. Quartz precipitated over the temperature interval of 70–90°C, kaolinite, 90–120°C, albite, *c.* 150°C and calcite over the full temperature range of diagenesis. See Milliken *et al.* (1981) for details.

recrystallization and re-equilibration during later, higher temperature diagenesis. The variable isotopic values of the carbonates illustrates the complex 'history' of cementation throughout the burial event.

Milliken *et al.* (1981) delineated three diagenetic zones (Fig. 8.14):
1 From the surface to *c.* 2600 m, spanning a temperature range up to *c.* 90°C. It is the zone of passive diagenesis, where quartz, carbonate, rarely feldspar

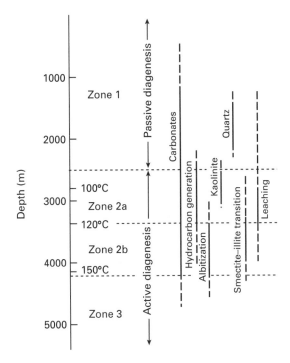

Fig. 8.14 Occurrence of diagenetic minerals as a function of depth in the Frio sandstone. Milliken *et al.* (1981) divided the diagenetic sequence into three zones. Temperatures are estimated from stable isotope thermometry. (From Milliken *et al.*, 1981.)

and some kaolinite cements form from pore fluids.
2 The beginning of active diagenesis is characterized by the reaction of aqueous fluids with unstable detrital components. The top of this zone is delineated by the beginning of the illite–smectite transition. Albitization and illite–smectite conversion provides major sources of chemical components for mineral formation. Hydrocarbon generation reaches its peak in zone 2.
3 The beginning of the end of illite–smectite conversion and albitization. This zone leads into what is considered very low-grade metamorphism.

One potential problem with the diagenetic reconstruction of Milliken *et al.* (1981) is that it is only valid if the $\delta^{18}O$ values of diagenetic fluids were constant over their entire diagenetic history. There are clearly cases where this assumption is not reasonable. Ion microprobe studies of authigenic quartz demonstrate that the $\delta^{18}O$ values of quartz cover a broad range, supporting a complex history of fluid mixing and changing $\delta^{18}O_{fluid}$ values.

Longstaffe (1987, 1989) and Longstaffe and Ayalon (1987) described detailed results of an isotopic and petrographic study of diagenetic minerals from a transitional marine to continental sequence. On the basis of textural and isotopic data, both temperatures and $\delta^{18}O$ values of pore waters were estimated, and there appeared to be significant temporal variability in the oxygen isotope composition of the porewater. Two units were studied—the Lower Cretaceous marine Viking Formation and the Upper Cretaceous marine to continental Basal Belly River sandstones from west-central Alberta. The relative timing of diagenetic mineral growth was inferred from textural considerations. For the mixed marine–continental Belly River sandstone, the progression is early chlorite and quartz, followed by kaolinite, and then illite–smectite and smectite. As is often the case (e.g. O'Neil, 1987), carbonates precipitated over a wide temperature interval. The predominantly marine Viking Formation underwent a different paragenetic sequence: very early kaolinite (kaolinite B), early siderite, chlorite and/or calcite (calcite B), followed by quartz, still later dolomite, calcite (calcite A) and ankerite and finally kaolinite (kaolinite A), illite and illite–smectite.

The $\delta^{18}O$ values of all phases were measured. At any given temperature, the $\delta^{18}O$ value of the porewater can be estimated from the measured mineral $\delta^{18}O$ value and the independently assumed temperature (Fig. 8.15). A $\delta^{18}O_{(pore\,water)}$ – temperature graph of all diagenetic phases was used to estimate the progression in both the $\delta^{18}O_{(pore\,water)}$ values and the temperature throughout the diagenetic history of the formation. Although there are a number of assumptions and inherent uncertainties in this study, Longstaffe concluded that the $\delta^{18}O$ value of the porewater changed by 8‰ over the course of diagenesis. Similar changes have been observed in deeply buried shales in the Gulf coast, where the calculated $\delta^{18}O$ values of porewaters change by over 10‰ (Yeh & Savin, 1977).

8.9.3 Low-grade metamorphism

8.9.3.1 Burial metamorphism of the Precambrian Belt Supergroup

Eslinger and Savin (1973b) measured the $\delta^{18}O$ values of mineral separates from the Precambrian Belt

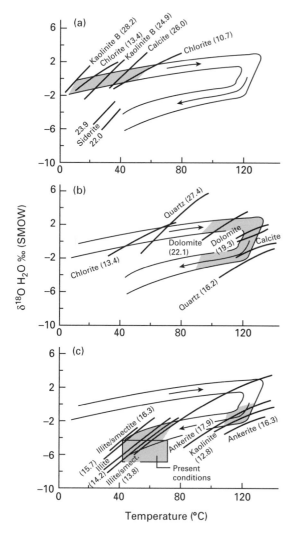

Fig. 8.15 Temperature-fluid variation throughout the diagenetic evolution of the Viking formation, Canada. The relative timing of different diagenetic minerals was determined from textural relations and temperatures are estimated from known formation temperatures of different diagenetic phases. For each phase (shaded areas); the range of oxygen isotope compositions of pore waters is plotted. In the earliest episode of diagenesis (Viking Stage 1; a), the $\delta^{18}O_{water}$ values are between −2 and 1‰. The $\delta^{18}O_{water}$ values are unchanged in Stages 2 and 3(b), even though temperatures are significantly higher. In Stages 4 and 5(c), the $\delta^{18}O_{water}$ values decrease to −4 to −6‰, reflecting a meteoric water contribution. (From Longstaffe, 1987, 1989.)

morphism. Samples were collected from various stratigraphic depths ranging from c. 700 m to over 2500 m. Whole rock $\delta^{19}O$ values decrease from 18‰ at the highest levels to 12‰ with increasing depth and then increase again by c. 2‰ in the lowest levels. The mineralogy of all samples is similar, and Eslinger and Savin attributed the $\delta^{18}O$ variations to isotopic exchange after deposition. In all samples the $\delta^{18}O$ values of the silicate minerals decrease from quartz to feldspar to clay, as expected from theoretical considerations. Temperature estimates from the quartz–illite thermometer (Eslinger & Savin, 1973a) increase progressively with depth from 225 to 310°C, and agree well with crude estimates on the basis of stratigraphic depth and an inferred geothermal gradient of 36°C km^{-1} (Fig. 8.16). The temperatures also correlate well with the percentage of illite in the samples (Fig. 8.17), supporting the conclusion that the 1M$_d$ polytype is transformed to the 2M polytype with increasing temperature and pressure

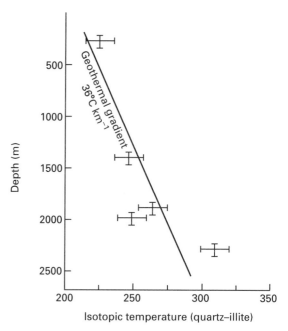

Fig. 8.16 Temperature estimates using the quartz–illite oxygen isotope thermometer (Eslinger & Savin, 1973a) as a function of estimated depth of burial for Precambrian Belt Supergroup rocks, Montana, USA. The data correspond well with an assumed geothermal gradient of 36°C km^{-1}. (After Eslinger & Savin, 1973b.)

Supergroup, Glacier National Park, Montana, USA in order to assess the degree of isotopic equilibrium attained during diagenesis and low-grade meta-

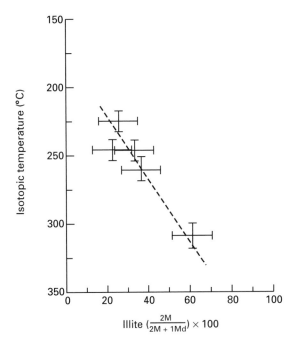

Fig. 8.17 Correlation between the inferred quartz–illite temperature and polytypism of illite for samples from the Precambrian Belt Supergroup, Montana, USA. (After Eslinger & Savin, 1973b.)

temperature of 170°C, compared to a quartz–illite temperature of 225°C, while a second gave a quartz–calcite temperature of 340°C, compared to 310°C for the quartz–illite thermometer. The differences may be related to retrograde exchange between calcite and the pore fluids. An uncertainty of only 0.51‰ in the $\Delta^{18}O_{(quartz–calcite)}$ value, however, is sufficient to bring the two thermometers into agreement. The quartz–calcite temperature estimates were made using a thermometric calibration based on experimental calcite–water fractionations of O'Neil *et al.* (1969) and the quartz–water fractionations of Clayton *et al.* (1972). Temperature estimates using a calibration based on more recent direct exchange experimental data (Clayton *et al.*, 1989; Clayton & Kieffer, 1991) are unreasonably low. On the basis of these data, either the calcite has undergone major exchange during retrogression or there is an error in the direct exchange experiments (Sharp & Kirschner, 1994).

8.9.3.2 Prograde metamorphism of the Central Swiss Alps

Sharp *et al.* (1995) measured the $\delta^{18}O$ and δD values of silicates and the $\delta^{18}O$ and $\delta^{13}C$ values of carbonates from a Triassic red-bed formation in the Central Swiss Alps. The section progresses from unmetamorphosed to mesozone (amphibolite) grade, and individual units can be traced throughout the metamorphic sequence. Hydrogen and oxygen isotope data were combined with mineralogy, water content and modal abundance in order to identify a number of processes that occurred during the metamorphic event. The $\delta^{18}O$ values of the carbonate and silicate fractions are not in equilibrium with each other in the unmetamorphosed and low-grade metamorphic samples. The $\delta^{18}O$ values of the carbonates are typical of marine sediments (30‰), while the lower $\delta^{18}O$ values of silicates (20‰) indicate a significant detrital component. Only at temperatures of about 400°C do the carbonates come into equilibrium with the far more abundant silicates (Fig. 8.18). If a fluid phase had been present during the early stages of metamorphism, it could not have been in equilibrium with both the carbonates and the silicates.

The hydrogen isotope data provide information on the fluid–rock interaction during metamorphism (Fig. 8.19). The δD values of the unmetamorphosed

(Maxwell & Hower, 1967). In contrast to the regular temperature progression from quartz–illite pairs, those obtained from quartz–feldspar pairs are scattered. Quartz–feldspar temperatures for two of the five samples were the same as those obtained with the quartz–illite thermometer; the other three samples gave lower values. Pure feldspar mineral separates could not be made; the $\delta^{18}O$ values of feldspar were inferred from the measured values of quartz–feldspar mixtures. The variable quartz–feldspar isotopic temperatures could therefore be due to errors in the calculated $\delta^{18}O_{feldspar}$ values. Eslinger and Savin (1973b) proposed as an alternative explanation that feldspar did not equilibrate with quartz, while the illite did. This explanation is reasonable if quartz and calcite recrystallized during mineral reactions, while feldspar was a passive phase.

Quartz–calcite temperatures were determined from two samples. They are in moderately good agreement with those obtained with the quartz–illite thermometer. One sample gave a quartz–calcite

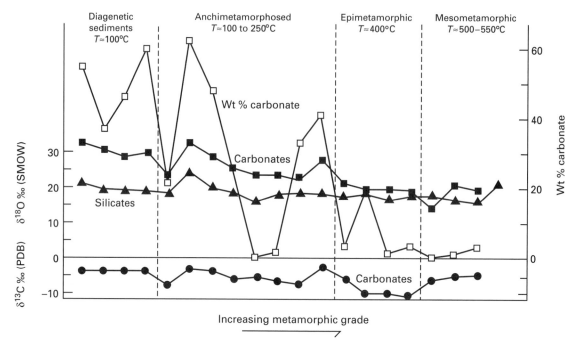

Fig. 8.18 Isotopic composition and wt % carbonate plotted against increasing metamorphic grade for the Triassic red beds of the Central Swiss Alps. The wt % carbonate is very variable, but generally decreases with increasing metamorphic grade. The $\delta^{18}O$ values of the diagenetic carbonates are typical marine, while the silicates have a detrital component. With increasing metamorphic grade (c. 400°C), the two come to isotopic equilibrium. The $\delta^{13}C$ of carbonates is variable and may reflect original detrital values or devolatilization processes. (From Sharp et al., 1995.)

samples (−70 to −80‰) indicate a large detrital component. They increase to values of −50‰ in the early metamorphic stages as a result of interaction with a deuterium-rich pore fluid. The δD values decrease again during the highest grades of metamorphism as a result of devolatilization reactions. During hydrogen isotope fractionation between a hydrous mineral and H_2O, the fluid phase concentrates deuterium. This is the reverse of oxygen, which at low temperatures is concentrated in the solid phase. Hence, during dehydration reactions, the δD of the minerals should decrease as a result of preferential removal of deuterium to the vapour phase. The lowering of the δD values in the Central Alps transect is expected; surprisingly it hasn't been investigated in sufficient detail in other metamorphic terranes to know whether or not it is a common process.

In the same sequence, the $\delta^{13}C_{carbonate}$ values are relatively constant at all metamorphic grades (Fig. 8.18). Small $\delta^{13}C$ variations may be primary, being related to sedimentary or early diagenetic

processes, and not a result of metamorphic reactions. Alternatively, a process of decarbonation and input of an oxidized organic matter component may be indicated (Fig. 8.20). At the highest metamorphic grades, total carbonate is depleted relative to the lower grade equivalents (Fig. 8.18), but $\delta^{13}C$ values are unchanged. If the carbon isotope fractionation between the carbonate and vapour was large at the temperature of decarbonation, the $\delta^{13}C$ values of the residual carbonate should be shifted from their original value. That this is not the case indicates the $\Delta^{13}O_{carbonate-CO_2}$ value was small.

The $\delta^{13}C$ values of organic material from the organic-rich Liassic units overlying the Triassic sequence have undergone significant carbon isotope shifts (Hoefs & Frey, 1976). The $\Delta^{13}C_{carbonate-graphite}$ values of these units become smaller with increasing temperature, roughly following the equilibrium fractionation curve (Fig. 8.21). The $\delta^{13}C$ values of the carbonates are relatively constant, which can be explained by simple mass balance — the carbonate is

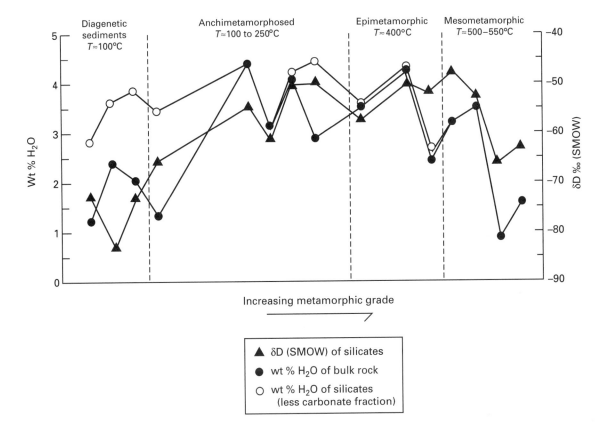

Fig. 8.19 Hydrogen isotope values of silicates and wt % water (with and without carbonate fraction) of samples from the Triassic red beds, Central Switzerland. (From Sharp *et al.*, 1995.)

the major source of carbon and acts as a buffered reservoir. In order to maintain the equilibrium $\Delta^{13}C_{carbonate-graphite}$ value with increasing temperature, the $\delta^{13}C$ value of the graphite must increase. Thus, in organic-rich, carbonate-bearing sediments, the carbon isotope values of the units are modified during metamorphism, while in closed-system, organic-free sediments, the $\delta^{13}C$ values are not changed during metamorphism, and their values can potentially be used as sedimentary markers. Overall, the stable isotope data can be used to constrain a number of fluid–rock interaction processes during prograde metamorphism of sediments.

8.9.3.3 Tracing formation waters in accreted terranes — an example from south-west Japan

The oxygen, and particularly hydrogen isotope composition of clay minerals is often directly related to the isotopic composition and source of the coexisting formation waters. Masuda *et al.* (1992) measured the $\delta^{18}O$ and δD values of two accreted sedimentary terranes in south-west Japan in order to determine the composition of the paleo-porewaters. Their results indicate that the two belts have distinctive isotopic compositions which are related to different tectonoenvironmental conditions of formation. The Mino-Tanba and Ryoke belts were accreted during the Jurassic to Early Cretaceous time, while the Sanbegawa and Shimanto belts farther to the south-east were accreted during the Cretaceous to Early Tertiary. The Chichibu belt is Jurassic in age, but overlies the younger Cretaceous–Tertiary belts due to late tectonic movement. All of the rocks analyzed were shales that have reached zeolite facies

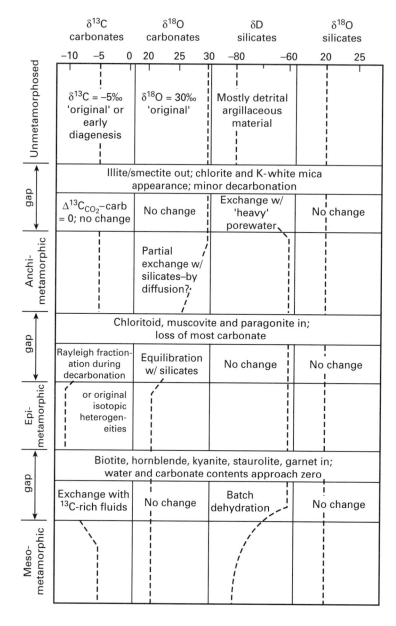

Fig. 8.20 Variations in the carbon, oxygen and hydrogen isotope compositions of phases from the Triassic red beds. Gaps represent regions where no outcrop is exposed. (From Sharp *et al.*, 1995.)

to chlorite-forming metamorphic grades. The major rock forming minerals are quartz, albite, chlorite and illite, the latter two being more concentrated in the <2-µm size fraction. Quartz and albite were considered to be predominantly of detrital origin, being deposited on a continental slope. Chlorite and illite are probably a mixture of detrital and authigenic in origin. The polytypes of illite were 1M and $2M_1$

for most samples. The lack of kaolinite, smectite and $1M_d$ illite indicates that surficial weathering was of minor importance.

The δD values for the different units are distinct, while the $\delta^{18}O$ values cover the same range for each unit (Fig. 8.22). The Jurassic–Early Cretaceous samples (Mino-Tanba and Ryoke belts) have δD values ranging from –110 to –70‰ (SMOW), while

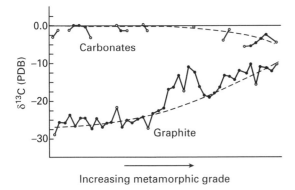

Fig. 8.21 Carbon isotope data of coexisting graphite and calcite as a function of metamorphic grade from the Central Swiss Alps. As temperatures increase, the carbon isotope fractionation between the two carbon-bearing phases decreases due to smaller equilibrium isotopic fractionations. (After Sharp *et al.*, 1995; data from Hoefs & Frey, 1976.)

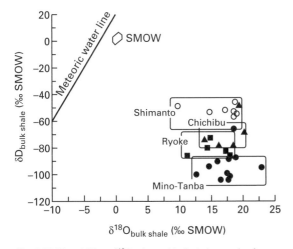

Fig. 8.22 Plot of δD vs. $\delta^{18}O$ values of bulk shale samples from different belts in the Kinki district, south-west Japan. Samples from different belts have distinctive hydrogen isotope ratios, indicative of their source fluids. (From Masuda *et al.*, 1992.)

the δD values for the younger Cretaceous–Early Tertiary units are –80‰ to –50‰. The δD values for illite in any one unit are tightly clustered, while the $\delta^{18}O$ values are scattered over a range of up to 10‰. That the samples do not parallel the meteoric water line indicates that they were not formed by surficial weathering or deep circulation of meteoric waters. Furthermore, the lack of correlation between illite

polytypes and δD values led Masuda *et al.* (1992) to conclude that the hydrogen isotope compositions were indicative of formation waters. The calculated formation water δD values for the older units are –10 to –60‰, depending upon the assumed temperature of equilibration and which fractionation factor is used for illite–water. Regardless of the uncertainties inherent in these calculations, it is clear that the water has a significant meteoric water component. In contrast, the calculated δD values for formation waters of the younger units are –5 to –35‰. The meteoric water component in these rocks was less, and there was a larger seawater-derived component.

Slightly different results are obtained from the δD values of chlorite. Chlorite poses an additional complication, because the Fe/(Fe + Mg) ratio has a very strong effect on the hydrogen isotope fractionation between hydrous phyllosilicates and water (Suzuoki & Epstein, 1976). Nevertheless, the calculated δD (water) values in equilibrium with chlorite are the following: –30 to –80‰ in the older Tanba and Ryoke belts and from –12 to as high as +19‰ in the Jurassic Chichibu belt and the Cretaceous Shimanto belts. In the case of chlorite, there is a distinct seawater–meteoric water signature for the different belts, but they are not strictly correlated with age. The chlorite from the Ryoke and Mino-Tanba belts equilibrated with meteoric water, while the Shimanto and Chichibu belts have a seawater signature. The authors concluded that the chlorite–water exchange ceased at higher temperatures than that of illite–water. Overall, the hydrogen isotope values give an indication of the paleohydrology of the different shale units. Unlike the hydrogen data, no relations between the $\delta^{18}O$ values and geological position exist.

8.10 Ocean basalts

There has been a great deal of interest in the metamorphism of ocean floor basalts, and many papers have been published on the isotopic compositions of seafloor metamorphosed oceanic crust (see Alt, Chapter 6). The chemical interactions that occur during metamorphism have wide-ranging importance and profound effects on other processes. Specifically, ocean floor metamorphism buffers the oxygen isotope composition of the oceans (Muehlenbachs

& Clayton, 1976; Muehlenbachs, 1987) and the fluids that are incorporated in subducted hydrated crustal material are the major source for subduction-related igneous activity. Our understanding of seafloor metamorphism has benefited greatly from advances in the ability to sample *in situ* material from the ocean depths; many of the samples that have been studied come from either dredge samples or core samples from deep-sea drilling projects, although there has certainly been some important observations made on obducted material (e.g. Gregory & Taylor, 1981).

Metamorphism of mid-ocean ridge basalts (MORB) can be subdivided into two types; high temperature hydrothermal metamorphism near spreading centres and low temperature seafloor 'weathering'. The $\delta^{18}O$ of the modern ocean is buffered to its present value of 0‰ by the combination of high- and low-temperature hydrothermal interaction with oceanic crust (Muehlenbachs, 1987). There is some speculation regarding the consistency of this value. Many authors have argued on the basis of secular isotopic variations of marine carbonates that the $\delta^{18}O$ values of the oceans have varied in the past, due to different rates of seafloor spreading and continental weathering (Veizer *et al.*, 1982; Walker & Lohmann, 1989), although data from ancient samples indicate that the ocean has had a constant oxygen and hydrogen isotope composition (Hoffman *et al.*, 1986; Holmden & Muehlenbachs, 1993; Lecuyer *et al.*, 1996; see Muehlenbachs, 1987, for a more complete discussion).

Low-temperature alteration occurs down to ambient seafloor temperatures and is a long-lived slow-acting process. Whole rock $\delta^{18}O$ values from dredged samples are very variable, ranging from 5.5 to over 17‰ (Muehlenbachs & Clayton, 1972). The rocks with the lowest water content (<0.5 wt %) have $\delta^{18}O$ values close to those of unmetamorphosed MORB (Fig. 8.23). There is a linear correlation between the $\delta^{18}O$ values of whole rock samples and the age (Muehlenbachs & Clayton, 1972), from *c.* 6.0‰ for modern samples to 10‰ for samples 16 Ma in age. The change in the $\delta^{18}O$ values of the basalts is not related to changing $\delta^{18}O$ values of the phenocrysts. Rather, it accompanies mineral alteration. Pyroxene and magnetite are the first to be altered, calcium-rich plagioclase is second, while fresh olivine is often preserved in samples over 16 Ma in age (Muehlenbachs & Clayton, 1972).

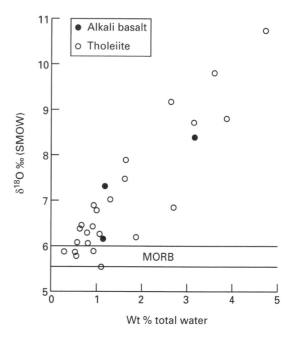

Fig. 8.23 Plot of the $\delta^{18}O$ values against total water content for submarine basalts. Low-temperature hydration causes the $\delta^{18}O$ values of the basalts to increase due to formation of low-temperature minerals. (From Muehlenbachs & Clayton, 1972; MORB $\delta^{18}O$ field from Ito *et al.*, 1987, and Harmon & Hoefs, 1995.)

Replacement minerals include montmorillonite and iron-rich chlorite. These low-temperature alteration phases have $\delta^{18}O$ values in excess of 27‰, thus explaining the dramatic increase in oxygen isotope values in altered basalts.

8.10.1 DSDP-ODP results

The combined effects of high- and low-temperature metamorphism in seafloor basalts inferred from dredged samples are confirmed by analyses of DSDP and ODP samples (Barrett & Friedrichsen, 1982; Friedrichsen, 1985; Alt *et al.*, 1986b) and from obducted ophiolites (Gregory & Taylor, 1981). Some general trends are observed from the deep DSDP Hole 504 B, Leg 111 (Fig. 8.24) (Barrett & Friedrichsen, 1982; Alt *et al.*, 1986b; Kusakabe *et al.*, 1989). The uppermost 300 m of pillow basalts are characterized by low-temperature (0–150°C) submarine weathering in an oxidizing environment. Smectite, celadonite, oxyhydroxide and zeolites characterize the alteration minerals after olivine. The

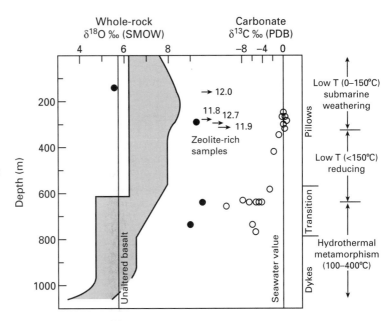

Fig. 8.24 Oxygen and carbon isotope variations in DSDP Hole 504B as a function of depth. In the upper section of the hole, the $\delta^{18}O$ values of the whole rocks (shown as shaded region — 113 analyses — with outliers shown as filled circles) are higher than those for unaltered basalt (5.8‰). The highest $\delta^{18}O$ values are found for samples that are rich in zeolites. In the region of hydrothermal metamorphism, the $\delta^{18}O$ values decrease to less than those of unaltered basalt. Carbonates in the upper section are typical for seawater, while those at depth have either a mantle component or a contribution from organic matter. (From Alt *et al.*, 1986b.)

$\delta^{18}O$ values of the rocks are higher than normal tholeiites due to the large fractionations between water ($\delta^{18}O$ value of 0‰) and basalt. Basalts of the lower pillow basalts and transition zone have a different mineralogy characterized by the presence of saponite and minor chlorite after olivine. The $\delta^{18}O$ values are slightly lower than the overlying pillow basalts, but still enriched in ^{18}O relative to unaltered material (and are still indicative of seawater). In the lithological transition zone and sheeted dykes below 600 m, there is an abrupt lowering of $\delta^{18}O$ to less than 5‰ (although samples with $\delta^{18}O$ values up to 7‰ persist). In the transition zone, the mineralogy and stable isotope data are consistent with superimposed greenschist and zeolite facies metamorphism (Alt *et al.*, 1986b). The whole rock $\delta^{18}O$ values of samples in the dyke section below 780 m are lower still, with an average of 5.4‰. Below 1000 m, the $\delta^{18}O$ values are only slightly above 5‰ (Alt *et al.*, 1986b; Kawahata *et al.*, 1987; Kusakabe *et al.*, 1989). Alteration minerals attain values as low as 1.1‰ (Friedrichsen, 1985; Alt *et al.*, 1986b), due to exchange with fluids with seawater composition at 200–250°C. Early near-axis high-temperature (>500°C) alteration of gabbros produced whole rock values of 3–6‰ (Früh-Green *et al.*, 1996).

The $\delta^{13}C$ values of carbonates in the upper section of Hole 504B are near 0‰, consistent with a seawater

carbon source to 600 m depth. In the transition zone, the $\delta^{13}C$ values are as low as –8‰ (Fig. 8.24). Alt *et al.* (1986b) and Hubberton (1982) interpreted this as a mantle or basalt-derived carbon signature, although they recognized the possibility of a contribution from organic carbon. Muehlenbachs (1987) interpreted $\delta^{13}C$ values of carbonates of –18‰ as clearly indicative of organic matter interaction.

The general alteration picture is one of low-temperature, freely circulating seawater in the upper oxidized zone of the pillow basalts. Water–rock ratios are between 10 and 100. Metamorphic temperatures in the transition zone are between 200 and 350°C, producing greenschist facies mineral assemblages. The water–rock ratios are lower and there is evidence for a mantle source of CO_2. The abrupt change in the $\delta^{18}O$ values at 620 m is thought to be due to mixing of hydrothermal upwelling waters with cooler seawater in the more permeable overlying section. This zone is characterized by a steep geothermal gradient of 2.5°C m^{-1}. The seawater–basalt interaction will cause the $\delta^{18}O$ values of the seawater to increase to as high as 4‰.

8.11 Veins

Veins can be assigned to specific structural events and temperatures can often be obtained independently from fluid inclusions. Monomineralic veins or porphyroblasts, such as quartz and calcite, can be used to infer the isotopic composition (and hence, the source) of fluids, while polymineralic veins can *occasionally* be used for thermometry. Intravein variations in the $\delta^{18}O$ values of quartz have been measured using *in situ* laser techniques to monitor variations in temperature/fluid composition during the life of the vein (Conrad & Chamberlain, 1992; Kirschner *et al.*, 1993). In the following discussion, several examples are discussed to illustrate the types of problems that can be addressed with stable isotope methods.

8.11.1 Tracing fluid origins: a combined stable isotope–fluid inclusion study of Variscan crustal fluids

Wilkinson *et al.* (1995) made a detailed fluid inclusion–stable isotope study of different generations of quartz veins from segments of Palaeozoic crust in south-west England. The veins formed over a period of >100 Ma, associated with multiple episodes of Variscan deformation, granitic emplacement and associated hydrothermal activity. The Rhenohercynian zone of the Variscan orogeny consists of Devonian rocks that underwent low-grade metamorphism during Carboniferous compressional

tectonics. This was followed by granitic intrusions over a period of 25 Ma until the Early Permian. Five deformation events were identified: D_1 (F_1 fluids) is associated with nappe emplacement during pumpellyite–actinolite grade regional metamorphism (320°C, 3.2 kbar); D_2 (F_2 fluids) occurred during continued Variscan shortening and is characterized by extensive quartz veining and local retrogression (*c.* 270°C); D_3 (F_3 fluids) is associated with the emplacement of granitic intrusions and the development of contact metamorphic zones; post-D_3 (F_4 and F_5 fluids) veining is related to late brittle extension and strike-slip faulting.

The δD and $\delta^{18}O$ values of fluid inclusions from quartz veins are plotted in Fig. 8.25. The F_1 fluids have the highest δD and $\delta^{18}O$ values of all samples, and are consistent with fluids derived from low-grade metamorphism of marine sediments and formation waters. The isotopic compositions of the F_2 fluids are slightly lower, but overlap with F_1. The isotopic values correlate with changes in salinity. Wilkinson *et al.* (1995) proposed that the shift in δD and $\delta^{18}O$ values was related to retrogressive hydration reactions during the waning stages of compressional tectonism. F_3 fluids are chemically intermediate between metamorphic and magmatic fluids. The conclusions, based on the stable isotope data, are that the F3 fluids were a combination of magmatic fluids and those

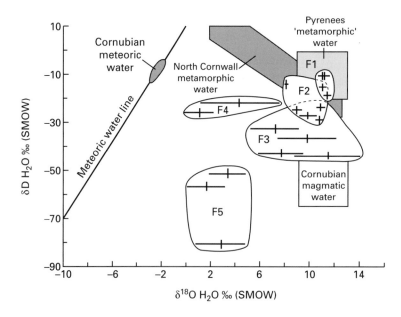

Fig. 8.25 Plot of the δD vs. $\delta^{18}O$ values of water from fluid inclusions from different generations of quartz veins from south Cornwall, UK. There is a general progression from early 'metamorphic' fluids (F1) to those with an increasingly larger magmatic and meteoric component (F4 and F5). Crosses are range of measured data. See text for details. (From Wilkinson *et al.*, 1995.)

derived locally from devolatilization of contact metamorphosed rocks. Potential reservoirs are pelitic and volcanic rocks. The F_4 fluids have much lower δD values and have low salinity, pointing to a meteoric component. Local meteoric waters in the Permo-Triassic period are inferred to have been −14 to −3‰ (the Cornubian meteoric water of Fig. 8.25). The lower δD values measured in the quartz veins (−25 to −21‰) can be explained if the local fluids were precipitated at higher altitude. Finally, the F_5 fluids have distinct δD and $\delta^{18}O$ values and high salinity, characteristic of basinal brines.

8.11.2 A combined structural-oxygen isotope thermometry study of nappe formation: an example from the Helvetic Alps

There are few quantitative thermometers available in low-grade metamorphic rocks. Other than mineral paragenesis and fluid inclusion homogenization temperatures, most 'thermometers' are qualitative (e.g. illite crystallinity, coal rank, conodont colour alteration index). Kirschner *et al.* (1995) used fibrous quartz–calcite veins to estimate the temperature of their formation. The quartz and calcite consist of interdigitated fibres, with the growth axis perpendicular to the plane of the vein. The similarity in curvature and orientation of the fibres indicates that both minerals formed (pene)contemporaneously from the same vein fluid. Simultaneous quartz–calcite formation may be related to a sudden pressure drop during vein extension. As such, they should have formed from the same fluid (same $\delta^{18}O$ value), and should be in oxygen isotope equilibrium.

Temperature estimates were made on samples from the Morcles nappe in the western Helvetic Alps, Switzerland. The Morcles nappe is a large recumbent fold with a maximum vertical extent of 6 km. The nappe was buried to depths of more than 10 km due to overthrusting of structurally higher nappes starting approximately at 30 Ma. Three generations of vein growth could be identified: S1 related to large scale folding, S2 to a second foliation and S3 related to crenulation cleavage. Temperature estimates from quartz–calcite fractionations were determined at single localities as a function of deformation histories, i.e. temperatures could be estimated for different episodes of vein growth. Due to the excellent exposure, temperature estimates could be made

throughout the nappe, providing a two-dimensional temperature deformation history. Index mineralogy, illite crystallinity, vitrinite reflectance and fluid inclusion chemistry are all consistent with increasing temperatures from anchizone conditions ($< 300°C$) in the foreland to epizone conditions ($> 300°C$) in the hinterland (Fig. 8.26).

The isotopic composition of quartz and calcite was analyzed from seven localities. At each locality, up to four different generations of veins could be identified and analyzed. Variations in the $\delta^{18}O$ values of individual veins were evaluated by microsampling; in general, the veins were found to be homogeneous throughout the thickness of the veins. The $\delta^{18}O$ and $\delta^{13}C$ calcite values of the veins are quite close to those of the host rock, indicating a rock-buffered system. Temperatures vary according to position in the nappe (Fig. 8.27). The deeper, hinterland localities range from close to 350°C for the pre S2 samples to 300°C for the later S3 veins. Temperature estimates for the shallower, foreland localities are more scattered, but show a similar trend, with an overall temperature that is about 50° cooler than observed in the deeper localities. There is an indication of a cooling anomaly during S2 foliation development, although the data are not conclusive. One sample (E) close to the Diablerets thrust actually has a higher temperature during the late period of nappe development. This may be a result of local heating from the thrust itself (passage of hot waters), an idea that is supported by other thermometric estimates, and causes the anchizone–epizone boundary to be deflected to higher elevations in the cross section (Fig. 8.26). By taking advantage of the combined structural–thermometric data inherent in veins, temperature variations related to structural events can be determined.

8.12 Retrograde metamorphism

Metamorphic petrologists, particularly those interested in high-grade rocks such as granulites and eclogites, generally ignore the postmetamorphic late-stage alteration. It is this late, hydrothermal interaction, however, that is most closely related to modern environmental concerns. Enormous amounts of water circulate in high-level hydrothermal systems, and the fluid–rock interaction will influence how polluted waters interact with rocks, how waters migrate to form aquifers and under what conditions

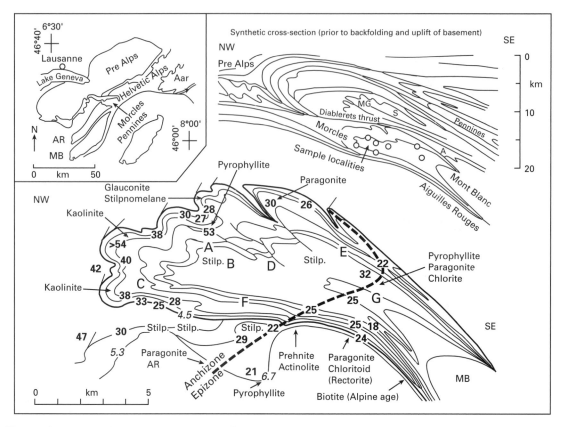

Fig. 8.26 Cross-section of the Morcles nappe, Western Swiss Alps. Shown are illite crystallinity values (bold), vitrinite reflectance (italic), index minerals and sample localities (A–G). Samples A–D are expected to be the lowest temperature samples, while E–G are more hinterland, higher temperature samples. The Diablerets thrust (see upper right inset) is situated over the top of the Morcles nappe, causing the upward inflection in the anchizone–epizone boundary (heavy dashed line on main figure). (After Kirschner *et al.*, 1995.)

deep waters will reappear at the surface. This last point is a major concern in terms of disposal of toxic wastes. Stable isotope data, particularly those of hydrogen and oxygen, provide a wealth of data regarding the interaction between rocks and fluid. Morrison and Valley (1988, 1991) demonstrated that postmetamorphic fluid infiltration of an anorthosite body in the Adirondack Mountains, USA was widespread, although locally patchy. The combined carbon and oxygen isotope data from retrograde calcites indicate fluid infiltration having occurred over a narrow temperature interval. They pointed out that the evidence for the retrograde alteration was cryptic. The widespread calcite veins were so small that they were visible only under cathodoluminescence.

A detailed study of retrograde alteration of granites and meta-igneous rocks from SW Connemara, Ireland was made by Jenkin *et al.* (1992). Numerous hydrous alteration phases are present in these rocks, involving sericitization and saussuritization of feldspar, and the formation of chlorite and epidote after biotite and hornblende. Stable mineral assemblages constrain the retrograde metamorphism to a peak temperature of *c.* 275°C at a pressure of ≤ 1.5 kbar. Oxygen isotope thermometry from coexisting quartz and epidote support these estimates. The authors also concluded that the fluids responsible for the retrogression remained in the rocks until temperatures dropped to 200–140°C. Calculated $\delta^{18}O$ values are –4 to 7‰ (SMOW). The oxygen isotope data of epidote correlate well with the f_{O_2} of the fluid, the higher

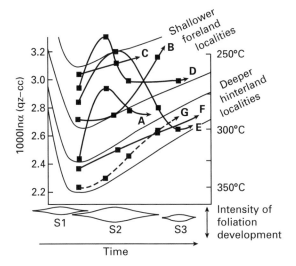

Fig. 8.27 Temperature estimates from quartz–calcite veins of the Morcles nappe. The deeper localities, F-G have recorded higher temperatures than the more frontal localities. Sample E is deflected to higher temperatures in the latest stages of deformation by local heating of the overlying Diablerets thrust. (After Kirschner *et al.*, 1995.)

f_{O_2} estimates being associated with the lower $\delta^{18}O$ values. Together, these data strongly argue for a meteoric water presence. The range of isotopic values is assumed to be due to differing amounts of exchange between fluid and rock.

In other studies of retrograde fluids, multiple episodes of fluid infiltration can be deduced by looking at different generations of minerals, or minerals with different closure temperature characteristics. In a study of alteration fluids from prehnite–pumpellyite facies Palaeozoic rocks from the English Lake District, Thomas *et al.* (1985) analyzed oxygen and hydrogen isotope compositions of primary and secondary minerals, veins and fluid inclusions. They recognized three different fluids that were responsible for late alteration: a synmetamorphic, deuterium-rich fluid in equilibrium with coexisting chlorite and epidote; a postdeformational, low-δD fluid related to quartz veining; and an intermediate-δD fluid that caused the alteration of granites. Fluid compositions were calculated from the δD–$\delta^{18}O$ values of each phase as a function of temperature. At temperatures between 250° and 350°C, the fluid compositions calculated from both chlorite and epidote are the same. The

lack of later retrograde exchange is interpreted to indicate that the fluid was removed from the system at the metamorphic temperatures. This first fluid phase is characteristic of metamorphic fluid.

The $\delta^{18}O$ values of the groundmass quartz from these rocks are not useful for constraining metamorphic fluid compositions. Metamorphic temperatures were never high enough to cause re-equilibration of the detrital material with the metamorphic fluid. However, the oxygen isotope fluid compositions calculated from the $\delta^{18}O$ values of nondeformed quartz veins are similar to those obtained from chlorite and epidote. The δD values of the fluid inclusions in the quartz are much lower than those estimated from the hydrous silicates. Thus, a second, postmetamorphic fluid source is considered to be responsible for the quartz vein formation. Thomas *et al.* (1985) proposed a mixed meteoric–magmatic fluid source (the later coming from syndeformational granites).

The final episode of fluid infiltration is recorded in the illites and is responsible for the granite alteration. The isotopic composition of this fluid is obtained from the measured δD values of the illite and the $\delta^{18}O$ whole rock values of the granites. The δD values are intermediate to the early metamorphic fluids and those responsible for quartz vein formation. A purely meteoric water source would have such a composition.

8.13 Summary

Stable isotope geochemistry is a powerful geochemical tool for constraining metamorphic conditions in very low-grade metamorphic rocks, but one that has not been used very extensively. Part of the problem may be related to the complexities of mineral separation and interpretation of such data. Mineral inheritance and the general complications associated with sluggish kinetic rates are daunting. Nevertheless, there have been a number of significant contributions using stable isotope data, and more are certain to follow. Source rocks, histories of fluid–rock interaction and temperatures of metamorphism can all be evaluated with stable isotope geochemistry. Careful strategies need to be adopted in order to maximize the opportunities for success. In the future, new microanalytical techniques, particularly the ion microprobe, will be used for determining oxygen and hydrogen isotope ratios of material with an extremely high spatial resolution. Because the variation in the isotopic ratios of minerals in low-grade rocks is so large, the very high analytical precision required in studies of high-temperature processes is not as important as in higher-grade equivalents. Even without specialized equipment, however, the combination of careful sampling strategies and effective mineral separation techniques provide important information in the 'fertile' field of very low-grade metamorphism.

References

Aagaard, P. & Jahren, J.S. (1992) Diagenetic illite–chlorite assemblages in arenites. II. Thermodynamic relations. *Clays and Clay Minerals* **40**, 547–554.

Abad-Ortega, M.M. & Nieto, F. (1995) Genetic and chemical relationships between berthierine, chlorite and cordierite in nodules associated to granitic pegmatites of Sierra Albarrana (Iberian Massif, Spain). *Contributions to Mineralogy and Petrology* **120**, 327–336.

Adams, C.J. & Robinson, P. (1993) Potassium–argon age studies of metamorphism uplift cooling in Haast Schist coastal section south of Dunedin, Otago, New Zealand. *New Zealand Journal of Geology and Geophysics* **36**, 317–325.

Adamson, A.C. & Richards, H.G. (1990) Low-temperature alteration of very young basalts from ODP Hole 648B: Serocki volcano, Mid-Atlantic Ridge. In: *Proceedings of ODP, Scientific Results v. 106/109* (eds R. Detrick, J. Honnorez, W. B. Bryan & T. Juteau), pp. 181–196. Ocean Drilling Program, College Station, TX.

Aguirre, L. (1993) Compositional variations of Cretaceous pumpellyites along the western margin of South America and their relation to an extensional geodynamic setting. *Journal of Metamorphic Geology* **11**, 437–448.

Aguirre, L. & Offler, R. (1985) Burial metamorphism in the Western Peruvian trough: its relation to Andean magmatism and tectonics. In: *Magmatism at a Plate Edge: the Peruvian Andes* (eds W. S. Pitcher, M. P. Atherton, E. J. Cobbing & R. D. Beckinsale), pp. 59–71. Blackie & Son Ltd, Glasgow.

Aguirre, L., Levi, B. & Offler, R. (1978) Unconformities as mineralogical breaks in the burial metamorphism of the Andes. *Contributions to Mineralogy and Petrology* **66**, 361–366.

Aguirre, L., Levi, B. & Nyström, J.O. (1989) The link between metamorphism, volcanism and geotectonic setting during the evolution of the Andes. In: *Evolution of Metamorphic Belts* (eds J. S. Daly, R. A. Cliff & B. W. D. Yardley), pp. 223–232. Geological Society of London Special Publication No. 43.

Ahn, J.H. & Peacor, D.R. (1986) Transmission and analytical electron microscopy of the smectite-to-illite transition. *Clays and Clay Minerals* **34**, 165–179.

Ahn, J.H., Peacor, D.R. & Essene, E J. (1986) Cation-diffusion-induced characteristic beam damage in transmission electron microscope images of micas. *Ultramicroscopy* **19**, 375–382.

Ahn, J.H., Peacor, D.R. & Coombs, D.S. (1988) Formation mechanism of illite, chlorite and mixed-layer illite–chlorite in Triassic volcanogenic sediments from the Southland Syncline, New Zealand. *Contributions to Mineralogy and Petrology* **99**, 82–89.

Aja, S.U. (1991) Illite equilibria in solutions: III. A reinterpretation of the data of Sass *et al.* (1987). *Geochimica et Cosmochimica Acta* **55**, 3431–3435.

Aja, S.U. & Rosenberg, P.E. (1992) The thermodynamic status of compositionally variable clay minerals: a discussion. *Clays and Clay Minerals* **40**, 292–299.

Aja, S.U. & Rosenberg, P.E. (1996) The thermodynamic status of compositionally variable clay minerals: discussion of 'Clay mineral thermometry—a critical perspective'. *Clays and Clay Minerals* **44**, 560–568.

Aja, S.U., Rosenberg, P.E. & Kittrick, J.A. (1991a) Illite equilibria in solutions: I. Phase relationships in the system $K_2O-Al_2O_3-SiO_2-H_2O$ between 25° and 250°C. *Geochimica et Cosmochimica Acta* **55**, 1353–1364.

Aja, S.U., Rosenberg, P.E. & Kittrick, J.A. (1991b) Illite equilibria in solutions: II. Phase relationships in the system $K_2O-MgO-Al_2O_3-SiO_2-H_2O$ between 25° and 250°C. *Geochimica et Cosmochimica Acta* **55**, 1365–1374.

Al-Aasm, I.S., Taylor, B.E. & South, B. (1990) Stable isotope analysis of multiple carbonate samples using selective acid extraction. *Chemical Geology* **80**, 119–125.

Allen, P.M., Cooper, D.C. & Fortey, N.J. (1987) Composite lava flows in the English Lake District. *Journal of the Geological Society of London* **144**, 945–960.

Alt, J.C. (1988) Hydrothermal oxide and nontronite deposits on seamounts in the eastern Pacific. *Marine Geology* **81**, 227–239.

Alt, J.C. (1993) Low-temperature alteration of basalts from the Hawaiian Arch, ODP leg 136. In: *Proceedings of ODP, Scientific Results 136* (eds R. H. Wilkens, J. Firth & J. Bender), pp. 133–146. Ocean Drilling Program, College Station, TX.

Alt, J.C. (1994) A sulfur isotopic profile through the Troodos Ophiolite, Cyprus: primary composition and the effects of seawater hydrothermal alteration. *Geochimica et Cosmochimica Acta* **58**, 1825–1840.

Alt, J.C. (1995) Subseafloor processes in mid-ocean ridge hydrothermal systems. In: *Seafloor Hydrothermal Systems: Physical, Chemical, and Biological Interactions, Geophysics Monograph No. 91* (eds S. Humphris, J. Lupton, L. Mullineaux & R. Zierenberg), pp. 85–114. American Geophysical Union, Washington, DC.

Alt, J.C. (1997) Hydrothermal alteration and mineralization of oceanic crust: mineralogy, geochemistry, and processes. In: *Volcanic Associated Massive Sulfide Deposits* (eds T. Barrie & M. Hannington). *Reviews in Economic Geology*, Vol. 8. Society of Economic Geologists, Chelsea, MI (in press).

Alt, J.C. & Anderson, T.F. (1991) The mineralogy and isotopic composition of sulfur in Layer 3 gabbros from the Indian Ocean, ODP Hole 735B. In: *Proceedings of ODP, Scientific Results 118* (eds P. T. Robinson & R. P. Von Herzen), pp. 113–125. Ocean Drilling Program, College Station, TX.

Alt, J.C. & Honnorez, J. (1984) Alteration of the upper oceanic crust DSDP Site 417: mineralogy and chemistry. *Contributions to Mineralogy and Petrology* **87**, 149–169.

Alt, J.C. & Teagle, D.A.H. (1998) Probing the TAG hydrothermal mound and stockwork: oxygen isotope profiles from deep ocean drilling. In: *Proceedings of ODP, Scientific Results 158* (eds P. M. Herzig, S. E. Humphris, D. J. Miller & R. A. Zierenberg), pp. 285–295. Ocean Drilling Program, College Station, TX.

Alt, J.C., Laverne, C. & Muehlenbachs, K. (1985) Alteration of the upper oceanic crust: mineralogy and processes in DSDP Hole 504B, Leg 83. In: *Initial Reports, DSDP v. 83* (eds R. N. Anderson, J. Honnorez & K. Becker), pp. 217–247. US Government Printing Office, Washington, DC.

Alt, J.C., Honnorez, J., Laverne, C. & Emmermann, R. (1986a) Hydrothermal alteration of a 1 km section through the upper oceanic crust, deep sea drilling project hole 504B: mineralogy, chemistry, and evolution of seawater–basalt interactions. *Journal of Geophysical Research* **91**, 10309–10335.

Alt, J.C., Muehlenbachs, K. & Honnorez, J. (1986b) An oxygen isotopic profile through the upper kilometer of the oceanic crust, DSDP Hole 504B. *Earth and Planetary Science Letters* **80**, 217–229.

Alt, J.C., Anderson, T.F. & Bonnell, L. (1989) The geochemistry of sulfur in a 1.3 km section of hydrothermally altered oceanic crust, DSDP Hole 504B. *Geochimica et Cosmochimica Acta* **53**, 1011–1023.

Alt, J.C., Lanord, C.F., Floyd, P.A., Castillo, P. & Galy, A. (1992) Low-temperature hydrothermal alteration of Jurassic ocean crust, Site 801. In: *Proceedings of ODP, Scientific Results 129* (eds R. L. Larson & Y. Lancelot), pp. 415–427. Ocean Drilling Program, College Station, TX.

Alt, J.C., Zuleger, E. & Erzinger, J. (1995) Stable isotopic compositions of hydrothermally altered lower sheeted dikes, ODP Hole 504B, Leg 140. In: *Proceedings of ODP, Scientific Results 140* (eds H. J. B. Dick, J. Erzinger & L. Stokking), pp. 155–166. Ocean Drilling Program, College Station, TX.

Alt, J.C., Laverne, C., Vanko, D. *et al.* (1996a) Hydrothermal alteration of a section of upper oceanic crust in the eastern equatorial Pacific: a synthesis of results from Site 504 (DSDP legs 69, 70, and 83, and ODP legs 111, 137, 140, and 148). In: *Proceedings of ODP, Scientific Results 148* (eds J. C. Alt, H. Kinoshita, L. Stokking & P. J. Michael), pp. 417–434. Ocean Drilling Program, College Station, TX.

Alt, J.C., Teagle, D.A.H., Bach, W., Halliday, A.N. & Erzinger, J. (1996b) Stable and strontium isotopic profiles through hydrothermally altered upper oceanic crust, ODP Hole 504B. In: *Proceedings of ODP, Scientific Results 148* (eds J. C. Alt, H. Kinoshita & L. Stokking), pp. 47–70. Ocean Drilling Program, College Station, TX.

Alt, J.C., Teagle, D.A.H., Laverne, C. *et al.* (1996c) Ridge flank alteration of upper ocean crust in the eastern Pacific: a synthesis of results for volcanic rocks of holes 504B and 896A. In: *Proceedings of ODP, Scientific Results 148* (eds J. C. Alt, H. Kinoshita, L. B. Stokking & P. J. Michael), pp. 434–452. Ocean Drilling Program, College Station, TX.

Alt, J.C., Teagle, D.A.H., Brewer, T.S., Shanks, W.C. & Halliday, A.N. (1998) Alteration and mineralization of an oceanic forearc and the ophiolite-ocean crust analogy. *Journal of Geophysical Research*, **103**, 12365–12380.

Altaner, S.P., Whitney, G. & Aronson, J.L. (1984) Model for K-bentonite formation: evidence from zoned K-bentonites in the disturbed belt, Montana. *Geology* **12**, 412–415.

Amouric, M. & Olives, J. (1991) Illitization of smectite as seen by high-resolution transmission electron microscopy. *European Journal of Mineralogy* **3**, 831–835.

Amstutz, G.C. (1977) Time- and stratabound features of the Michigan copper deposits (USA). In: *Time- and Strata-bound Ore Deposits* (eds D. D. Klemm & H.-J. Schneider), pp. 123–140. Springer, Berlin, Heidelberg, New York.

Anderson, T.B. & Oliver, G.J.H. (1986) The Orlock Bridge Fault: a major Late Caledonian sinistral fault in the Southern Uplands terrane, British Isles. *Transactions of the Royal Society of Edinburgh: Earth Sciences* **77**, 203–222.

Andrews, A.J. (1977) Low temperature fluid alteration of oceanic Layer 2 basalts. *Canadian Journal of Earth Science* **14**, 911–926.

Andrews, A.J. (1979) On the effect of low-temperature seawater–basalt interaction on the distribution of sulfur in oceanic crust, Layer 2. *Earth and Planetary Science Letters* **46**, 68–80.

Andrews, A.J. (1980) Saponite and celadonite in layer 2

basalts, DSDP Leg 37. *Contributions to Mineralogy and Petrology* **73**, 323–340.

Andrews, A.J., Dollase, W.A. & Fleet, M.E. (1983) A Mossbauer study of saponite occurring in Layer 2 basalts, DSDP Leg 69. In: *Initial Reports, DSDP 69* (eds J. R. Cann, M. G. Langseth, J. Honnorez, R. P. Von Herzen & S. M. White), pp. 585–588. US Government Printing Office, Washington, DC.

Arch, J., Stevenson, E. & Maltman, A. (1996) Factors affecting the containment properties of natural clays. In: *Engineering Geology of Waste Disposal* (ed. S. P. Bentley), pp. 259–265. Geological Society, Engineering Geology Special Publication No. 11.

Árkai, P. (1973) Pumpellyite–prehnite–quartz facies Alpine metamorphism in the Middle Triassic volcanogenic-sedimentary sequence of the Bükk Mountains, Northeast Hungary. *Acta Geologica Academiae Scientiarum Hungaricae* **17**, 67–83.

Árkai, P. (1977) Low-grade metamorphism of Paleozoic formations of the Szendrö Mountains (NE-Hungary). *Acta Geologica Academiae Scientiarum Hungaricae* **21**, 53–80.

Árkai, P. (1983) Very low- and low-grade Alpine regional metamorphism of the Palaeozoic and Mesozoic formations of the Bükkium, NE Hungary. *Acta Geologica Hungarica* **26**, 83–101.

Árkai, P. (1991) Chlorite crystallinity: an empirical approach and correlation with illite crystallinity, coal rank and mineral facies as exemplified by Palaeozoic and Mesozoic rocks of northeast Hungary. *Journal of Metamorphic Geology* **9**, 723–734.

Árkai, P. (1995) *Alpine regional metamorphism in the Dinaric-type tectonic domains of the Pannonian Basin and innermost Western Carpathians, Hungary.* Geological Society of Greece, Special Publication No. 4, pp. 585–589.

Árkai, P. & Ghabrial, D.S. (1997) Chlorite crystallinity as an indicator of metamorphic grade of low-temperature meta-igneous rocks: a case study from the Bükk Mountains, northeast Hungary. *Clay Minerals* **32**, 205–222.

Árkai, P. & Tóth, N.M. (1983) Illite crystallinity: combined effects of domain size and lattice distortion. *Acta Geologica Hungarica* **26**, 341–358.

Árkai, P., Horváth, Z.A. & Tóth, M. (1981) Transitional very low- and low-grade regional metamorphism of the Paleozoic formations, Uppony Mountains, NE-Hungary: mineral assemblages, illite-crystallinity, -b_0 vitrinite reflectance data. *Acta Geologica Academiae Scientiarum Hungaricae* **24**, 265–294.

Árkai, P., Balogh, K. & Dunkl, I. (1995) Timing of low-temperature metamorphism and cooling of the Paleozoic and Mesozoic formations of the Bükkium, innermost Western Carpathians, Hungary. *Geologische Rundschau* **84**, 334–344.

Árkai, P., Merriman, R.J., Roberts, B., Peacor, D.R. &

Tóth, M. (1996) Crystallinity, crystallite size and lattice strain of illite–muscovite and chlorite: comparison of XRD and TEM data for diagenetic and epizonal pelites. *European Journal of Mineralogy* **8**, 1119–1137.

Árkai, P., Balogh, K. & Frey, M. (1997) The effects of tectonic strain on crystallinity, apparent mean crystallite size and lattice strain of phyllosilicates in low-temperature metamorphic rocks. A case study from the Glarus overthrust, Switzerland. *Schweizerische Mineralogische und Petrographische Mitteilungen* **77**, 27–40.

Arnorsson, S., Gronvold, K. & Sigurdsson, S. (1978) Aquifer chemistry of four high-temperature geothermal systems in Iceland. *Geochimica et Cosmochimica Acta* **42**, 523–536.

Aronson, J.L. & Douthitt, C.B. (1986) K/Ar systematics of an acid-treated illite/smectite: implications for evaluating age and crystal structure. *Clays and Clay Minerals* **34**, 473–482.

Aronson, J.L. & Hower, J. (1976) Mechanism of burial metamorphism of argillaceous sediment: 2. Radiogenic argon evidence. *Geological Society of America Bulletin* **87**, 738–743.

Auzende, J.-M., Bideau, D., Bonatti, E. *et al.* (1989) Direct observation of a section through slow-spreading oceanic crust. *Nature* **337**, 726–729.

Awan, M.A. & Woodcock, N.H. (1991) A white mica crystallinity study of the the Berwyn Hills, North Wales. *Journal of Metamorphic Geology* **9**, 765–773.

Awwiller, D.N. (1993) Illite/smectite formation and potassium mass transfer during burial diagenesis of mudrocks: a study from the Texas Gulf Coast Paleocene–Eocene. *Journal of Sedimentary Petrology* **63**, 501–512.

Ayalon, A. & Longstaffe, F.J. (1990) Isolation of diagenetic minerals in clastic sedimentary rocks for oxygen isotope analysis: a summary of methods. *Israeli Journal of Earth Sciences* **39**, 139–148.

Baadsgaard, H. (1987) Rb–Sr and K–Ca isotope systematics in minerals from potassium horizons in the Prairie Evaporite formation, Saskatchewan, Canada. *Chemical Geology* **66**, 1–15.

Baertschi, P. & Silverman, S.R. (1951) The determination of relative abundances of the oxygen isotopes in silicate rocks. *Geochimica et Cosmochimica Acta* **1**, 4–6.

Bailey, S.W. (1980) Structures of layer silicates. In: *Crystal Structures of Clay Minerals and Their X-Ray Identification* (eds G. W. Brindley & G. Brown), Monograph 5, pp. 1–123. Mineralogical Society, London.

Bailey, S.W. (1988) X-ray diffraction identification of the polytypes of mica, serpentine, and chlorite. *Clays and Clay Minerals* **36**, 193–213.

Bailey, S.W. & Brown, B.E. (1962) Chlorite polytypism: I. Regular and semi-random one-layer structures. *American Mineralogist* **47**, 819–850.

Bally, A.W. (1975) A geodynamic scenario for

hydrocarbon occurrences. *Proceedings of the Ninth World Petroleum Congress, Tokyo* **2**, 33–44.

Barker, A.J. (1990) *Introduction to Metamorphic Textures and Microstructures*. Blackie, Glasgow and London.

Barnes, R.P. (1998) The geology of the Kirkcowan–Wigtown district—a concise account of the geology. *Memoir of the British Geological Survey*, Sheets 4E, 4W and part of 2 (Scotland).

Barnes, R.P., Lintern, B.C. & Stone, P. (1989) Timing and regional implications of deformation in the Southern Uplands of Scotland. *Journal of the Geological Society of London* **146**, 905–908.

Baronnet, A. (1982) Ostwald ripening in solution. The case of calcite and mica. *Estudios Geol* **38**, 185–198.

Baronnet, A. (1992) Polytypism and stacking disorder. In: *Minerals and Reactions at the Atomic Scale: Transmission Electron Microscopy* (ed. P. R. Buseck), Mineralogical Society of America. *Reviews in Mineralogy* **27**, 231–288.

Barr, T.D., Dahlen, F.A. & McPhail, D.C. (1991) Brittle frictional mountain building 3. Low-grade metamorphism. *Journal of Geophysical Research* **96**, 10319–10338.

Barrett, T.J. & Friedrichsen, H. (1982) Strontium and oxygen isotopic composition of some basalts from Hole 504B, Costa Rica Rift, DSDP legs 69 and 70. *Earth and Planetary Science Letters* **60**, 27–38.

Barrow, G. (1893) On an intrusion of muscovite–biotite gneiss in the Southeast Highlands of Scotland and its accompanying metamorphism. *Quarterly Journal of the Geological Society* **49**, 330–358.

Barrow, G. (1912) On the geology of lower Dee-side and the southern Highland Border. *Proceedings of the Geologists Association* **23**, 274–290.

Bass, M.N. (1976) Secondary minerals in oceanic basalts, with special reference to Leg 34, DSDP. In: *Initial Reports DSDP 34* (eds R. S. Yeats & S. R. Hart), pp. 393–431. US Government Printing Office, Washington, DC.

Bath, A.H. (1974) New isotopic age data on rocks from the Long Mynd, Shropshire. *Journal of the Geological Society, London* **130**, 567–574.

Bath, A.H. (1977) Experimental observation of exchange of Rb and Sr between clays and solution. In: *2nd International Symposium of Water–Rock Interactions and International Association of Geochemistry and Cosmochemistry*, Strasbourg, IV. pp. 244–249.

Beaufort, D. & Meunier, A. (1994) Saponite, corrensite and chlorite–saponite mixed-layers in the Sancerre-Couy deep drill-hole (France). *Clay Minerals* **29**, 47–61.

Bechtel, A. & Hoernes, S. (1990) Oxygen isotope fractionation between oxygen of different sites in illite minerals; a potential single-mineral thermometer. *Contributions to Mineralogy and Petrology* **104**, 463–470.

Becker, K., Sakai, H., Adamson, A. *et al.* (1989) Drilling deep into young oceanic crust, Hole 504B, Costa Rica Rift. *Reviews of Geophysics* **27**, 79–102.

Bednarz, U. & Schmincke, H.U. (1989) Mass transfer during sub-seafloor alteration of the upper Troodos crust (Cyprus). *Contributions to Mineralogy and Petrology* **102**, 93–101.

Beiersdorfer, R.E. (1993) Metamorphism of a Late Jurassic volcano-plutonic arc, northern California, USA. *Journal of Metamorphic Geology* **11**, 415–428.

Beiersdorfer, R.E. & Day, H.W. (1983) Pumpellyite–actinolite facies metamorphism in the Smartville Complex, Northern Sierra Nevada. *Geological Society of America, Abstracts with Programs* **15**, 436.

Beiersdorfer, R.E. & Day, H.W. (1995) Mineral paragenesis of pumpellyite in low-grade mafic rocks. In: *Low-grade Metamorphism of Mafic Rocks* (eds P. Schiffman & H. W. Day). Geological Society of America, Boulder, CO. Special Paper No. 296, pp. 5–28.

Belmar, M., Schmidt, S.Th., Frey, M., Vergara, M. & Aguirre, M. (1997) Low-grade metamorphic patterns in the Chilean Andes between 34° and 35° South—preliminary results. *Terra Abstracts* **9**, 575.

Bennett, R.H., O'Brien, N.R. & Hulbert, M. (1991) Determinants of clay and shale microfabric signatures: processes and mechanisms. In: *Microstructure of Fine-Grained Sediments: From Mud to Shale* (eds R. H. Bennett, W. R. Bryant & M. H. Hulbert), pp. 5–32. Springer-Verlag.

Berman, R.G. (1988) Internally-consistent thermodynamic data for stoichiometric minerals in the system $Na_2O-K_2O-CaO-MgO-FeO-Fe_2O_3-Al_2O_3-SiO_2-TiO_2-H_2O-CO_2$. *Journal of Petrology* **29**, 445–522.

Berman, R.G. (1991) Thermobarometry using multiequilibrium calculations: a new technique with petrologic applications. *Canadian Mineralogist* **29**, 833–855.

Berman, R.G., Brown, T.H. & Perkins, E.H. (1987) GE0-CALC: software for calculation and display of pressure–temperature–composition phase diagrams. *American Mineralogist* **72**, 861–862.

Berndt, M.E. & Seyfried, W.E. (1986) B, Li, and associated trace element chemistry of alteration minerals, Holes 597B and 597C. In: *Initial Reports DSDP 92* (eds M. Leinen & D. K. Rea), pp. 491–497, US Government Printing Office, Washington, DC.

Bertin, E.P. (1978) *Introduction to X-ray Spectrometric Analysis*. Plenum Press, New York.

Bettison, L.A. & Schiffman, P. (1988) Compositional and structural variations of phyllosilicates from the Point Sal ophiolite, California. *American Mineralogist* **73**, 62–76.

Bettison-Varga, L. & MacKinnon, I.D.R. (1997) The role of randomly mixed-layer chlorite/smectite in the transformation of smectite to chlorite. *Clays and Clay Minerals* **45**, 506–516.

Bettison-Varga, L., Mackinnon, I.D.R. & Schiffman, P. (1991) Integrated TEM, XRD and electron microprobe investigation of mixed-layer chlorite–smectite from the Point Sal ophiolite, California. *Journal of Metamorphic Geology* **9**, 697–710.

Bettison-Varga, L., Varga, R.J. & Schiffman, P. (1992) Relation between ore-forming hydrothermal systems and extensional deformation in the Solea graben spreading center, Troodos ophiolite, Cyprus. *Geology* **20**, 987–990.

Bevington, P.R. & Robinson, D.K. (1992) *Data Reduction and Error Analysis for the Physical Sciences.* McGraw-Hill, New York.

Bevins, R.E. (1978) Pumpellyite-bearing basic igneous rocks from the Lower Ordovician of North Pembrokeshire, Wales. *Mineralogical Magazine* **42**, 81–83.

Bevins, R.E. & Merriman, R.J. (1988) Compositional controls on coexisiting prehnite–actinolite and prehnite–pumpellyite facies assemblages in the Tal y Fan metabasite intrusion, North Wales: implications for Caledonian metamorphic field gradients. *Journal of Metamorphic Geology* **6**, 17–39.

Bevins, R.E. & Robinson, D. (1988) Low grade metamorphism in the Welsh Basin Lower Palaeozoic succession: an example of diastathermal metamorphism? *Journal of the Geological Society, London* **145**, 363–366.

Bevins, R.E. & Robinson, D. (1993) Parageneses of Ordovician sub-greenschist to greenschist facies metabasites from Wales, UK. *European Journal of Mineralogy* **5**, 925–935.

Bevins, R.E. & Robinson, D. (1994) A review of low grade metabasite parageneses. In: *Very Low Grade Metamorphism: Mechanisms and Geological Applications* (eds W. Hanquan, T. Bai & L. Yiqun), pp. 1–8. The Seismological Press, Beijing, China.

Bevins, R.E. & Robinson, D. (1995) Regional low-grade polygenetic metamorphism and inversion in the northern part of the Eastern Belt, Northern Sierra Nevada, California. In: *Low-grade Metamorphism of Mafic Rocks* (eds P. Schiffman & H. W. Day). Geological Society of America, Boulder, CO. Special Paper No. 296, pp. 29–50.

Bevins, R.E. & Rowbotham, G. (1983) Low grade metamorphism within the Welsh sector of the paratectonic Caledonides. *Geological Journal* **18**, 141–167.

Bevins, R.E., Robinson, D., Gayer, R.A. & Allman, S. (1986) Low-grade metamorphism and its relationship to thrust tectonics in the Caledonides of Finnmark, North Norway. *Norges Geologiske Undersøkelse Bulletin* **404**, 31–42.

Bevins, R.E., Robinson, D. & Rowbotham, G. (1991a) Compositional variations in mafic phyllosilicates from regional low-grade metabasites and application of the chlorite geothermometer. *Journal of Metamorphic Geology* **9**, 711–721.

Bevins, R.E., Rowbotham, G. & Robinson, D. (1991b)

Zeolite to prehnite–pumpellyite facies metamorphism of the late Proterozoic Zig-Zag Dal Basalt Formation, eastern North Greenland. *Lithos* **27**, 155–165.

Bevins, R.E., White, S.C. & Robinson, D. (1996) The South Wales Coalfield: low grade metamorphism in a foreland basin setting? *Geological Magazine* **133**, 739–749.

Bickle, M.J. & Teagle, D.A.H. (1992) Strontium alteration in the Troodos ophiolite: implications for fluid fluxes and geochemical transport in mid-ocean ridge hydrothermal systems. *Earth and Planetary Science Letters* **113**, 219–237.

Bigeleisen, J., Perlman, M.L. & Prosser, H.C. (1952) Conversion of hydrogenic materials to hydrogen for isotopic analysis. *Analytical Chemistry* **24**, 1356–1357.

Bird, D.K., Schiffman, P., Elders, W.A., Williams, A.E. & McDowell, S.D. (1984) Calc-silicate mineralization in active geothermal systems. *Economic Geology* **79**, 671–695.

Bird, M.I., Longstaffe, F.J. & Fyfe, W.S. (1993) Oxygen isotope fractionation in titanium-oxide minerals at low temperature. *Geochimica et Cosmochimica Acta* **57**, 3083–3091.

Bish, D.L. & Aronson, J.L. (1993) Paleogeothermal and paleohydrologic conditions in silicic tuff from Yucca Mountain, Nevada. *Clays and Clay Minerals* **41**, 148–161.

Blanco, J.A., Corrochano, A., Montigny, R. & Thuizat, R. (1982) Sur l'âge du début de la sédimentation dans le bassin tertiaire du Duero (Espagne). Attribution au Paléocène par datation isotopique des alunites de l'unité inférieure. *Comptes Rendus de l'Académie des Sciences, Paris, série II* **295**, 259–262.

Blattner, P. (1975) Oxygen isotopic composition of fissure-grown quartz, adularia, and calcite from Broadlands geothermal field, New Zealand, with an appendix on quartz–K-feldspar–calcite–muscovite oxygen isotope geothermometers. *American Journal of Science* **275**, 785–800.

Blenkinsop, T.G. (1988) Definition of low-grade metamorphic zones using illite crystallinity. *Journal of Metamorphic Geology* **6**, 623–636.

Boettcher, A.L. & O'Neil, J.R. (1980) Stable isotope, chemical and petrographic studies of high-pressure amphiboles and micas: evidence for metasomatism in the mantle source regions of alkali basalt and kimberlites. *American Journal of Science* **280A**, 594–621.

Boger, P.D. & Faure, G. (1974) Strontium-isotope stratigraphy of a Red Sea core. *Geology* **2**, 181–183.

Bohlke, J.K., Honnorez, J. & Honnorez-Guerstein, M.B. (1980) Alteration of basalts from Site 396B, DSDP: petrographic and mineralogic studies. *Contributions to Mineralogy and Petrology* **73**, 341–364.

Bohlke, J.K., Honnorez, J., Honnorez-Guerstein, B.M., Muehlenbachs, K. & Petersen, N. (1981) Heterogeneous alteration of the upper oceanic crust:

correlation of rock chemistry, magnetic properties, and O isotope ratios with alteration patterns in basalts from Site 396B, DSDP. *Journal of Geophysical Research* **86**, 7935–7950.

Bohlke, J.K., Alt, J.C. & Muehlenbachs, K. (1984) Oxygen isotope–water relations in altered deep-sea basalts: low-temperature mineralogical controls. *Canadian Journal of Earth Science* **21**, 67–77.

Boles, J.R. & Franks, S.G. (1979) Clay diagenesis in Wilcox sandstones of southwest Texas: implications of smectite diagenesis on sandstone cementation. *Journal of Sedimentary Petrology* **49**, 55–70.

Bons, A.-J. (1988) Intracrystalline deformation and slaty cleavage development in very low-grade slates from the Central Pyrénées. *Geological Ultraiectina* **56**, 173 pp.

Borradaile, G.J., Bayly, M.B. & Powell, C.Mc.A. (eds) (1982) *Atlas of Deformational and Metamorphic Rock Fabrics.* Springer-Verlag.

Bostick, N.H., Cashman, S.M., McCulloch, T.H. & Waddell, C.T. (1978) Gradients of vitrinite reflectance and present temperature in the Los Angeles and Ventura Basins, California. In: *Low Temperature Metamorphism of Kerogens and Clay Minerals* (ed. D. F. Oltz), pp. 65–96. Society of Economic Paleontologists and Mineralogists. Pacific Section Symposium in Geochemistry.

Bottinga, Y. (1969) Carbon isotope fractionation between graphite, diamond and carbon dioxide. *Earth and Planetary Science Letters* **5**, 301–307.

Bottinga, Y. & Javoy, M. (1973) Comments on oxygen isotope geothermometry. *Earth and Planetary Science Letters* **20**, 250–265.

Bottinga, Y. & Javoy, M. (1975) Oxygen isotope partitioning among the minerals in igneous and metamorphic rocks. *Reviews of Geophysics and Space Physics* **13**, 401–418.

Bottrell, S.H., Greenwood, P.B., Yardley, B.W.D., Shepherd, T.J. & Spiro, B. (1990) Metamorphic and post-metamorphic fluid flow in the low-grade rocks of the Harlech Dome, North Wales. *Journal of Metamorphic Geology* **8**, 131–143.

Bowen, R. (1966) *Paleotemperature Analysis.* Elsevier, Amsterdam and New York.

Boyd, F.R., Finger, L.W. & Chayes, F. (1967) Computer reduction of electron microprobe data. *Carnegie Institution of Washington Yearbook* **67**, 210–215.

Bray, C.J., Spooner, E.T.C., Hall, C.M., York, D., Bills, T.M. & Krueger, H.W. (1987) Laser probe $^{40}Ar/^{39}Ar$ and conventional K–Ar dating of illites associated with the McClean unconformity-related uranium deposits, north Saskatchewan, Canada. *Canadian Journal of Earth Sciences* **24**, 10–23.

Brearley, A.J. (1988) Chloritoid from low-grade pelitic rocks in North Wales. *Mineralogical Magazine* **52**, 394–396.

Breitschmid, A. (1982) Diagenese und schwache Metamorphose inden sedmentären Abfolgen der Zentralschweizer Alpen (Vierwaldstätter See, Urirotstock). *Eclogae Geologica Helvetica* **75**, 331–380.

Brindley, G.W. & Brown, G. (eds) (1980) *Crystal structures of clay minerals and their X-ray identification.* Mineralogical Society, London, Monograph No. 5.

British Geological Survey (1992a) Kirkcowan, Scotland. Sheet 4W. Solid 1:50000. Nottingham, British Geological Survey.

British Geological Survey (1992b) Wigtown, Scotland. Sheet 4E. Solid 1:50000. Nottingham, British Geological Survey.

British Geological Survey (1993a) The Rhins of Galloway, Scotland. Sheets 1 and 3 with parts of 7 and 4W. Solid Geology 1:50000. Keyworth, Nottingham, British Geological Survey.

British Geological Survey (1993b) Rhayader, England and Wales. Sheet 179. Solid 1:50000. Nottingham, British Geological Survey.

British Geological Survey (1993c) Llanilar, England and Wales. Sheet 178. Solid with Drift Geology 1:50000. Keyworth, Nottingham.

British Geological Survey (1993d) Kirkcudbright, Scotland. Sheet 5W. Solid Geology 1:50000. Keyworth, Nottingham.

British Geological Survey (1994) Geological map of Wales. Solid 1:250000, 1st edn. Nottingham, British Geological Survey.

British Geological Survey (1996) Tectonic map of Britain, Ireland and adjacent areas. Compilers T. C. Pharaoh, J. H. Morris, C. B. Long & P. D. Ryan. 1:1500000. Nottingham, British Geological Survey.

British Geological Survey (1997) New Galloway, Scotland. Sheet 9W. Solid Geology 1:50000. Keyworth, Nottingham.

Brockamp, O., Zuther, M. & Clauer, N. (1987) Epigenetic-hydrothermal origin of the sediment-hosted Mullenbach Uranium deposit, Baden-Baden, W-Germany. *Monography Series of Mineral Deposits* **27**, 87–98.

Bröcker, M. & Day, H.W. (1995) Low-grade blueschist metamorphism in metagraywackes, Eastern Franciscan Belt, Northern California. *Journal of Metamorphic Geology* **13**, 61–78.

Broecker, W.S. & Oversby, V.M. (1971) *Chemical Equilibria in the Earth.* McGraw-Hill, New York.

Brookins, D.G. (1980) Geochronologic studies in the Grants mineral belt. *New Mexico Bureau of Mines and Mineral Resources Memoire* **38**, 52–58.

Bros, R., Stille, P., Gauthier-Lafaye, F., Weber, F. & Clauer, N. (1992) Sm–Nd isotopic dating of Proterozoic clay materials: an example from the Francevillian sedimentary series (Gabon). *Earth and Planetary Science Letters* **113**, 207–218.

Brown, E.H. (1968) The Si^{+4} content of natural

phengites: a discussion. *Contributions to Mineralogy and Petrology* **17**, 78–81.

Brown, E.H. (1975) A petrogenetic grid for reactions producing biotite and other Al–Fe–Mg silicates in the greenschist facies. *Journal of Petrology* **16**, 258–271.

Brown, E.H. (1977) Phase equilibria among pumpellyite, lawsonite, epidote and associated minerals in low-grade metamorphic rocks. *Contributions to Mineralogy and Petrology* **87**, 149–169.

Browne, P.R.L. (1978) Hydrothermal alteration in active geothermal fields. *Annual Review of Earth and Planetary Sciences* **6**, 279–250.

Bruce, C.H. (1984) Smectite dehydration—its relation to structural development and hydrocarbon accumulation in Northern Gulf of Mexico basin. *American Association of Petroleum Geologists Bulletin* **68**, 673–683.

Bryant, W.R., Bennett, R.H., Burkett, P.J. & Rack, F.R. (1991) Microfabric and physical characteristics of a consolidated clay section: ODP Site 697, Weddell Sea. In: *Microstructure of Fine-Grained Sediments: From Mud to Shale* (eds R. H. Bennett, W. R. Bryant & M. H. Hulbert), pp. 73–92. Springer-Verlag.

Buatier, M.D., Peacor, D.R. & O'Neil, J.R. (1992) Smectite–illite transition in Barbados accretionary wedge sediments: TEM and AEM evidence for dissolution/crystallization at low temperature. *Clays and Clay Minerals* **40**, 65–80.

Buatier, M.D., Fruh-Green, G.L. & Karpoff, A.M. (1995) Mechanism of Mg-phyllosilicate formation in a hydrothermal system at a sedimented ridge (Middle Valley, Juan de Fuca). *Contributions to Mineralogy and Petrology* **122**, 134–151.

Buatier, M.D., Travé, A., Labaume, P. & Potdevin, J.L. (1997) Dickite related to fluid–sediment interaction and deformation in Pyrenean thrust-fault zones. *European Journal of Mineralogy* **9**, 875–888.

Bucher, K. & Frey, M. (1994) *Petrogenesis of Metamorphic Rocks*, 6th edn. Springer-Verlag, Berlin, Heidelberg.

Bucher-Nurminen, K. (1987) A recalibration of the chlorite–biotite–muscovite geobarometer. *Contributions to Mineralogy and Petrology* **96**, 519–522.

Buckley, H.A., Bevan, J.C., Brown, K.M., Johnson, L.R. & Farmer, V.C. (1978) Glauconite and celadonite: two separate mineral species. *Mineralogical Magazine* **42**, 373–382.

Burkhard, D.J.M. & O'Neil, J.R. (1988) Contrasting serpentinization processes in the eastern Central Alps. *Contributions to Mineralogy and Petrology* **99**, 498–506.

Burkhard, M. (1988) L'Helvétique de la bordure occidentale du massif de l'Aar (évolution tectonique et métamorphique). *Eclogae Geologicae Helvetiae* **81**, 63–114.

Burkhard, M. & Kerrich, R. (1988) Fluid regimes in the deformation of the Helvetic nappes, Switzerland, as inferred from stable isotope data. *Contributions to Mineralogy and Petrology* **99**, 416–429.

Burns, S.J., Baker, P.A. & Elderfield, H. (1992) Timing of carbonate mineral precipitation and fluid flow in seafloor basalts, northwest Indian Ocean. *Geology* **20**, 255–258.

Burst, J.F. (1959) Post diagenetic clay mineral-environmental relationships in the Gulf Coast Eocene in clays and clay minerals. *Clays and Clay Minerals* **6**, 327–341.

Burst, J.F. (1969) Diagenesis of Gulf Coast clayey sediments and its possible relation to petroleum migration. *Bulletin of the American Association for Petroleum Geology* **53**, 73–93.

Butterfield, D.A. & Massoth, G.J. (1994) Geochemistry of north Cleft segment vent fluids: temporal changes in chlorinity and their possible relation to recent volcanism. *Journal of Geophysical Research* **99**, 4951–4969.

Butterfield, D.A., Massoth, G.J., McDuff, R.E., Lupton, J.E. & Lilley, M.D. (1990) Geochemistry of hydrothermal fluids from Axial Seamount hydrothermal emissions study vent field, Juan de Fuca Ridge: subseafloor boiling and subsequent fluid–rock interaction. *Journal of Geophysical Research* **95**, 12895–12922.

Carmichael, D.M. (1991) Metamorphism, metasomatism and Archean lode gold deposits. In: *Nuna Conference on Greenstone Gold Evolution* (eds F. Robert & S. B. Green). Geological Association of Canada, Mineral Deposits Division.

Carr, A. (1998) A vitrinite reflectance kinetic model incorporating overpressure retardance. *Marine and Petroleum Geology*, in press.

Caruso, L.J., Bird, D.K., Cho, M. & Liou, J.G. (1988) Epidote bearing veins in the State 2–14 drill hole: implications for hydrothermal fluid composition. *Journal of Geophysical Research* **93**, 13123–13133.

Cathelineau, M. (1988) Cation site occupancy in chlorites and illites as a function of temperature. *Clay Minerals* **23**, 471–485.

Cathelineau, M. & Nieva, D. (1985) A chlorite solid solution geothermometer. The Los Azufres (Mexico) geothermal system. *Contributions to Mineralogy and Petrology* **91**, 235–244.

Cerling, T.E. & Sharp, Z.D. (1996) Stable carbon and oxygen isotope analysis of fossil tooth enamel using laser ablation. *Palaeogeography, Palaeoclimatology, Palaeoecology* **126**, 173–186.

Chacko, T., Hu, X., Mayeda, T., Clayton, R.N. & Goldsmith, J.R. (1996) Oxygen isotope fractionations in muscovite, phlogopite, and rutile. *Geochimica et Cosmochimica Acta* **60**, 2595–2608.

Chamberlain, C.P. & Conrad, M.E. (1991) Oxygen isotope zoning in garnet. *Science* **254**, 403–406.

Chamberlain, C.P. & Conrad, M.E. (1993) Oxygen-isotope zoning in garnet; a record of volatile transport. *Geochimica et Cosmochimica Acta* **57**, 2613–2629.

Chamberlain, C.P., Zeitler, P.K., Barnett, D.E. *et al.* (1995) Active hydrothermal systems during the recent uplift of Nanga Parbat, Pakistan Himalaya. *Journal of Geophysical Research, B, Solid Earth and Planets* **100**, 439–453.

Charef, A. & Sheppard, S.M.F. (1984) Carbon and oxygen isotope analysis of calcite or dolomite associated with organic matter. *Chemical Geology* **46**, 325–333.

Chaudhuri, S. & Brookins, D.G. (1969) The isotopic age of the Flathead Sandstone (Middle Cambrian), Montana. *Journal of Sedimentary Petrology* **39**, 364–366.

Chaudhuri, S., Stille, P. & Clauer, N. (1992) Sm–Nd isotopes in fine-grained clastic sedimentary materials: clues to sedimentary processes and recycling growth of the continental crust. In: *Isotopic Signatures and Sedimentary Records* (eds N. Clauer & S. Chaudhuri). *Lecture Notes in Earth Sciences* **43**, 287–319. Springer-Verlag, Berlin, Heidelberg.

Chiba, H., Chacko, T., Clayton, R.N. & Goldsmith, J.R. (1989) Oxygen isotope fractionations involving diopside, forsterite, magnetite, and calcite; application to geothermometry. *Geochimica et Cosmochimica Acta* **53**, 2985–2995.

Cho, M. & Liou, J.G. (1988) Pressure dependence of the Fe and Mg partitioning between actinolite and chlorite. *Geological Society of America Abstracts with Programs* **20**, 150.

Cho, M., Liou, J.G. & Maruyama, S. (1986) Transition from zeolite to prehnite–pumpellyite facies in the Karmutsen metabasites, Vancouver Island, British Columbia. *Journal of Petrology* **27**, 467–494.

Cho, M., Liou, J.G. & Bird, D.K. (1988) Prograde phase relations in the State 2-14 well metasandstones, Salton Sea geothermal field, California. *Journal of Geophysical Research* **93**, 13081–13103.

Church, W.R., Pinet, N. & Tremblay, A. (1996) Is the Taconian orogeny of southern Quebec the result of an Oman-type obduction? Comment and Reply. *Geology* **24**, 285–287.

Clauer, N. (1974) Utilisation de la méthode rubidium–strontium pour la datation d'une schistosité de sédiments peu métamorphisés: application au Précambrien II de la boutonnière de Bou Azzer-El Graara (Anti-Atlas). *Earth and Planetary Science Letters* **22**, 404–412.

Clauer, N. & Chaudhuri, S. (1995) *Clays in Crustal Environments. Isotope Dating and Tracing*. Springer-Verlag, Berlin, Heidelberg.

Clauer, N., Vidal, P. & Auvray, B. (1985) Differential behavior of the Rb–Sr and K–Ar systems of spilitic flows and interbedded metasediments: the spilitic group of Erquy (Brittany, France). Paleomagnetic implications. *Contributions to Mineralogy and Petrology* **89**, 81–89.

Clauer, N., O'Neil, J.R., Bonnot, C.C. & Holtzapffel, T. (1990) Morphological, chemical, and isotopic evidence for an early diagenetic evolution of detrital smectite in marine sediments. *Clays and Clay Minerals* **38**, 33–46.

Clauer, N., Chaudhuri, S., Kralik, M. & Bonnot-Courtois, C. (1993) Effects of experimental leaching on Rb–Sr and K–Ar isotopic systems and REE contents of diagenetic illite. *Chemical Geology* **103**, 1–16.

Clauer, N., Rais, N., Schaltegger, U. & Piqué, A. (1995) K–Ar systematics of clay-to-mica minerals in a multi-stage low-grade metamorphic evolution. *Chemical Geology, Isotope Geoscience Section* **124**, 305–316.

Clauer, N., Środoń, J., Francu, J. & Sucha, V. (1997) K–Ar dating of illite fundamental particles separated from illite/smectite. *Clay Minerals* **32**, 181–196.

Clayton, R.N. & Kieffer, S.W. (1991) Oxygen isotopic thermometer calibrations. In: *Stable Isotope Geochemistry: A Tribute to Samuel Epstein* (eds H. P. J. Taylor, J. R. O'Neil & I. R. Kaplan), pp. 3–10. Lancaster Press, Inc., San Antonio.

Clayton, R.N. & Mayeda, T.K. (1963) The use of bromine pentafluoride in the extraction of oxygen in oxides and silicates for isotopic analysis. *Geochimica et Cosmochimica Acta* **27**, 43–52.

Clayton, R.N. & Steiner, A. (1975) Oxygen isotope studies of the geothermal system at Wairakei, New Zealand. *Geochimica et Cosmochimica Acta* **39**, 1179–1186.

Clayton, R.N., Muffler, L.J.P. & White, D.E. (1968) Oxygen isotope study of calcite and silicates of the River ranch No. 1 well, Salton Sea geothermal field, California. *American Journal of Science* **266**, 968–979.

Clayton, R.N., O'Neil, J.R. & Mayeda, T.K. (1972) Oxygen isotope exchange between quartz and water. *Journal of Geophysical Research* **77**, 3057–3066.

Clayton, R.N., Jackson, M.L. & Sridhar, K. (1978) Resistance of quartz silt to isotopic exchange under burial and intense weathering conditions. *Geochimica et Cosmochimica Acta* **42**, 1517–1522.

Clayton, R.N., Goldsmith, J.R. & Mayeda, T.K. (1989) Oxygen isotope fractionation in quartz, albite, anorthite and calcite. *Geochimica et Cosmochimica Acta* **53**, 725–733.

Cliff, G. & Lorimer, G.W. (1975) The quantitative analysis of thin specimens. *Journal of Microscopy* **103**, 203–207.

Cloos, M. (1983) Comparative study of melange matrix and metashales from the Franciscan subduction complex with the basal Great Valley sequence, California. *Journal of Geology* **91**, 291–306.

Cocker, J.D., Griffin, B.J. & Muehlenbachs, K. (1982) Oxygen and carbon isotope evidence for seawater-hydrothermal alteration of the Macquarie Island ophiolite. *Earth and Planetary Science Letters* **61**, 112–122.

Coish, R.A. (1977) Ocean floor metamorphism in the Betts Cove ophiolite, Newfoundland. *Contributions to Mineralogy and Petrology* **60**, 225–270.

Coleman, M.L., Walsh, J.N. & Benmore, R.A. (1989) Determination of both chemical and stable isotope composition in milligramme-size carbonate samples. In: *Zoned Carbonate Cements; Techniques, Applications and Implication* (ed. B. W. Sellwood), pp. 233–238. Reading.

Collier, J.S. & Sinha, M.C. (1992) Seismic mapping of a magma chamber beneath the Valu-Fa Ridge, Lau Basin. *Journal of Geophysical Research* **97**, 14031–14053.

Colten-Bradley, V.A. (1987) Role of pressure in smectite dehydration—effects on geopressure and smectite-to-illite transformation. *American Association of Petroleum Geologists Bulletin* **71**, 1414–1427.

Compston, W. & Pidgeon, R.T. (1962) Rb–Sr dating of shales by the total-rock method. *Journal of Geophysical Research* **67**, 3493–3502.

Conrad, M.E. & Chamberlain, C.P. (1992) Laser-based, *in situ* measurements of fine-scale variations in the delta ^{18}O values of hydrothermal quartz. *Geology* **20**, 812–816.

Coombs, D.S. (1954) The nature and alteration of some Triassic sediments from Southland, New Zealand. *Transactions of the Royal Society of New Zealand* **82**, 65–109.

Coombs, D.S. (1960) Lower grade mineral facies in New Zealand. In: *21st International Geological Congress*, Vol. 13, pp. 339–351. Copenhagen, 1960.

Coombs, D.S. (1961) Some recent work on the lower grades of metamorphism. *Australian Journal of Science* **24**, 203–215.

Coombs, D.S. & Cox, S.C. (1991) Low- and very low grade metamorphism in southern New Zealand and its geological setting. *Geological Society of New Zealand Miscellaneous Publications* **58**, 88 pp.

Coombs, D.S., Ellis, A.J., Fyfe, W.S. & Taylor, A.M. (1959) The zeolite facies, with comments on the interpretation of hydrothermal syntheses. *Geochimica et Cosmochimica Acta* **17**, 53–107.

Coombs, D.S., Nakamura, Y. & Vuagnat, M. (1976) Pumpellyite–actinolite facies schists of the Taveyanne Formation near Loèche, Valais, Switzerland. *Journal of Petrology* **17**, 440–471.

Coombs, D.S., Kawachi, Y. & Ford, P.B. (1996) Porphyroblastic manganaxinite metapelites with incipient garnet in prehnite–pumpellyite facies, near Meyers Pass, Torlesse Terrane, New Zealand. *Journal of Metamorphic Geology* **14**, 125–142.

Cooper, A.F. (1980) Retrograde alteration of chromian kyanite in metachert and amphibolite whiteschist from the Southern Alps, New Zealand, with implications for uplift on the Alpine Fault. *Contributions to Mineralogy and Petrology* **75**, 153–164.

Cooper, A.H. & Molyneux, S.G. (1990) The age and correlation of the Skiddaw Group (early Ordovician) sediments in the Cross Fell inlier (northern England). *Geological Magazine* **127**, 147–157.

Coplen, T.B., Kendall, C. & Hopple, J. (1983) Comparison of stable isotope reference samples. *Nature* **302**, 236–238.

Cordani, U.G., Kawashita, K. & Thomaz-Filho, A. (1978) Applicability of the Rb–Sr method to shales and related rocks. *American Association of Petroleum Geologists Special Publication* **6**, 93–117.

Cordani, U.G., Thomaz-Filho, A., Brito-Neves, B.B. & Kawashita, K. (1985) On the applicability of the Rb–Sr method to argillaceous sedimentary rocks: some examples from Precambrian sequences of Brazil. *Giornal of Geologica (Bologna)* **47**, 253–280.

Cornides, J. & Kusakabe, M. (1977) Preparation of carbon dioxide from magnesite for isotopic analysis. *Fresenius Zeitschrift fuer Analytische Chemie* **287**, 310–311.

Cornwall, H.R. (1951a) Differentiation in lavas of the Michigan Keweenawan series and the origin of the copper deposits. *Geological Society of America Bulletin* **62**, 159–202.

Cornwall, H.R. (1951b) Differentiation in magmas of the Keweenawan series. *Journal of Geology* **59**, 151–172.

Cornwall, H.R. (1951c) Ilmenite, magnetite, hematite, and copper in lavas of the Keweenawan series. *Economic Geology* **46**, 51–67.

Cornwall, H.R. (1956) A summary of ideas on the origin of native copper deposits. *Bulletin of the Society of Economic Geology* **51**, 615–631.

Craig, H. (1961) Isotopic variations in meteoric waters. *Science* **133**, 1702–1703.

Craig, H. & Gordon, L.I. (1965) Deuterium and oxygen-18 variations in the ocean and the marine atmosphere. In: *Symposium of Marine Geochemistry*, pp. 277–374. University of Rhode Island, Narragansett Marine Laboratory.

Crespo-Blanc, A., Masson, H., Sharp, Z.D., Cosca, M. & Hunziker, J. (1995) A stable and $^{40}Ar/^{39}Ar$ isotope study of a major thrust in the Helvetic nappes. *Geological Society of America Bulletin* **107**, 1129–1144.

Crowe, D.E., Valley, J.W. & Baker, K.L. (1990) Micro-analysis of sulfur-isotope ratios and zonation by laser microprobe. *Geochimica et Cosmochimica Acta* **54**, 2075–2092.

Crowley, K.D. (1985) Thermal significance of fission-track lengths distributions. *Nuclear Tracks Radiation Measurements* **10**, 311–322.

Curry, J.R. & Moore, D.G. (1982) *Initial Reports DSDP 64*. US Government Printing Office, Washington DC.

Dalla Torre, M., de Capitani, C., Frey, M., Underwood, M.B., Mullis, J. & Cox, R. (1996a) Very low-temperature metamorphism of metashales from the Diablo Range, Franciscan Complex, California, USA: new constraints on the exhumation path. *Geological Society of America Bulletin* **108**, 578–601.

Dalla Torre, M., Livi, K.J.T. & Frey, M. (1996b) Chlorite textures and compositions from high-pressure/low-temperature metashales and metagreywackes, Franciscan Complex, Diablo Range, California, USA. *European Journal of Mineralogy* **8**, 825–846.

Dalla Torre, M., Livi, K.J.T., Veblen, D.R. & Frey, M. (1996c) White K-mica evolution from phengite to muscovite in shales and shale matrix melange, Diablo Range, California. *Contributions to Mineralogy and Petrology* **123**, 390–405.

Dalla Torre, M., Ferreiro Mählmann, R. & Ernst, W.G. (1997) Experimental study on the pressure dependence of vitrinite maturation. *Geochimica et Cosmochemica Acta* **61**, 2921–2928.

Dallmeyer, R.D. & Wright, T.O. (1992) Diachronous cleavage development in the Robertson Bay terrane, Northern Victoria Land, Antarctica—Tectonic implications. *Tectonics* **11**, 437–448.

Dansgaard, W. (1964) Stable isotopes in precipitation. *Tellus* **16**, 436–468.

Darwin, C. (1846) *Geological Observations on South America.* Smith, Elder & Co., London.

Date, J., Watanabe, Y. & Inoue, T. (1983) Zonal alteration around the Fukazawa Kuroko deposits, Akita prefecture, northern Japan. In: *The Kuroko and Related Volcanogenic Massive Sulfide Deposits* (eds H. Ohmoto & B. J. Skinner), pp. 365–386. Economic Geology, New Haven, CT.

Davies, J.R., Fletcher, C.J.N., Waters, R.A., Wilson, D., Woodhall, D.G. & Zalasiewicz, J.A. (1997) Geology of the country around Llanilar and Rhayader. Memoir for 1: 50,000 Geological Sheets 178 & 179 (England & Wales). The Stationery Office for British Geological Survey, London.

Davis, E.E., Lister, C.R.B., Wade, U.S. & Hyndman, R.D. (1980) Detailed heat flow measurements over the Juan de Fuca Ridge system. *Journal of Geophysical Research* **85**, 299–310.

Davis, E.E., Chapman, D.S., Mottl, M.J. *et al.* (1992a) FlankFlux: an experiment to study the nature of hydrothermal circulation in young oceanic crust. *Canadian Journal of Earth Science* **29**, 925–952.

Davis, E.E., Mottl, M.J. & Fisher, A.T. (1992b) *Proceedings of ODP Initial Reports 139.* Ocean Drilling Program, College Station, TX.

Davis, E.E., Chapman, D.S. & Forster, C.B. (1996) Observations concerning the vigor of hydrothermal circulation in young oceanic crust. *Journal of Geophysical Research* **101**, 2927–2942.

Day, H.W. (1972) Geometrical analysis of phase equilibria in Ternary systems of six phases. *American Journal of Science* **272**, 711–734.

Day, H.W. (1976) A working model of some equilibria in the system alumina–silica–water. *American Journal of Science* **276**, 1254–1284.

Day, H.W., Schiffman, P. & Moores, E.M. (1988) Metamorphism and tectonics of the northern Sierra Nevada. In: *Metamorphism and Crustal Evolution of the Western United States (Rubey Volume VII)* (ed. W. G. Ernst), pp. 1–28. Prentice-Hall, Englewood Cliffs, New Jersey.

De Caritat, P., Hutcheon, I. & Walshe, J.L. (1993) Chlorite geothermometry: a review. *Clays and Clay Minerals* **41**, 219–239.

Deer, W.A., Howie, R.A. & Zussman, J. (1992) *An Introduction to the Rock Forming Minerals.* Longman, Harlow.

Delgado, A. & Reyes, E. (1996) Oxygen and hydrogen isotope compositions in clay minerals: a potential single-mineral geothermometer. *Geochimica et Cosmochimica Acta* **60**, 4285–4289.

Deming, D. (1994) Overburden rock, temperature, and heat flow. In: *The Petroleum System—From Source to Trap* (eds L. B. Magoon & D. G. Dow). *American Association of Petroleum Geologists Memoirs* **60**, 165–186.

DePaolo, D.J. (1988) *Neodymium Isotope Geochemistry.* Springer-Verlag, Berlin, Heidelberg, New York.

Detrick, R.S., Buhl, P., Vera, E. *et al.* (1987) Multichannel seismic imaging of a crustal magma chamber along the East Pacific Rise. *Nature* **326**, 35–41.

Detrick, R.S., Harding, A.J., Kent, G., Orcutt, J., Mutter, J. & Buhl, P. (1993) Seismic structure of the southern East Pacific Rise. *Science* **259**, 499–503.

Dettman, D.L. & Lohmann, K.C. (1995) Microsampling carbonates for stable isotope and minor element analysis: physical separation of samples on a 20 micrometer scale. *Sedimentary Research* **A65**, 566–569.

DeWolf, C.P. & Halliday, A.N. (1991) U–Pb dating of a remagnetized Paleozoic limestone. *Geophysical Research Letters* **18**, 1445–1448.

Digel, S. & Ghent, E.D. (1994) Fluid–mineral equilibrium in prehnite–pumpellyite to greenschist facies metabasites near Flin Flon, Manitoba, Canada: implications for petrogenetic grids. *Journal of Metamorphic Geology* **12**, 467–477.

Digel, S.G. & Gordon, T.M. (1995) Phase relations in metabasites and pressure–temperature relations in prehnite–pumpellyite to greenschist facies transition, Flin Flon, Manitoba, Canada. In: *Low-grade Metamorphism of Mafic Rocks* (eds P. Schiffman & H. W. Day). Geological Society of America, Boulder, CO. Special Paper No. 296, pp. 67–80.

Doe, B.R. (1970) Evaluation of U–Th–Pb whole-rock dating on Phanerozoic sedimentary rocks. *Eclogae Geologicae Helvetiae* **63**, 79–82.

Dong, H. & Peacor, D.R. (1996) TEM observations of coherent stacking relations in smectite, I/S and illite of shales: evidence for MacEwan crystallites and dominance of $2M_1$ polytypism. *Clays and Clay Minerals* **44**, 257–275.

Dong, H., Hall, C.M., Peacor, D.R. & Halliday, A.N. (1995) Mechanisms of argon retention in clays revealed by laser $^{40}Ar–^{39}Ar$ dating. *Science* **267**, 355–359.

Dong, H., Hall, C.M., Halliday, A.N., Peacor, D.R., Merriman, R.J. & Roberts, B. (1997a) $^{40}Ar/^{39}Ar$ dating of Late Caledonian (Acadian) metamorphism and cooling of K-bentonites and slates from the Welsh Basin, U.K. *Earth and Planetary Science Letters* **150**, 337–351.

Dong, H., Peacor, D.R. & Freed, R.L. (1997b) Phase relations among smectite, R1 illite–smectite and illite. *American Mineralogist* **82**, 379–391.

Donnelly, T.W., Pritchard, R.A., Emmermann, R. & Puchelt, H. (1979) The aging of oceanic crust: synthesis of mineralogical and chemical results of DSDP Legs 51–53. In: *Initial Reports DSDP* (ed. T. W. Donnelly), part 2, pp. 1563–1577. US Government Printing Office, Washington, DC.

Drits, V., Środoń, J. & Eberl, D.D. (1997) XRD measurements of mean crystallite thickness of illite and illite/smectite: reappraisal of the Kübler index and the Scherrer equation. *Clays and Clay Minerals* **45**, 461–475.

Drits, V.A., Eberl, D.D. & Środoń, J. (1998) XRD measurement of mean thickness, thickness distribution and strain for illite and illite–smectite crystallites by the Bertaut–Warren–Averbach technique. *Clays and Clay Minerals*, **46**, 38–50.

Duba, D. & Williams-Jones, A.E. (1983) The application of illite crystallinity, organic matter reflectance, and isotopic techniques to mineral exploration: a case study in southwestern Gaspé, Quebec. *Economic Geology* **78**, 1350–1363.

Dunoyer de Segonzac, G. (1969) Les minéraux argileux dans la diagenèse. Passage au métamorphisme. *Sciences Géologiques Mémoire, Strasbourg* **29**, 320 p.

Dunoyer de Segonzac, G. (1970) The transformation of clay minerals during diagenesis and low-grade metamorphism: a review. *Sedimentology* **15**, 281–346.

Dunoyer de Segonzac, G. & Bernoulli, D. (1976) Diagenèse et métamorphisme des argiles dans le Rhétien Sud-alpin et Austro-alpin (Lombardie et Grisons). *Bulletin Société Géologique de la France* **18**, 1283–1293.

Durney, D.W. & Kisch, H.J. (1994) A field classification and intensity scale for first-generation cleavages. *Journal of Australian Geology and Geophysics* **15**, 257–295.

Eberl, D.D. (1993) Three zones for illite formation during burial diagenesis and metamorphism. *Clays and Clay Minerals* **41**, 26–37.

Eberl, D.D. & Środoń, J. (1988) Ostwald ripening and interparticle-diffraction effects from illite crystals. *American Mineralogist* **73**, 1335–1345.

Eberl, D.D., Środoń, J., Lee, M., Nadeau, P.H. & Northrop, H.R. (1987) Sericite from the Silverton caldera, Colorado: correlation among structure, composition, origin, and particle thickness. *American Mineralogist* **72**, 914–934.

Eberl, D.D., Środoń, J., Kralik, M., Taylor, B.E. & Peterman, Z.E. (1990) Ostwald ripening of clays and metamorphic minerals. *Science* **248**, 474–477.

Eberl, D.D., Drits, V., Środoń, J. & Nuesch, R. (1996) Mudmaster: a program for calculating crystallite size distributions and strain from the shapes of X-ray diffraction peaks. USGS Open File Report 96–171, 44 pp.

Eberl, D.D., Nuesch, R., Sucha, V. & Tsipursky, S. (1998) Measurement of fundamental illite particle thicknesses by X-ray diffraction using PVP-10 intercalation. *Clays and Clay Minerals*, **46**, 89–97.

Edmond, J., Campbell, A.C., Palmer, M.R. *et al.* (1995) Time-series of vent fluids from the TAG and MARK sites (1986, 1990), Mid-Atlantic Ridge, and a mechanism for Cu/Zn zonation in massive sulfide orebodies. In *Hydrothermal Vents and Processes* (eds L. M. Parson *et al.*), Special Publication No. 87, pp. 77–86. Geological Society, London.

Edmond, J.M., Measures, C., McDuff, R.E. *et al.* (1979a) Ridge crest hydrothermal activity and the balances of the major and minor elements in the ocean: the Galapagos data. *Earth and Planetary Science Letters* **46**, 1–18.

Edmond, J.M., Measures, C., Magnum, B. *et al.* (1979b) On the formation of metal-rich deposits at ridge crests. *Earth and Planetary Science Letters* **46**, 19–30.

Edmonds, H.N. & Edmond, J.M. (1995) A three-component mixing model for ridge-crest hydrothermal fluids. *Earth and Planetary Science Letters* **134**, 53–67.

Eggleton, R.A. & Keller, J. (1982) The palagonitization of limburgite glass—a TEM study. *Neues Jahrbuch für Mineralogie Montashefte* **7**, 321–336.

Ehrenberg, S.N., Aagaard, P., Wilson, M.J., Fraser, A.R. & Duthie, D.M.L. (1993) Depth-dependent transformation of kaolinite to dickite in sandstones of the Norwegian continental shelf. *Clay Minerals* **28**, 325–352.

Elderfield, H. & Greaves, M.J. (1981) Strontium isotope geochemistry of Icelandic hydrothermal systems and implications for sea water chemistry. *Geochimica et Cosmochimica Acta* **45**, 2201–2212.

Elders, W.A., Bird, D.K., Williams, A.E. & Schiffman, P. (1984) Hydrothermal flow regime and magmatic heat source of the Cerro Prieto geothermal system, Baja California, Mexico. *Geothermics* **13**, 27–47.

Elliott, W.C. & Aronson, J.L. (1987) Alleghanian episode of K-bentonite illitization in the southern Appalachian Basin. *Geology* **15**, 735–739.

Elliott, W.C. & Matisoff, G. (1996) Evaluation of kinetic models for the smectite to illite transformation. *Clays and Clay Minerals* **44**, 77–87.

Elsenheimer, D. & Valley, J.W. (1992) *In situ* oxygen isotope analysis of feldspar and quartz by Nd: YAG laser microprobe. *Chemical Geology* **101**, 21–42.

Embley, R.W., Jonasson, I.R., Perfit, M.R. *et al.* (1988) Submersible investigation of an extinct hydrothermal system on the Galapagos Ridge: sulfide mounds, stockwork zone, and differentiated lavas. *Canadian Mineralogist* **26**, 517–540.

Engel, M.H., Macko, S.A. & Silfer, J.A. (1990) Carbon isotope composition of individual amino acids in the Murchison Meteorite. *Nature* **348**, 47–49.

Eslinger, E.V. (1971) *Mineralogy and oxygen isotope ratios of hydrothermal and low-grade metamorphic argillaceous rocks.* Unpublished PhD Thesis, Case Western University, Cleveland, OH.

Eslinger, E.V. & Glasmann, J.R. (1993) Geothermometry and geochronology using clay minerals—an introduction. *Clays and Clay Minerals* **41**, 117–118.

Eslinger, E.V. & Savin, S.M. (1973a) Mineralogy and oxygen isotope geochemistry of the hydrothermally altered rocks of the Ohaki-Broadlands, New Zealand, geothermal area. *American Journal of Science* **273**, 240–267.

Eslinger, E.V. & Savin, S.M. (1973b) Oxygen isotope geothermometry of the burial metamorphic rocks of the Precambrian Belt Supergroup, Glacier National Park, Montana. *Geological Society of America Bulletin* **84**, 2549–2560.

Eslinger, E.V., Savin, S.M. & Yeh, H.W. (1979) Oxygen isotope geothermometry of diagenetically altered shales. *SEPM Special Publication* **26**, 113–124.

Essene, E.J. (1982) Geologic thermometry and barometry. In *Characterization of Metamorphism Through Mineral Equilibria* (ed. J. M. Ferry), Mineralogical Society of America. *Reviews in Mineralogy* **10**, 153–206.

Essene, E.J. (1989) The current status of thermobarometry in metamorphic rocks. In *Evolution of Metamorphic Belts* (eds J. S. Daly, R. A. Cliff & B. W. D. Yardley). *Geological Society Special Publication* **43**, 1–44.

Essene, E.J. & Fyfe, W.S. (1967) Omphacite in California metamorphic rocks. *Contributions to Mineralogy and Petrology* **15**, 1–23.

Essene, E.J. & Peacor, D.R. (1995) Clay mineral thermometry—a critical perspective. *Clays and Clay Minerals* **43**, 540–553.

Essene, E.J. & Peacor, D.R. (1997) Illite and smectite: metastable, stable, or unstable? Further discussion and a correction. *Clays and Clay Minerals* **45**, 116–122.

Evans, B.W. (1990) Phase relations of epidote blueschists. *Lithos* **25**, 3–23.

Evans, J.A. (1989) A note on Rb–Sr whole-rock ages from cleaved mudrocks in the Welsh Basin. *Journal of the Geological Society, London* **146**, 901–904.

Evans, J.A. (1991) Resetting of Rb–Sr whole-rock ages during Acadian low-grade metamorphism in North Wales. *Journal of the Geological Society, London* **148**, 703–710.

Evans, J.A. (1996) Dating the transition of smectite to illite in Palaeozoic mudrocks using the Rb–Sr whole-rock technique. *Journal of the Geological Society, London* **153**, 101–108.

Evarts, R.C. & Schiffman, P. (1983) Submarine hydrothermal metamorphism of the Del Puerto ophiolite, California. *American Journal of Science* **283**, 289–340.

Exley, R.A. (1982) Electron microprobe studies of Iceland Research Drilling Project high temperature mineral geochemistry. *Journal of Geophysical Research* **87**, 6547–6558.

Fagan, T. & Day, H.W. (1997) Formation of amphibole after clinopyroxene by dehydration reactions: implications for pseudomorphic reactions and mass fluxes. *Geology* **25**, 395–398.

Fairchild, I.J., Hendry, G., Quest, M. & Tucker, M. (1988) Chemical analysis of sedimentary rocks. In: *Techniques in Sedimentology* (ed. M. Tucker), pp. 274–354. University of Durham, Durham, UK.

Fallick, A.E., Macaulay, C.I. & Haszeldine, R.S. (1993) Implications of linearly correlated oxygen and hydrogen isotopic compositions for kaolinite and illite in the Magnus Sandstone, North Sea. *Clays and Clay Minerals* **41**, 184–190.

Faure, G. (1986) *Principles of Isotope Geology*, 2nd edn. John Wiley, New York.

Feeley, T.C. & Sharp, Z.D. (1996) Chemical and hydrogen isotope evidence for *in situ* dehydrogenation of biotite in silicic magma chambers. *Geology* **24**, 1021–1024.

Fehn, U., Green, K.E., Von Herzen, R.P. & Cathles, L.M. (1983) Numerical models for the hydrothermal field at the Galapagos Spreading Center. *Journal of Geophysical Research* **88**, 1033–1048.

Fein, J.B., Graham, C.M., Holness, M.B., Fallick, A.E. & Skelton, A.D.L. (1994) Controls on the mechanisms of fluid infiltration and front advection during regional metamorphism: a stable isotope and textural study of retrograde Dalradian rocks of the SW Scottish Highlands. *Journal of Metamorphic Geology* **12**, 249–260.

Ferrell, R.E. & Carpenter, P.K. (1990) Application of the electron microprobe and the electron microprobe in the study of clays. In: *Electron-Optical Methods in Clay Sciences* (eds I. D. R. Mackinnon & F. A. Mumpton), pp. 108–132. Clay Minerals Society, Boulder, Colorado.

Fettes, D.J., Graham, C.M., Sassi, F.P. & Scolari, A. (1976) The basal spacing of potassic white micas and facies series variation across the Caledonides. *Scottish Journal of Geology* **3**, 227–236.

Field, D. & Raheim, A. (1973) A geologically meaningless Rb–Sr total-rock isochron. *Nature* **282**, 497–499.

Findlay, R.H. (1992) The age of cleavage development in the Ross orogen, northern Victoria Land, Antarctica: evidence from $^{40}Ar/^{39}Ar$ whole-rock slate ages: discussion. *Journal of Structural Geology* **14**, 887–890.

Fisher, A.T., Becker, K., Narasimhan, T.N., Langseth, M.G. & Mottl, M.J. (1990) Passive off-axis convection through the southern flank of the Costa Rica Rift. *Journal of Geophysical Research* **95**, 9343–9370.

Fisher, A.T., Becker, K. & Narasimhan, K. (1994) Off-axis hydrothermal circulation: parametric tests of a refined model of processes at DSDP/ODP site 504. *Journal of Geophysical Research* **99**, 3097–3123.

Fisher, G.W. (1989) Matrix analysis of metamorphic mineral assemblages and reactions. *Contributions to Mineralogy and Petrology* **102**, 69–77.

Fitch, F.J., Miller, J.A., Evans, A.L., Grasty, R.N.H. & Meneisy, M.Y. (1969) Isotopic determinations on rocks from Wales and the Welsh borders. In: *The Pre-Cambrian and Lower Palaeozoic Rocks of North Wales* (ed. A. Wood), pp. 23–45. University of Wales, Cardiff.

Foland, K.A., Hubacher, F.A. & Arehart, G.B. (1992) $^{40}Ar/^{39}Ar$ dating of very fine-grained samples: an encapsulated-vial procedure to overcome the problem of ^{39}Ar recoil loss. *Chemical Geology, Isotope Geoscience Section* **102**, 269–276.

Fortey, N.J. (1989) Low grade metamorphism in the Lower Ordovician Skiddaw Group of the Lake District, England. *Proceedings of the Yorkshire Geological Society* **47**, 325–337.

Fortey, N.J., Roberts, B. & Hirons, S.R. (1993) Relationship between metamorphism and structure in the Skiddaw Group, English Lake District. *Geological Magazine* **130**, 631–638.

Fortey, N.J., Roberts, B. & Hirons, S.R. (1998) Metamorphism. In: *Geology of the Ambleside District* (compiler D. Millward). Memoir of the British Geological Survey, Sheet 38 (England and Wales).

Francheteau, J., Armijo, R., Cheminee, J.L., Hekinian, R., Lonsdale, P. & Blum, N. (1992) 1 Ma East Pacific Rise oceanic crust and uppermost mantle exposed by rifting in Hess Deep (equatorial Pacific Ocean). *Earth and Planetary Science Letters* **101**, 281–295.

Freed, R.L. & Peacor, D.R. (1989) Variability in temperature of the smectite/illite reaction in Gulf Coast sediments. *Clay Minerals* **24**, 171–180.

Freed, R.L. & Peacor, D.R. (1992a) Diagenesis and the formation of authigenic illite-rich I/S crystals in Gulf Coast shales: TEM study of clay separates. *Journal of Sedimentary Petrology* **62**, 220–234.

Freed, R.L. & Peacor, D.R. (1992b) Geopressured shale and sealing effect of smectite to illite transition. *American Association of Petroleum Geologists Bulletin* **73**, 1223–1232.

Freeman, K.H. (1991) *The carbon isotopic compositions of individual compounds from ancient and modern depositional environments*. Unpublished Doctoral Thesis, Indiana University.

Frey, M. (1969a) Die Metamorphose des Keupers vom Tafeljura bis zum Lukmanier-Gebiet. *Beiträge zur geologischen Karte der Schweiz*, NF 131, 112 p.

Frey, M. (1969b) A mixed-layer paragonite/phengite of low-grade metamorphic origin. *Contributions to Mineralogy and Petrology* **14**, 63–65.

Frey, M. (1970) The step from diagenesis to metamorphism in pelitic rocks during Alpine orogenesis. *Sedimentology* **15**, 261–279.

Frey, M. (1974) Alpine metamorphism of pelitic and marly rocks of the Central Alps. *Schweizerische Mineralogische und Petrographische Mitteilungen* **54**, 489–506.

Frey, M. (1986) Very low-grade metamorphism of the Alps—an introduction. *Schweizerische Mineralogische und Petrographische Mitteilungen* **66**, 13–27.

Frey, M. (1987a) Very low-grade metamorphism of clastic sedimentary rocks. In: *Low Temperature Metamorphism* (ed. M. Frey). Blackie and Son Ltd, Glasgow.

Frey, M. (1987b) The reaction-isograd kaolinite + quartz = pyrophyllite + H_2O, Helvetic Alps, Switzerland. *Schweizerische Mineralogische und Petrographische Mitteilungen* **67**, 1–11.

Frey, M. (1988) Discontinuous inverse metamorphic zonation, Glarus Alps, Switzerland: evidence from illite 'crystallinity' data. *Schweizerische Mineralogische und Petrographische Mitteilungen* **68**, 171–183.

Frey, M. & Kisch, H.J. (1987) Scope of subject. In: *Low Temperature Metamorphism* (ed. M. Frey), pp. 1–8. Blackie and Son Ltd, Glasgow,

Frey, M., Hunziker, J.C., Roggwiler, P. & Schindler, C. (1973) Progressive niedriggradige Metamorphose glaukonitführender Horizonte in den helvetischen Alpen der Ostschweiz. *Contributions to Mineralogy and Petrology* **39**, 185–218.

Frey, M., Teichmüller, M., Teichmüller, R. *et al.* (1980) Very low grade metamorphism in external parts of the Central Alps: illite crystallinity, coal rank and fluid inclusion data. *Eclogae Geologica Helvetica* **73**, 173–203.

Frey, M., Hunziker, J.C., Jäger, E. & Stern, W.B. (1983) Regional distribution of white K-mica polymorphs and their phengite content in the Central Alps. *Contributions to Mineralogy and Petrology* **83**, 185–197.

Frey, M., de Capitani, D. & Liou, J.G. (1991) A new petrogenetic grid for low-grade metabasites. *Journal of Metamorphic Geology* **9**, 497–509.

Fridleifsson, G.O. (1984) Mineralogical evolution of a hydrothermal system: heat source–fluid interactions. *Transactions of the Geothermal Resources Council* **8**, 119–123.

Friedman, I. (1953) Deuterium content of natural water and other substances. *Geochimica et Cosmochimica Acta* **4**, 89–103.

Friedrichsen, H. (1985) Strontium, oxygen, and hydrogen isotope studies on primary and secondary minerals in basalts from the Costa Rica Rift, Deep Sea Drilling Project Hole 504B, Leg 83. In: *Initial Reports of the Deep Sea Drilling Project, Covering Leg 504b* (eds R. N. Anderson, J. Honnorez, A. C. Adamson *et al.*), pp. 289–295. Palisades, NY.

Früh-Green, G.L., Plas, A. & Dell'Angelo, L.N. (1996) Mineralogic and stable isotope record of polyphase alteration of upper crustal gabbros of the East Pacific Rise (Hess Deep, Site 894). In: *Proceedings of the Ocean Drilling Program, Scientific Results* (eds C. Mével, K. M. Gillis, J. F. Allan & P. S. Meyer), pp. 235–254. College Station.

Fryer, P., Pearce, J.A. & Stokking, L.B. (1990) *Proceedings of ODP Initial Reports 125.* Ocean Drilling Program, College Station, TX.

Gallahan, W.E. & Duncan, R.A. (1994) Spatial and temporal variability in crystallization of celadonites within the Troodos ophiolite, Cyprus: implications for low-temperature alteration of the oceanic crust. *Journal of Geophysical Research* **99**, 3147–3162.

Gamo, T., Okamura, K., Charlou, J.L. *et al.* (1997) Acidic and sulfate-rich hydrothermal fluids from the Manus back-arc basin, Papua New Guinea. *Geology* **25**, 139–142.

Garcia-Lopez, S., Brime, C., Bastida, F. & Sarmiento, G.N. (1997) Simultaneous use of thermal indicators to analyse the transition from diagenesis to metamorphism: an example from the Variscan Belt of northwest Spain. *Geological Magazine* **134**, 323–334.

Garlick, G.D. & Epstein, S. (1967) Oxygen isotope ratios in coexisting minerals of regionally metamorphosed rocks. *Geochimica et Cosmochimica Acta* **31**, 181–214.

Garlick, G.D., MacGregor, I.D. & Vogel, D.E. (1971) Oxygen isotope ratios in eclogites from kimberlites. *Science* **172**, 1025–1027.

Gaudemer, Y., Jaupart, C. & Tapponier, P. (1988) Thermal control on post-orogenic extension in collision belts. *Earth and Planetary Science Letters* **89**, 48–62.

Gauthier-Lafaye, F., Bros, R. & Stille, P. (1996) Pb–Pb isotope systematics on diagenetic clays: an example from Proterozoic black shales of the Franceville basin (Gabon). *Chemical Geology* **133**, 243–250.

Gebauer, D. & Grünenfelder, M. (1974) Rb–Sr whole-rock dating of late diagenetic to anchimetamorphic Paleozoic sediments in southern France (Montagne Noire). *Contributions to Mineralogy and Petrology* **47**, 113–130.

Gee, J., Staudigel, H., Tauxe, L., Pick, T. & Gallet, Y. (1993) Magnetization of the La Palma seamount series: implications for seamount paleopoles. *Journal of Geophysical Research* **98**, 11743–11767.

Gharrabi, M. & Velde, B. (1995) Clay mineral evolution in the Illinois Basin and its causes. *Clay Minerals* **30**, 353–364.

Giaramita, M.J. & Day, H.W. (1990) Error propagation in calculations of structural formulas. *American Mineralogist* **75**, 170–182.

Giaramita, M.J. & Day, H.W. (1991) The four-phase AFM assemblage staurolite–aluminium silicate–biotite–garnet: extra components and implications for staurolite-out isograds. *Journal of Petrology* **32**, 1203–1229.

Gieskes, J.M., Kastner, M., Einsele, G., Kelts, K. & Niemitz, J. (1982) Hydrothermal activity in the Guaymas Basin, Gulf of California: a synthesis. In: *Initial Reports of the Deep Sea Drilling Project* (eds J. R. Curray & D. G. Moore), pp. 1159–1167. US Government Printing Office, Washington, DC.

Gillis, K.M. (1995) Controls on hydrothermal alteration in a section of fast-spreading oceanic crust. *Earth and Planetary Science Letters* **134**, 473–489.

Gillis, K.M. & Robinson, P.T. (1985) Low-temperature alteration of the extrusive sequence, Troodos Ophiolite, Cyprus. *Canadian Mineralogist* **23**, 431–444.

Gillis, K.M. & Robinson, P.T. (1988) Distribution of alteration zones in the upper oceanic crust. *Geology* **16**, 262–266.

Gillis, K.M. & Robinson, P.T. (1990) Patterns and processes of alteration in the lavas and dikes of the Troodos ophiolite, Cyprus. *Journal of Geophysical Research* **95**, 21523–21548.

Gillis, K.M. & Thompson, G. (1993) Metabasalts from the Mid-Atlantic Ridge: new insights into hydrothermal systems in slow-spreading crust. *Contributions to Mineralogy and Petrology* **113**, 502–523.

Girard, J.-P. & Deynoux, M. (1991) Oxygen isotope study of diagenetic quartz overgrowths from the upper Proterozoic quartzites of western Mali, Taoudeni Basin; implications for conditions of quartz cementation. *Journal of Sedimentary Petrology* **61**, 406–418.

Girard, J.-P. & Savin, S.M. (1996) Intracrystalline fractionation of oxygen isotopes between hydroxyl and non-hydroxyl sites in kaolinite measured by thermal dehydroxylation and partial fluorination. *Geochimica et Cosmochimica Acta* **60**, 469–487.

Gleadow, A.J.W., Duddy, I.R., Green, P.F. & Lovering, J.F. (1986) Confined fission track lengths in apatite: a diagnostic tool for thermal history analysis. *Contributions to Mineralogy and Petrology* **94**, 405–415.

Goodge, J.W. (1995) Pre-Middle Jurassic accretionary metamorphism in the southern Klamath Mountains of northern California, USA. *Journal of Metamorphic Geology* **13**, 93–110.

Gordon, T.M., Ghent, E.D. & Stout, M.Z. (1991) Algebraic analysis of the biotite–sillimanite isograd in the File Lake area, Manitoba. *Canadian Mineralogist* **29**, 673–686.

Graham, C.M., Valley, J.W. & Winter, B.L. (1996) Ion microprobe analysis of $^{18}O/^{16}O$ in authigenic and detrital quartz in the St. Peter Sandstone, Michigan Basin and Wisconsin Arch, USA; contrasting diagenetic histories. *Geochimica et Cosmochimica Acta* **60**, 5101–5116.

Green, J.C. (1982) *Geology of the Keweenawan extrusive rocks.* Geological Society of America Memoir 156.

Green, P.F., Duddy, I.R., Laslett, G.M., Hegarty, K.A., Gleadow, A.J.W. & Lovering, J.F. (1989) Thermal annealing of fission tracks in apatite. 1—quantitative modelling techniques and extension to geological time-scales. *Chemical Geology, Isotope Geoscience Section* **59**, 237–253.

Greenwood, H.J. (1967) The N-dimensional tie-line problem. *Geochimica et Cosmochimica Acta* **31**, 465–490.

Greenwood, H.J. (1968) Matrix methods and the phase rule in petrology. In: *23rd International Geological Congress*, Prague Vol. 6, pp. 267–279.

Greenwood, H.J. (1975) Thermodynamically valid projections of extensive phase relationships. *American Mineralogist* **60**, 1–8.

Gregory, R.T. (1991) Oxygen isotope history of seawater revisited; timescales for boundary event changes in the oxygen isotope composition of seawater. In: *Stable Isotope Geochemistry: A Tribute to Samuel Epstein* (eds H. P. Taylor, J. R. O'Neil & I. R. Kaplan), pp. 65–76. Geochemical Society, San Antonio.

Gregory, R.T. & Taylor, H.P. (1981) An oxygen isotope profile in a section of Cretaceous oceanic crust, Samail ophiolite, Oman: evidence for ^{18}O buffering of the oceans by deep (> 5 km) seawater-hydrothermal circulation at mid-ocean ridges. *Journal of Geophysical Research* **86**, 2737–2755.

Grubb, S.M.B., Peacor, D.R. & Jiang, W.-T. (1991) Transmission electron microscope observations of illite polytypism. *Clays and Clay Minerals* **39**, 540–550.

Guidotti, C.V. (1984) Micas in metamorphic rocks. In: *Micas* (ed. S.W. Bailey). *Reviews in Mineralogy* **13**, 357–467.

Guidotti, C.V. & Sassi, F.P. (1986) Classification and correlation of metamorphic facies series by means of muscovite bo data from low-grade metapelites. *Neues Jahrbuch für Mineralogie Abhandlungen* **153**, 363–380.

Guidotti, C.V., Sassi, F.P. & Blencoe, J.G. (1989) Compositional controls on the a and b cell dimensions of $2M_1$ muscovite. *European Journal of Mineralogy* **1**, 71–84.

Guthrie, G.D. & Veblen, D.R. (1989a) High resolution transmission electron microscopy of mixed-layer illite/smectite: computer simulations. *Clays and Clay Minerals* **37**, 1–11.

Guthrie, G.D. & Veblen, D.R. (1989b) High resolution transmission electron microscopy applied to clay minerals. In: *Spectroscopic Characterization of Minerals and Their Surfaces* (eds L. M. Coyne, D. F. Blake & S. McKeever), pp. 75–93. American Chemical Society Symposium Serial No. 415.

Gutiérrez-Alonso, G. & Nieto, F. (1996) White-mica 'crystallinity', finite strain and cleavage development across a large Variscan structure, NW Spain. *Journal of the Geological Society, London* **153**, 287–299.

Hamza, M.S. & Epstein, S. (1980) Oxygen isotopic fractionation between oxygen of different sites in hydroxyl-bearing silicate minerals. *Geochimica et Cosmochimica Acta* **44**, 173–182.

Hannington, M., Jonasson, I.R., Herzig, P.M. & Petersen, S. (1995) Physical, chemical and microbial processes affecting sulfide deposition. In: *Seafloor Hydrothermal Systems, Physical, Chemical, and Biological Interactions* (eds S. Humphris, J. Lupton, L. Mullineaux & R. Zierenberg), pp. 115–157. Geophysical Monograph 91. AGU, Washington DC.

Hanson, R.B. (1995) The hydrodynamics of contact metamorphism. *Geological Society of America Bulletin* **107**, 595–611.

Harmon, R.S. & Hoefs, J. (1995) Oxygen isotope heterogeneity of the mantle deduced from global ^{18}O systematics of basalts from different geotectonic settings. *Contributions to Mineralogy and Petrology* **120**, 95–114.

Harper, C.T. (1970) Graphic solution to the problem of $^{40}Ar^*$ loss from metamorphic minerals. *Eclogae Geologicae Helvetica* **63**, 119–140.

Harper, G.D. (1995) Pumpellyosite and prehnitite associated with epidosite in the Josephine ophiolite-Ca metasomatism during upwelling of hydrothermal fluids at a spreading axis. In: *Low-Grade Metamorphism of Mafic Rocks* (eds P. Schiffman & H. W. Day). Geological Society of America, Boulder, CO. Special Paper No. 296, pp. 101–122.

Harper, G.D. & Tartarotti, P. (1996) Structural evolution of upper layer 2, Hole 896A. In: *Proceedings of ODP Scientific Results 148* (eds J. C. Alt, H. Kinoshita, L. Stokking & P. J. Michael), pp. 245–259. Ocean Drilling Program, College Station, TX.

Harper, G.D., Bowman, J.R. & Kuhns, R. (1988) A field, chemical, and stable isotope study of subseafloor metamorphism of the Josephine ophiolite, California-Oregon. *Journal of Geophysical Research* **93**, 4625–4656.

Harrison, T.M., Armstrong, R.L., Naeser, C.W. & Harakal, J.E. (1979) Geochronology and thermal history of the Coast Plutonic Complex, near Prince Rupert, British Columbia. *Canadian Journal of Earth Sciences* **16**, 400–410.

Hart, R. (1973) Chemical exchange between sea water and deep ocean basalts. *Earth and Planetary Science Letters* **9**, 269–279.

Hart, S.R. & Staudigel, H. (1978) Oceanic crust: age of hydrothermal alteration. *Geophysical Research Letters* **5**, 1009–1012.

Hart, S.R. & Staudigel, H. (1982) The control of alkalis and uranium in seawater by ocean crust alteration. *Earth and Planetary Science Letters* **58**, 202–212.

Hart, S.R. & Staudigel, H. (1986) Ocean crust vein mineral deposition: Rb/Sr ages, U–Th–Pb geochemistry, and duration of circulation at DSDP Sites 261, 462, and 516. *Geochimica et Cosmochimica Acta* **50**, 2751–2761.

Hart, S.R., Erlank, A.J. & Kable, E.J.D. (1974) Sea floor basalt alteration: some chemical and Sr isotopic effects. *Contributions to Mineralogy and Petrology* **44**, 219–230.

Hart, S.R., Blusztajn, J., Dick, H.J.B. & Lawrence, J.R. (1994) Fluid circulation in the oceanic crust: contrast between volcanic and plutonic regimes. *Journal of Geophysical Research* **99**, 3163–3174.

Harte, B. & Graham, C.M. (1975) The graphical analysis of greenschist to amphibolite facies mineral assemblages in metabasites. *Journal of Petrology* **16**, 347–370.

Hay, R.L. & Iijima, A. (1968) Nature and origin of palagonitic tuffs of the Honolulu Group on Oahu, Hawaii. *Geological Society of America Memoir* **116**, 338–376.

Hayes, J.M., Freeman, K.H., Popp, B.N. & Hoham, C.H. (1990) Compound-specific isotopic analyses; a novel tool for reconstruction of ancient biogeochemical processes. In: *14th International Meeting on Organic Geochemistry, Paris* (eds B. Durand & F. Behar). *Advances in Organic Geochemistry*, 1115–1128.

Haymon, R.M., Koski, R.A. & Abrams, M.J. (1989) Hydrothermal discharge zones beneath massive sulfide deposits mapped in the Oman ophiolite. *Geology* **17**, 531–535.

Heaton, T.H.E. (1987) The $^{15}N/^{14}N$ ratios of plants in South Africa and Namibia: relationship to climate and coastal/saline environments. *Oecologia* **74**, 236–246.

Henley, R.W. & Ellis, A.J. (1983) Geothermal systems ancient and modern: a geochemical review. *Earth Science Reviews* **19**, 1–50.

Hervig, R.L., Williams, L.B., Kirkland, I.K. & Longstaffe, F.J. (1995) Oxygen isotope microanalyses of diagenetic quartz: possible low temperature occlusion of pores. *Geochimica et Cosmochimica Acta* **59**, 2537–2543.

Hesse, R. & Dalton, E. (1991) Diagenetic and low-grade metamorphic terranes of the Gaspé Peninsula related to the geological structure of the Taconian and Acadian orogenic belts, Quebec Appalachians. *Journal of Metamorphic Geology* **9**, 775–790.

Hey, M.H. (1954) A new review of chlorites. *Mineralogical Magazine* **30**, 277–292.

Hillier, S. (1993a) *High gradient magnetic separation: principles, application to clay minerals and laboratory manual.* Internal Report, Geological Institute, University of Bern.

Hillier, S. (1993b) Origin, diagenesis, and mineralogy of chlorite minerals in Devonian lacustrine mudrocks, Orcadian Basin, Scotland. *Clays and Clay Minerals* **41**, 240–259.

Hillier, S. (1995) Mafic phyllosilicates in low-grade metabasites. Characterization using deconvolution analysis—Discussion. *Clay Minerals* **30**, 67–73.

Hillier, S. & Clayton, T. (1992) Cation exchange 'staining' of clay minerals in thin-section for electron microscopy. *Clay Minerals* **27**, 379–384.

Hillier, S. & Marshall, J.E.A. (1992) Organic maturation, thermal history and hydrocarbon generation in the Orcadian basin, Scotland. *Journal of the Geological Society, London* **149**, 491–502.

Hillier, S. & Velde, B. (1991) Octahedral occupancy and the chemical composition of diagenetic (low-temperature) chlorites. *Clay Minerals* **26**, 149–168.

Hillier, S., Mátyás, J., Matter, A. & Vasseur, G. (1995) Illite/smectite diagenesis and its variable correlation with vitrinite reflectance in the Pannonian Basin. *Clays and Clay Minerals* **43**, 174–183.

Hirons, S.R. (1997) Temporal and spatial relationships between metamorphism and structure in the Windermere Supergroup, English Lake District. In: *Clay Mineral Evolution, Basin Maturity and Mudrock Properties* (eds R. J. Merriman & S. J. Kemp), p 57. British Geological Survey, Technical Report WG/97/45.

Hirons, S.R., Roberts, B. & Merriman, R.J. (1997) *Metamorphism of the Lower Palaeozoic rocks of the Carrick-Loch Doon region, southern Scotland.* British Geological Survey, Technical Report WG/97/25.

Ho, N.-C., Peacor, D.R. & Van der Pluijm, B.A. (1995) Reorientation mechanisms of phyllosilicates in the mudstone-to-slate transition at Lehigh Gap, Pennsylvania. *Journal of Structural Geology* **17**, 345–356.

Ho, N.-C., Peacor, D.R. & Van der Pluijm, B.A. (1996) Contrasting roles of detrital and authigenic phyllosilicates during slaty cleavage development. *Journal of Structural Geology* **18**, 615–623.

Ho, N.-C., Peacor, D.R. & Van der Pluijm, B.A. (1998) Preferred orientation of phyllosilicates in Gulf Coast mudstones and relation to the smectite–illite transition. *Clays and Clay Minerals*, in review.

Hodgson, C.J., Hamilton, J.V. & Piroscho, D.W. (1990) Structural setting of gold deposits and the tectonic evolution of the Timmins-Kirkland Lake area, southwestern Abitibi greenstone belt. In: *Gold and Base-Metal Mineralization in the Abitibi Province, Canada with Emphasis on the Quebec Segment* (eds S. E. Ho, F. Robert & D. I. Groves), pp. 101–120.

Hoefs, J. & Frey, M. (1976) The isotopic composition of carbonaceous matter in a metamorphic profile from the Swiss Alps. *Geochimica et Cosmochimica Acta* **40**, 945–951.

Hoffman, J. & Hower, J. (1979) Clay mineral assemblages as low grade metamorphic geothermometers: application to the thrust faulted Disturbed Belt of Montana, U.S.A. *Society of Economic Paleontologists & Mineralogists Special Publication* **26**, 55–79.

Hoffman, S.E., Wilson, M. & Stakes, D.S. (1986) Inferred oxygen isotope profile of Archaean oceanic crust, Onverwacht Group, South Africa. *Nature* **321**, 55–58.

Holder, M.T. & Leveridge, B.E. (1994) *A framework for the European Variscides.* British Geological Survey, Technical Report WA/94/24R.

Holland, T.J.B. & Blundy, J.D. (1994) Non-ideal interactions in calcic amphiboles and their bearing on amphibole-plagioclase thermometry. *Contributions to Mineralogy and Petrology* **116**, 433–447.

Holland, T.J.B. & Powell, R. (1990) An enlarged and updated internally consistent thermodynamic dataset with uncertainties and correlations: the system K_2O-Na_2O-CaO-MgO-MnO-FeO-Fe_2O_3-Al_2O_3-TiO_2-SiO_2-C-H_2-O_2. *Journal of Metamorphic Geology* **8**, 89–124.

Holmden, C. & Muehlenbachs, K. (1993) The $^{18}O/^{16}O$ ratio of 2-billion-year-old seawater inferred from ancient oceanic crust. *Science* **259**, 1733–1736.

Honnorez, J. (1981) The aging of the oceanic crust at low temperature. In: *The Sea Vol. 7. The Oceanic Lithosphere* (ed. C. Emiliani), pp. 525–587. John Wiley & Sons, New York.

Honnorez, J., Von Herzen, R.P., Barrett, T.J. *et al.* (1981) Hydrothermal mounds and young ocean crust of the Galapagos: Preliminary Deep Sea Drillling results, Leg 70. *Geological Society of America Bulletin* **92**, 457–472.

Honnorez, J., Laverne, C., Hubberten, H.W., Emmermann, R. & Muehlenbachs, K. (1983) Alteration processes of layer 2 basalts from DSDP Hole 504B, Costa Rica Rift. In: *Initial Reports DSDP, 70* (eds J. R. Cann, J. Honnorez, M. G. Langseth, R. P. Von Herzen & S. M. White), pp. 509–546. US Government Printing Office, Washington, DC.

Honnorez, J., Alt, J.C., Honnorez, B.M. *et al.* (1985) Stockwork-like sulfide mineralization in young oceanic crust: DSDP Hole 504B. In: *Initial Reports DSDP, 83* (eds J. Honnorez, R. N. Anderson & K. Becker), pp. 263–282. US Government Printing Office, Washington, DC.

Honnorez, J., Alt, J.C. & Humphris, S.E. (1997) Vivisection and autopsy of active and fossil hydrothermal alterations of basalt beneath and within the TAG hydrothermal mound. In: *Proceedings of ODP, Scientific Results 158* (eds P. M. Herzig, S. E. Humphris, D. J. Miller & R. A. Zierenberg), pp. 231–254. Ocean Drilling Program, College Station, TX.

Horseman, S.T. (1997) *Thermal aspects of performance assessment for HLW disposal in clays and mudrocks.* British Geological Survey, Technical Report WE/97/36R.

Horseman, S.T. & Volckaert, G. (1996) Disposal of radioactive wastes in argillaceous formations. In: *Engineering Geology of Waste Disposal* (ed. S. P. Bentley), pp. 179–191. Geological Society Engineering Geology Special Publication No. 11.

Horsfield, B. & Rullkötter, J. (1994) Diagenesis, catagenesis, and metagenesis of organic matter. In: *The Petroleum System—From Source to Trap.* (eds L. B. Magoon & D. G. Dow). *American Association of Petroleum Geologists Memoir* **60**, 189–199.

Hover, V.C., Peacor, D.R. & Walker, L.M. (1996) STEM/AEM evidence for preservation of burial diagenetic fabrics in Devonian shales: implications for fluid/rock interaction in cratonic basins (U.S.A.). *Journal of Sedimentary Research* **66**, 519–530.

Howells, M.F., Reedman, A.J. & Campbell, S.D.G. (1991) *Ordovician (Caradoc) Marginal Basin Volcanism in Snowdonia (North-West Wales).* London, HMSO for the British Geological Survey.

Hower, J., Eslinger, E.V., Hower, M.E. & Perry, E.A. (1976) Mechanism of burial metamorphism of argillaceous sediment: 1. Mineralogical and chemical evidence. *Geological Society of America Bulletin* **87**, 725–737.

Huang, P.Y. & Solomon, S.C. (1988) Centroid depths of mid-ocean ridge earthquakes: dependence on spreading rate. *Journal of Geophysical Research* **93**, 13445–13477.

Huang, W.-L., Longo, J.M. & Pevear, D.R. (1993) An experimentally derived kinetic model for smectite-to-illite conversion and its use as a geothermometer. *Clays and Clay Minerals* **41**, 162–177.

Hubberton, H.-W. (1982) The isotopic composition of carbonate carbon from deep-sea basalts. *Geochemical Journal* **16**, 99–105.

Hubberton, H.-W. (1983) Sulfur content and sulfur isotopes of basalts from the Costa Rica Rift (Hole 504B, DSDP Legs 69 and 70). In: *Initial Reports DSDP, 69* (eds J. R. Cann, J. Honnorez, M. G. Langseth, R. P. Von Herzen & S. M. White), pp. 629–635. US Government Printing Office, Washington, DC.

Huff, W.D., Whiteman, J.A. & Curtis, C.D. (1988) Investigation of a K-bentonite by X-ray powder diffraction and analytical transmission electron microscopy. *Clays and Clay Minerals* **36**, 83–93.

Huggett, J.M. (1995) Formation of authigenic illite in Palaeocene mudrocks from the central North Sea: a study by high resolution electron microscopy. *Clays and Clay Minerals* **43**, 682–692.

Humphris, S., Herzig, P., Miller, J. *et al.* (1995) The internal stucture of an active sea-floor massive sulphide deposit. *Nature* **377**, 713–716.

Humphris, S.E., Melson, W.G. & Thompson, R.N. (1980) Basalt weathering on the East Pacific Rise and Galapagos Spreading Center, DSDP Leg 54. In: *Initial Reports DSDP 54* (eds B. R. Rosendahl & R. Hekinian), pp. 773–787. US Government Printing Office, Washington, DC.

Hunziker, J.C., Frey, M., Clauer, N. *et al.* (1986) The evolution from illite to muscovite: mineralogical and isotopic data from the Glarus Alps, Switzerland. *Contributions to Mineralogy and Petrology* **92**, 157–180.

Hunziker, J.C., Frey, M., Clauer, N. & Dallmeyer, R.D. (1987) Reply to the comment on the evolution of illite to muscovite by J. R. Glasmann. *Contributions to Mineralogy and Petrology* **96**, 74–77.

Huon, S., Cornée, J.J., Piqué, A., Rais, N., Clauer, N. & Zayane, R. (1993) Mise en évidence au Maroc d'évènements thermiques d'âge triasico-liasique liés à l'ouverture de l'Atlantique. *Bulletin Société géologique de France* **164**, 165–176.

Huon, S., Burkhard, M. & Hunziker, J.-C. (1994) Mineralogical, K–Ar, stable and Sr isotope systematics of K-white micas during very low grade metamorphism of limestones (Helvetic nappes, western Switzerland). *Chemical Geology, Isotope Geoscience Section* **113**, 347–376.

Hurford, A.J. & Green, P.F. (1982) A users' guide to fission track dating calibration. *Earth and Planetary Science Letters* **59**, 343–354.

Hurford, A.J., Hunziker, J.C. & Stökert, B. (1991) Constraints on the late thermotectonic evolution of the wetern Alps: evidence for episodic rapid uplift. *Tectonics* **10**, 758–769.

Hutcheon, I. (1990) Clay-carbonate reactions in the Venture area, Scotia Shelf, Nova Scotia, Canada. In: *Fluid–Mineral Interactions: A Tribute to H. P. Eugster* (eds R. J. Spencer & I.-M. Chou). *The Geochemical Society Special Publication* **2**, 199–212.

IAEA-TECDOC-825 (1995) *Reference and intercomparison materials for stable isotopes of light elements.* International Atomic Energy Agency, Vienna.

Inoue, A. (1987) Conversion of smectite to chlorite by hydrothermal and diagenetic alterations, Hokuroku Kuroko mineralization area, northeast Japan. In: *Proceedings of the International Clay Conference, Denver* (eds L. G. Shultz, H. van Olphen & F. A. Mumpton), pp. 158–164. Clay Minerals Society, Bloomington, MA.

Inoue, A. & Utada, M. (1991) Smectite to chlorite transformation in thermally metamorphosed volcanoclastic rocks in the Kamikita area, northern Honshu, Japan. *American Mineralogist* **76**, 628–640.

Inoue, A., Utada, M., Nagata, H. & Watanabe, T. (1984) Conversion of trioctahedral smectite to interstratified chlorite/smectite in Pliocene acidic pyroclastic sediments of the Ohyu district, Akita prefecture, Japan. *Clay Science* **6**, 103–116.

Inoue, A., Watanabe, T., Kohyama, N. & Brusewitz, A.M. (1990) Characterization of illitization of smectite in bentonite beds at Kinnekulle, Sweden. *Clays and Clay Minerals* **38**, 241–249.

Ishii, K. (1988) Grain growth and re-orientation of phyllosilicate minerals during the development of slaty cleavage in the South Kitakami Mountains, northeast Japan. *Journal of Structural Geology* **10**, 145–154.

Ishikawa, T. & Nakamura, E. (1992) Boron isotope geochemistry of the oceanic crust from DSDP/ODP Hole 504B. *Geochimica et Cosmochimica Acta* **56**, 1633–1639.

Ishizuka, H. (1985) Prograde metamorphism of the Horokanai ophiolite in the Kamuikotan zone, Hokkaido, Japan. *Journal of Petrology* **26**, 391–417.

Ishizuka, H. (1989) Mineral paragenesis of altered basalts from Hole 504B, ODP Leg 111. In: *Proceedings of ODP Scientific Results* (eds K. Becker & H. Sakai), pp. 61–76. Ocean Drilling Program, College Station, TX.

Islam, S., Hesse, R. & Chagnon, A. (1982) Zonation of diagenesis and low-grade metamorphism in Cambro-Ordovician flysch of Gaspé Peninsula, Quebec Appalachians. *Canadian Mineralogist* **20**, 155–167.

Ito, E., White, W.M. & Göpel, C. (1987) The O, Sr, Nd and Pb isotope geochemistry of MORB. *Chemical Geology* **62**, 157–176.

Jackson, M.L. (1979) *Soil Chemical Analysis — Advanced Course.* Author, Madison.

Jahn, B.M. (1988) Pb–Pb dating of young marbles from Taïwan. *Nature* **332**, 429–432.

Jahn, B.M., Bertrand-Sarfati, J., Morin, N. & Macé, J. (1990) Direct dating of stromatolitic carbonates from the Schmidtdrif Formation (Transvaal dolomite), South Africa, with implications on the age of the Ventersdorp Supergroup. *Geology* **18**, 1211–1214.

Jahn, B.M., Chi, W.R. & Yui, T.F. (1992) A Late Permian formation of Taïwan marbles from Chia-Li well no. 1: Pb–Pb isochron and Sr isotopic evidence, and its regional and geological significance. *Journal of Geological Society of China* **35**, 193–218.

Jahren, J.S. & Aagaard, P. (1989) Compositional variations in diagenetic chlorites and illites, and relationships with formation-water chemistry. *Clay Minerals* **24**, 157–170.

Jakobsen, H.J., Nielsen, N.C. & Lindgree, H. (1995) Sequences of charged sheets in rectorite. *American Mineralogist* **80**, 247–252.

Jakobsson, S.P. & Moore, J.G. (1986) Hydrothermal alteration minerals and alteration rates at Surtsey Volcano, Iceland. *Geological Society of America Bulletin* **97**, 648–659.

James, H.L. (1955) Zones of regional metamorphism in the Precambrian of northern Michigan. *Bulletin of the Geological Society of America* **66**, 1455–1488.

James, R.H. & Elderfield, H. (1996) Chemistry of ore-forming fluids and mineral formation rates in an active hydrothermal sulfide deposit on the Mid-Atlantic Ridge. *Geology* **24**, 1147–1150.

Jamieson, R.A., Beaumont, C., Hamilton, I. & Fullsack, P. (1996) Tectonic assembly of inverted metamorphic sequences. *Geology* **24**, 839–842.

Jayko, A.S., Blake, M.C. & Brothers, R.N. (1986) Blueschist metamorphism of the Eastern Franciscan belt, northern California. *Geological Society of America Memoir* **164**, 107–123.

Jeans, C.V., Fallick, A.E., Fisher, M.J., Merriman, R.J., Corfield, R.M. & Manighetti, B. (1997) Clay- and zeolite-bearing Triassic sediments at Kaka Point, New Zealand: evidence of microbially influenced mineral formation from earliest diagenesis into the lowest grade of metamorphism. *Clay Minerals* **32**, 373–423.

Jenkin, G.R.T., Fallick, A.E. & Leake, B.E. (1992) A stable isotope study of retrograde alteration in SW Connemara, Ireland. *Contributions to Mineralogy and Petrology* **110**, 269–288.

Jiang, W.-T. & Peacor, D.R. (1993) Formation and modification of metastable intermediate sodium potassium mica, paragonite and muscovite in hydrothermally altered metabasites from northern Wales. *American Mineralogist* **78**, 782–793.

Jiang, W.-T. & Peacor, D.R. (1994a) Prograde transitions of corrensite and chlorite in low-grade pelitic rocks from the Gaspe Peninsula, Quebec. *Clays and Clay Minerals* **42**, 497–517.

Jiang, W.-T. & Peacor, D.R. (1994b) Formation of corrensite, chlorite and chlorite-mica stacks by replacement of detrital biotite in low-grade pelitic rocks. *Journal of Metamorphic Geology* **12**, 867–884.

Jiang, W.-T. & Peacor, D.R. (1994c) Chlorite geothermometry? — Contamination and apparent octahedral vacancies. *Clays and Clay Minerals* **42**, 593–605.

Jiang, W.-T., Peacor, D.R., Merriman, R.J. & Roberts, B. (1990a) Transmission and analytical electron microscopic study of mixed-layer illite/smectite formed as an apparent replacement product of diagenetic illite. *Clays and Clay Minerals* **38**, 449–468.

Jiang, W.-T., Peacor, D.R. & Essene, E.J. (1990b) Transmission electron microscopic study of coexisting pyrophyllite and muscovite: direct evidence for the metastability of illite. *Clays and Clay Minerals* **38**, 225–240.

Jiang, W.-T., Nieto, F. & Peacor, D.R. (1992a) Composition of diagenetic illite as defined by analytical electron microscope analyses: implications for smectite–illite–muscovite transitions. In: *29th International Geological Congress*, abstract 100, Kyoto, Japan.

Jiang, W.-T., Peacor, D.R. & Slack, J.F. (1992b) Microstructures, mixed layering, and polymorphism of chlorite and retrograde berthierine in the Kidd Creek massive sulfide deposit, Ontario. *Clays and Clay Minerals* **40**, 501–514.

Jiang, W.-T., Peacor, D.R. & Essene, E.J. (1994) Analytical and transmission electron microscopic study of clay minerals in sandstone of Kettleman North Dome, California: implications for the metastability of illite. *Clays and Clay Minerals* **42**, 35–45.

Jiang, W.-T., Peacor, D.R., Árkai, P., Tóth, M. & Kim, J.-W. (1997) TEM and XRD determination of crystallite size and lattice strain as a function of illite crystallinity in pelitic rocks. *Journal of Metamorphic Geology* **15**, 267–281.

Johnson, D.M. (1979) Crack distribution in the upper oceanic crust and its effects upon seismic velocity, seismic structure, formation permeability, and fluid circulation. In: *Initial Reports DSDP 51–53* (eds T. Donnelly, J. Francheteau, W. Bryan, P. Robinson, M. Flower & M. Salisbury), pp. 1473–1478. US Government Printing Office, Washington, DC.

Jolly, W.T. (1974) Behaviour of Cu, Zn and Ni during prehnite–pumpellyite rank metamorphism of the Keweenawan basalts, northern Michigan. *Economic Geology* **69**, 1118–1125.

Jolly, W.T. & Smith, R.E. (1972) Degradation and metamorphic differentiation of the Keweenawan tholeiitic lavas of Northern Michigan, USA. *Journal of Petrology* **13**, 272–309.

Jowett, E.C. (1991) *Fitting iron and magnesium into the hydrothermal chlorite geothermometer.* GAC/MAC/SEG Joint Annual Meeting, Toronto, Program with Abstracts, Vol. 16, A62.

Juteau, T., Bingol, F., Noack, Y. *et al.* (1979) Preliminary results: mineralogy and geochemistry of alteration products in Leg 45 basement samples. In: *Initial Reports DSDP 45* (eds W. G. Melson & P. D. Rabinowitz), pp. 613–645. US Government Printing Office, Washington, DC.

Kadko, D., Baross, J. & Alt, J.C. (1995) Hydrothermal fluxes and global change. In: *Seafloor Hydrothermal Systems, Physical, Chemical, and Biological Interactions* (eds S. Humphris, J. Lupton, L. Mullineaux & R. Zierenberg), pp. 446–466. Geophysical Monograph No. 91. AGU, Washington, DC.

Karlsson, H.R. & Clayton, R.N. (1990) Oxygen and hydrogen isotope geochemistry of zeolites. *Geochimica et Cosmochimica Acta* **54**, 1369–1386.

Karson, J.A. (1990) Seafloor spreading on the Mid-Atlantic ridge: implications for the structure of ophiolites and oceanic lithosphere produced in slow spreading environments. In: *Ophiolites, Oceanic Crustal Analogues, Proceedings of the Symposium 'Troodos 1987'*, pp. 547–555. Geological Survey Department, Ministry of Agriculture and Human Resources, Nicosia, Cyprus.

Kastner, M. & Gieskes, J.M. (1976) Interstitial water profiles and sites of diagenetic reactions, Leg 35, Bellingshausen abyssal plain. *Earth and Planetary Science Letters* **33**, 11–20.

Kawahata, H., Kusakabe, M. & Kikuchi, Y. (1987) Strontium, oxygen, and hydrogen isotope geochemistry of hydrothermally altered and weathered rocks in DSDP Hole 504B, Costa Rica Rift. *Earth and Planetary Science Letters* **85**, 343–355.

Kelley, S.P. & Fallick, A.E. (1990) High precision spatially resolved analysis of $d^{34}S$ in sulphides using a laser extraction technique. *Geochimica et Cosmochimica Acta* **54**, 883–888.

Kelley, S.P., Arnaud, N.O. & Turner, S.P. (1994) High spatial resolution $^{40}Ar/^{39}Ar$ investigations using an ultra-violet laser probe extraction technique. *Geochimica et Cosmochimica Acta* **58**, 3519–3525.

Kelso, P.R., Banerjee, S.K. & Worm, H.-U. (1991) The effect of low-temperature hydrothermal alteration on the remanent magnetization of synthetic titanomagnetites; a case for acquisition of chemical remanent magnetization. *Journal of Geophysical Research* **96**, 19545–19553.

Kemp, A.E.S., Oliver, G.J.H. & Baldwin, J.R. (1985) Low-grade metamorphism and accretion tectonics:

Southern Uplands terrain, Scotland. *Mineralogical Magazine* **49**, 335–344.

Kemp, S.J. & Rochelle, C.A. (1998) *A mineralogical, geochemical, petrographic and physical testing study of the thermal alteration of smectite-bearing mudstones from the Isle of Skye, Inner Hebrides. (I) Lùb Score, below Bealach Iochdarach.* British Geological Survey, Technical Report WG/98/3R.

Kempe, D.R.C. (1974) The petrology of the basalts, Leg 26. In: *Initial Reports DSDP 26* (eds T. A. Davies & B. P. Luyendyk), pp. 465–504. US Government Printing Office, Washington, DC.

Kendall, C. & Coplen, T.B. (1985) Multisample conversion of water to hydrogen by zinc for stable isotope determination. *Analytical Chemistry* **57**, 1437–1440.

Kennedy, W.Q. (1948) On the significance of the thermal structure of the Scottish Highlands. *Geological Magazine* **85**, 229–234.

Kerrich, R. (1987) Stable isotope studies of fluids in the crust. In: *Short Course in Stable Isotope Geochemistry of Low Temperature Processes* (ed. T. K. Kyser), pp. 258–286. Mineralogical Society of Canada, Saskatoon, SK.

Kerrick, D.M. (1991) Overview of contact metamorphism. In: *Contact Metamorphism* (ed. D. M. Kerrick), Mineralogical Society of America. *Reviews in Mineralogy* **26**, 1–12.

Kieffer, S.W. (1982) Thermodynamics and lattice vibrations in minerals: 5. Applications to phase equilibria, isotopic fractionation, and high-pressure thermodynamic properties. *Reviews of Geophysics and Space Physics* **20**, 827–849.

Kim, J.-W., Peacor, D.R., Teyssier, D. & Elsass, F. (1995) A technique for maintaining texture and permanent expansion of smectite interlayers for TEM observations. *Clays and Clay Minerals* **43**, 51–57.

Kirschner, D.L., Sharp, Z.D. & Teyssier, C. (1993) Vein growth mechanisms and fluid sources revealed by oxygen isotope laser microprobe. *Geology* **21**, 85–88.

Kirschner, D.L., Sharp, Z.D. & Masson, H. (1995) Oxygen isotope thermometry of quartz–calcite veins: unraveling the thermal-tectonic history of the subgreenschist facies Morcles nappe (Swiss Alps). *Geological Society of America Bulletin* **107**, 1145–1156.

Kisch, H.J. (1974) Anthracite and meta-anthracite coal ranks associated with 'anchimetamorphism' and 'very-low-stage' metamorphism. I, II, III. *Koninkl Nederland Akademie Wetenschappen Amsterdam, Proceedings Series B* **77**, 81–118.

Kisch, H.J. (1980) Incipient metamorphism of Cambro-Silurian clastic rocks from the Jämtland Supergroup, central Scandinavian Caledonides, western Sweden: illite crystallinity and 'vitrinite' reflectance. *Journal of the Geological Society, London* **137**, 271–288.

Kisch, H.J. (1981) Coal rank and illite crystallinity associated with the zeolite facies of Southland and the pumpellyite-bearing facies of Otago, southern New

Zealand. *New Zealand Journal of Geology and Geophysics* **24**, 349–360.

Kisch, H.J. (1983) Mineralogy and petrology of burial diagenesis (burial metamorphism) and incipient metamorphism in clastic rocks. In *Diagenesis in Sediments and Sedimentary Rocks*, Vol. 2 (eds G. Larsen & G. V. Chilingar), pp. 289–493, 513–541. Elsevier, Amsterdam. (Appendix B-literature published since 1976.)

Kisch, H.J. (1987) Correlation between indicators of very-low-grade metamorphism. In: *Low Temperature Metamorphism* (ed. M. Frey), pp. 227–300. Blackie & Son, Glasgow.

Kisch, H.J. (1989) Discordant relationship between degree of very-low grade metamorphism and the development of slaty cleavage. In: *Evolution of Metamorphic Belts* (eds J. S. Daly, R. A. Cliff & B.W.D. Yardley). *Geological Society Special Publication* **43**, 173–185.

Kisch, H.J. (1990) Calibration of the anchizone: a critical comparison of illite 'crystallinity' scales used for definition. *Journal of Metamorphic Geology* **8**, 31–46.

Kisch, H.J. (1991a) Illite crystallinity: recommendations on sample preparation, X-ray diffraction settings and interlaboratory standards. *Journal of Metamorphic Geology* **6**, 665–670.

Kisch, H.J. (1991b) Development of slaty cleavage and degree of very low-grade metamorphism. *Journal of Metamorphic Geology* **6**, 735–750.

Kisch, H.J. (1994) X-ray diffraction intensity ratios of phyllosilicate reflections in cleavage- and bedding-parallel slabs: incipient development of slaty cleavage in the Caledonides of Jämtland, western central Sweden. *Revista Geológica de Chile* **21**, 253–267.

Kisch, H.J. & Frey, M. (1987) Appendix: Effects of sample preparation on the measured 10Å peak width of illite (illite 'crystallinity'). In: *Low Temperature Metamorphism* (ed. M. Frey), pp. 301–304. Blackie & Son, Glasgow.

Kita, I. & Honda, S. (1987) Oxygen isotopic difference between hydrothermally and diagenetically altered rocks from the Tsugaru-Yunosawa area, Aomori, Japan. *Geochemical Journal* **21**, 35–41.

Klug, H.P. & Alexander, L.E. (1974) *X-ray Diffraction Procedures*, 2nd edn. Wiley, New York.

Kneller, B.C. & Bell, A.M. (1993) An Acadian mountain front in the English Lake District: the Westmorland Monocline. *Geological Magazine* **130**, 203–213.

Kneller, B.C., King, L.M. & Bell, A.M. (1993) Foreland basin development and tectonics on the northwest margin of eastern Avalonia. *Geological Magazine* **130**, 691–697.

Knipe, R.J. (1979) Chemical changes during slaty cleavage development. *Bulletin de Minéralogie* **105**, 206–209.

Knipe, R.J. (1981) The interaction of deformation and metamorphism in slates. *Tectonophysics* **78**, 249–272.

Knipe, R.J. (1986) Faulting mechanisms in slope sediments: examples from Deep Sea Drilling Project cores. In: *Structural Fabrics in Deep Sea Drilling Project Cores From Forearcs* (ed. J. C. Moore). *Geological Society of America Memoir* **166**, 45–54.

Knipe, R.J. & White, S.H. (1977) Microstructural variations of an axial plane cleavage around a fold—H.V.E.M. study. *Tectonophysics* **39**, 255–380.

Kokelaar, B.P., Howells, M.F., Bevins, R.E., Roach, R.A. & Dunkley, P.N. (1984) The Ordovician marginal basin of Wales. In: *Marginal Basin Geology: Volcanic and Associated Sedimentary and Tectonic Processes in Modern and Ancient Marginal Basins* (eds B. P. Kokelaar & M. F. Howells). *Special Publication of the Geological Society, London* **16**, 245–269.

Kolodny, Y. & Epstein, S. (1974) Stable isotope geochemistry of deep sea cherts. *Geochimica et Cosmochimica Acta* **40**, 1195–1209.

Kong, L.S., Solomon, S.C. & Purdy, G.M. (1992) Microearthquake characteristics of a Mid-Ocean ridge along-axis high. *Journal of Geophysical Research* **97**, 1659–1685.

Kossovskaya, A.G. & Shutov, V.D. (1961) The correlation of zones of regional epigenesis and metagenesis in terrigenous and volcanic rocks. *Doklady Akaademii Nauk SSSR, Earth Science Section* **139** (1963), 732–736 [in Russian].

Kossovskaya, A.G. & Shutov, V.D. (1970) Main aspects of the epigenesis problem. *Sedimentology* **15**, 11–40.

Kossovskaya, A.G., Logvinenko, N.V. & Shutov, V.D. (1957) Stages of formation and alteration in terrigenous rocks. *Doklady Akaademii Nauk SSSR, Earth Science Section* **116** (2), 293–296 [in Russian].

Kranidiotis, P. & MacLean, W.H. (1987) Systematics of chlorite alteration at the Phelps Dodge massive sulfide deposit, Matagami, Quebec. *Economic Geology* **82**, 1898–1911.

Kristmannsdóttir, H. (1975) Hydrothermal alteration of basaltic rocks in Icelandic geothermal areas. In: *Proceedings of the Second UN Symposium on the Development and Use of Geothermal Resources*, pp. 441–445. Lawrence Berkeley Laboratory, CA.

Kristmannsdóttir, H. (1977) Types of clay minerals in hydrothermally altered basaltic rocks, Reykjanes, Iceland. *Jokull* **26**, 30–39.

Kristmannsdóttir, H. (1978) Alteration of basaltic rocks by hydrothermal activity at 100–300°C. In: *International Clay Conference 1978* (eds M. M. Mortland & V. C. Farmer). Elsevier, Amsterdam.

Kristmannsdóttir, H. (1979) Alteration of basaltic rocks by hydrothermal activity at 100–300°C. *Developments in Sedimentology* **27**, 359–367.

Kristmannsdóttir, H. (1982) Alteration in the IRDP drill hole compared with other drill holes in Iceland. *Journal of Geophysical Research* **87**, 6525–6531.

Kristmannsdóttir, H. & Tómasson, J. (1978) Zeolite zones in geothermal areas in Iceland. In: *Natural Zeolite Occurrence, Properties and Use* (eds L. B. Sand & F. M. Mumpton), pp. 277–284. Pergamon Press, Oxford.

Kröner, A., Byerly, G.R. & Lowe, D.R. (1991) Chronology of Early Archean granite–greenstone evolution in the Barberton Mountain Land, South Africa, based on precise dating by single zircon evaporation. *Earth and Planetary Science Letters* **103**, 41–54.

Krumm, S. (1997) Crystallite-size distributions, peak shape, and their implications for XRD domain-size determinations. In: *Clay Mineral Evolution, Basin Maturity and Mudrock Properties* (eds R. J. Merriman & S. J. Kemp), pp. 17–18 (abstract). British Geological Survey, Technical Report WG/97/45.

Krumm, S. & Buggisch, W. (1991) Sample preparation effects on illite crystallinity measurements: grain-size gradation and particle orientation. *Journal of Metamorphic Geology* **9**, 671–678.

Krumm, S., Kisch, H.J. & Warr, L.N. (1994) Inter-laboratory study of the effects of sample preparation on illite 'crystallinity': a progress report. XIII. Conference on clay mineralogy and petrology. *Acta University Caolinae Geologica* **38**, 263–270.

Kübler, B. (1964) Les argiles, indicateurs de métamorphisme. *Revue Institué de la Français de Pétrole* **19**, 1093–1112.

Kübler, B. (1967a) La cristallinité de l'illite et les zones tout á fait supérieures du métamorphisme. In: *Etages Tectoniques, Colloque de Neuchâtel 1966*, pp. 105–121. Université Neuchâtel, à la Baconnière, Neuchâtel, Switzerland.

Kübler, B. (1967b) Anchimetamorphisme et schistosité. *Bulletin Centre Recherche Pau-SNPA* **1**, 259–278.

Kübler, B. (1968) Evaluation quantitative du métamorphism par la cristallinité de l'illite. *Bulletin Centre Recherche Pau-SNPA* **2**, 385–397.

Kübler, B. (1975) *Diagenese-anchimétamorphisme et métamorphisme*. Institut national de la recherche scientifque-Pétrole, Quebec.

Kübler, B. (1984) Les indicateurs des transformations physiques et chimiques dans la diagenése, température et calorimétrie. In: *Thérmométrie et barométrie géologiques* (ed. M. Lagache), pp. 489–596. Sociétié de Français Minéralogie et Cristallographie, Paris.

Kulla, J.B. & Anderson, T.F. (1978) Experimental oxygen isotope fractionation between kaolinite and water. In: *Short Papers of the 4th International Conference of Geochronology, Cosmochronology, and Isotope Geology* (ed. R. E. Zartman), pp. 234–235. US Geological Survey, Open File Report.

Kuniyoshi, S. & Liou, J.G. (1976) Burial metamorphism of the Karmutsen Volcanic rocks, northeastern Vancouver Island, British Columbia. *American Journal of Science* **276**, 1096–1119.

Kunk, M.J. & Sutter, J.F. (1984) $^{40}Ar/^{39}Ar$ age spectrum dating of biotites from Middle Ordovician bentonites, eastern North America. In: *Aspects of the Ordovician System* (ed. D. L. Bruton). Paleontological Contributions from the University of Oslo.

Kusakabe, M., Shibata, T., Yamamoto, M. *et al.* (1989) Petrology and isotope characteristics (H, O, S, Sr, and Nd) of basalts from Ocean Drilling Program Hole 504B, Leg 111, Costa Rica Rift. In: *Proceedings of the Ocean Drilling Program, Scientific Results* (eds E. K. Mazzullo, K. Becker, H. Sakai *et al.*), pp. 47–60. Ocean Drilling Program, College Station, TX.

Kyser, T.K. (1987) Equilibrium fractionation factors for stable isotopes. In: *Short Course in Stable Isotope Geochemistry of Low Temperature Fluids* (ed. T. K. Kyser), pp. 1–84. Mineralogical Society of Canada, Saskatoon, SK.

Labeyrie, L. (1972) Composition isotopique de l'oxygene de la silice biogenique. *Academie Science, CR, Serie D* **274**, 1605–1608.

Labeyrie, L.D. & Jullet, J. (1982) Oxygen isotopic exchangeability of diatom valve silica; interpretation and consequences for paleoclimatic studies. *Geochimica et Cosmochimica Acta* **46**, 967–975.

Laird, J. (1980) Phase equilibria in mafic schists from Vermont. *Journal of Petrology* **21**, 1–38.

Land, L.S. (1984) Frio sandstone diagenesis, Texas Gulf Coast: a regional isotopic study. In: *Clastic Diagenesis* (eds R. C. Surdam & D. A. MacDonald), pp. 47–62. American Association Petroleum Geologists Memoir.

Land, L.S. & Dutton, S.P. (1978) Cementation of a Pennsylvanian Deltaic Sandstone: isotopic data. *Journal of Sedimentary Petrology* **48**, 1167–1176.

Landis, C.A. (1971) Graphitization of dispersed carbonaceous material in metamorphic rocks. *Contributions to Mineralogy and Petrology* **30**, 34–45.

Lang, H.M. & Rice, J.M. (1985) Regression modeling of metamorphic reactions in metapelites, Snow, Peak, Northern Idaho. *Journal of Petrology* **26**, 889–924.

Langley, K.M. (1978) Dating sediments by a K–Ar method. *Nature* **276**, 56–57.

Langseth, M.G., Mottl, M.J., Hobart, M. & Fisher, A. (1988) The distribution of geothermal and geochemical gradients near site 501/504. In: *Proceedings of ODP, Initial Reports, 111* (eds K. Becker & H. Sakai), pp. 23–32. Ocean Drilling Program, College Station, TX.

Lanson, B. (1990) *Mise en évidence des mécanismes de transformation des interstratifiés illite/smectite au cours de la diagenése.* PhD Thesis, University of Paris, France.

Lanson, B. & Besson, G. (1992) Characterization of the end of smectite-to-illite transformation: decomposition of X-ray patterns. *Clays and Clay Minerals* **40**, 40–52.

Lanson, B. & Champion, D. (1991) The I/S-to-illite reaction in the late stage diagenesis. *American Journal of Science* **291**, 473–596.

Lanson, B. & Velde, B. (1992) Decomposition of X-ray diffraction patterns: a convenient way to describe complex I/S diagenetic evolution. *Clays and Clay Minerals* **40**, 629–643.

Laverne, C. (1993) Occurrence of siderite and ankerite in young basalts from the Galapagos Spreading Center (DSDP Holes 605G and 507B). *Chemical Geology* **106**, 27–46.

Laverne, C. & Vivier, G. (1983) Petrographical and chemical study of basement from the Galapagos Spreading Center, Leg 70. In: *Initial Reports DSDP, 70* (eds R. P. Von Herzen & J. Honnorez), pp. 375–390. US Government Printing Office, Washington, DC.

Laverne, C., Vanko, D.A., Tartarotti, P. & Alt, J.C. (1995) Chemistry and geothermometry of secondary minerals from the deep sheeted dike complex, DSDP/ODP Hole 504B. In: *Proceedings of ODP, Scientific Results, 140* (eds H. J. B. Dick, J. Erzinger & L. Stokking), pp. 167–190. Ocean Drilling Program, College Station, TX.

Lawrence, J.R. (1979) Temperatures of formation of calcite veins in the basalts from DSDP Holes 417A and 417D. In: *Initial Reports DSDP 51–53* (eds T. Donnelly, J. Francheteau, W. Bryan *et al.*), pp. 1183–1184. US Government Printing Office, Washington, DC.

Lawrence, J.R. (1991) Stable isotopic composition of pore waters and calcite veins. In: *Proceedings of ODP, Scientific Results 121* (eds J. Weissel, J. Pierce, E. Taylor & J. Alt), pp. 1–6. Ocean Drilling Program, College Station, TX.

Lawrence, J.R. & Drever, J.I. (1981) Evidence for cold water circulation at DSDP Site 395: isotopes and chemistry of alteration products. *Journal of Geophysical Research* **86**, 5125–5133.

Lawrence, J.R. & Taylor, H.P. (1972) Hydrogen and oxygen isotope systematics in weathering profiles. *Geochimica et Cosmochimica Acta* **36**, 1377–1393.

Lecuyer, C., Gruau, G., Frueh-Green, G.L. & Picard, C. (1996) Hydrogen isotope composition of early Proterozoic seawater. *Geology* **24**, 291–294.

Lee, H.J., Ahn, J.H. & Peacor, D.R. (1985) Textures in layered silicates: progressive changes through diagenesis and low-temperature metamorphism. *Journal of Sedimentary Petrology* **55**, 532–540.

Lee, H.J., Peacor, D.R., Lewis, D.D. & Wintsch, R.P. (1986) Evidence for syntectonic crystallization for the mudstone to slate transition at Lehigh Gap, Pennsylvania, U.S.A. *Journal of Structural Geology* **8**, 767–780.

Lee, M. & Savin, S.M. (1985) Isolation of diagenetic overgrowths on quartz sand grains for oxygen isotopic analysis. *Geochimica et Cosmochimica Acta* **49**, 497–501.

Lee, M.K. (1986) A new gravity survey of the Lake District and three-dimensional model of the granite batholith. *Journal of the Geological Society, London* **143**, 425–436.

Lee, T., Mayeda, T.K. & Clayton, R.N. (1980) Oxygen isotopic anomalies in Allende inclusion HAL. *Geophysical Research Letters* **7**, 493–496.

Lee, Y.I. (1987) Isotopic aspects of thermal and burial diagenesis of sandstones at DSDP Site 445, Daito Ridge, Northwest Pacific Ocean. *Chemical Geology* **65**, 95–102.

Leggett, J.K., McKerrow, W.S. & Eales, M.H. (1979) The Southern Uplands of Scotland: a Lower Palaeozoic accretionary prism. *Journal of the Geological Society, London* **136**, 755–770.

Lelkes-Felvári, G., Árkai, P. & Sassi, F.P. (1996) Main features of the regional metamorphic events in Hungary: a review. *Geologica Carpathica* **47**, 257–270.

LeMaitre, R.W. (1976) The chemical variability of some common igneous rocks. *Journal of Petrology* **17**, 589–637.

Leveridge, B.E., Holder, M.T., Goode, A.J.J., Scivener, R.C., Jones, N.S. & Merriman, R.J. (1998) *The Plymouth and south-east Cornwall area—a concise account of the geology.* Memoir of the British Geological Survey, Sheets 348 (England and Wales), in press.

Levi, B. (1969) Burial metamorphism of a Cretaceous volcanic sequence west from Santiago, Chile. *Contributions to Mineralogy and Petrology* **24**, 30–49.

Levi, B., Aguirre, L. & Nyström, J.O. (1982) Metamorphic gradients in burial metamorphosed vesicular lavas: comparison of basalt and spilite in Cretaceous basic flows from central Chile. *Contributions to Mineralogy and Petrology* **80**, 49–58.

Levi, B., Aguirre, L., Nyström, J.O., Padilla, H. & Vergara, M. (1989) Low-grade regional metamorphism in the Mesozoic–Cenozoic volcanic sequences of the Central Andes. *Journal of Metamorphic Geology* **7**, 487–495.

Lezzerini, M., Sartori, R. & Tamponi, M. (1995) Effect of amount of material used on sedimentation slides in the control of illite 'crystallinity' measurements. *European Journal of Mineralogy* **7**, 819–823.

Li, G., Peacor, D.R., Merriman, R.J., Roberts, B. & Van der Pluijm, B.A. (1994a) TEM and AEM constraints on the origin and significance of chlorite-mica stacks in slates: an example from Central Wales, U.K. *Journal of Structural Geology* **16**, 1139–1157.

Li, G., Peacor, D.R., Merriman, R.J. & Roberts, B. (1994b) The diagenetic to low grade metamorphic evolution of matrix white micas in the system muscovite–paragonite in a mudrock from Central Wales, U.K. *Clays and Clay Minerals* **42**, 369–381.

Li, G., Peacor, D.R. & Coombs, D.S. (1997) Transformation of smectite to illite in bentonite and associated sediments from Kaka Point, New Zealand: contrast in rate and mechanism. *Clays and Clay Minerals* **45**, 54–67.

Li, G., Peacor, D.R., Buseck, P.R. & Árkai, P. (1998) Modification of illite–muscovite crystallite size distributions by sample preparation for powder XRD analysis. *Canadian Mineralogist*, in press.

Lichtenstein, U. & Hoernes, S. (1992) Oxygen isotope fractionation between grossular-spessartine garnet and water: an experimental investigation. *European Journal of Mineralogy* **4**, 239–249.

Liewig, N., Clauer, N. & Sommer, F. (1987) Rb–Sr and K–Ar dating of clay diagenesis in Jurassic sandstone reservoirs, North Sea. *American Association of Petroleum Geologists Bulletin* **71**, 1467–1474.

Lindgreen, H. & Hansen, P.L. (1991) Ordering of illite–smectite in upper Jurassic claystones from the North Sea. *Clay Minerals* **26**, 105–125.

Lindgreen, H., Garneaes, J., Besenbacher, F., Laegsgaard, E. & Stensgaard, I. (1992) Illite–smectite from the North Sea investigated by scanning tunnelling microscopy. *Clay Minerals* **27**, 331–342.

Lindqvist, J.-E. & Andréasson, P.-G. (1987) Illite crystallinity and prograde metamorphism in thrust zones of the Scandinavian Caledonides. *Sciences Géologiques Bulletin* **40**, 217–230.

Lindsley, D.H. & Anderson, D.J. (1983) A two-pyroxene thermometer. *Journal of Geophysical Research* **88**, A887–A906.

Lintern, B.C. & Floyd, J.D. (1997) *The Kirkcudbright-Dalbeattie district—a concise account of the geology.* Memoir of the British Geological Survey, Sheets 5W, 5E and part of 6W (Scotland).

Liou, J.G. (1979) Zeolite facies metamorphism of basaltic rocks from the east Taiwan ophiolite. *American Mineralogist* **64**, 1–14.

Liou, J.G. & Ernst, W.G. (1979) Ocean ridge metamorphism of the Taiwan ophiolite. *Contributions to Mineralogy and Petrology* **68**, 335–348.

Liou, J.G., Kim, H.S. & Maruyama, S. (1983) Prehnite-epidote equilibria and their petrologic applications. *Journal of Petrology* **24**, 321–342.

Liou, J.G., Maruyama, S. & Cho, M. (1985a) Phase equilibria and mineral paragenesis of metabasites in low-grade metamorphism. *Mineralogical Magazine* **49**, 321–333.

Liou, J.G., Seki, Y., Guillemette, R.N. & Sakai, H. (1985b) Compositions and parageneses of secondary minerals in the Onikobe geothermal system, Japan. *Chemical Geology* **49**, 1–20.

Liou, J.G., Maruyama, S. & Cho, M. (1987) Very low-grade metamorphism of volcanic and volcaniclastic rocks—mineral assemblages and mineral facies. In: *Low Temperature Metamorphism* (ed. M. Frey), pp. 59–113. Blackie & Son, Glasgow.

Liou, J.G., de Capitani, C. & Frey, M. (1991) Zeolite equilibria in the system $CaAl_2Si_2O_8$-$NaAlSi_3O_8$-SiO_2-H_2O. *New Zealand Journal of Geology and Geophysics* **34**, 293–301.

Lippmann, F. (1981) Stability diagrams involving clay minerals. In: *8th Conference on Clay Mineralogy and Petrology Teplice, 1979* (ed. J. Konata), pp. 153–171. University of Karlova, Prague, Czechoslovakia.

Lippmann, F. (1982) The thermodynamic status of clay minerals. In: *Proceedings of the 7th International Clay*

Conference, Bologna, Pavia, 1981 (eds H. van Olphen & F. Veniale), pp. 475–485. Elsevier, New York.

Lister, C.R.B. (1972) On the thermal balance of a mid-ocean ridge. *Geophysical Journal of the Royal Astronomical Society* **26**, 515–535.

Lister, C.R.B. (1982) 'Active' and 'passive' hydrothermal systems in the ocean crust. Predicted physical conditions. In: *The Dynamic Environment of the Ocean Floor* (eds K. A. Fanning & F. T. Manheim), pp. 441–470. D. C. Heath, Lexington, MA.

Livi, K.J.T., Ferry, J.M., Veblen, D.R. & Frey, M. (1997a) Reactions and physical conditions during metamorphism of the Liassic aluminous black shales, Central Switzerland. *Journal of Metamorphic Geology* **15**, 323–344.

Livi, K.J.T., Veblen, D.R., Ferry, J.M. & Frey, M. (1997b) Evolution of 2:1 layered silicates in low-grade metamorphosed Liassic shales of Central Switzerland. *Journal of Metamorphic Geology* **15**, 323–344.

Longstaffe, F.J. (1983) Stable isotope studies of diagenesis in clastic rocks. *Geosciences Canada* **10**, 43–58.

Longstaffe, F.J. (1987) Stable isotope studies of diagenetic processes. In: *Short Course in Stable Isotope Geochemistry of Low Temperature Processes* (ed. T. Kyser), pp. 187–257. Mineralogical Society of Canada, Saskatoon, SK.

Longstaffe, F.J. (1989) Stable isotopes as tracers in clastic diagenesis. In: *Short Course in Burial Diagenesis* (ed. I. E. Hutcheon), pp. 201–277. Mineralogical Association of Canada, Montreal.

Longstaffe, F.J. & Ayalon, A. (1987) Oxygen-isotope studies of clastic diagenesis in the Lower Cretaceous Viking Formation, Alberta; implications for the role of meteoric water. In: *Diagenesis of Sedimentary Sequences* (ed. J. D. Marshall), pp. 277–296. Geological Society Special Publication, Liverpool.

Longstaffe, F.J. & Schwarcz, H.P. (1977) $^{18}O/^{16}O$ of Archean clastic metasedimentary rocks: a petrogenetic indicator for Archean gneisses? *Geochimica et Cosmochimica Acta* **41**, 1303–1312.

Lonker, S.W. & Fitzgerald, J.D. (1990) Formation of coexisting 1M and 2M polytypes in illite from an active hydrothermal system. *American Mineralogist* **75**, 1282–1289.

Lonker, S.W., Fitzgerald, J.D., Hedenquist, J.W. & Walshe, J. (1990) Mineral–fluid interactions in the Broadlands-Ohaaki geothermal system, New Zealand. *American Journal of Science* **290**, 995–1068.

Loucks, R.R. (1992) The bound interlayer H_2O content of potassic white micas: muscovite–hydromuscovite–hydropyrophyllite solutions. *American Mineralogist* **76**, 1563–1579.

Ludden, J.N. & Thompson, G. (1979) An evaluation of the behavior of the rare earth elements during the weathering of sea floor basalt. *Earth and Planetary Science Letters* **43**, 85–92.

Lundberg, N. & Moore, J.C. (1986) Macroscopic structural features in Deep Sea Drilling Project cores from forearc regions. In: *Structural Fabrics in Deep Sea Drilling Project Cores from Forearcs* (ed. J. C. Moore). *Geological Society of America Memoir* **166**, 13–44.

Macaulay, C.I., Fallick, A.E. & Haszeldine, R.S. (1993a) Textural and isotopic variations in diagenetic kaolinite from the Magnus oilfield sandstones. *Clay Minerals* **28**, 625–639.

Macaulay, C.I., Haszeldine, R.S. & Fallick, A.E. (1993b) Distribution, chemistry, isotopic composition and origin of diagenetic carbonates; Magnus Sandstone, North Sea. *Journal of Sedimentary Petrology* **63**, 33–43.

MacGregor, I.D. & Manton, W.I. (1986) Roberts Victor eclogites; ancient oceanic crust. *Journal of Geophysical Research B* **91**, 14063–14079.

Maekawa, H., Shozui, M., Ishii, T., Saboda, K.L. & Ogawa, Y. (1992) Metamorphic rocks from the serpentinite seamounts in the Marina and Izu-Ogasawara forearcs. In: *Proceedings of ODP, Scientific Results 125* (eds P. Fryer, J. Pearce & L. Stokking), pp. 415–430. Ocean Drilling Program, College Station, TX.

Magaritz, M. & Taylor, H.P. (1976) Oxygen, hydrogen and carbon isotope studies of the Franciscan Formation, Coast Ranges, California. *Geochimica et Cosmochimica Acta* **40**, 215–234.

Manning, C.E. & Bird, D.K. (1995) Porosity, permeability and basalt metamorphism. In: *Low-Grade Metamorphism of Mafic Rocks* (eds P. Schiffman & H. W. Day). Geological Society of America, Boulder, CO. Special Paper No. 296, pp. 123–140.

Manning, C.E., Weston, P.E. & Mahon, K.I. (1996) Rapid high-temperature metamorphism of the East Pacific Rise gabbros from Hess Deep. *Earth and Planetary Science Letters* **144**, 123–132.

Marumo, K., Longstaffe, F.J. & Matsubaya, O. (1995) Stable isotope geochemistry of clay minerals from fossil and active hydrothermal systems, southwestern Hokkaido, Japan. *Geochimica et Cosmochimica Acta* **59**, 2545–2559.

Maruyama, S. & Liou, J.G. (1985) The stability of Ca–Na pyroxene in low grade metabasites of high pressure intermediate facies series. *American Mineralogist* **70**, 16–29.

Maruyama, S., Suzuki, K. & Liou, J.G. (1983) Greenschist-amphibolite transition at low pressures. *Journal of Petrology* **24**, 583–604.

Maruyama, S., Liou, J.G. & Terabayashi, M. (1996) Blueschists and eclogites of the world and their exhumation. *International Geological Review* **38**, 485–594.

Massonne, H.J. & Schreyer, W. (1987) Phengite geobarometry based on the limiting assemblage with K-feldspar, phlogopite and quartz. *Contributions to Mineralogy and Petrology* **96**, 212–224.

Masuda, H., Sakai, H., Matsuhisa, Y. & Nakamura, T.

(1986) Stable isotopic and mineralogical studies of hydrothermal alteration at Arima Spa, Japan. *Geochimica et Cosmochimica Acta* **50**, 19–28.

Masuda, H., Kusakabe, M. & Sakai, H. (1992) Hydrogen and oxygen isotope ratios of shales and characteristics of formation waters in sedimentary complexes accreted at different times, Kinki District, Southwest Japan. *Geochimica et Cosmochimica Acta* **56**, 3505–3511.

Masuda, H., O'Neil, J.R., Jiang, W.-T. & Peacor, D.R. (1996) Relation between interlayer composition of authigenic smectite, mineral assemblages, I/S reaction rate and fluid composition in silicic ash of the Nankai Trough. *Clays and Clay Minerals* **44**, 460–469.

Matheney, R.K. & Knauth, L.P. (1989) Oxygen-isotope fractionation between marine biogenic silica and seawater. *Geochimica et Cosmochimica Acta* **53**, 3207–3214.

Matthews, A., Beckinsale, R.D. & Durham, J.J. (1979) Oxygen isotope fractionation between rutile and water and geothermometry of metamorphic eclogites. *Mineralogical Magazine* **43**, 405–413.

Matthews, D.E. & Hayes, J.M. (1978) Isotope-ratio-monitoring gas chromatography-mass spectrometry. *Analytical Chemistry* **50**, 1465–1473.

Maxwell, D.T. & Hower, J. (1967) High grade diagenesis and low-grade metamorphism of illite in the Precambrian Belt series. *American Mineralogist* **52**, 843–857.

May, H.M., Kinniburgh, D.G., Helmke, P.A. & Jackson, M.L. (1986) Aqueous dissolution, solubilities and thermodynamic stabilities of common aluminosilicate clay minerals: kaolinite and smectites. *Geochimica et Cosmochimica Acta* **50**, 1667–1677.

McConville, P., Kelley, S.P. & Turner, S.P. (1988) Laser probe ^{40}Ar–^{39}Ar studies of the Peace River shocked L6 chondrite. *Geochimica et Cosmochimica Acta* **52**, 2487–2499.

McCrea, J.M. (1950) On the isotopic chemistry of carbonates and a paleotemperature scale. *Journal of Chemical Physics* **18**, 849–857.

McDougall, I. & Harrison, T.M. (1988) *Geochronology and Thermochronology by the $^{40}Ar/^{39}Ar$ Method*. Oxford University Press, Oxford, New York.

McDowell, S.D. & Elders, W.A. (1980) Authigenic layer silicate minerals in borehole Elmore 1, Salton Sea geothermal field, California, USA. *Contributions to Mineralogy and Petrology* **74**, 293–310.

McKenzie, D.P. (1981) The variation of temperature with time and hydrocarbon maturation in sedimentary basins formed by extension. *Earth and Planetary Science Letters* **40**, 25–32.

McKerrow, W.S., Leggett, J.K. & Eales, M.H. (1977) Imbricate thrust model of the Southern Uplands of Scotland. *Nature* **267**, 237–239.

McKinney, C.R., McCrea, J.M., Epstein, S., Allen, H.A. & Urey, H.C. (1950) Improvements in mass spectrometers for the measurement of small differences in isotope abundance ratios. *Review of Scientific Instruments* **21**, 724–730.

McMillan, A.A. (1998) *A concise account of the geology around New Galloway and Thornhill*. Memoir of the British Geological Survey, sheets 9W and 9E (Scotland), in press.

McPowell, A. (1979) A morphological classification of rock cleavage. *Tectonophysics* **58**, 21–34.

Mehegan, J.M., Robinson, P.T. & Delaney, J.R. (1982) Secondary mineralization and hydrothermal alteration in the Reydarfjordur drill core, eastern Iceland. *Journal of Geophysical Research* **87**, 6511–6524.

Melson, W.G. & Thompson, G. (1973) Glassy abyssal basalts, Atlantic sea floor near St. Paul's Rocks: petrography and composition of secondary clay minerals. *Geological Society of America Bulletin* **84**, 703–716.

Merrihue, C.M. & Turner, G. (1966) Potassium–argon dating by activation with fast neutrons. *Journal of Geophysical Research* **71**, 2852–2857.

Merriman, R.J. & Kemp, S.J. (1996) Clay minerals and sedimentary basin maturity. *Mineralogical Society Bulletin* **111**, 7–8.

Merriman, R.J. & Roberts, B. (1985) A survey of white mica crystallinity and polytypes in pelitic rocks of Snowdonia and Llyn, N. Wales. *Mineralogical Magazine* **49**, 305–319.

Merriman, R.J. & Roberts, B. (1990) Metabentonites in the Moffat Shale Group, Southern Uplands of Scotland: geochemical evidence of ensialic marginal basin volcanism. *Geological Magazine* **127**, 259–271.

Merriman, R.J. & Roberts, B. (1995) Low-grade metamorphism of the Lower Palaeozoic sequence. In: *The Geology of the Rhins of Galloway District* (ed. P. Stone), pp. 67–70. Memoir of the British Geological Survey, sheets 1 and 3 (Scotland).

Merriman, R.J. & Roberts, B. (1996) Metamorphism of the Lower Palaeozoic rocks. In: *Geology in South-West Scotland: An Excursion Guide* (ed. P. Stone). British Geological Survey, Keyworth, Nottingham.

Merriman, R.J., Bevins, R.E. & Ball, T.K. (1987) Petrological and geochemical variations within the Tal y Fan intrusion; a study of element mobility during low-grade metamorphism with implications for petrotectonic modelling. *Journal of Petrology* **27**, 1409–1436.

Merriman, R.J., Roberts, B. & Peacor, D.R. (1990) A transmission electron microscope study of white mica crystallite size distribution in a mudstone to slate transitional sequence, North Wales, U.K. *Contributions to Mineralogy and Petrology* **106**, 27–40.

Merriman, R.J., Roberts, B. & Hirons, S.R. (1992) *Regional low grade metamorphism in the central part of the Lower Palaeozoic Welsh Basin — an account of the Llanilar and Rhayader Districts, BGS 1:50K sheets 178 and 179*. British Geological Survey Technical Report, WG/92/16.

Merriman, R.J., Pharaoh, T.C., Woodcock, N.H. & Daly, P. (1993) The metamorphic history of the concealed Caledonides of eastern England and their foreland. *Geological Magazine* **130**, 613–620.

Merriman, R.J., Rex, D.C., Soper, N.J. & Peacor, D.R. (1995a) The age of Acadian cleavage in northern England, U.K.: K–Ar and TEM analysis of a Silurian metabentonite. *Proceedings of the Yorkshire Geological Society* **50**, 255–265.

Merriman, R.J., Roberts, B., Peacor, D.R. & Hirons, S.R. (1995b) Strain-related differences in the crystal growth of white mica and chlorite: a TEM and XRD study of the development of metapelite microfabrics in the Southern Uplands thrust terrane, Scotland. *Journal of Metamorphic Geology* **13**, 559–576.

Merritt, D.A., Freeman, K.H., Ricci, M.P., Studley, S.A. & Hayes, J.M. (1995) Isotope-ratio-monitoring gas chromatography-mass spectrometry: optimization of the combustion interface and evaluation of analytical performance. *Analytical Chemistry* **67**, 2461–2473.

Metcalfe, R., Bevins, R.E. & Robinson, D. (1994) Theoretical constraints on calcium silicate domain formation and fluid flow during alteration of basic igneous rocks. *Mineralogical Magazine* **58A**, 601–602.

Meunier, A., Inoue, A. & Beaufort, D. (1991) Chemiographic analysis of trioctahedral smectite to chlorite conversion series from the Ohyu caldera, Japan. *Clays and Clay Minerals* **39**, 409–415.

Mevel, C. (1981) Occurrence of pumpellyite in hydrothermally altered basalts from the Vena fracture zone. *Contributions to Mineralogy and Petrology* **76**, 386–393.

Mevel, C. & Cannut, M. (1991) Lithospheric stretching and hydrothermal processes in oceanic gabbros from slow-spreading ridges. In: *Ophiolite Genesis and Evolution of the Oceanic Lithosphere* (eds T. Peters, A. Nicholas & R. G. Coleman), pp. 293–312. Kluwer Academic, Dordrecht.

Mevel, C., Gillis, K.M., Allan, J.F. & Meyer, P.S. (1996) *Proceedings of ODP, Scientific Results, 147*. Ocean Drilling Program, College Station, TX.

Miller, W., Alexander, R., Chapman, N., McKinley, I. & Smellie, J. (1994) *Natural Analogue Studies in the Geological Disposal of Radioactive Wastes. Studies in Environmental Science 57*. Elsevier.

Milliken, K.L., Land, L.S. & Loucks, R.G. (1981) History of burial diagenesis determined from isotopic geochemistry, Frio Formation, Brazoria County, TX. *AAPG Bulletin* **65**, 1397–1413.

Milnes, A.G. & Pfiffner, O.A. (1977) Structural development of the infrahelvetic complex, eastern Switzerland. *Eclogae Geologica Helvetica* **70**, 83–95.

Milodowski, A.E. & Zalasiewicz, J.A. (1991) The origin, depositional and prograde evolution of chlorite–mica stacks in Llandovery sediments of the central Wales Basin. *Geological Magazine* **128**, 263–278.

Miyashiro, A. (1961) Evolution of metamorphic belts. *Journal of Petrology* **2**, 277–311.

Miyashiro, A., Shido, F. & Ewing, M. (1971) Metamorphism in the Mid-Atlantic Ridge near 24° and 30°N. *Philosophical Transactions of the Royal Society of London* **A268**, 589–603.

Mohr, D.W., Fritz, S.J. & Eckert, J.O. (1990) Estimation of elemental microvariation within minerals analyzed by the microprobe: use of model population estimates. *American Mineralogist* **75**, 1406–1414.

Molyneux, S.G. (1979) New evidence for the age of the Manx Group, Isle of Man. In: *The Caledonides of the British Isles—Reviewed* (eds A. J. Harris, C. H. Holland & B. E. Leake). *Geological Society London, Special Publication* **8**, 415–422.

Moorbath, S., Stewart, A.D., Lawson, D.E. & Williams, G.E. (1967) Geochronological studies on the Torridonian sediments of north-west Scotland. *Scottish Journal of Geology* **3**, 389–412.

Moorbath, S., Taylor, P.N., Orpen, J.L., Treloar, P. & Wilson, J.F. (1987) First direct radiometric dating of Archean stromatolitic limestone. *Nature* **326**, 865–867.

Moore, D.M. & Reynolds, R.C. (1989) *X-ray Diffraction and Identification of Clay Minerals*. Oxford University Press, Oxford.

Moore, J.C., Roeske, S., Lundberg, N. *et al.* (1986) Scaly fabrics from Deep Sea Drilling project cores from forearcs. In: *Structural Fabrics in Deep Sea Drilling Project Cores from Forearcs* (ed. J. C. Moore). *Geological Society of America Memoir* **166**, 55–73.

Moore, J.C., Diebold, J., Fisher, M.A. *et al.* (1991) EDGE deep seismic reflection transect of the eastern Aleutian arc-trench layered lower crust reveals underplating and continental growth. *Geology* **19**, 420–424.

Morata, D., Vergara, M., Aguirre, L., Cembrano, J. & Puga, E. (1997) Primary rock composition as a controlling factor in the genesis of low-grade metamorphic minerals in Cretaceous metabasites from central Chile. *Terra Abstracts* **9**, 578.

Morrison, J. & Valley, J.W. (1988) Post-granulite facies fluid infiltration in the Adirondack Mountains. *Geology* **16**, 513–516.

Morrison, J. & Valley, J.W. (1991) Retrograde fluids in granulites; stable isotope evidence of fluid migration. *Journal of Geology* **99**, 559–570.

Morse, J.S. & Casey, W.H. (1988) Ostwald processes and mineral paragenesis in sediments. *American Journal of Science* **288**, 537–560.

Morton, J.L. & Sleep, N.H. (1985) A mid-ocean ridge thermal model: constraints on the volume of axial hydrothermal heat flux. *Journal of Geophysical Research* **90**, 11345–11353.

Mottl, M.J. (1992) Pore waters from serpentinite seamounts in the Mariana and Izu-Bonin forearcs, Leg 125: evidence for volatiles from the subducting slab. In: *Proceedings of ODP, Scientific Results 125* (eds P. Fryer,

J. Pearce & L. Stokking), pp. 373–386. Ocean Drilling Program, College Station, TX.

Mottl, M.J. & Wheat, C.G. (1994) Hydrothermal circulation through mid-ocean ridge flanks: fluxes of heat and magnesium. *Geochimica et Cosmochimica Acta* **58**, 2225–2237.

Muehlenbachs, K. (1979) The alteration and aging of the basaltic layer of the seafloor: oxygen isotope evidence from DSDP/IPOD Legs 51, 52, and 53. In: *Initial Reports DSDP, 51–53* (eds T. Donnelly, J. Francheteau, W. Bryan, P. Robinson, M. Flower & M. Salisbury), pp. 1159–1167. US Government Printing Office, Washington, DC.

Muehlenbachs, K. (1987) Oxygen isotope exchange during weathering and low temperature alteration. In: *Stable Isotope Geochemistry of Low Temperature Processes* (ed. T. K. Kyser), pp. 187–257. Mineralogical Society of Canada, Saskatoon, SK.

Muehlenbachs, K. & Clayton, R.N. (1972) Oxygen isotope studies of fresh and weathered submarine basalts. *Canadian Journal of Earth Science* **9**, 172–184.

Muehlenbachs, K. & Clayton, R.N. (1976) Oxygen isotope composition of the oceanic crust and its bearing on seawater. *Journal of Geophysical Research* **81**, 4365–4369.

Mullis, J. (1979) The system methane–water as a geologic thermometer and barometer from the external part of the Central Alps. *Bulletin de Minéralogie* **102**, 526–536.

Mullis, J. (1987) Fluid inclusion studies during very low grade metamorphism. In: *Low Temperature Metamorphism* (ed. M. Frey), pp. 162–199. Blackie & Son, Glasgow and London.

Mullis, J., Stern, W.B., de Capitani, C. *et al.* (1993) Correlation of fluid inclusion temperatures with illite, smectite and chlorite 'crystallinity' data and smear slide chemistry in sedimentary rocks from the external parts of the Central Alps (Switzerland). In: *IGCP Project 294, Very Low-Grade Metamorphism Symposium*, November 1993, Santiago de Chile.

Murata, K.J., Friedman, I. & Gleason, J.D. (1977) Oxygen isotope relations between diagenetic silica minerals in Monterey Shale, Temblor Range, California. *American Journal of Science* **277**, 259–272.

Murphey, M.B. & Nier, A.O. (1941) Variations in the relative abundances of the carbon isotopes. *Physical Review* **59**, 771–772.

Nabelek, P.I., O'Neil, J.R. & Papike, J.J. (1983) Vapor phase exsolution as a controlling factor in hydrogen isotope variation in granitic rocks; the Notch Peak granitic stock, Utah. *Earth and Planetary Science Letters* **66**, 137–150.

Nabelek, P.I., Labotka, T.C., O'Neil, J.R. & Papike, J.J. (1984) Contrasting fluid/rock interaction between the Notch Peak granitic intrusion and argillites and limestones in western Utah; evidence from stable isotopes and phase assemblages. *Contributions to Mineralogy and Petrology* **86**, 25–34.

Nadeau, P.H. (1985) The physical dimensions of fundamental clay particles. *Clay Minerals* **20**, 499–514.

Nadeau, P.H. (1998) The fundamental particle model: a clay mineral paradigm. In: *Proceedings of the 11th International Clay Conference*, Ottawa, Canada (in press).

Nadeau, P.H. & Bain, D.C. (1986) Composition of some smectites and diagenetic illitic clays and implications for their origin. *Clays and Clay Minerals* **34**, 455–464.

Nadeau, P.H. & Reynolds, R.C. (1981) Burial and contact metamorphism in the Mancos Shale. *Clays and Clay Minerals* **29**, 249–259.

Nadeau, P.H., Tait, J.M., McHardy, W.J. & Wilson, M.J. (1984a) Interstratified XRD characteristics of physical mixtures of elementary clay particles. *Clay Minerals* **19**, 67–76.

Nadeau, P.H., Wilson, M.J., McHardy, W.J. & Tait, J.M. (1984b) Interstratified clays as fundamental particles. *Science* **225**, 923–925.

Nadeau, P.H., Wilson, M.J., McHardy, W.J. & Tait, J.M. (1984c) Interparticle diffraction: a new concept for interstratified clays. *Clay Minerals* **19**, 757–769.

Naeser, C.W. (1979) Thermal history of sedimentary basins: fission-track dating of subsurface rocks. In: *Aspects of Diagenesis* (eds P. A. Scholle & P. R. Schluger). *Society of Economic Paleontologists and Mineralogists Special Publication* **26**, 109–112.

Naeser, N.D. & McCulloh, T.H. (eds) (1989) *Thermal History of Sedimentary Basins. Methods and Case Histories.* Springer-Verlag, Berlin, Heidelberg, New York.

Naeser, N.D., Naeser, C.W. & McCulloh, T.H. (1989) The application of fission-track dating to the depositional and thermal history of rocks in sedimentary basins. In: *Thermal History of Sedimentary Basins. Methods and Case Histories* (eds N. D. Naeser & T. H. McCulloh), pp. 157–180. Springer-Verlag, Berlin, Heidelberg, New York.

Natland, J. & Hekinian, R. (1981) Hydrothermal alteration of basalts and sediments at DSDP Site 456, Mariana Trough. In: *Initial Reports DSDP 60* (eds D. M. Hussong & S. Ueda), pp. 759–769. US Government Printing Office, Washington, DC.

Natland, J.H. & Mahoney, J.J. (1981) Alteration of igneous rocks at DSDP Sites 458 and 459, Mariana forearc region: relationship to basement structure. In: *Initial Reports DSDP 60* (eds D. M. Hussong & S. Ueda), pp. 769–788. US Government Printing Office, Washington, DC.

Nehlig, P. & Juteau, T. (1988) Flow porosities, permeabilities and preliminary data on fluid inclusions and fossil thermal gradients in the crustal sequence of the Sumail ophiolite (Oman). *Tectonophysics* **151**, 199–221.

Nehlig, P., Juteau, T., Bendel, V. & Cotten, J. (1994) The root zones of oceanic hydrothermal systems:

constraints from the Samail ophiolite (Oman). *Journal of Geophysical Research* **99**, 4703–4713.

Nesbitt, B.E. & Muehlenbachs, K. (1995) Geochemical studies of the origins and effects of synorogenic crustal fluids in the southern Omineca Belt of British Columbia, Canada. *Geological Society of America Bulletin* **107**, 1033–1050.

Neuhoff, P.S., Watt, W.S., Bird, D.K. & Pedersen, A.K. (1997) Timing and structural relations of regional zeolite zones in basalts of the East Greenland continental margin. *Geology* **25**, 803–806.

Nicholls, G.D. (1959) Autometasomatism in the Lower Spilites of the Builth Volcanic Series. *Quarterly Journal of the Geological Society of London* **114**, 137–162.

Nier, A.O. (1950) A redetermination of the relative abundances of the isotopes of carbon, nitrogen, oxygen, argon and potassium. *Physical Review* **77**, 789–793.

Nieto, F. & Sánchez-Navas, A. (1994) A comparative XRD and TEM study of the physical meaning of the white mica 'crystallinity' index. *European Journal of Mineralogy* **6**, 611–621.

Nieto, F., Velilla, N., Peacor, D.R. & Huertas, M.O. (1994) Regional retrograde alteration of sub-greenschist facies chlorite to smectite. *Contributions to Mineralogy and Petrology* **115**, 243–252.

Nieto, F., Ortega-Huertas, M., Peacor, D.R. & Arostegui, J. (1996) Evolution of illite/smectite from early diagenesis through incipient metamorphism in sediments of the Basque-Cantabrian Basin. *Clays and Clay Minerals* **46**, 304–323.

Norton, D. & Knupp, R. (1977) Transport phenomena in hydrothermal systems: the nature of porosity. *American Journal of Science* **277**, 913–936.

Obradovich, J.D. & Peterman, Z.E. (1968) Geochronology of the Belt series, Montana. *Canadian Journal of Earth Sciences* **5**, 737–747.

O'Brien, N.R. & Slatt, R.M. (1990) *Argillaceous Rock Atlas*. Springer-Verlag.

Odin, G.S. (1982) How to measure glaucony ages. In: *Numerical Dating in Stratigraphy* (ed. G. S. Odin), pp. 387–403. John Wiley & Sons, New York.

Odin, G.S. & Bonhomme, M.G. (1982) Argon behaviour in clays and glauconies during preheating experiments. In: *Numerical Dating in Stratigraphy* (ed. G. S. Odin), pp. 333–343. John Wiley & Sons, New York.

Oertel, G. (1983) The relationship of strain and preferred orientation of phyllosilicate grains in rocks—a review. *Tectonophysics* **100**, 413–447.

Oertel, G., Engelder, T. & Evans, K. (1989) A comparison of the strain of crinoid columns with that of their enclosing silty and shaly matrix on the Appalachian Plateau, New York. *Journal of Structural Geology* **11**, 975–993.

Offler, R., Aguirre, L., Levi, B. & Child, S. (1980) Burial metamorphism in rocks of the western Andes of Peru. *Lithos* **13**, 31–42.

Ohr, M., Halliday, A.N. & Peacor, D.R. (1994) Mobility and fractionation of rare earth elements in argillaceous sediments: implications for dating diagenesis and low-grade metamorphism. *Geochimica et Cosmochimica Acta* **58**, 289–312.

Oki, Y., Hirano, T. & Suzuki, T. (1974) Hydrothermal metamorphism and vein minerals of the Yugawara geothermal area, Japan. In: *Water–Rock Interaction*, pp. 81–94. Geological Survey, Prague.

Oliver, G.J.H. & Leggett, J.K. (1980) Metamorphism in an accretionary prism: prehnite–pumpellyite facies metamorphism of the Southern Uplands of Scotland. *Transactions of the Royal Society of Edinburgh, Earth Sciences* **71**, 235–246.

Oliver, G.J.H., Smellie, J.L., Thomas, L.J. *et al.* (1984) Early Palaeozoic metamorphic history of the Midland Valley, the Southern Uplands-Longford Down massif and the Lake District, British Isles. *Transactions of the Royal Society of Edinburgh, Earth Sciences* **75**, 259–273.

O'Neil, J.R. (1986a) Terminology and standards. In: *Stable Isotopes in High Temperature Geological Processes* (eds J. W. Valley, H. P. Taylor & J. R. O'Neil), pp. 561–570. Mineralogical Society of America, Chelsea.

O'Neil, J.R. (1986b) Theoretical and experimental aspects of isotopic fractionation. In: *Stable Isotopes in High Temperature Geological Processes* (eds J. W. Valley, H. P. Taylor & J. R. O'Neil), pp. 1–40. Mineralogical Society of America, Chelsea.

O'Neil, J.R. (1987) Preservation of H, C, and O isotopic ratios in the low temperature environment. In: *Stable Isotope Geochemistry of Low Temperature Fluids* (ed. T. K. Kyser), pp. 85–128. Mineralogical Society of America, Saskatoon, SK.

O'Neil, J.R. & Chappell, B.W. (1977) Oxygen and hydrogen isotope relations in the Berridale batholith. *Journal of the Geological Society of London* **133**, 559–571.

O'Neil, J.R. & Hay, R.L. (1973) $^{18}O/^{16}O$ ratios in cherts associated with the saline lake deposits of East Africa. *Earth and Planetary Science Letters* **19**, 257–266.

O'Neil, J.R. & Kharaka, Y.K. (1976) Hydrogen and oxygen isotope exchange reactions between clay minerals and water. *Geochimica et Cosmochimica Acta* **40**, 241–246.

O'Neil, J.R. & Taylor, H.P. (1967) The oxygen isotope and cation exchange chemistry of feldspars. *American Mineralogist* **52**, 1414–1437.

O'Neil, J.R., Clayton, R.N. & Mayeda, T.K. (1969) Oxygen isotope fractionation between divalent metal carbonates. *Journal of Chemical Physics* **51**, 5547–5558.

O'Nions, R.K., Oxburgh, E.R., Hawkesworth, C.J. & MacIntyre, R.M. (1973) New isotopic and stratigraphical evidence on the age of the Ingletonian: probable Cambrian of northern England. *Journal of the Geological Society of London* **129**, 445–452.

Ongley, J.S., Basu, A.R. & Kyser, T.K. (1987) Oxygen isotopes in coexisting garnets, clinopyroxenes and phlogopites of Roberts Victor eclogites; implications for petrogenesis and mantle metasomatism. *Earth and Planetary Science Letters* **83**, 80–84.

Osborne, M.J. & Swarbrick, R.E. (1997) Mechanisms for generating overpressure in sedimentary basins: a reevaluation. *American Association of Petroleum Geologists Bulletin* **81**, 1023–1041.

Ostwald, W.Z. (1897) Studien uber die Bildung und Umwandlung fester Körper. 1. *Abhandlung: Bersstattigung und Begrundlung Zeitung Physik Chemie* **22**, 289–330.

Ostwald, W.Z. (1900) Über die vermeintliche Isomeric des roten und gelben Quecksilberoxyds und die Oberflächenspannung fester. *Körper Zeitung Physik Chemie Stoechiom Verwandtschaftsl* **34**, 495–503.

Padan, A., Kisch, H.J. & Shagam, R. (1982) Use of the lattice parameter b_0 of dioctahedral illite/muscovite, for the characterization of P/T gradients of incipient metamorphism. *Contributions to Mineralogy and Petrology* **79**, 85–95.

Pagel, M., Braun, J.J., Disnar, J.R., Martinez, L., Renac, C. & Vasseur, G. (1997) Thermal history constraints from studies of organic matter, clay minerals, fluid inclusions, and apatite fission tracks at the Ardèche paleo-margin (BA1 drill hole, GPF Program), France. *Journal of Sedimentary Research* **67**, 235–245.

Palache, C. & Vassar, H.E. (1925) Some minerals of the Keweenawan copper deposits: pumpellyite, a new mineral; sericite; saponite. *American Mineralogist* **10**, 412.

Palmason, G., Arnorsson, S., Fridleifsson, I.B. *et al.* (1979) The Iceland crust: evidence from drillhole data on structure and processes. In: *Deep Drilling in the Atlantic Ocean, Ocean Crust* (eds M. Talwani, C. G. A. Harrison & D. E. Hayes), pp. 43–65. American Geophysical Union, Washington, DC.

Pamplin, C.F. (1990) A model for the tectono-thermal evolution of north Cornwall. *Proceedings of the Ussher Society* **7**, 206–211.

Park, K.H. & Staudigel, H. (1990) Radiogenic isotope ratios and initial seafloor alteration in submarine Serocki Volcano basalts. In: *Proceedings of ODP, Scientific Results, 106/109* (eds R. Detrick, J. Honnorez, W. B. Bryan & T. Juteau), pp. 117–122. Ocean Drilling Program, College Station, TX.

Parsons, L., Hawkins, J. & Allan, J. (1992) *Proceedings of ODP, Initial Reports 135*. Ocean Drilling Program, College Station, TX.

Passaglia, E. & Gottardi, G. (1973) Crystal chemistry and nomenclature of pumpellyites and julgoldites. *Canadian Mineralogist* **12**, 219–223.

Patrick, B.E. & Day, H.W. (1989) Controls on the first appearance of jadeitic pyroxene, northern Diablo range, California. *Journal of Metamorphic Geology* **7**, 629–639.

Pattison, D.R.M. & Tracy, R.J. (1991) Phase equilibria and thermobarometry of metapelites. In: *Contact Metamorphism* (ed. D. M. Kerrick), Mineralogical Society of America. *Reviews in Mineralogy* **26**, 105–206.

Paull, R.K., Campbell, J.D. & Coombs, D.S. (1996) New information on the age and thermal history of a probable Early Triassic siltstone near Kaka Point, South Island, New Zealand. *New Zealand Journal of Geology and Geophysics* **39**, 581–584.

Peacock, M.A. (1926) The petrology of Iceland. Part 1, the basic tuffs. *Transactions of the Royal Society of Edinburgh* **55**, 154–159.

Peacock, S.M. (1990) Fluid processes in subduction zones. *Science* **248**, 329–337.

Peacor, D.R. (1992a) Diagenesis and low-grade metamorphism of shales and slates. In: *Minerals and Reactions at the Atomic Scale: Transmission Electron Microscopy* (ed. P. R. Buseck), Mineralogical Society of America. *Reviews in Mineralogy* **27**, 335–380.

Peacor, D.R. (1992b) Analytical electron microscopy: X-ray analysis. In: *Minerals and Reactions at the Atomic Scale: Transmission Electron Microscopy*. (ed. P. R. Buseck), Mineralogical Society of America. *Reviews in Mineralogy* **27**, 113–140.

Peacor, D.R. (1998) Fundamental particles: fact or fiction? *Canadian Mineralogist*, in press.

Peresson, H. & Decker, K. (1997) Far-field effects of Late Miocene subduction in the Eastern Carpathians: E-W compression and inversion of structures in the Alpine–Carpathian–Pannonian region. *Tectonics* **16**, 38–56.

Perry, E. & Hower, J. (1970) Burial diagenesis in Gulf Coast pelitic sediments. *Clays and Clay Minerals* **18**, 167–177.

Perry, E.A. & Turekian, K.K. (1974) The effects of diagenesis on the redistribution of strontium isotopes in shales. *Geochimica et Cosmochimica Acta* **38**, 929–935.

Peterson, C., Duncan, R. & Scheidegger, K.F. (1986) The sequence and longevity of basalt alteration at Deep Sea Drilling Project Site 597. In: *Initial Reports, DSDP 92* (eds M. Leinen & D. K. Rea), pp. 491–497. US Government Printing Office, Washington, DC.

Petterson, M.G., Beddoe-Stephens, B., Millward, D. & Johnson, E.W. (1992) A pre-caldera plateau-andesite field in the Borrowdale Volcanic Group of the English Lake District. *Journal of the Geological Society, London* **149**, 889–906.

Pevear, D.R. & Vrolijk, P.J. (1996) Separation, preparation and characterization of clay minerals for unstable isotope analysis. In: *Clay Minerals Society Workshop, 33rd Annual Meeting*, pp. 4B1–4B17. June 15, Gatlinburg, TN.

Pfiffner, O.A. (1986) Evolution of the north Alpine foreland basin in the Central Alps. *Special Publication of the International Association of Sedimentologists* **8**, 219–228.

Pfiffner, O.A., Frei, W., Valasek, P. *et al.* (1990) Crustal shortening in the Alpine Orogen: results from deep seismic reflection profiling in the eastern Swiss Alps, line NFP 20-east. *Tectonophysics* **9**, 1327–1355.

Pflumio, C. (1991) Evidences for polyphased oceanic alteration of the extrusive sequence of the Semail ophiolite from the Salahi block (Northern Oman). In: *Ophiolite Genesis and Evolution of Oceanic Lithosphere* (eds T. Peters, A. Nicolas & R. G. Coleman), pp. 313–352. Kluwer.

Pharaoh, T.C., Merriman, R.J., Webb, P.C. & Beckinsale, R.D. (1987) The concealed Caledonides of eastern England: preliminary results of a multidisciplinary study. *Proceedings of the Yorkshire Geological Society* **46**, 355–369.

Pharaoh, T.C., Merriman, R.J., Evans, J.A., Brewer, T.S., Webb, P.C. & Smith, N.J.P. (1991) Early Palaeozoic arc-related volcanism in the concealed Caledonides of southern Britain. *Annales de la Société de Belgique* **114**, 63–91.

Phillips, E.R., Barnes, R.P., Boland, M.P., Fortey, N.J. & McMillan, A.A. (1995) The Moniaive Shear Zone: a major zone of sinistral strike-slip deformation in the Southern Uplands of Scotland. *Scottish Journal of Geology* **31**, 139–149.

Pickering, K.T. (1987) Deep-marine foreland basin and forearc sedimentation: a comparative study from the Lower Palaeozoic Northern Appalachians, Quebec and Newfoundland. In: *Marine Clastic Sediments* (eds J. K. Leggett & G. G. Zuffa), pp. 190–211. Graham and Trotman, London.

Pinet, N. & Tremblay, A. (1995) Is the Taconian orogeny of southern Quebec the result of an Oman-type obduction? *Geology* **23**, 121–124.

Pollastro, R.M. (1990) The illite/smectite geothermometer—concepts, methodology and application to basin history and hydrocarbon generation. In: *Applications of Thermal Maturity Studies to Energy Exploration* (eds V. F. Nuccio & C. E. Barker). SEPM 1–18.

Pollastro, R.M. (1993) Considerations of the illite–smectite geothermometer in hydrocarbon-bearing rocks of Miocene to Mississippian age. *Clays and Clay Minerals* **41**, 119–133.

Potts, P.J., Tindle, A.G. & Isaacs, M.C. (1983) On the precision of electron microprobe data: a new test for the homogeneity of mineral standards. *American Mineralogist* **68**, 1237–1242.

Powell, C.Mc.A. (1979) A morphological classification of rock cleavage. *Tectonophysics* **58**, 21–34.

Powell, R. & Holland, T.J.B. (1988) An internally consistent data set with uncertainties and correlations: 3. Applications to geobarometry, worked examples and a computer program. *Journal of Metamorphic Geology* **6**, 173–204.

Powell, W.G., Carmichael, D.M. & Hodgson, C.J. (1993)

Thermobarometry in a subgreenschist to greenschist transition in metabasites of the Abitibi greenstone belt, Superior Province, Canada. *Journal of Metamorphic Geology* **11**, 165–178.

Powers, M.C. (1959) Adjustment of clays to chemical change and the concept of the equivalence level. *Clays and Clay Minerals* **6**, 309–326.

Powers, M.C. (1967) Fluid-release mechanisms in compacting marine mudrocks and their importance in oil exploration. *American Association of Petroleum Geologists Bulletin* **51**, 1240–1254.

Preiss, W.V. (1995) Rb/Sr dating of differentiated cleavage from the Upper Adelaidean metasediments at Hallett-Cove, southern Adelaide fold belt—Discussion. *Journal of Structural Geology* **17**, 1797–1800.

Price, K.L. & McDowell, S.D. (1993) Illite/smectite geothermometry of the Proterozoic Oronto group, Midcontinent rift system. *Clays and Clay Minerals* **41**, 134–147.

Price, P.B. & Walker, R.M. (1963) Fossil tracks of charged particles in mica and the age of minerals. *Journal of Geophysical Research* **68**, 4847–4862.

Pritchard, R.G. (1979) Alteration of basalts from DSDP Legs 51, 52, and 53, Holes 417A and 418A. In: *Initial Reports, DSDP 51–53, part 2* (eds T. Donnelly, J. Francheteau, W. Bryan *et al.*), pp. 1185–1199. US Government Printing Office, Washington DC.

Pumpelly, R. (1871) The paragenesis and derivation of copper and its associates on Lake Superior. *American Journal of Science* **2**, 188–198, 243–258, 347–355.

Purdy, G.M. & Detrick, R.S. (1986) Crustal structure of the mid-atlantic ridge at 23°N from seismic refraction studies. *Journal of Geophysical Research* **91**, 3739–3762.

Ragnarsdottir, K.V., Walther, J. & Arnorsson, S. (1984) Description and interpretation of the composition of fluid and alteration mineralogy in the geothermal system at Svartsengi, Iceland. *Geochimica et Cosmochimica Acta* **48**, 1535–1553.

Rahn, M., Mullis, J., Erdelbrock, K. & Frey, M. (1994) Very low-grade metamorphism of the Taveyanne greywacke, Glarus Alps, Switzerland. *Journal of Metamorphic Geology* **12**, 625–641.

Rahn, M., Mullis, J., Erdelbrock, K. & Frey, M. (1995) Alpine metamorphism in the North Helvetic Flysch of the Glarus Alps, Switzerland. *Eclogae Geologica Helvetica* **88**, 157–178.

Rahn, M.K., Hurford, A.J. & Frey, M. (1997) Rotation and exhumation of a thrust plane: apatite fission track data from the Glarus thrust, Switzerland. *Geology* **25**, 599–602.

Ransom, B. & Helgeson, H.C. (1993) Compositional end members and thermodynamic components of illite and dioctahedral aluminous smectite solid solutions. *Clays and Clay Minerals* **41**, 537–550.

Ravenhurst, C.E. & Donelick, R.A. (1992) Fission track thermochronology. In: *Low Temperature*

Thermochronology (eds M. Zentilli & P. H. Reynolds), Mineralogical Association of Canada. *Short Course Handbook* **20**, 21–42.

Redfield, A.C. & Friedman, I. (1965) Factors affecting the distribution of deuterium in the ocean, in symposium on marine geochemistry, 1964. *Rhode Island University Narragansett Marine Laboratory Occasional Publication* **3**, 149–168.

Reed, F.R.C. (1895) The geology of the country around Fishguard. *Quarterly Journal of the Geological Society of London* **51**, 149–195.

Reed, S.J.B. (1996) *Electron Microprobe Analysis and Scanning Electron Microscopy in Geology*. Cambridge University Press, Cambridge.

Rees, C.E. (1978) Sulfur isotope measurements using SO_2 and SF_6. *Geochimica et Cosmochimica Acta* **42**, 383–390.

Rees, C.E. & Holt, B.D. (1991) The isotopic analysis of sulphur and oxygen. In: *Stable Isotopes: Natural and Anthropogenic Sulphur in the Environment* (eds H. R. Krouse & V. A. Grinenko), pp. 43–64. John Wiley & Sons, Chichester.

Reuter, A. (1987) Implications of K–Ar ages of whole-rock and grain-size fractions of metapelites and intercalated metatuffs within an anchizonal terrane. *Contributions to Mineralogy and Petrology* **97**, 105–115.

Reuter, A. & Dallmeyer, R.D. (1989) K–Ar and $^{40}Ar/^{39}Ar$ dating of cleavage formed during very low-grade metamorphism: a review. In: *Evolution of Metamorphic Belts* (eds J. S. Daly, R. A. Cliff & B. W. D. Yardley), pp. 161–172. Geological Society of London Special Publication. Blackwell, Oxford.

Reynolds, R.C. (1985) *NEWMOD™, a Computer Program for the Calculation of One-Dimensional Diffraction Patterns of Mixed-Layer Clays*.

Reynolds, R.C. (1988) Mixed layer chlorite minerals. In: *Hydrous Phyllosilicates (Exclusive of Micas)* (ed. S. W. Bailey), Mineralogical Society of America, Washington, DC. *Reviews in Mineralogy* **19**, 601–629.

Reynolds, R.C. (1992) X-ray diffraction studies of illite/smectite from rocks, <1 μm randomly oriented powders, and <1 μm oriented powder aggregates: the absence of laboratory-induced artifacts. *Clays and Clay Minerals* **40**, 387–396.

Reynolds, R.C. & Hower, J. (1970) The nature of interlayering in mixed-layer illite–montmorillonites. *Clays and Clay Minerals* **18**, 25–36.

Rice, A.N.H., Bevins, R.E., Robinson, D. & Roberts, D. (1989) Evolution of low-grade metamorphic zones in the Caledonides of Finnmark, N Norway. In: *The Caledonian Geology of Scandinavia* (ed. R. A. Gayer), pp. 177–191. Graham and Trotman, Norwell, MA.

Richards, H.G., Cann, J.R. & Jensenius, J. (1989) Mineralogical zonation and metasomatism of the alteration pipes of Cyprus sulfide deposits. *Economic Geology* **84**, 91–115.

Richardson, C.J., Cann, J.R., Richards, H.G. & Cowan, J.G. (1987) Metal-depleted root zones of the Troodos ore-forming hydrothermal systems, Cyprus. *Earth and Planetary Science Letters* **84**, 243–253.

Richardson, S.H., Hart, S.R. & Staudigel, H. (1980) Vein mineral ages of old oceanic crust. *Journal of Geophysical Research* **85**, 7195–7200.

Roberson, H.E. (1988) *Random mixed-layer chlorite–smectite: does it exist?* Abstract for Clay Mineral Society 25th Annual Meeting, Grand Rapids, MI, USA.

Roberts, B. (1979) *The Geology of Snowdonia and Llŷn*. Adam Hilger Ltd, Bristol.

Roberts, B. (1981) Low grade and very low grade regional Ordovician metabasic rocks of Llŷn and Snowdonia, Gwynedd, North Wales. *Geological Magazine* **118**, 189–200.

Roberts, B. & Merriman, R.J. (1985) The distinction between Caledonian burial and regional metamorphism in metapelites from North Wales: an analysis of isocryst patterns. *Journal of the Geological Society, London* **142**, 615–624.

Roberts, B. & Merriman, R.J. (1990) Cambrian and Ordovician metabentonites and their relevance to the origins of associated mudrocks in the northern sector of the Lower Palaeozoic Welsh marginal basin. *Geological Magazine* **127**, 31–43.

Roberts, B., Evans, J.A., Merriman, R.J. & Smith, M. (1989) Discussion on low grade metamorphism of the Welsh Basin Lower Palaeozoic succession: an example of diastathermal metamorphism? *Journal of the Geological Society, London* **146**, 885–890.

Roberts, B., Morrison, C. & Hirons, S. (1990) Low grade metamorphism of the Manx Group, Isle of Man: a comparative study of white mica 'crystallinity' techniques. *Journal of the Geological Society, London* **147**, 271–277.

Roberts, B., Merriman, R.J. & Pratt, W. (1991) The relative influences of strain, lithology and stratigraphical depth on white mica (illite) crystallinity in mudrocks from the district centred on the Corris Slate Belt, Gwynedd–Powys. *Geological Magazine* **128**, 633–645.

Roberts, B., Merriman, R.J., Hirons, S.R., Fletcher, C.J.N. & Wilson, D. (1996) Synchronous very low grade metamorphism, contraction and inversion in the central part of the Welsh Lower Palaeozoic Basin. *Journal of the Geological Society, London* **153**, 277–286.

Roberts, D. & Gee, D.G. (1985) An introduction to the structure of the Scandinavian Caledonides. In: *The Caledonian Orogen-Scandinavia and Related Areas* (eds D. A. Gee & B. A. Sturt), pp. 55–68. John Wiley, New York.

Robin, P.-Y.F. (1978) Pressure solution at grain-to-grain contacts. *Geochimica et Cosmochimica Acta* **42**, 1383–1389.

Robinson, D. (1987) Transition from diagenesis to metamorphism in extensional and collision settings. *Geology* **15**, 866–869.

Robinson, D. (1993) Pre- and post-orogenic extensional settings for burial-type metamorphism. *The Island Arc* **2**, 280–287.

Robinson, D. & Bevins, R.E. (1986) Incipient metamorphism in the Lower Palaeozoic marginal basin of Wales. *Journal of Metamorphic Geology* **4**, 101–113.

Robinson, D. & Bevins, R.E. (1989) Diastathermal (extensional) metamorphism at very low grades and possible high grade analogues. *Earth and Planetary Science Letters* **92**, 81–88.

Robinson, D. & Bevins, R.E. (1994) Mafic phyllosilicates in low-grade metabasites. Characterization using deconvolution analysis. *Clay Minerals* **29**, 223–237.

Robinson, D. & Santana de Zamora, A. (1998) The transition from smectite to chlorite in the Chipilapa geothermal system, El Salvador. *American Mineralogist*, in press.

Robinson, D., Warr, L.N. & Bevins, R.E. (1990) The illite 'crystallinity' technique: a critical appraisal of its precision. *Journal of Metamorphic Geology* **8**, 333–344.

Robinson, D., Bevins, R.E. & Rowbotham, G. (1993) The characterization of mafic phyllosilicates in low grade metabasalts from eastern North Greenland. *American Mineralogist* **78**, 377–390.

Robinson, D., Schmidt, S.Th. & Santana de Zamora, A. (1997a) The smectite to chlorite transition in Icelandic and El Salvador geothermal systems: evidence for non-equilibrium reaction pathways. In: *Clay Mineral Evolution, Basin Maturity and Mudrock Properties* (eds R. J. Merriman & S. J. Kemp), pp. 37–38. British Geological Survey, Technical Report WG/97/45.

Robinson, D., Bevins, R.E., Aguirre, L. & Vergara, M. (1997b) Mafic phyllosilicate variation with depth of burial in the Valle Nevado section of the Miocene Farellones Formation. In: *VIII Congreso Geológico Chileno, Antofagasta*, Actas Vol. II, pp. 1483–1487.

Robinson, D., Reverdatto, V.V. & Bevins, R.E. (1998a) Low grade metamorphism in the Welsh Basin: a model for burial-type metamorphism. *Journal of Geophysical Research*, in review.

Robinson, D., Schmidt, S.Th. & Santana de Zamora, A. (1998b) Non-equilibrium reaction pathways in the smectite-to-chlorite transition. *Geology*, in review.

Roden, M.K., Elliott, W.C., Aronson, J.L. & Miller, D.S. (1993) A comparison of fission-track ages of apatite and zircon to the K–Ar ages of illite–smectite (I/S) from Ordovician K-bentonites, southern Appalachian Basin. *Journal of Geology* **101**, 633–641.

Rohr, K., Mildreit, B. & Yorath, C.J. (1988) Asymmetric deep crustal structure across the Juan de Fuca Ridge. *Geology* **16**, 533–537.

Rose, N.M. (1995) Geochemical consequences of fluid flow in porous basaltic crust containing permeability contrasts. *Geochimica et Cosmochimica Acta* **59**, 4381–4392.

Rose, N.M. & Bird, D.K. (1987) Prehnite–epidote phase relations in the Nordre Aputiteq and Kruuse Fjord layered gabbros, East Greenland. *Journal of Petrology* **28**, 1193–1218.

Rose, N.M. & Bird, D.K. (1994) Hydrothermally altered dolerite dykes in East Greenland: implications for Ca-metasomatism of basaltic protoliths. *Contributions to Mineralogy and Petrology* **116**, 420–432.

Roux, J. & Hovis, G.L. (1996) Thermodynamic mixing models for muscovite–paragonite solutions based on solution calorimetric and phase equilibrium data. *Journal of Petrology* **37**, 1241–1254.

Royden, L., Sclater, J.G. & Von Herzen, R.P. (1980) Continental margin subsidence and heat flow: important parameters in formation of petroleum hydrocarbons. *American Association of Petroleum Geologists Bulletin* **64**, 173–187.

Ruiz Cruz, M.D. & Andreo, B. (1996) Genesis and transformation of dickite in Permo-Triassic sediments (Betic Cordilleras, Spain). *Clay Minerals* **31**, 133–152.

Ruiz Cruz, M.D. & Moreno Real, L. (1993) Diagenetic kaolinite/dickite (Betic Cordilleras, Spain). *Clays and Clay Minerals* **41**, 570–579.

Rumble, D.I. & Sharp, Z.D. (1998) Laser microanalysis of silicates for $^{18}O/^{17}O/^{16}O$ and of carbonates for $^{18}O/^{16}O$ and $^{13}C/^{12}C$. In: *Applications of Microanalytical Techniques to Understanding Mineralizing Processes* (eds M. A. McKibben & W. C. Shanks III). *Reviews of Economic Geology*, in press.

Saccocia, P.J. & Seyfried, W.E. (1994) The solubility of chlorite solid solutions in 3.2 wt% NaCl fluids from 300–400°C, 500 bars. *Geochimica et Cosmochimica Acta* **58**, 567–585.

Sakakibara, M., Ofuka, H., Kimura, G. *et al.* (1997) Metamorphic evolution of the Susunai metabasites in southern Sakhalin, Russian Republic. *Journal of Metamorphic Geology* **15**, 564–580.

Sample, J.C. & Moore, J.C. (1987) Structural style and kinematics of an underplated slate belt, Kodiak and adjacent islands, Alaska. *Geological Society of America Bulletin* **99**, 7–20.

Sass, B.M., Rosenberg, P.E. & Kittrick, J.A. (1987) The stability of illite/smectite during diagenesis: an experimental study. *Geochimica et Cosmochimica Acta* **51**, 2103–2115.

Sassi, F.P. (1972) The petrological and geological significance of the b_0 values of potassic white micas in low-grade metamorphic rocks. An application to the Eastern Alps. *Tschermaks Mineralogische und Petrographische Mitteilungen* **18**, 105–113.

Sassi, F.P. & Scolari, A. (1974) The b_0 value of the potassic white mica as a barometric indicator in low-grade metamorphism of pelitic schists. *Contributions to Mineralogy and Petrology* **45**, 143–152.

Sassi, R., Árkai, P., Lantai, C. & Venturini, C. (1995) Location of the boundary between the metamorphic Southalpine basement and Paleozoic sequences of the

Carnic Alps: illite 'crystallinity' and vitrinite reflectance data. *Schweizerische Mineralogische und Petrographische Mitteilungen* **75**, 399–412.

Savin, S.M. (1980) Oxygen and hydrogen isotope effects in low-temperature mineral–water interactions. In: *Handbook of Environmental Isotope Geochemistry* (eds A. P. Fritz & J. C. Fontes), pp. 283–327. Elsevier, Amsterdam.

Savin, S.M. & Epstein, S. (1970a) The oxygen and hydrogen isotope geochemistry of clay minerals. *Geochimica et Cosmochimica Acta* **34**, 35–42.

Savin, S.M. & Epstein, S. (1970b) The oxygen and hydrogen isotope geochemistry of ocean sediments and shales. *Geochimica et Cosmochimica Acta* **34**, 43–63.

Savin, S.M. & Epstein, S. (1970c) The oxygen isotopic compositions of coarse grained sedimentary rocks and minerals. *Geochimica et Cosmochimica Acta* **34**, 323–329.

Savin, S.M. & Lee, M. (1988) Isotopic studies of phyllosilicates. In: *Hydrous Phyllosilicates* (ed. S. W. Bailey), pp. 189–223. Mineralogical Society of America, Chelsea.

Savin, S.M. & Yeh, H.-W. (1981) Stable isotopes in ocean sediments. In: *The Oceanic Lithosphere* (ed. C. Emiliani), pp. 1521–1554. John Wiley & Sons, New York.

Schaltegger, U., Stille, P., Rais, N., Piqué, A. & Clauer, N. (1994) Nd and Sr isotopic dating of diagenesis and low-grade metamorphism of argillaceous sediments. *Geochimica et Cosmochimica Acta* **58**, 1471–1481.

Scheidegger, K.F. & Stakes, D.S. (1977) Mineralogy, chemistry, and crystallization sequence of clay minerals in altered tholeiitic basalts from the Peru Trench. *Earth and Planetary Science Letters* **36**, 413–422.

Schiffman, P. (1995) Low grade metamorphism of mafic rocks. *Review of Geophysics* (Suppl.), 81–86.

Schiffman, P. & Day, H.W (eds) *Low-Grade Metamorphism of Mafic Rocks.* Geological Society of America, Boulder, CO. Special Paper No. 296, 187 pp.

Schiffman, P. & Fridleifsson, G.O. (1991) The smectite–chlorite transition in drillhole NJ-15, Nesjavellir geothermal field, Iceland: XRD, BSE, and electron microprobe investigations. *Journal of Metamorphic Geology* **9**, 679–696.

Schiffman, P. & Smith, B.M. (1988) Petrology and oxygen isotope geochemistry of a fossil seawater hydrothermal system within the Solea Graben, northern Troodos Ophiolite, Cyprus. *Journal of Geophysical Research B, Solid Earth and Planets* **93**, 4612–4624.

Schiffman, P. & Southard, R.J. (1996) Cation exchange capacity of layer silicates and palagonitized glass in mafic volcanic rocks: a comparative study of bulk extraction and *in situ* techniques. *Clays and Clay Minerals* **44**, 624–634.

Schiffman, P. & Staudigel, H. (1994) Hydrothermal alteration of a seamount complex on La Palma, Canary Islands: implications for metamorphism in accreted terranes. *Geology* **22**, 151–154.

Schiffman, P. & Staudigel, H. (1995) The smectite to chlorite transition in a fossil seamount hydrothermal system: the basement complex of La Palma, Canary Islands. *Journal of Metamorphic Geology* **13**, 487–498.

Schiffman, P., Elders, W.A., Williams, A.E., McDowell, S.D. & Bird, D.K. (1984) Active metasomatism in the Cerro Prieto geothermal system, Baja California, Mexico: a telescoped low-pressure, low-temperature metamorphic facies series. *Geology* **12**, 12–15.

Schiffman, P., Smith, B.M., Varga, R.J. & Moores, E.M. (1987) Geometry, conditions and timing of off-axis hydrothermal metamorphism and ore deposition in the Solea Graben. *Nature* **325**, 423–425.

Schiffman, P., Evarts, R.C., Williams, A.E. & Pickthorn, W.P. (1991) Hydrothermal metamorphism in oceanic crust from the coast range ophiolite of California: fluid–rock interaction in a rifted arc. In: *Ophiolite Genesis and Evolution of Oceanic Lithosphere* (eds T. Peters, A. Nicolas & R. G. Coleman), pp. 399–426. Kluwer.

Schmid, S.M. (1975) The Glarus overthrust: field evidence and mechanical control. *Eclogae Geologica Helvetica* **68**, 247–280.

Schmidt, D. & Livi, K.J.T. (1998) HRTEM and SAED investigations of polytypism, stacking disorder, crystal growth, and vacancies in chlorite from subgreenschist facies outcrops. *American Mineralogist*, in review.

Schmidt, D., Livi, K.J.T. & Frey, M. (1999) Reaction progress in chloritic material: an electron microbeam study of the Taveyanne greywacke, Switzerland. *Journal of Metamorphic Geology*, in press.

Schmidt, S.T. (1990) Alteration under conditions of burial metamorphism in the North Shore Volcanic Group, Minnesota—mineralogical and geochemical zonation. *Heidelberger Geowissenschaftliche Abhandlungen* **41**, 309 pp.

Schmidt, S.Th. (1993) Regional and local patterns of low grade metamorphism in the North Shore Volcanic Group, Minnesota, USA. *Journal of Metamorphic Geology* **11**, 401–414.

Schmidt, S.Th. & Robinson, D. (1997) Metamorphic grade and porosity/permeability controls on mafic phyllosilicate distributions in a regional zeolite to greenschist facies transition of the North Shore Volcanic Group, Minnesota. *Bulletin of the Geological Society of America* **109**, 683–697.

Schmidt, S.Th. & Sharp, Z.D. (1993) Conventional and *in situ* stable isotopes of calcite and quartz in amygdules of low-grade metamorphosed basalts: evidence for non-equilibrium conditions? *Terra Abstracts* **5**, 377.

Schreinemakers. (1915–1925) In-, mono-, and divariant equilibria: Knnikl. In: *Nederlandse Akad. Wetensch. Proc.* Reprinted as papers by F. A. H. Schreinemakers, Vol. 2. The Pennsylvania State University, University Park, PA. June, 1965.

Scott, R.B. & Hajash, A. (1976) Initial submarine alteration of basaltic pillow lavas: a microprobe study. *American Journal of Science* **276**, 480–501.

Seki, Y. & Liou, J.G. (1981) Recent study of low-grade metamorphism. *Geological Society of China Memoir* **4**, 207–228.

Séranne, M. & Séguret, M. (1987) The Devonian Basins of western Norway: tectonics and kinematics of an extending crust. In: *Continental Extensional Tectonics* (eds M. P. Coward, J. F. Dewey & P. L. Hancock), pp. 537–548. Geological Society of London Special Publication No. 28.

Seyfried, W.E. (1987) Experimental and theoretical constraints on hydrothermal alteration processes at mid-ocean ridges. *Annual Review of Earth and Planetary Science* **15**, 317–335.

Seyfried, W.E. & Ding, K. (1995) Phase equilibria in seafloor hydrothermal systems: a review of the role of redox, temperature, pH and dissolved Cl on the chemistry of hot spring fluids at mid-ocean ridges. In: *Seafloor Hydrothermal Systems: Physical, Chemical, Biological and Geological Interactions* (eds S. E. Humphris, R. A. Zierenberg, L. S. Millineaux & R. E. Thompson), pp. 248–272. American Geophysical Union, Washington, DC.

Seyfried, W.E., Mottl, M.J. & Bischoff, J.L. (1978a) Seawater/basalt ratio effects on the chemistry and mineralogy of spilites from the ocean floor. *Nature* **275**, 211–213.

Seyfried, W.E., Shanks, W.C. & Dibble, W.E. (1978b) Clay mineral formation in DSDP Leg 34 basalts. *Earth and Planetary Science Letters* **41**, 265–276.

Shanin, L.L., Ivanov, I.B. & Shipulin, F.K. (1968) The possible use of alunite in K–Ar geochronology. *Geokhimyia* **1**, 109–111.

Sharp, Z.D. (1990) A laser-based microanalytical method for the *in situ* determination of oxygen isotope ratios of silicates and oxides. *Geochimica et Cosmochimica Acta* **54**, 1353–1357.

Sharp, Z.D. (1992) *In situ* laser microprobe techniques for stable isotope analysis. *Chemical Geology* **101**, 3–19.

Sharp, Z.D. (1995) Oxygen isotope geochemistry of the Al_2SiO_5 polymorphs. *American Journal of Science* **295**, 1058–1076.

Sharp, Z.D. & Cerling, T.E. (1996) A laser GC-IRMMS technique for *in situ* stable isotope analyses of carbonates and phosphates. *Geochimica et Cosmochimica Acta* **60**, 2909–2916.

Sharp, Z.D. & Kirschner, D.L. (1994) Quartz–calcite oxygen isotope thermometry; a calibration based on natural isotopic variations. *Geochimica et Cosmochimica Acta* **58**, 4491–4501.

Sharp, Z.D., Essene, E.J. & Hunziker, J.C. (1993) Stable isotope geochemistry and phase equilibria of coesite-bearing whiteschists, Dora Maira Massif, Western Alps. *Contributions to Mineralogy and Petrology* **114**, 1–12.

Sharp, Z.D., Frey, M. & Livi, K.J.T. (1995) Stable isotope variations (H, C, O) in a prograde metamorphic Triassic red bed formation, Central Swiss Alps. *Schweizerische Mineralogische und Petrographische Mitteilungen* **5**, 147–161.

Sharpe, D. (1847) On slaty cleavage. *Quarterly Journal of the Geological Society* **5**, 111–129.

Shau, Y.-H. & Peacor, D.R. (1992) Phyllosilicates in hydrothermally altered basalts from DSDP Hole 504B, Leg-83—a TEM and AEM study. *Contributions to Mineralogy and Petrology* **12**, 119–133.

Shau, Y.-H., Peacor, D.R. & Essene, E.J. (1990) Corrensite and mixed-layer chlorite/corrensite in metabasalt from Northern Taiwan: TEM/AEM, EMPA, XRD and optical studies. *Contributions to Mineralogy and Petrology* **105**, 123–142.

Shau, Y.-H., Feather, M.E., Essene, E.J. & Peacor, D.R. (1991) Genesis and solvus relations of submicroscopically intergrown paragonite and phengite in a blueschist from northern California. *Contributions to Mineralogy and Petrology* **106**, 367–378.

Sheppard, S.M.F. (1986) Characterization and isotopic variations in natural waters. In: *Stable Isotopes in High Temperature Geological Processes* (eds J. W. Valley, H. P. Taylor & J. R. O'Neil), pp. 165–183. Mineralogical Society of America, Chelsea.

Sheppard, S.M.F. & Gilg, H.A. (1996) Stable isotope geochemistry of clay minerals. *Clay Minerals* **31**, 1–24.

Sheppard, S.M.F. & Schwarcz, H.P. (1970) Fractionation of carbon and oxygen isotopes and magnesium between coexisting metamorphic calcite and dolomite. *Contributions to Mineralogy and Petrology* **26**, 161–198.

Shido, F., Miyashiro, A. & Ewing, M. (1974) Compositional variation in pillow lavas from the Mid-Atlantic Ridge. *Marine Geology* **16**, 177–190.

Shieh, Y. (1983) Oxygen isotope study of Precambrian granites from the Illinois Deep Hole Project. *Journal of Geophysical Research B* **88**, 7300–7304.

Shikazono, N. (1984) Compositional variations in epidote from geothermal areas. *Geochemical Journal* **18**, 181–187.

Shutov, V.D., Aleksandrova, A.V. & Losievskaya, S.A. (1970) Genetic interpretation of the polymorphism of the kaolinite group in sedimentary rocks. *Sedimentology* **15**, 69–82.

Siddans, A.W.B. (1972) Slaty cleavage—a review of research since 1815. *Earth Science Reviews* **8**, 205–232.

Siegenthaler, U. & Oeschger, H. (1980) Correlation of ^{18}O in precipitation with temperature and altitude. *Nature* **288**, 314–316.

Silverman, S.R. (1951) The isotope geology of oxygen. *Geochimica et Cosmochimica Acta* **2**, 26–42.

Singh, B. (1996) Why does halloysite roll?—a new model. *Clays and Clay Minerals* **44**, 191–196.

Singh, B. & Gilkes, R.J. (1992) An electron optical investigation of the alteration of kaolinite to halloysite. *Clays and Clay Minerals* **40**, 212–229.

Sinha, M.C., Navin, D.A., MacGregor, L.M. *et al.* (1997) Evidence for accumulated melt beneath the slow-spreading mid-Atlantic ridge. *Philosophical Transactions of the Royal Society of London A* **355**, 233–253.

Sinton, J.M. & Detrick, R.S. (1992) Mid-ocean ridge magma chambers. *Journal of Geophysical Research* **97**, 197–216.

Small, J.S. (1993) Experimental determination of the rates of precipitation of authigenic illite and kaolinite in the presence of aqueous oxalate and comparison to the K/Ar ages of authigenic illite in reservoir sandstones. *Clays and Clay Minerals* **41**, 191–208.

Small, J.S. (1994) Fluid composition, mineralogy and morphological changes associated with the smectite-to-illite reaction: an experimental investigation of the effect of organic acid anions. *Clay Minerals* **29**, 539–554.

Smellie, J.L., Roberts, B. & Hirons, S.R. (1996) Very low- and low-grade metamorphism in the Trinity Peninsula Group (Permo-Triassic) of northern Graham Land, Antarctic Peninsula. *Geological Magazine* **133**, 583–594.

Smith, P.E. & Farquhar, R.M. (1989) Direct dating of Phanerozoic sediments by the $^{238}U^{206}Pb$ method. *Nature* **341**, 518–521.

Smith, P.E., Farquhar, R.M. & Hancock, R.G. (1991) Direct radiometric age determination of carbonate diagenesis using U–Pb in secondary calcite. *Earth and Planetary Science Letters* **105**, 474–491.

Smith, R.E. (1968) Redistribution of major elements in the alteration of some basic lavas during burial metamorphism. *Journal of Petrology* **9**, 191–219.

Solomon, S.C. & Toomey, D.R. (1992) The structure of mid-ocean ridges. *Annual Review of Earth and Planetary Science* **20**, 329–364.

Soper, N.J. (1980) Non-metamorphic Caledonides of Great Britain. *Episodes* **1980** (1), 17–18.

Soper, N.J. & Woodcock, N.J. (1990) Silurian collision and sediment dispersal patterns in southern Britain. *Geological Magazine* **127**, 527–542.

Soper, N.J., Webb, B.C. & Woodcock, N.J. (1987) Late Caledonian (Acadian) transpression in North West England: timings, geometry and geotectonic significance. *Proceedings of the Yorkshire Geological Society* **46**, 175–192.

Sorby, H.C. (1853) On the origin of slaty cleavage. *New Philosophical Journal, Edinburgh* **55**, 137–148.

Spangenberg, J., Sharp, Z.D. & Fontboté, L. (1995) Apparent stable isotope heterogeneities in gangue carbonates of the Mississippi Valley-type Zn-Pb deposit of San Vicente, Central Peru. *Mineralium Deposita* **30**, 67–74.

Spangenberg, J.E. & Macko, S.A. (1996) Molecular and isotopic organic geochemistry of bitumens and kerogens in the Mississippi Valley-type zinc–lead district of San Vicente, Central Peru. *Geological Society of America Abstracts with Programs* **28**, A-84.

Spear, F.S. (1993) *Metamorphic Phase Equilibria and Pressure–Temperature–Time Paths.* Mineralogical Society of America Monograph, Washington, DC.

Spivack, A.J. & Staudigel, H. (1994) Low-temperature alteration of the upper oceanic crust and the alkalinity budget of seawater. *Chemical Geology* **115**, 239–247.

Spooner, E.T.C. & Fyfe, W.S. (1973) Sub-seafloor metamorphism, heat, and mass transfer. *Contributions to Mineralogy and Petrology* **42**, 278–304.

Spooner, E.T.C., Beckinsdale, R.D., England, P.C. & Senior, A. (1977) Hydration, ^{18}O-enrichment, and oxidation during ocean floor metamorphism of ophiolitic metabasic rocks from E. Liguria, Italy. *Geochimica et Cosmochimica Acta* **41**, 857–871.

Spötl, C., Houseknecht, D.W. & Jaques, R. (1993) Clay mineralogy and illite crystallinity of the Atoka Formation, Arkoma Basin, and frontal Ouachita Mountains. *Clays and Clay Minerals* **41**, 745–754.

Springer, R.K., Day, H.W. & Beiersdorfer, R.E. (1992) Prehnite–pumpellyite to greenschist facies transition, Smartville Complex, near Auburn, California. *Journal of Metamorphic Geology* **10**, 147–170.

Środoń, J. (1984) X-ray identification of illitic materials. *Clays and Clay Minerals* **32**, 337–349.

Środoń, J. & Eberl, D.D. (1984) Illite. In: *Micas* (ed. S. W. Bailey). Mineralogical Society of America. *Reviews in Mineralogy* **13**, 495–544.

Środoń, J., Morgan, D.J., Eslinger, E.V., Eberl, D.D. & Karlinger, M.R. (1986) Chemistry of illite/smectite and end-member illite. *Clays and Clay Minerals* **34**, 368–378.

Środoń, J., Elsass, W.J., McHardy, W.J. & Morgan, D.J. (1992) Chemistry of illite–smectite inferred from TEM measurements of fundamental particles. *Clay Minerals* **27**, 137–158.

Stacey, J.S. & Kramers, J.D. (1975) Approximation of terrestrial lead isotope evolution by a two-stage model. *Earth and Planetary Science Letters* **26**, 206–221.

Stakes, D.S. & O'Neil, J.R. (1982) Mineralogy and stable isotope geochemistry of hydrothermally altered oceanic rocks. *Earth and Planetary Science Letters* **57**, 285–304.

Stakes, D.S. & Scheidegger, K.F. (1981) Temporal variations in secondary minerals from Nazca Plate basalts. In: *Nazca Plate: Crustal Formation and Andean Convergence* (ed. L. D. Kulm), pp. 109–130. Geological Society of America, Boulder, CO.

Stakes, D.S. & Schiffman, P. (1996) Hydrothermal alteration of basalts in the sedimented ridge environment of Middle Valley. *EOS Transactions American Geophysical Union* **77**, F727 (abstract).

Stallard, M.L. & Boles, J.R. (1989) Oxygen isotope measurements of albite–quartz–zeolite mineral assemblages, Hokonui Hills, Southland, New Zealand. *Clays and Clay Minerals* **37**, 409–418.

Staudigel, H. & Hart, S.R. (1983) Alteration of basaltic glass: mechanisms and significance for the oceanic crust-seawater budget. *Geochimica et Cosmochimica Acta* **47**, 337–350.

Staudigel, H. & Hart, S.R. (1985) Dating of ocean crust hydrothermal alteration: strontium isotope ratios from Hole 504B carbonates and the re-interpretation of Sr isotope data from Deep Sea Drilling Project Sites 105, 332, 417, and 418. In: *Initial Reports, DSDP 83* (eds R.N. Anderson, J. Honnorez & K. Becker), pp. 297–303. US Government Printing Office, Washington, DC.

Staudigel, H., Hart, S.R. & Richardson, S.H. (1981a) Alteration of the oceanic crust: processes and timing. *Earth and Planetary Science Letters* **52**, 311–327.

Staudigel, H., Muehlenbachs, K., Richardson, S.H. & Hart, S.R. (1981b) Agents of low temperature ocean crust alteration. *Contributions to Mineralogy and Petrology* **77**, 150–157.

Staudigel, H., Gillis, K. & Duncan, R. (1986) K/Ar and Rb/Sr ages of celadonites from the Troodos ophiolite, Cyprus. *Geology* **14**, 72–75.

Staudigel, H.R., Hart, S.R., Schmincke, H.U. & Smith, B.M. (1989) Cretaceous ocean crust at DSDP Sites 417 and 418: carbon uptake from weathering versus loss by magmatic outgassing. *Geochimica et Cosmochimica Acta* **53**, 3091–309.

Staudigel, H., Davies, G.R., Hart, S.R., Marchant, K.M. & Smith, B.M. (1995) Large scale Sr, Nd, and O isotopic anatomy of altered oceanic crust: DSDP sites 417/418. *Earth and Planetary Science Letters* **130**, 169–185.

Stein, C.A. & Stein, S. (1994) Constraints on hydrothermal heat flux through the oceanic lithosphere from global heat flow. *Journal of Geophysical Research* **99**, 3081–3096.

Steiner, A. (1968) Clay minerals in hydrothermally altered rocks at Waireki, New Zealand. *Clays and Clay Minerals* **16**, 193–213.

Stern, C. & Elthon, D. (1979) Vertical variations in the effects of hydrothemal metamorphism in Chilean ophiolites: their implications for ocean floor metamorphism. *Tectonophysics* **55**, 179–213.

St Julien, P. & Hubert, C. (1975) Evolution of the Taconian orogen in the Quebec Appalachians. *American Journal of Science* **275A**, 337–362.

Stoiber, R.E. & Davidson, E.S. (1959) Amygdule mineral zoning in the Portage Lake Lava Series, Michigan copper district. *Economic Geology* **54**, 1250–1277.

Stone, P. (1995) *The Geology of the Rhins of Galloway District*. Memoir of the British Geological Survey, sheets 1and 3 (Scotland).

Stone, P., Kimbell, G.S. & Henney, P.J. (1997) Basement control on the location of strike-slip shear in the Southern Uplands of Scotland. *Journal of the Geological Society, London* **154**, 141–144.

Súcha, V., Kraus, I., Gerthorfferová, H., Petes, J. & Serekova, M. (1993) Smectite to illite conversion in bentonites and shales of the East Slovak Basin. *Clay Minerals* **28**, 243–253.

Sudo, T. & Shimoda, S. (1978) *Clays and Clay Minerals of Japan*. Elsevier, Amsterdam.

Sutter, J.F. & Hartung, J.B. (1984) Laser microprobe ^{40}Ar/^{39}Ar dating of mineral grains *in situ*. In: *Scanning Electron Microscopy*, pp. 1525–1529. Society of Economic Mineralogy Inc., Chicago.

Suzuoki, T. & Epstein, S. (1976) Hydrogen isotope fractionation between OH-bearing minerals and water. *Geochimica et Cosmochimica Acta* **40**, 1229–1240.

Swart, P.K., Burns, S.J. & Leder, J.J. (1991) Fractionation of the stable isotopes of oxygen and carbon in carbon dioxide during the reaction of calcite with phosphoric acid as a function of temperature and technique. *Chemical Geology* **86**, 89–96.

Sweeney, J.J. & Burnham, A.K. (1990) Evaluation of a simple model of vitrinite reflectance based on chemical kinetics. *American Association of Petroleum Geologists Bulletin* **74**, 1559–1570.

Syers, J.K., Chapman, S.L., Jackson, M.L., Rex, R.W. & Clayton, R.N. (1968) Quartz isolation from rocks, sediments and soils for the determination of oxygen isotopes composition. *Geochimica et Cosmochimica Acta* **32**, 1022–1025.

Tagami, T. & Shimada, C. (1996) Natural long-term annealing of the zircon fission track system around a granitic pluton. *Journal of Geophysical Research* **101**, B4, 8245–8255.

Tagami, T., Ito, H. & Nishimura, S. (1990) Thermal annealing characteristics of spontaneous fission tracks in zircon. *Chemical Geology, Isotope Geoscience Section* **80**, 159–169.

Taieb, R. (1990) *Les isotopes de l'hydrogene, du carbone, et de l'oxygene dans les sédiments argileux et les eaux de formation*. Unpublished PhD Thesis, Nancy.

Taylor, E., Burkett, P.J., Wackler, J.D. & Leonard, J.N. (1991) Physical properties and microstructural response of sediments to accretion-subduction: Barbados forearc. In: *Microstructure of Fine-Grained Sediments: From Mud to Shale* (eds R. H. Bennett, W. R. Bryant & M. H. Hulbert), pp. 213–228. Springer-Verlag.

Taylor, H.P. (1971) Oxygen isotope evidence for large-scale interaction between meteoric ground waters and Tertiary granodiorite intrusions, Western Cascade Range, Oregon. *Journal of Geophysical Research* **76**, 7855–7874.

Taylor, H.P. (1978) Oxygen and hydrogen isotope studies of plutonic granitic rocks. *Earth and Planetary Science Letters* **38**, 177–210.

Taylor, H.P. (1980) The effects of assimilation of country rocks by magmas on (18)O/(16)O and (87)Sr/(86)Sr systematics in igneous rocks. *Earth and Planetary Science Letters* **47**, 243–254.

Taylor, H.P. & Epstein, S. (1962) Relationship between

O^{18}/O^{16} ratios in coexisting minerals of igneous and metamorphic rocks, Part 2. Application to petrologic problems. *Geological Society of America Bulletin* **73**, 675–694.

Taylor, H.P. & Forester, R.W. (1971) Low-O^{18} igneous rocks from the intrusive complexes of Skye, Mull and Ardnamurchan, Western Scotland. *Journal of Petrology* **12**, 465–497.

Taylor, J.R. (1982) *An Introduction to Error Analysis.* University Science Books, Oxford University Press.

Taylor, P.N. & Kalsbeek, F. (1990) Dating the metamorphism of Precambrian marbles: examples from Proterozoic mobile belts in Greenland. *Chemical Geology* **86**, 21–28.

Teagle, D.A.H., Alt, J.C., Bach, W., Halliday, A.N. & Erzinger, J. (1996) Alteration of upper ocean crust in a ridge-flank hydrothermal upflow zone: mineral, chemical and isotopic constraints from ODP Hole 896A. In: *Proceedings ODP, Scientific Results 148* (eds J. C. Alt, H. Kinoshita & L. Stokking), pp. 119–150. Ocean Drilling Program, College Station, TX.

Teagle, D.A.H., Alt, J.C., Chiba, H. & Halliday, A.N. (1997) Dissecting an active hydrothermal deposit: the strontium and oxygen isotopic anatomy of the TAG hydrothermal mound. *Chemical Geology*, **149**, 1–24.

Teichmüller, M. (1987) Organic material and very low-grade metamorphism. In: *Low Temperature Metamorphism* (ed. M. Frey), pp. 114–161. Blackie, Glasgow and London.

Tellier, K.E., Hluchy, M.M., Walker, J.R. & Reynolds, R.C. (1988) Applications of high gradient magnetic separation (HGMS) to structural and compositional studies of clay mineral mixtures. *Journal of Sedimentary Petrology* **58**, 761–763.

Terabayashi, M. (1988) Actinolite-forming reaction at low pressure and the role of Fe^{2+}–Mg substitution. *Contributions to Mineralogy and Petrology* **100**, 268–280.

Thode, H.G. (1949) Natural variations in the isotopic content of sulphur and their significance. *Canadian Journal of Research* **27B**, 361.

Thode, H.G. & Rees, C.E. (1971) Measurement of sulphur concentrations and the isotope ratios (33)S/(32)S, (34)S/(32)S and (36)S/(32)S in Apollo 12 samples. *Earth and Planetary Science Letters* **12**, 434–438.

Thode, H.G., Macnamara, J. & Fleming, W.H. (1953) Sulphur isotope fractionation in nature and geological and biological time scales. *Geochimica et Cosmochimica Acta* **3**, 235–243.

Thomas, L.J. (1986) *Low grade metamorphism of the Lake District, England.* PhD Thesis, University of St Andrews, UK.

Thomas, L.J., Harmon, R.S. & Oliver, G.J.H. (1985) Stable isotope compositions of alteration fluids in low-grade lower Palaeozoic rocks, English Lake District. *Mineralogical Magazine* **49**, 425–434.

Thompson, G. (1973) A geochemical study of the low-temperature interaction of seawater and oceanic igneous rocks. *EOS Transactions American Geophysical Union* **54**, 1015–1019.

Thompson, G. (1983) Basalt–seawater interaction. In: *Hydrothermal Processes at Seafloor Spreading Centers* (eds P. A. Rona, K. Bostrom & K. L. Smith), pp. 225–278. Plenum, New York.

Thompson, G.R. & Hower, J. (1973) An explanation for low radiometric ages from glauconite. *Geochimica et Cosmochimica Acta* **37**, 1473–1491.

Thompson, J.B. (1957) The graphical analysis of mineral assemblages in pelitic schists. *American Mineralogist* **42**, 842–858.

Thorpe, R.S., Beckinsale, R.D., Patchett, P.D., Piper, J.D.A., Davies, G.R. & Evans, J.A. (1984) Crustal growth and late Precambrian-early Palaeozoic plate tectonic evolution of England and Wales. *Journal of the Geological Society, London* **141**, 521–536.

Tivey, M.K. (1995) Modeling chimney growth and associated fluid flow at seafloor hydrothermal vent sites. In: *Seafloor Hydrothermal Systems, Physical, Chemical, and Biological Interactions* (eds S. Humphris, J. Lupton, L. Mullineaux & R. Zierenberg), pp. 157–177. Geophysical Monograph No. 91. AGU, Washington, DC.

Tivey, M.K., Humphris, S.E., Thompson, G., Hannington, M.D. & Rona, P.A. (1995) Deducing patterns of fluid flow and mixing within the active TAG hydrothermal mound using mineralogical and geochemical data. *Journal of Geophysical Research* **100**, 12527–12555.

Tómasson, J. & Kristmannsdóttir, H. (1972) High temperature alteration minerals and thermal brines, Reykjanes, Iceland. *Contributions to Mineralogy and Petrology* **36**, 123–134.

Toomey, D.R., Solomon, S. & Purdy, G.M. (1988) Microearthquakes beneath the median valley of the mid-Atlantic ridge near 23°N: tomography and plate tectonics. *Journal of Geophysical Research* **93**, 9093–9112.

Toulkeridis, T. & Clauer, N. (1997) Effect of sequential leaching on the Rb–Sr, Sm–Nd and Pb–Pb systematics, and on the REE distribution of Archean carbonates. In: *South-American Symposium on Isotope Geology*, Sao Paulo, June 15–18, pp. 317–319.

Toulkeridis, T., Goldstein, S.L., Clauer, N., Kröner, A. & Lowe, D.R. (1994) Sm–Nd dating of fig tree clays: implications for the thermal history of the Barberton Greenstone Belt, South Africa. *Geology* **22**, 199–202.

Toulkeridis, T., Clauer, N., Chaudhuri, S. & Goldstein, S.L. (1998) Multi-method (K–Ar, Rb–Sr, Sm–Nd) dating of bentonite minerals from eastern United States. *Basin Research*, in press.

Trümpy, R. (1980) *Geology of Switzerland, Part A.* Wepf & Co., Basel, New York.

Tucker, M.E. (1981) *Sedimentary Petrology: an Introduction.* Blackwell Science Ltd, Oxford.

Turner, C.E. & Fishman, N.S. (1991) Jurassic Lake

Toodichi: a large alkaline saline lake, Morrison Formation, eastern Colorado Plateau. *Geological Society of America Bulletin* **103**, 538–558.

Turner, G. & Cadogan, P.H. (1974) Possible effects of ^{39}Ar recoil in ^{40}Ar–^{39}Ar dating. *Geochimica et Cosmochimica Acta* XX, 1601–1615.

Turner, S., Sandiford, M., Flottmann, T. & Foden, J. (1994) Rb/Sr dating of differentiated cleavage from the Upper Adelaidean metasediments at Hallett-Cove, southern Adelaide fold belt. *Journal of Structural Geology* **16**, 1233–1241.

Turner, S., Sandiford, M., Flottmann, T. & Foden, J. (1995) Rb/Sr dating of differentiated cleavage from the Upper Adelaidean metasediments at Hallett-Cove, southern Adelaide fold belt: reply. *Journal of Structural Geology* **17**, 1801–1803.

Underwood, B.M., Laughland, M.M., Shelton, K.L. & Sedlock, R.L. (1995) Thermal-maturity trends within the Franciscan rocks near Big Sur, California: implications for offset along the San Gregorio–San Simeon–Hsogri fault zone. *Geology* **23**, 839–842.

Ustinov, V.I. & Grinenko, V.A. (1985) Intrastructural isotope distribution during mineral formation. *Geochemistry International* **22**, 143–149.

Vali, H. & Hesse, R. (1990) Alkylammonium ion treatment of clay minerals in ultrathin section: a new method for HRTEM examination of expandable layers. *American Mineralogist* **75**, 1443–1446.

Vali, H. & Koster, H.M. (1986) Expanding behaviour, structural disorder, regular and random irregular interstratification of 2:1 layer-silicates studied by high-resolution images of transmission electron microscopy. *Clay Minerals* **24**, 827–859.

Valley, J.W. (1986) Stable isotope geochemistry of metamorphic rocks. In: *Stable Isotopes in High Temperature Geological Processes* (eds J. W. Valley, H. P. Taylor & J. R. O'Neil), pp. 445–489. Mineralogical Society of America, Chelsea.

Van der Pluijm, B.A., Lee, J.H. & Peacor, D.R. (1988) Analytical electron microscopy and the problem of potassium diffusion. *Clays and Clay Minerals* **36**, 498–504.

Van der Pluijm, B.A., Ho, N.-C. & Peacor, D.R. (1994) High-resolution X-ray texture goniometry. *Journal of Structural Geology* **16**, 1029–1032.

Van der Pluijm, B.A., Ho, N.-C., Peacor, D.R. & Merriman, R.J. (1998) Contradictions of slate formation resolved. *Nature* **392**, 348.

Van Everdingen, D.A. (1995) Fracture characteristics of the sheeted dike complex, Troodos ophiolite, Cyprus: implications for permeability of oceanic crust. *Journal of Geophysical Research* **100**, 19957–19972.

Vanko, D.A. & Stakes, D.S. (1991) Fluids in oceanic layer 3: evidence from veined rocks, Hole 735B, SW Indian Ridge. In: *Proceedings of ODP, Scientific Results 118* (eds R. P. von Herzen & P. T. Robinson), pp. 181–218. Ocean Drilling Program, College Station, TX.

Vanko, D.A., Laverne, C., Tartarotti, P. & Alt, J.C. (1996) Chemistry and origin of secondary minerals from the deep sheeted dikes cored during ODP Leg 148, Hole 504B. In: *Proceedings of ODP, Scientific Results 148* (eds J. C. Alt, H. Kinoshita, L. Stokking & P. Michael), pp. 71–86. Ocean Drilling Program, College Station, TX.

Veblen, D.R., Guthrie, G.D., Livi, K.J.T. & Reynolds, R.C. (1990) High-resolution transmission electron microscopy and electron diffraction of mixed-layer illite/smectite: experimental results. *Clays and Clay Minerals* **38**, 1–13.

Veizer, J., Compston, W., Hoefs, J. & Nielsen, H. (1982) Mantle buffering of the early oceans. *Naturwissenschaften* **69**, 173–180.

Velde, B. (1965) Phengitic micas: synthesis, stability, and natural occurrence. *American Journal of Science* **263**, 886–913.

Velde, B. (1967) Si^{+4} content of natural phengites. *Contributions to Mineralogy and Petrology* **14**, 250–258.

Velde, B. (1968) The effects of chemical reduction on the stability of pyrophyllite and kaolinite in pelitic rocks. *Journal of Sedimentary Petrology* **38**, 13–16.

Velde, B. (1985) *Clay Minerals: A Physico-chemical Explanation of Their Occurrence.* Elsevier, Amsterdam.

Velde, B. (1992) The stability of clays. In: *The Stability of Minerals* (eds G. D. Price & N. L. Ross), pp. 329–335. Chapman & Hall, London, New York.

Velde, B. & Hower, J. (1963) Petrological significance of illite polytypism in Paleocene sedimentary rocks. *American Mineralogist* **48**, 1239–1254.

Velde, B. & Lanson, B. (1993) Comparison of I/S transformations and maturity of organic matter at elevated temperatures. *Clays and Clays Minerals* **41**, 178–183.

Vennemann, T.W. & O'Neil, J.R. (1993) A simple and inexpensive method of hydrogen isotope and water analyses of minerals and rocks based on zinc reagent. *Chemical Geology* **103**, 227–234.

Vennemann, T.W. & Smith, H.S. (1990) The rate and temperature of reaction of ClF_3 with silicate minerals, and their relevance to oxygen isotope analysis. *Chemical Geology* **86**, 83–88.

Vergara, M. & Drake, R. (1979) Edades K/Ar en secuencias volcanicas continentales post-neocomianas de Chile central; su depositacion en cuencas intermontanas restringidas. *Revista de la Asociación Geológica Argentina* **34**, 42–52.

Vicente, M.A., Elsass, F., Molina, E. & Robert, M. (1997) Palaeoweathering in slates from the Iberian Hercynian Massif (Spain): investigation by TEM of clay mineral signatures. *Clay Minerals* **32**, 435–451.

Vidal, P. (1980) L'évolution polyorogénique du Massif Armoricain: apport de la géochronologie et de la géochimie isotopique du strontium. *Mémoire de la Société de Géologie et Minéralogie de Bretagne* **21**, 162P.

Videtich, P.E. (1981) A method for analyzing dolomite

for stable isotopic composition. *Journal of Sedimentary Petrology* **51**, 661–662.

Viereck, L.G., Griffin, B.J., Schmincke, H.U. & Pritchard, R. (1982) Volcaniclastic rocks of the Reydarfjordur drill hole, Eastern Iceland 2, Alteration. *Journal of Geophysical Research* **87**, 6459–6476.

Von Damm, K.L. (1988) Systematics of and postulated controls on submarine hydrothermal solution chemistry. *Journal of Geophysical Research* **93**, 4551–4561.

Von Damm, K.L. (1995) Controls on the chemistry and temporal variability of fluids. In: *Seafloor Hydrothermal Systems, Physical, Chemical, and Biological Interactions* (eds S. Humphris, J. Lupton, L. Mullineaux & R. Zierenberg), pp. 222–247. Geophysical Monograph No. 91. AGU, Washington, DC.

Von Damm, K.L., Oosting, S.E., Kozlowski, R. *et al.* (1995) Evolution of East Pacific Rise hydrothermal vent fluids following a volcanic eruption. *Nature* **375**, 47–50.

Von Gosen, W., Buggisch, W. & Krumm, S. (1991) Metamorphism and deformation in the Sierras Australes fold and thrust belt (Buenos Aires Province, Argentina). *Tectonophysics* **185**, 335–356.

Von Herzen, R.P. & Robinson, P.T. (eds) *Proceedings of ODP, Initial Reports 118.* Ocean Drilling Program, College Station, TX.

Wada, H. (1988) Microscale isotopic zoning in calcite and graphite crystals in marble. *Nature* **331**, 61–63.

Waggoner, D.G. (1993) The age and alteration of central Pacific oceanic crust near Hawaii, Site 843. In: *Proceedings of ODP 136* (eds R. H. Wilkens, J. Firth & J. Bender), pp. 119–132. Ocean Drilling Program, College Station, TX.

Wagner, G. & Van den Haute, P. (1992) *Fission-Track Dating.* Kluwer Academic Publishers, Dordrecht, The Netherlands.

Wakabayashi, J. (1992) Nappes, tectonics of oblique plate convergence, and metamorphic evolution related to 140 million years of continuous subduction, Franciscan Complex, California. *Journal of Geology* **100**, 19–40.

Wakabayashi, J. & Unruh, J.R. (1995) Tectonic wedging, blueschist metamorphism, and exposure of blueschists: are they compatible? *Geology* **23**, 85–88.

Walker, G.P.L. (1960a) Zeolite zones and dike distributions in relation to the structure of the basalts of eastern Iceland. *Journal of Geology* **68**, 515–528.

Walker, G.P.L. (1960b) The amygdale minerals in the Tertiary basalts of Ireland, III. Regional distribution. *Mineralogical Magazine* **32**, 503–527.

Walker, G.P.L. (1971) The distribution of amygdale minerals in Mull and Morvern (Western Scotland). In: *Studies in Earth Sciences, West Commemorative Volume* (eds T. V. V. G. R. K. Murty & S. S. Rao), pp. 181–194.

Walker, J.C.G. & Lohmann, K.C. (1989) Why the oxygen isotopic composition of sea water changes with time. *Geophysical Research Letters* **16**, 323–326.

Walker, J.R. (1989) Polytypism of chlorite in very low grade metamorphic rocks. *American Mineralogist* **74**, 738–743.

Walker, J.R. (1993) Chlorite polytype geothermometry. *Clays and Clay Minerals* **41**, 260–267.

Walker, J.R. & Murphy, M.P. (1995) Chloritic minerals from prehnite–pumpellyite facies rocks of the Winterville Formation, Aroostook County, Maine. In: *Low-Grade Metamorphism of Mafic Rocks* (eds P. Schiffman & H. W. Day). Geological Society of America, Boulder, CO. Special Paper No. 296, pp. 141–156.

Walshe, J.L. (1986) A six-component chlorite solid solution model and the conditions of chlorite formation in hydrothermal and geothermal systems. *Economic Geology* **81**, 681–703.

Wang, H., Frey, M. & Stern, W.B. (1996) Diagenesis and metamorphism of clay minerals in the Helvetic Alps of Eastern Switzerland. *Clays and Clay Minerals* **44**, 96–112.

Waples, D.W. (1994) Maturity modelling: thermal indicators, hydrocarbon generation, and oil cracking. In: *The Petroleum System—From Source to Trap* (eds L. B. Magoon & D. G. Dow). American Association of Petroleum Geologists Memoir **60**, 285–306.

Warr, L.N. (1991) Basin inversion and foreland basin development in the Rhenohercynian of southwest England. In: *The Rhenohercynian and Subvariscan Fold Belts* (eds R. A. Gayer, R. Greiling & A. Vogel), pp. 197–224. Earth Evolution Series. Vieweg and Sohn.

Warr, L.N. (1995) A reconnaissance study of very low-grade metamorphism in south Devon. *Proceedings of the Ussher Society* **8**, 405–410.

Warr, L.N. (1996) Standardized clay mineral crystallinity data from the very low-grade metamorphic facies rocks of southern New Zealand. *European Journal of Mineralogy* **8**, 115–127.

Warr, L.N. & Rice, A.H.N. (1994) Interlaboratory standardization and calibration of clay mineral crystallinity and crystallite size data. *Journal of Metamorphic Geology* **12**, 141–152.

Warr, L.N. & Robinson, D. (1990) The application of isocryst mapping to geological interpretation: a case study from north Cornwall. *Proceedings of the Ussher Society* **7**, 223–227.

Warr, L.N., Primmer, T.J. & Robinson, D. (1991) Variscan very low-grade metamorphism in southwest England: a diastathermal and thrust-related origin. *Journal of Metamorphic Geology* **9**, 751–764.

Warr, L.N., Greiling, R.O. & Zachrisson, E. (1996) Thrust related very low grade metamorphism in the marginal part of an orogenic wedge, Scandinavian Caledonides. *Tectonics* **15**, 1213–1229.

Warren, B.E. & Averbach, B.L. (1950) The effects of cold-work distortion on X-ray patterns. *Journal of Applied Physics* **21**, 595–599.

Weaver, C.E. (1959) The clay petrology of sediments. In: *Proceedings of the 6th National Conference*, pp. 15–187. *Clays and Clay Minerals.*

Weaver, C.E. (1960) Possible uses of clay minerals in search for oil. *American Association of Petroleum Geologists Bulletin* **44**, 1505–1518.

Weaver, C.E. (1984) Response of physilites to temperature and tectonics. In: *Shale-Slate Metamorphism in the Southern Appalachians* (ed. C. E. Weaver). *Developments in Petrology* **10**, 185–199.

Weaver, C.E. & Broekstra, B.R. (1984) Illite–mica. In: *Shale-Slate Metamorphism in the Southern Appalachians* (ed. C. E. Weaver). *Developments in Petrology* **10**, 67–97.

Weber, F., Dunoyer de Segonzac, G. & Economou, C. (1976) Une nouvelle expression de la 'crystallinité' de l'illite et des micas. Notion d'epaisseur des cristallites. *Compte Rendu Sommaire de Seances de la Societé Géologique de France* **5**, 225–227.

Weber, K. (1972a) Note on the determination of illite crystallinity. *Neues Jahrbuch für Mineralogie Monatshefte* **6**, 267–276.

Weber, K. (1972b) Kristallinität des Illites in tonschiefern un andere Kriteerien schwacher Metamorphose in nordöstlichen Rheinischen Schiefergebirge. *Neues Jahrbuch für Geologie und Palaontologie Abhandlugen* **141**, 333–363.

Weber, K. (1981) Kinematic and metamorphic aspects of cleavage formation in very low-grade metamorphic slates. *Tectonophysics* **78**, 291–306.

Wenner, D.B. & Taylor, H.P. (1973) D/H and O^{18}/O^{16} studies of serpentinization of ultramafic rocks. *Geochimica et Cosmochimica Acta* **38**, 1255–1286.

Wenner, D.B. & Taylor, H.P. (1974) Oxygen and hydrogen isotope studies of the serpentinization of ultramafic rocks in oceanic environments and continental ophiolite complexes. *American Journal of Science* **273**, 207–239.

White, R.S., McKenzie, D. & O'Nions, R.K. (1992) Oceanic crustal thickness from seismic measurements and rare earth element inversions. *Journal of Geophysical Research* **97**, 19683–19715.

White, S.H. & Knipe, R.J. (1978) Microstructure and cleavage development in selected slates. *Contributions to Mineralogy and Petrology* **66**, 165–174.

Whitney, G. & Velde, B. (1993) Changes in particle morphology during illitization: an experimental study. *Clays and Clay Minerals* **41**, 209–218.

Wiechert, U. & Hoefs, J. (1995) An excimer laser-based micro analytical preparation technique for *in-situ* oxygen isotope analysis of silicate and oxide minerals. *Geochimica et Cosmochimica Acta* **59**, 4093–4101.

Wilkinson, J.J., Jenkin, G.R.T., Fallick, A.E. & Foster, R.P. (1995) Oxygen and hydrogen isotopic evolution of Variscan crustal fluids, south Cornwall, U.K. *Chemical Geology* **123**, 239–254.

Willett, S.D. (1992) Modelling thermal annealing of fission tracks in apatite. In: *Low Temperature Thermochronology* (eds M. Zentilli & P. H. Reynolds). Mineralogical Association of Canada, Short Course Handbook Vol. 20, pp. 43–72.

Wilson, A.J.C. (1963) *Mathematical Theory of X-ray Powder Diffractometry.* Philips Technical Library, Eindhoven.

Winkler, H.G.F. (1967) *Petrogenesis of Metamorphic Rocks,* 2nd edn. Springer-Verlag, Berlin, New York.

Woodcock, N.H. (1984) Early Palaeozoic sedimentation and tectonics in Wales. *Proceedings of the Geologists' Association* **95**, 323–335.

Woodcock, N.H. & Pharaoh, T.C. (1993) Silurian facies beneath East Anglia. *Geological Magazine* **130**, 681–690.

Wright, T.O. & Dallmeyer, R.D. (1991) The age of cleavage development in the Ross orogen, northern Victoria Land, Antarctica: evidence from Ar-40 Ar-39 whole-rock slate ages. *Journal of Structural Geology* **13**, 677–690.

Wright, T.O. & Dallmeyer, R.D. (1992) The age of cleavage development in the Ross orogen, northern Victoria Land, Antarctica: evidence from Ar-40 Ar-39 whole-rock slate ages — reply. *Journal of Structural Geology* **14**, 891–892.

Yamada, R., Tagami, T. & Nishimura, S. (1995) Confined fission-track length measurement of zircon: assessment of factors affecting the paleotemperature estimate. *Chemical Geology* **122**, 249–258.

Yang, H.Y., Huang, T.M. & Lo, Y.M. (1994) The pyrophyllite isograde in the metamorphic terrain of Taiwan. *Bulletin of the Central Geological Survey, Taiwan* **9**, 123–135.

Yeh, H.W. (1980) D/H ratios and late-stage dehydration of shales during burial. *Geochimica et Cosmochimica Acta* **44**, 341–352.

Yeh, H.W. & Epstein, S. (1978) Hydrogen isotope exchange between clay minerals and sea water. *Geochimica et Cosmochimica Acta* **42**, 140–143.

Yeh, H.W. & Eslinger, E.V. (1986) Oxygen isotopes and the extent of diagenesis of clay minerals during sedimentation and burial in the sea. *Clays and Clay Minerals* **34**, 403–406.

Yeh, H.W. & Savin, S.M. (1976) The extent of oxygen isotope exchange between clay minerals and sea water. *Geochimica et Cosmochimica Acta* **40**, 743–748.

Yeh, H.W. & Savin, S.M. (1977) The mechanism of burial metamorphism of argillaceous sediments, 3. Oxygen isotopic evidence. *Geological Society of America Bulletin* **88**, 1321–1330.

Yoder, H.S. & Eugster, H.P. (1955) Synthetic and natural muscovites. *Geochimica et Cosmochimica Acta* **6**, 157–185.

York, D., Masliwec, H., Hall, C.M. *et al.* (1981) The direct dating of ore minerals. *Ontario Geological Survey* **98**, 334–340.

Yuasa, M., Watanabe, T., Kuwajima, T., Hirama, T. & Fujioka, K. (1992) Prehnite–pumpellyite facies metamorphism in oceanic arc basement from Site 791 in the Sumisu Rift, western Pacific. In: *Proceedings of ODP, Initial Reports 126* (eds B. Taylor & K. Fujioka), pp. 185–193. Ocean Drilling Program, College Station, TX.

Zane, A., Sassi, F.P. & Sassi, R. (1996) New data on chlorites as petrogenetic indicator mineral in metamorphic rocks. *Plinius* **16**, 212–213.

Zen, E-an (1966) Construction of pressure–temperature diagrams for multicomponent systems after the method of Schreinemakers—a geometric approach. *US Geological Survey Bulletin* **1225**, 1–56.

Zen, E-an & Roseboom, E.H., Jr (1972) Some topological relationships in multisystems of n + 3 phases III. Ternary systems. *American Journal of Science* **272**, 677–710.

Zhao, G., Peacor, D.R. & McDowell, S.D. (1998) 'Retrograde diagenesis' of clay minerals of the Freda Sandstone, Wisconsin. *Clays and Clay Minerals*, in press.

Zheng, Y.F. (1993) Calculation of oxygen isotope fractionation in hydroxyl-bearing silicates. *Earth and Planetary Science Letters* **120**, 247–263.

Zheng, Y.F. (1994) Oxygen isotope fractionation in metal monoxides. *Mineralogical Magazine* **58a**, 1000–1001.

Zhou, Z., Fyfe, W., Tazaki, K. & van der Gaast, S.J. (1992) The structural characteristics of palagonite from DSDP Site 335. *Canadian Mineralogist* **30**, 75–81.

Zierenberg, R.A., Shanks, W.C., Seyfried, W.E., Koski, R.A. & Strickler, M.D. (1988) Mineralization, alteration and hydrothermal metamorphism of the ophiolite-hosted Turner–Albright sulfide deposits, southwest Oregon. *Journal of Geophysical Research* **93**, 4657–4674.

Zimmermann, J.L. & Odin, G.S. (1982) Kinetics of the release of argon and fluids from glauconies. In: *Numerical Dating in Stratigraphy* (ed. G. S. Odin), pp. 345–362. John Wiley & Sons, New York.

Index

Please note: page numbers in *italics* refer to figures; those in **bold** to tables